Methods in Molecular Biology™ • 58

Basic DNA and RNA Protocols

Edited by

Adrian J. Harwood

MRC Laboratory for Molecular Cell Biology, London, UK

Stafford Library
Columbia College
1001 Rogers Street
Columbia, Missouri 65216

Humana Press ✱ Totowa, New Jersey

© 1996 Humana Press Inc.
999 Riverview Drive, Suite 208
Totowa, New Jersey 07512

All rights reserved. No part of this book may be reproduced, stored in a retrieval system, or transmitted in any form or by any means, electronic, mechanical, photocopying, microfilming, recording, or otherwise without written permission from the Publisher. Methods in Molecular Biology™ is a trademark of the Humana Press Inc.

All authored papers, comments, opinions, conclusions, or recommendations are those of the author(s) and do not necessarily reflect the views of the publisher.

This publication is printed on acid-free paper. ∞
ANSI Z39.48-1984 (American National Standards Institute)
Permanence of Paper for Printed Library Materials

Cover illustration: Fig. 2B from Chapter 58, "One Step Purification of Recombinant Proteins with the 6xHis Tag and Ni-NTA Resin," by Joanne Crowe, Brigitte Steude Masone, and Joachim Ribbe.

For additional copies, pricing for bulk purchases, and/or information about other Humana titles, contact Humana at the above address or at any of the following numbers: Tel.: 201-256-1699; Fax: 201-256-8341; E-mail: humana@interramp.com

Photocopy Authorization Policy:
Authorization to photocopy items for internal or personal use, or the internal or personal use of specific clients, is granted by Humana Press Inc., provided that the base fee of US $5.00 per copy, plus US $00.25 per page, is paid directly to the Copyright Clearance Center at 222 Rosewood Drive, Danvers, MA 01923. For those organizations that have been granted a photocopy license from the CCC, a separate system of payment has been arranged and is acceptable to Humana Press Inc. The fee code for users of the Transactional Reporting Service is: [0-89603-331-7/96 $5.00 + $00.25].

Printed in the United States of America. 10 9 8 7 6 5 4 3 2

Library of Congress Cataloging in Publication Data
Basic DNA and RNA protocols/edited by Adrian J. Harwood.
    p.   cm.—(Methods in molecular biology™; 58)
  ISBN 0-89603-402-X (hc: alk. paper).—ISBN 0-89603-331-7 (comb: alk. paper)
  1. Nucleic acids—Laboratory manuals. I. Harwood, Adrian J. II. Series: Methods in molecular biology™ (Totowa, NJ); 58
QP620.B38   1996
574.87'328'0724—dc20
                                                                      95-50258
                                                                       CIP

Preface

Molecular genetics, or "genetic engineering" as it is sometimes described, has had a profound effect on the study of biology. It is now 12 years since the publication of *Methods in Molecular Biology™, vol. 2: Nucleic Acids*, which contained what were then the current molecular and genetic techniques. The methodology is rapidly evolving and any volume of these basic techniques needs constant revision. This was last carried out in this series in 1988 *(*vol. 4: *New Nucleic Acids Methods)*. Since then, the descriptions of many new, but nonetheless basic, techniques have been incorporated into the more specialized volumes of the *Methods in Molecular Biology* series. It is the aim of *Basic DNA and RNA Protocols* to bring together all of these core techniques in a single volume.

In the last eight years, there have been both dramatic new developments and a myriad of minor improvements to existing techniques. The polymerase chain reaction (PCR), for example, has led to a completely new approach to gene cloning and recombinant DNA manipulation. On the other hand, relatively minor changes, such as the development of Sequenase™, have made plasmid DNA sequencing commonplace. *Basic DNA and RNA Protocols* covers both these new techniques even as it revises and updates the older ones. Since we are still in a period of rapid technical development, a number of methods are presented that, although sufficiently established for inclusion in this volume, may point in the direction of further progress. In particular, the development of nonisotopic detection methods and techniques is becoming increasingly important and such methods as nonisotopic DNA labeling, silver staining, and automatic sequencing are described.

More than ever before, industrial companies are playing an important part in development of new techniques. A number of these innovative companies have contributed to *Basic DNA and RNA Protocols*, and their efforts are greatly appreciated. Commercial reagents are greatly beneficial for the ease and speed with which they allow us to carry out our research. There is a danger, however, that the novice may miss the principles on which the methods are based through excessive use of "ready-to-use" kits. In an attempt to counter this trend, we have also included some methods for producing certain reagents that are most commonly bought from commercial sources.

Basic DNA and RNA Protocols has been prepared in the same manner as the previous volumes of the *Methods in Molecular Biology* series. Each chapter has been contributed by an author, or authors, who have had considerable laboratory experience of the technique and in some cases were instrumental in the initial development. Each chapter was written in a recipe-like format designed for direct practical use within the laboratory, and not only includes a detailed description of a basic method, but in the Notes section discusses the problems (and solutions) that may be encountered and the variant versions that may be tried.

Basic DNA and RNA Protocols is divided into seven sections: DNA analysis, RNA analysis, gene cloning, subcloning methods, PCR, DNA sequencing, and site-directed mutagenesis and protein expression. These groupings are for convenience of use and should not be taken as discrete units. All the methods are interlocking and reinforce one another. Accordingly, extensive cross-referencing between the individual chapters is provided to demonstrate their interrelationship.

Basic DNA and RNA Protocols is primarily aimed at the competent scientist who is new to these techniques. I hope, however, that it contains a sufficient number of tips and hints to assure its interest to even the most experienced aficionado.

I would like to thank the editors of the volumes from which I plundered many useful chapters. I refer the reader to these more detailed volumes for further methods on a more specific topic. I am particularly grateful to all of the authors for their contributions and revisions. Finally, I would especially like to thank Janet for the many ways in which she contributed to this work.

Adrian J. Harwood

Contents

Preface v
Contributors xi

PART I. DNA ANALYSIS 1
CH. 1. The Simultaneous Isolation of RNA and DNA from Tissues
and Cultured Cells,
*Frank Merante, Sandeep Raha, Juta K. Reed,
and Gerald Proteau* 3
CH. 2. Restriction Endonuclease Digestion of DNA,
Duncan R. Smith 11
CH. 3. Agarose Gel Electrophoresis,
Duncan R. Smith 17
CH. 4. Capillary Blotting of Agarose Gels,
Duncan R. Smith and David Murphy 23
CH. 5. Random Primed ^{32}P-Labeling of DNA,
Duncan R. Smith 27
CH. 6. Hybridization and Competition Hybridization of Southern Blots,
Rosemary E. Kelsell 31
CH. 7. Utilization of DNA Probes with Digoxigenin-Modified Nucleotides
in Southern Hybridizations,
Tim Helentjaris and Tom McCreery 41
CH. 8. The Preparation of Fluorescein-Labeled Nucleic Acid Probes
and Their Detection Using Alkaline Phosphatase-Catalyzed
Dioxetane Chemiluminescence,
Martin Cunningham and Martin Harris 53
CH. 9. Monitoring Incorporation of Fluorescein into Nucleic Acid Probes
Using a Rapid Labeling Assay,
Martin Cunningham 65
CH. 10. The Preparation of Horseradish Peroxidase Labeled Nucleic Acid
Probes and Their Detection Using Enhanced Chemiluminescence,
Bronwen Harvey, Ian Durrant, and Martin Cunningham 67
CH. 11. 3'-End Labeling of Oligonucleotides with Fluorescein-11-dUTP
and Enhanced Chemiluminescent Detection,
Martin Cunningham, Ian Durrant, and Bronwen Harvey 77

Ch. 12.	The Preparation of Riboprobes, *Dominique Belin*	83
Ch. 13.	Native Polyacrylamide Gel Electrophoresis, *Adrian J. Harwood*	93
Ch. 14.	Use of Silver Staining to Detect Nucleic Acids, *Lloyd G. Mitchell, Angelika Bodenteich, and Carl R. Merrill*	97
Ch. 15.	End-Labeling of DNA Fragments, *Adrian J. Harwood*	105
Part II.	RNA Analysis	111
Ch. 16.	Northern Blot Analysis, *Robb Krumlauf*	113
Ch. 17.	RNA Slot Blotting, *David Murphy*	129
Ch. 18.	The RNase Protection Assay, *Dominique Belin*	131
Ch. 19.	Primer Extension Analysis of mRNA, *Mark W. Leonard and Roger Patient*	137
Ch. 20.	S1 Mapping Using Single-Stranded DNA Probes, *Stéphane Viville and Roberto Mantovani*	147
Ch. 21.	Nonradioactive *In Situ* Hybridization for Cells and Tissues, *Ian Durrant*	155
Part III.	Gene Cloning	169
Ch. 22.	In Vitro Packaging of DNA, *Jeremy W. Dale and Peter J. Greenaway*	171
Ch. 23.	Construction of Mammalian Genomic Libraries Using λ Replacement Vectors, *Alan N. Bateson and Jeffrey W. Pollard*	177
Ch. 24.	The Production of Double-Stranded Complementary DNA for Use in Making Libraries, *Steve Mayall and Jane Kirk*	191
Ch. 25.	Construction of cDNA Libraries, *Michael M. Burrell*	199
Ch. 26.	Screening λ Libraries, *Janet C. Harwood and Adrian J. Harwood*	211
Part IV.	Subcloning Methods	219
Ch. 27.	Subcloning Strategies and Protocols, *Danielle Gioioso Taghian and Jac A. Nickoloff*	221
Ch. 28.	Purification of DNA Fragments from Agarose Gels Using Glass Beads, *Etienne Joly*	237
Ch. 29.	Transformation of *E. coli*, *Fiona M. Tomley*	241

Contents

CH. 30.	Transformation of Bacteria by Electroporation, *Lucy Drury*	249
CH. 31.	Preparation of Plasmid DNA by Alkaline Lysis, *Etienne Joly*	257
CH. 32.	The Rapid Boiling Method for Small-Scale Preparation of Plasmid DNA, *Adrian J. Harwood*	265
CH. 33.	Plasmid Preparations with Diatomaceous Earth, *Laura Machesky*	269
PART V.	PCR TECHNIQUES	273
CH. 34.	Polymerase Chain Reaction: *Basic Protocols*, *Beverly C. Delidow, John P. Lynch, John J. Peluso, and Bruce A. White*	275
CH. 35.	Inverse Polymerase Chain Reaction, *Daniel L. Hartl and Howard Ochman*	293
CH. 36.	Polymerase Chain Reaction with Degenerate Oligonucleotide Primers to Clone Gene Family Members, *Gregory M. Preston*	303
CH. 37.	Cloning PCR Products Using T-Vectors, *Michael K. Trower and Greg S. Elgar*	313
CH. 38.	Direct Radioactive Labeling of Polymerase Chain Reaction Products, *Tim McDaniel and Stephen J. Meltzer*	325
CH. 39.	A Rapid PCR-Based Colony Screening Protocol for Cloned Inserts, *Michael K. Trower*	329
CH. 40.	Use of Polymerase Chain Reaction to Screen Phage Libraries, *Lei Yu and Laura J. Bloem*	335
PART VI.	DNA SEQUENCING	341
CH. 41.	Cloning into M13 Bacteriophage Vectors, *Qingzhong Yu*	343
CH. 42.	Ordered Deletions Using Exonuclease III, *Denise Clark and Steven Henikoff*	349
CH. 43.	M13 Phage Growth and Single-Stranded DNA Preparation, *Fiona M. Tomley*	359
CH. 44.	Preparation of ssDNA from Phagemid Vectors, *Michael K. Trower*	363
CH. 45.	A Rapid Plasmid Purification Method for Dideoxy Sequencing, *Annette M. Griffin and Hugh G. Griffin*	367
CH. 46.	DNA Sequencing Using Sequenase Version 2.0 T7 DNA Polymerase, *Carl W. Fuller, Bernard F. McArdle, Annette M. Griffin, and Hugh G. Griffin*	373
CH. 47.	Pouring Linear and Buffer-Gradient Sequencing Gels, *Paul Littlebury*	389

CH. 48. Electrophoresis of Sequence Reaction Samples,
Alan T. Bankier 393
CH. 49. Direct Sequencing of PCR Products,
Janet C. Harwood and Geraldine A. Phear 403
CH. 50. Thermal Cycle Dideoxy DNA Sequencing,
Barton E. Slatko 413
CH. 51. Using the Automated DNA Sequencer,
Zijin Du and Richard K. Wilson 425
CH. 52. Terminal Labeling of DNA for Maxam and Gilbert Sequencing,
Eran Pichersky 441
CH. 53. DNA Sequencing by the Chemical Method,
Eran Pichersky 447

PART VII. SITE-DIRECTED MUTAGENESIS AND PROTEIN SYNTHESIS 453

CH. 54. Site-Directed Mutagenesis of Double-Stranded Plasmids, Domain Substitution, and Marker Rescue by Comutagenesis of Restriction Enzyme Sites,
Jac A. Nickoloff, Win-Ping Deng, Elizabeth M. Miller, and F. Andrew Ray 455
CH. 55. A Protocol for Site-Directed Mutagenesis Employing a Uracil-Containing Phagemid Template,
Michael K. Trower 469
CH. 56. In Vitro Translation of Messenger RNA in a Rabbit Reticulocyte Lysate Cell-Free System,
Louise Olliver and Charles D. Boyd 477
CH. 57. In Vitro Translation of Messenger RNA in a Wheat Germ Extract Cell-Free System,
Louise Olliver, Anne Grobler-Rabie, and Charles D. Boyd 485
CH. 58. One-Step Purification of Recombinant Proteins with the 6xHis Tag and Ni-NTA Resin,
Joanne Crowe, Brigitte Steude Masone, and Joachim Ribbe 491

Index 511

Contributors

ALAN T. BANKIER • *MRC Laboratory of Molecular Biology, Cambridge, UK*
ALAN N. BATESON • *Department of Pharmacology, University of Alberta, Edmonton, Canada*
DOMINIQUE BELIN • *Department of Pathology, University of Geneva Medical School, Geneva, Switzerland*
LAURA J. BLOEM • *Department of Medical and Molecular Genetics, Indiana University School of Medicine, Indianapolis, IN*
ANGELIKA BODENTEICH • *Laboratory of Biochemical Genetics, NIMH Neuroscience Center at St. Elizabeth's Hospital, Washington, DC*
CHARLES D. BOYD • *University of Medicine and Dentistry of New Jersey, New Brunswick, NJ*
MICHAEL M. BURRELL • *Advanced Technologies, Cambridge, UK*
DENISE CLARK • *Department of Biology, University of New Brunswick, Fredericton, Canada*
JOANNE CROWE • *QIAGEN GmbH, Hilden, Germany*
MARTIN CUNNINGHAM • *Research and Development, Amersham International, Amersham, Bucks, UK*
JEREMY W. DALE • *School of Biological Sciences, University of Surrey, UK*
BEVERLY C. DELIDOW • *Department of Anatomy, University of Connecticut Health Center, Farmington, UK*
WIN-PING DENG • *Department of Cancer Biology, Harvard University School of Public Health, Boston, MA*
LUCY DRURY • *ICRF Clare Hall Laboratories, South Mimms, UK*
ZIJIN DU • *Genome Sequencing Center, Washington University School of Medicine, St. Louis, MO*
IAN DURRANT • *Research and Development, Amersham International, Amersham, Bucks, UK*

GREG S. ELGAR • *Molecular Genetics Unit, MRC Centre, Cambridge, UK*
CARL W. FULLER • *Institute of Food Research, Norwich Research Park, Colney, UK*
PETER J. GREENAWAY • *Department of Health, London, UK*
ANNETTE M. GRIFFIN • *Institute of Food Research, Norwich Research Park, Colney, UK*
HUGH G. GRIFFIN • *Institute of Food Research, Norwich Research Park, Colney, UK*
ANNE GROBLER-RABIE • *MRC Unit for Molecular and Cellular Cardiology, University of Stellenbosch Medical School, Tygerberg, South Africa*
MARTIN HARRIS • *Research and Development, Amersham International, Amersham, Bucks, UK*
DANIEL L. HARTL • *Department of Organismic and Evolutionary Biology, Harvard University, Cambridge, UK*
BRONWEN HARVEY • *Research and Development, Amersham International, Amersham, Bucks, UK*
ADRIAN J. HARWOOD • *MRC Laboratory for Molecular Cell Biology and Department of Biology, University College, London, UK*
JANET C. HARWOOD • *Imperial Cancer Research Fund, Clare Hall Laboratories, South Mimms, UK*
TIM HELENTJARIS • *Pioneer Hi-Bred Institute, Inc., Johnston, IA*
STEVEN HENIKOFF • *Howard Hughes Medical Institute, Fred Hutchinson Cancer Research Center, Seattle, WA*
ETIENNE JOLY • *Department of Immunology, AFRC Babraham Institute, Cambridge, UK*
ROSEMARY E. KELSELL • *Institute of Ophthalmology, Department of Molecular Genetics, London, UK*
JANE KIRK • *Imperial Cancer Research Fund, Clare Hall Laboratories, Potters Bar, UK*
ROBB KRUMLAUF • *National Institute for Medical Research, Laboratory of Eukaryotic Molecular Genetics, London, UK*
MARK W. LEONARD • *Genetics Institute, Andover, MA*
PAUL LITTLEBURY • *Department of Biology, University of York, UK*
JOHN P. LYNCH • *Department of Anatomy, University of Connecticut Health Center, Farmington, CT*

Contributors

Laura Machesky • *MRC Laboratory for Molecular Cell Biology, Cambridge, UK*
Roberto Mantovani • *Faculté de Médecine, DOI/LGME, Strasbourg Cedex, France*
Brigitte Steude Masone • *Qiagen, Inc., Chatsworth, CA*
Steve Mayall • *Imperial Cancer Research Fund, Clare Hall Laboratories, Potters Bar, UK*
Bernard F. McArdle • *United States Biochemical Corp., Cleveland, OH*
Tom McCreery • *Department of Plant Sciences, University of Arizona, Tucson, AZ*
Tim McDaniel • *University of Maryland Hospital, Baltimore, MD*
Stephen J. Meltzer • *University of Maryland Hospital, Baltimore, MD*
Frank Merante • *Department of Genetics, Research Institute, The Hospital for Sick Children, Toronto, Ontario, Canada*
Carl R. Merril • *Laboratory of Biochemical Genetics, NIMH Neuroscience Center at St. Elizabeth's Hospital, Washington, DC*
Elizabeth M. Miller • *Department of Cancer Biology, Harvard University School of Public Health, Boston, MA*
Lloyd G. Mitchell • *Laboratory of Biochemical Genetics, NIMH Neuroscience Center at St. Elizabeth's Hospital, Washington, DC*
David Murphy • *Neuropeptide Laboratory, Institute of Molecular and Cell Biology, National University of Singapore, Republic of Singapore*
Jac A. Nickoloff • *Department of Cancer Biology, Harvard University School of Public Health, Boston, MA*
Howard Ochman • *Department of Genetics, Washington University School of Medicine, St. Louis, MO*
Louise Olliver • *MRC Unit for Molecular and Cellular Cardiology, University of Stellenbosch Medical School, Tygerberg, South Africa*
Roger Patient • *Developmental Biology Research Centre, Kings College, University of London, UK*
John J. Peluso • *Department of Anatomy, University of Connecticut Health Center, Farmington, CT*

GERALDINE A. PHEAR • *Department of Radiation Oncology, Health Sciences Center, University of Utah, Salt Lake City, UT*
ERAN PICHERSKY • *Department of Biology, University of Michigan, Ann Arbor, MI*
JEFFREY W. POLLARD • *Albert Einstein College of Medicine, Yeshiva University, Bronx, NY*
GREGORY M. PRESTON • *Department of Medicine and Biological Chemistry, Johns Hopkins University School of Medicine, Baltimore, MD*
GERALD PROTEAU • *College Universitaire de Saint-Boniface, Canada*
SANDEEP RAHA • *Department of Genetics Research Institute, The Hospital for Sick Children, Toronto, Canada*
F. ANDREW RAY • *Life Sciences Division, Los Alamos National Laboratory, Los Alamos, NM*
JUTA K. REED • *Departments of Biochemistry and Chemistry, Erindale College, University of Toronto, Ontario, Canada*
JOACHIM RIBBE • *Qiagen GmbH, Hilden, Germany*
BARTON E. SLATKO • *New England Biolabs, Inc., Beverly, MA*
DUNCAN R. SMITH • *Department of Colorectal Surgery, Singapore General Hospital, Republic of Singapore*
DANIELLE GIOIOSO TAGHIAN • *Department of Cancer Biology, Harvard University School of Public Health, Boston, MA*
FIONA M. TOMLEY • *Institute of Animal Health, Compton, UK*
MICHAEL K. TROWER • *Molecular Pathology Department, Glaxo Research and Development Inc., Greenford, UK*
STÉPHANE VIVILLE • *Faculty of Medicine, Institute of Genetics and Molecular and Cellular Biology, University of Louis Pasteur, CNRS/INSERM, France*
BRUCE A. WHITE • *Department of Anatomy, University of Connecticut Health Center, Framington, CT*
RICHARD K. WILSON • *Genome Sequencing Center, Washington University School of Medicine, St. Louis, MO*
LEI YU • *Department of Medical and Molecular Genetics, Indiana University School of Medicine, Indianapolis, IN*
QINGZHONG YU • *Department of Microbiology, University of Alabama at Birmingham, AL*

PART I
DNA Analysis

CHAPTER 1

The Simultaneous Isolation of RNA and DNA from Tissues and Cultured Cells

Frank Merante, Sandeep Raha, Juta K. Reed, and Gerald Proteau

1. Introduction

Many techniques are currently available that allow the isolation of DNA *(1–7)* or RNA *(8–23)*, but such methods allow only the purification of one type of nucleic acid at the expense of the other. Frequently, when cellular material is limiting, it is desirable to isolate both RNA and DNA from the same source. Such is the case for biopsy specimens, primary cell lines, or manipulated embryonic stem cells.

Although several procedures have been published that address the need to simultaneously purify both RNA and DNA from the same source *(24–31)*, most methods are simply a modification of the original procedure of Chirgwin et al. *(8)*. Such procedures utilize strong chaotropic agents, such as guanidinium thiocyanate and cesium trifluoroacetate *(25,27)*, to simultaneously disrupt cellular membranes and inactivate potent intracellular RNases *(26,29,30)*. The limitations of such techniques are the need for ultracentrifugation *(26–28,30)* and long processing times (ranging 16–44 h).

Methods for isolating both RNA and DNA that circumvent the ultracentrifugation step take advantage of the fact that phenol *(1,32)* can act as an efficient deproteinization agent quickly disrupting cellular integrity and denaturing proteins *(24,31)*. The method presented here takes advantage of the qualities offered by phenol extraction when it is coupled with a suitable extraction buffer and a means for selectively separating high-mol-wt DNA from RNA *(31)*.

The method utilizes an initial phenol extraction coupled with two phenol:chloroform extractions to simultaneously remove proteins and lipids from nucleic acid containing solutions. In addition, the constituents of the aqueous extraction buffer are optimized to increase nucleic acid recovery, as discussed by Wallace *(33)*. For example, the pH of the buffer (pH 7.5), the presence of detergent (0.2% SDS), and relatively low salt concentration (100 mM LiCl) allow the efficient partitioning of nucleic acids into the aqueous phase and the dissociation of proteins. In addition, the presence of 10 mM EDTA discourages the formation of protein aggregates *(33)* and chelates Mg^{2+}, thereby inhibiting the action of magnesium dependent nucleases *(34)*.

This method differs from that presented by Krieg et al. *(24)* in that the lysis and extraction procedure is gentle enough to allow the selective removal of high-mol-wt DNA by spooling onto a hooked glass rod *(2,34,35)* following ethanol precipitation. This avoids additional LiCl precipitation steps following the recovery of total nucleic acids. Finally, the procedure can be scaled up or down to accommodate various sample sizes, hence allowing the processing of multiple samples at one time. The approximate time required for the isolation of total cellular RNA and DNA is 2 h. Using this method nucleic acids have been isolated from PC12 cells and analyzed by Southern and Northern blotting techniques *(31)*.

2. Materials

Molecular biology grade reagents should be utilized whenever possible. Manipulations were performed in disposable, sterile polypropylene tubes whenever possible, otherwise glassware that had been previously baked at 280°C for at least 3 h was used.

2.1. Nucleic Extraction from Nonadherent Tissue Culture Cells

1. PBS: 0.137M NaCl, 2.68 mM KCl, 7.98 mM Na_2HPO_4, 1.47 mM KH_2PO_4, pH 7.2.
2. DEPC-treated water: Diethylpyrocarbonate (DEPC)-treated water is prepared by adding 1 mL DEPC to 1 L of double-distilled water (0.1% DEPC v/v) and stirring overnight. The DEPC is inactivated by autoclaving at 20 psi for 20 min (*see* Note 1).

3. STEL buffer: 0.2% SDS, 10 mM Tris-HCl, pH 7.5, 10 mM EDTA, and 100 mM LiCl. The buffer is prepared in DEPC-treated water by adding the Tris-HCl, EDTA, and LiCl components first, autoclaving, and then adding an appropriate volume of 10% SDS. The 10% SDS stock solution is prepared by dissolving 10 g SDS in DEPC-treated water and incubating at 65°C for 2 h prior to use.
4. Phenol: Phenol is equilibrated as described previously *(34)*. Ultrapure, redistilled phenol, containing 0.1% hydroxyquinoline (as an antioxidant), is initially extracted with 0.5M Tris-HCl, pH 8.0, and then repeatedly extracted with 0.1M Tris-HCl, pH 8.0, until the pH of the aqueous phase is 8.0. Then equilibrate with STEL extraction buffer twice prior to use. This can be stored at 4°C for at least 2 mo.
5. Phenol:chloroform mixture: A 1:1 mixture was made by adding an equal volume of chloroform to STEL-equilibrated phenol. Can be stored at 4°C for at least 2 mo. Phenol should be handled with gloves in a fume hood.
6. 5M LiCl: Prepare in DEPC-treated water and autoclave.
7. TE: 10 mM Tris-HCl, pH 8.0, 1 mM EDTA, pH 8.0. Prepare in DEPC-treated water and autoclave.
8. RNA guard, such as RNasin (Promega; Madison, WI).

2.2. Variations for Adherent Cell Cultures and Tissues

9. Trypsin: A 0.125% solution in PBS. For short term store at 4°C; for long term freeze.

3. Methods

3.1. Nucleic Extraction from Nonadherent Tissue Culture Cells

In this section we detail nucleic extraction from nonadherent tissue culture cells. Section 3.2. describes variations of this protocol for adherent cell cultures and tissue.

To prevent RNase contamination from skin, disposable gloves should be worn throughout the RNA isolation procedure. In addition, it is advisable to set aside equipment solely for RNA analysis; for example, glassware, pipets, and an electrophoresis apparatus.

1. Cultured cells (1 × 10^7) should be cooled on ice (*see* Note 2). Transfer to 15-mL polypropylene tubes and pellet by centrifugation at 100g for 5 min. Wash the cells once with 10 mL of ice-cold PBS and repellet. The pelleted cells may be left on ice to allow processing of other samples.
2. Simultaneously add 5 mL of STEL-equilibrated phenol and 5 mL of ice-cold STEL buffer to the pelleted cells. Gently mix the solution by inver-

sion for 3–5 min, ensuring the cellular pellet is thoroughly dissolved (*see* Notes 3 and 4).
3. Centrifuge the mixture at 10,000*g* for 5 min at 20°C to separate the phases. Transfer the aqueous (upper) phase to a new tube using a sterile polypropylene pipet and re-extract twice with an equal volume of phenol:chloroform (*see* Note 5).
4. Transfer the aqueous phase to a 50-mL Falcon tube. Differentially precipitate high-mol-wt DNA from the RNA component by addition of 0.1 vol of ice-cold 5*M* LiCl and 2 vol of ice-cold absolute ethanol. The DNA will precipitate immediately as a threaded mass.
5. Gently compact the mass by mixing and remove the DNA by spooling onto a hooked glass rod. Remove excess ethanol from the DNA by touching onto the side of the tube. Remove excess salts by rinsing the DNA with 1 mL of ice-cold 70% ethanol while still coiled on the rod. Excess ethanol can be removed by carefully washing the DNA with 1 mL of ice-cold TE, pH 8.0 (*see* Note 6).
6. The DNA is then resolubilized by transferring the glass rod into an appropriate volume of TE, pH 8.0 and storing at 4°C.
7. The RNA is precipitated by placing the tube with the remaining solution at –70°C, or in an ethanol/dry ice bath for 30 min.
8. Collect the RNA by centrifuging at 10,000*g* for 15 min. Gently aspirate the supernatant and rinse the pellet with ice-cold 70% ethanol. Recentrifuge for 5 min and remove the supernatant. Dry the RNA pellet under vacuum and resuspend in DEPC-treated water.
9. For storage as aqueous samples add 5–10 U of RNasin (RNase inhibitor) according to manufacturer's instructions. Alternatively, the RNA can be safely stored as an ethanol/LiCl suspension (*see* Notes 7 and 8).

3.2. Variations for Adherent Cell Cultures and Tissues

3.2.1. Adherent Cells

1. Remove the culture medium from the equivalent of 1×10^7 cells by aspiration and wash the cells once with 10 mL of PBS at 37°C.
2. Add 1 mL of trypsin solution and incubate plates at 37°C until the cells have been dislodged. This should take approx 10 min. Dilute the trypsin solution by addition of ice-cold PBS.
3. Follow Section 3.1., steps 2–9.

3.2.2. Procedure for Tissue

1. Rinse approx 500 mg of tissue free of blood with ice-cold PBS. Cool and mince into 3–5-cm cubes with a sterile blade.

Isolation of RNA and DNA

2. Gradually add the tissue to a mortar containing liquid nitrogen and ground to a fine powder.
3. Slowly add the powdered tissue to an evenly dispersed mixture of 5 mL of phenol and 5 mL of STEL. This is best accomplished by gradually stirring the powdered tissue into the phenol:STEL emulsion with a baked glass rod. Mix the tissue until the components are thoroughly dispersed. Continue mixing by gentle inversion for 5 min.
4. Follow Section 3.1., steps 3–9.

4. Notes

1. DEPC is a suspected carcinogen and should be handled with gloves in a fume hood. Because it acts by acylating histidine and tyrosine residues on proteins, susceptible reagents, such as Tris solutions, should not be directly treated with DEPC. Sensitive reagents should simply be made up in DEPC-treated water as outlined.
2. The integrity of the nucleic acids will be improved by maintaining harvested cells or tissues cold.
3. The success of this procedure hinges on the ability to gently disrupt cellular integrity while maintaining DNA in an intact, high-mol-wt form. Thus, mixing of the STEL:phenol should be performed by gentle inversion, which minimizes shearing forces on the DNA.
4. The proteinaceous interface that partitions between the aqueous (upper) and phenol phase following the initial phenol extraction (Section 3.1., step 3) can be re-extracted with phenol:chloroform to improve DNA recovery.
5. Chloroform is commonly prepared as a 24:1 (v/v) mixture with isoamyl alcohol, which acts as a defoaming agent. We have found that foaming is not a problem if extractions are performed by gentle inversion or on a rotating wheel and routinely omit isoamyl alcohol from the mixture.
6. The DNA may be air dried, but will then take longer to resuspend.
7. Following the selective removal of high-mol-wt DNA, the remaining RNA is sufficiently free of DNA contamination such that DNA is not detected by ethidium bromide staining *(31)*. If the purified RNA is to be used for PCR procedures it is strongly recommended that a RNase-treated control be performed to ensure the absence of contaminating DNA. This recommendation extends to virtually any RNA purification procedure, particularly those involving an initial step in which the DNA is sheared.
8. Typical yields of total cellular RNA range between 60–170 µg when using approx 1.5–2×10^7 cells with A_{260}/A_{280} values of approx 1.86 *(31)*. These values compare favorably with those obtained using guanidinium thiocyanate CsCl centrifugation methods *(8)*.

References

1. Graham, D. E. (1978) The isolation of high molecular weight DNA from whole organisms of large tissue masses. *Anal. Biochem.* **85**, 609–613.
2. Bowtell, D. D. (1987) Rapid isolation of eukaryotic DNA. *Anal. Biochem.* **162**, 463–465.
3. Longmire, J. L., Albright, K. L., Meincke, L. J., and Hildebrand, C. E. (1987) A rapid and simple method for the isolation of high molecular weight cellular and chromosome-specific DNA in solution without the use of organic solvents. *Nucleic Acids Res.* **15**, 859.
4. Owen, R. J. and Borman, P. (1987) A rapid biochemical method for purifying high molecular weight bacterial chromosomal DNA for restriction enzyme analysis. *Nucleic Acids Res.* **15**, 3631.
5. Reymond, C. D. (1987) A rapid method for the preparation of multiple samples of eukaryotic DNA. *Nucleic Acids Res.* **15**, 8118.
6. Miller, S. A. and Polesky, H. F. (1988) A simple salting out procedure for extracting DNA from human nucleated cells. *Nucleic Acids Res.* **16**, 1215.
7. Grimberg, J., Nawoschik, S., Belluscio, L., McKee, R., Turck, A., and Eisenberg, A. (1989) A simple and efficient non-organic procedure for the isolation of genomic DNA from blood. *Nucleic Acids Res.* **17**, 8390.
8. Chirgwin, J. M., Przybyla, A. E., MacDonald, R. J., and Rutter, W. J. (1979) Isolation of biologically active ribonucleic acid from sources enriched in ribonuclease. *Biochemistry* **18**, 5294–5299.
9. Auffray, C. and Rougeon, F. (1980) Purification of mouse immunoglobulin heavy-chain messenger RNAs from total myeloma tumour RNA. *Eur. J. Biochem.* **107**, 303–314.
10. Elion, E. A. and Warner, J. R. (1984) The major promoter element of rRNA transcription in yeast lies 2 Kb upstream. *Cell* **39**, 663–673.
11. Chomczynski, P. and Sacchi, N. (1987) Single-step method of RNA extraction by acid guanidinium thiocyanate-phenol-chloroform extraction. *Anal. Biochem.* **162**, 156–159.
12. Hatch, C. L. and Bonner, W. M. (1987) Direct analysis of RNA in whole cell and cytoplasmic extracts by gel electrophoresis. *Anal. Biochem.* **162**, 283–290.
13. Emmett, M. and Petrack, B. (1988) Rapid isolation of total RNA from mammalian tissues. *Anal. Biochem.* **174**, 658–661.
14. Gough, N. M. (1988) Rapid and quantitative preparation of cytoplasmic RNA from small numbers of cells. *Anal. Biochem.* **173**, 93–95.
15. Meier, R. (1988) A universal and efficient protocol for the isolation of RNA from tissues and cultured cells. *Nucleic Acids Res.* **16**, 2340.
16. Wilkinson, M. (1988) RNA isolation: a mini-prep method. *Nucleic Acids Res.* **16**, 10,933.
17. Wilkinson, M. (1988) A rapid and convenient method for isolation of nuclear, cytoplasmic and total cellular RNA. *Nucleic Acids Res.* **16**, 10,934.
18. Ferre, F. and Garduno, F. (1989) Preparation of crude cell extract suitable for amplification of RNA by the polymerase chain reaction. *Nucleic Acids Res.* **17**, 2141.
19. McEntee, C. M. and Hudson, A. P. (1989) Preparation of RNA from unspheroplasted yeast cells *(Saccharomyces cerevisiae). Anal. Biochem.* **176**, 303–306.

20. Nemeth, G. G., Heydemann, A., and Bolander, M. E. (1989) Isolation and analysis of ribonucleic acids from skeletal tissues. *Anal. Biochem.* **183,** 301–304.
21. Verwoerd, T. C., Dekker, B. M. M., and Hoekema, A. (1989) A small-scale procedure for the rapid isolation of plant RNAs. *Nucleic Acids Res.* **17,** 2362.
22. Schmitt, M. E., Brown, T. A., and Trumpower, B. L. (1990) A rapid and simple method for preparation of RNA from *Saccharomyces cerevisiae*. *Nucleic Acids Res.* **18,** 3091.
23. Tavangar, K., Hoffman, A. R., and Kraemer, F. B. (1990) A micromethod for the isolation of total RNA from adipose tissue. *Anal. Biochem.* **186,** 60–63.
24. Krieg, P., Amtmann, E., and Sauer, G. (1983) The simultaneous extraction of high-molecular-weight DNA and of RNA from solid tumours. *Anal. Biochem.* **134,** 288–294.
25. Mirkes, P. E. (1985) Simultaneous banding of rat embryo DNA, RNA and protein in cesium trifluroracetate gradients. *Anal. Biochem.* **148,** 376–383.
26. Meese, E. and Blin, N. (1987) Simultaneous isolation of high molecular weight RNA and DNA from limited amounts of tissues and cells. *Gene Anal. Tech.* **4,** 45–49.
27. Zarlenga, D. S. and Gamble, H. R. (1987) Simultaneous isolation of preparative amounts of RNA and DNA from Trichinella spiralisby cesium trifluoroacetate isopycnic centrifugation. *Anal. Biochem.* **162,** 569–574.
28. Chan, V. T.-W., Fleming, K. A., and McGee, J. O. D. (1988) Simultaneous extraction from clinical biopsies of high-molecular-weight DNA and RNA: comparative characterization by biotinylated and ^{32}P-labeled probes on Southern and Northern blots. *Anal. Biochem.* **168,** 16–24.
29. Karlinsey, J., Stamatoyannopoulos, G., and Enver, T. (1989) Simultaneous purification of DNA and RNA from small numbers of eukaryotic cells. *Anal. Biochem.* **180,** 303–306.
30. Coombs, L. M., Pigott, D., Proctor, A., Eydmann, M., Denner, J., and Knowles, M. A. (1990) Simultaneous isolation of DNA, RNA and antigenic protein exhibiting kinase activity from small tumour samples using guanidine isothiocyanate. *Anal. Biochem.* **188,** 338–343.
31. Raha, S., Merante, F., Proteau, G., and Reed, J. K. (1990) Simultaneous isolation of total cellular RNA and DNA from tissue culture cells using phenol and lithium chloride. *Gene Anal. Tech.* **7,** 173–177.
32. Kirby, K. S. (1957) A new method for the isolation of deoxyribonucleic acids: evidence on the nature of bonds between deoxyribonucleic acid and protein. *Biochem. J.* **66,** 495–504.
33. Wallace, D. M. (1987) Large and small scale phenol extractions, in *Methods in Enzymology, vol. 152: Guide to Molecular Cloning Techniques* (Berger, S. L. and Kimmel, A. R., eds.), Academic, Orlando, FL, pp. 33–41.
34. Sambrook, J., Fritsch, E. F., and Maniatis, T. (1989) *Molecular Cloning: A Laboratory Manual,* 2nd ed. Cold Spring Harbor Laboratory, Cold Spring, Harbor, NY.
35. Davis, L. G., Dibner, M. D., and Battey, J. F. (1986) *Basic Methods in Molecular Biology.* Elsevier, New York.

CHAPTER 2

Restriction Endonuclease Digestion of DNA

Duncan R. Smith

1. Introduction

The ability to cleave DNA at specific sites is one of the cornerstones of today's methods of DNA manipulation. Restriction endonucleases are bacterial enzymes that cleave duplex DNA at specific target sequences with the production of defined fragments. These enzymes can be purchased from the many manufacturers of biotechnology products. The nomenclature of enzymes is based on a simple system, proposed by Smith and Nathans *(1)*. The name of the enzyme (such as *Bam*HI, *Eco*RI, and so on) tells us about the origin of the enzyme, but does not give us any information about the specificity of cleavage (*see* Note 1). This has to be determined for each individual enzyme. The recognition site for most of the commonly used enzymes is a short palindromic sequence, usually either 4, 5, or 6 bp in length, such as AGCT (for *Alu*I), GAATTC (for *Eco*RI), and so on. Each enzyme cuts the palindrome at a particular site, and two different enzymes may have the same recognition sequence, but cleave the DNA at different points within that sequence. The cleavage sites fall into three different categories, either flush (or blunt) in which the recognition site is cut in the middle, or with either 5'- or 3'-overhangs, in which case unpaired bases will be produced on both ends of the fragment. For a comprehensive review of restriction endonucleases, *see* Fuchs and Blakesley *(2)*.

2. Materials

1. A 10X stock of the appropriate restriction enzyme buffer (*see* Note 2).
2. DNA to be digested (*see* Notes 3 and 4) in either water or TE (10 mM Tris-HCl, pH 8.3, 1 mM EDTA).
3. Bovine serum albumin (BSA) at a concentration of 1 mg/mL (*see* Note 5).
4. Sterile distilled water (*see* Note 6).
5. The correct enzyme for the digest (*see* Note 7).
6. 5X loading buffer: 50% (v/v) glycerol, 100 mM Na$_2$EDTA, pH 8, 0.125% (w/v) bromophenol blue (6pb), 0.125% (w/v) xylene cyanol.
7. 100 mM Spermidine (*see* Note 8).

3. Methods

1. Thaw all solutions, with the exception of the enzyme, and then place on ice.
2. Decide on a final volume for the digest, usually between 10 and 50 µL (*see* Note 9), and then into a sterile Eppendorf tube, add 1/10 vol of reaction buffer, 1/10 vol of BSA, between 0.5 and 1 µg of the DNA to be digested (*see* Note 3), and sterile distilled water to the final volume.
3. Take the restriction enzyme stock directly from the –20°C freezer, and remove the desired units of enzyme (*see* Notes 7 and 10) with a clean sterile pipet tip. Immediately add the enzyme to the reaction and mix (*see* Note 11).
4. Incubate the tube at the correct temperature (*see* Note 12) for approx 1 h. Genomic DNA can be digested overnight.
5. An aliquot of the reaction (usually 1–2 µL) may be mixed with a 5X concentrated loading buffer and analyzed by gel electrophoresis (*see* Chapter 3).

4. Notes

1. Enzymes are named according to the system proposed by Smith and Nathans *(1)* in which enzymes are named according to the bacteria from which they are first purified. Therefore, for example, a restriction enzyme purified from *Providencia stuartii*, would be identified by the first letter of the genus name (in this case *Providencia* and hence *P*) and the first two letters of the specific epithet (in this case *stuartii* and hence *st*) joined together to form a three-letter abbreviation—*Pst*. The first restriction enzyme isolated from this source of bacteria would therefore be called *Pst*I (with the number in Roman numerals), and the second *Pst*II, and so on. Note, however, that the name of the enzyme gives no information about the specificity of cleavage, which must be determined from one of the numerous lists of enzymes and cleavage specificities (the catalog of most suppliers of restriction enzymes will provide extensive information about

Restriction Endonuclease Digestion

restriction enzymes, such as specificity of cleavage, optimal reaction conditions, number of cleavage sites in common DNA templates, and so on, and these catalogs should be treated as valuable sources of information).

2. Each enzyme has an optimal reaction buffer. The recommended reaction conditions are normally to be found on the manufacturer's assay sheet. In practice, many enzymes share common conditions, and it is possible to make up reaction buffers that are suitable for a number of enzymes. The vast majority of enzymes will work in one of three buffers, either a high-, low-, or medium-salt buffer, recipes for which are given below. These buffers are normally made as a 10X stock and then 1/10 final vol is added to each digest. Great care must be taken in matching the buffer to the enzyme, since the wrong buffer can give either a dramatically reduced activity, altered specificity, or no activity at all. Several manufacturers of restriction enzymes now provide the correct buffer with their enzymes as an added benefit, and it is recommended that where these buffers are provided, they should be used.
   a. High-salt buffer (1X): 100 mM NaCl, 50 mM Tris-HCl, pH 7.5, 10 mM MgCl$_2$, 1 mM DTT.
   b. Medium-salt buffer (1X): 50 mM NaCl, 10 mM Tris-HCl, pH 7.5, 10 mM MgCl$_2$, 1 mM DTT.
   c. Low-salt buffer (1X): 10 mM Tris-HCl, pH 7.5, 10 mM MgCl$_2$, 1 mM DTT.
   In addition, two "universal buffers" are occasionally used, which are buffers in which all restriction enzymes have activity, although in some cases, activity can be reduced to only 20% of optimal activity. These are the potassium-glutamate *(3)* and potassium-acetate *(4)* buffers. These buffers can be particularly useful when a piece of DNA must be digested by two enzymes having very different optimal buffers.
3. The amount of DNA to be digested depends on subsequent steps. A reasonable amount for a plasmid digestion to confirm the presence of an insertion would be 500 ng to 1 µg, depending on the size of the insert. The smaller the insert, the more DNA should be digested to enable visualization of the insert after agarose gel analysis.
4. The DNA to be digested should be relatively pure and free from reagents, such as phenol, chloroform, alcohols, salts, detergents, and chelating agents. Any trace amounts of these chemicals will inhibit or inactivate the restriction endonuclease.
5. BSA is routinely included in restriction digests to stabilize low protein concentrations and to protect against factors that cause denaturation.
6. Good-quality sterile distilled water should be used in restriction digests. Water should be free of ions and organic compounds, and must be detergent-free.

7. An enzyme unit is defined as the amount of enzyme required to digest 1 µg of a standard DNA in 1 h under optimal temperature and buffer conditions. The standard DNA used is normally λ DNA. Hence, for *Eco*RI (for example), there are five sites for this enzyme in λ. If one is digesting pBR322, which has one site with 1 U of enzyme for 1 h, this is actually a fivefold overdigestion.
8. Digests of genomic DNA are dramatically improved by the inclusion of spermidine in the digest mixture to a final concentration of 1 m*M*, since the polycationic spermidine binds negatively charged contaminants. Note that spermidine can cause precipitation of DNA at low temperatures, so it should not be added while the reaction is kept on ice.
9. The smallest practical volume in which to undertake a restriction digest is 10 µL. Below this, pipeting errors can introduce significant errors in the reaction conditions. This volume also allows the entire digest to be loaded onto a small agarose gel after the addition of the stop/loading buffer. If the stock DNA concentration is too dilute to give 0.5–1 µg in 5–6 µL, then the reaction can be scaled up to 20–50 µL. If double digestion is to be undertaken (i.e., digestion with two different enzymes), then 20 µL is the recommended minimum volume, 1 µL of each enzyme can be added, and the glycerol concentration is kept low (*see* Note 10).
10. Many enzymes are susceptible to the presence of glycerol. The majority of stock enzymes are provided in approx 50% (v/v) glycerol. A reaction digest in which more than approx 10% (v/v) glycerol is present can give cleavage at different sites from the normal (the so-called star activity). For this reason, it is advisable to keep the enzyme total reaction volume ratio at 1:10 or lower. Similar star activity can result from incorrect salt concentrations.
11. Stock restriction enzymes are very heat-labile and so should be removed from −20°C storage for as short a time as possible and placed on ice.
12. Note that the incubation temperature for the vast majority of restriction endonucleases is 37°C, but that this is not true for all enzymes. Other enzymes, such as *Taq*I and *Sma*I, require different optimal temperatures (in this case 65 and 25°C, respectively). It is wise, therefore, to check new or unfamiliar enzymes before use.
13. If large-scale preparative digests are to be undertaken (100–500 µL reaction mixes), then the reaction is scaled up accordingly. However, care must be taken to ensure that the reaction components are fully mixed, especially with regard to the viscous constituents, such as DNA solutions and stock restriction enzymes. For all volume digests, vortexing should be avoided, since this can significantly reduce the activity of the enzyme. For small volumes, mixing can be achieved by tapping or gently flicking the tube

with a finger (often followed by a brief 1–5 s spin in an Eppendorf centrifuge to deposit the reaction at the bottom of the tube). For larger volumes, mixing can be achieved by gentle pipeting, taking liquid from the bottom of the reaction volume and mixing at the top of the reaction volume until a homogenous solution is obtained.

References

1. Smith, H. O. and Nathans, D. (1973) A suggested nomenclature for bacterial host modification and restriction systems and their enzymes. *J. Mol. Biol.* **81,** 419–423.
2. Fuchs, R. and Blakesley, R. (1983) Guide to the use of Type II restriction endonucleases. *Methods Enzymol.* **100,** 3–38.
3. McClelland, M., Hanish, J., Nelson, M., and Patel, Y. (1988) KGB: a single buffer for all restriction endonucleases. *Nucleic Acids Res.* **16,** 364.
4. O'Farrell, P. H., Kutter, E., and Nakanishe, M. (1980) A restriction map of the bacteriophage T4 genome. *Mol. Gen. Genet.* **170,** 411–435.

CHAPTER 3

Agarose Gel Electrophoresis

Duncan R. Smith

1. Introduction

After digestion of DNA with a restriction enzyme (*see* Chapter 2), it is usually necessary, for both preparative and analytical purposes, to separate and visualize the products. In most cases, where the products are between 200 and 20,000 bp long, this is achieved by agarose gel electrophoresis. Agarose is a linear polymer that is extracted from seaweed and sold as a white powder that is melted in buffer and allowed to cool, whereby the agarose forms a gel by hydrogen bonding. The hardened matrix contains pores, the size of which depends on the concentration of agarose. The concentration of agarose is referred to as a percentage of agarose to volume of buffer (w/v), and agarose gels are normally in the range of 0.3–3%. Many different apparatus arrangements have been devised to run agarose gels. For example, they can be run horizontally or vertically, and the current can be conducted by wicks or the buffer solution. However, today, the "submarine" gel system is almost universally used. In this method, the agarose gel is formed on a supporting plate, and then the plate is submerged into a tank containing a suitable electrophoresis buffer. Wells are preformed in the agarose gel with the aid of a "comb" that is inserted into the cooling agarose before it has gelled. Into these wells is loaded the sample to be analyzed, which has been mixed with a dense solution (a loading buffer) to ensure that the sample sinks into the wells.

Electrophoresis apparatus is arguably one of the most vital pieces of equipment in the laboratory. It consists of four main parts: a power supply (capable of at least 100 V and currents of up to 100 mA), an electrophoresis tank, a casting plate, and a well-forming comb. Apparatus is available from many commercial suppliers, but tends to be fairly expensive. Alternatively, apparatus can be "home made" with access to a few sheets of perspex and minor electrical fittings. The construction of such apparatus is outside the scope of this chapter, but can be found in refs. *1–3*.

The essence of electrophoresis is that when DNA molecules within an agarose gel matrix are subjected to a steady electric field, they first orient in an end-on position *(4,5)* and then migrate through the gel at rates that are inversely proportional to the $\log_{10}$ of the number of base pairs *(6)*. This is because larger molecules migrate more slowly than smaller molecules because of their higher frictional drag and greater difficulty in "worming" through the pores of the gel *(1)*. This relationship only applies to linear molecules. Circular molecules, such as plasmids, migrate much more quickly than their molecular weight would imply because of their smaller apparent size with respect to the gel matrix. The migration rate also depends on other factors, such as the composition and ionic strength of the electrophoresis buffer as well as the percentage of agarose in the gel. The gel percentage presents the best way to control the resolution of agarose gel electrophoresis (*see* Table 1). An excellent treatment of the theory of gel electrophoresis can be found in Sambrook et al. *(1)*.

2. Materials

1. Molecular-biology grade agarose (high melting point, *see* Table 1).
2. Running buffer at 1X and 10X concentrations (*see* Table 2 for choice).
3. Sterile distilled water.
4. A heating plate or microwave oven.
5. Suitable gel apparatus and power pack: *see* Section 1.
6. Ethidium bromide: Dissolve in water at 10 mg/mL (*see* Note 1). Ethidium bromide is both carcinogenic and mutagenic and therefore must be handled with extreme caution.
7. An ultraviolet (UV) light transilluminator (long wave, 365 nm).
8. 5X loading buffer (*see* Note 2): Many variations exist, but this one is fairly standard: 50% (v/v) glycerol, 50 mM EDTA, pH 8.0, 0.125% (w/v) bromophenol blue, 0.125% (w/v) xylene cyanol.
9. A size marker: a predigested DNA sample for which the product band sizes are known. Many such markers are commercially available.

Agarose Gel Electrophoresis

Table 1
Resolution of Agarose Gels

Agarose, %	Mol wt range, kb	Comments
0.2	5–40	Gel very weak; separation in 20–40 kb range improved by increase in ionic strength of running buffer (i.e., Loenings E); only use high-melting point agarose
0.4	5–30	With care can use low-melting point agarose
0.6	3–10	Essentially as above, but with greater mechanical strength
0.8	1–7	General-purpose gel separation not greatly affected by choice of running buffer; bromophenol blue runs at about 1 kb
1	0.5–5	As for 0.8%
1.5	0.3–3	As for 0.8%, bromophenol blue runs at about 500 bp
2	0.2–1.5	Do not allow to cool to 50°C before pouring
3	0.1–1	Can separate small fragments differing from each other by a small amount; must be poured rapidly onto a prewarmed glass plate

3. Methods

1. An appropriate amount of powdered agarose (Table 1) is weighed carefully into a conical flask.
2. One-tenth of the final volume of 10X concentrated running buffer is added (Table 2), followed by distilled water to the final volume.
3. The contents of the flask are mixed by swirling, and placed on a hot plate or in a microwave until the contents just start to boil and all the powdered agarose is melted.
4. The contents are cooled to approx 50°C and ethidium bromide solution added to give a final concentration of 5 µg/mL. The gel mixture can then be poured into the gel apparatus.
6. A "comb" is inserted into the apparatus to form the wells, and the gel is left to solidify.
7. When the gel has set, the comb can be carefully removed and the solidified gel, still on its gel plate support, placed into the running apparatus. Fill with 1X running buffer to just cover the wells.

Table 2
Commonly Used Agarose Gel Electrophoresis Running Buffers

Buffer	Description	Solution
Loenings E	High ionic strength, and not recommended for preparative gels	For 5 L of 10X: 218 g of Tris base, 234 g of $NaH_2PO_4 \cdot 2H_2O$, and 18.6 g of $Na_2EDTA \cdot 2H_2O$
Glycine	Low ionic strength, very good for preparative gels, but can also be used for analytical gels	For 2 L 10X: 300 g of glycine, 300 mL of $1M$ NaOH (or 12 g pellets), and 80 mL of $0.5M$ EDTA, pH 8.0
Tris-borate EDTA (TBE)	Low ionic strength can be used for both preparative and analytical gels	For 5 L of 10X: 545 g of Tris, 278 g boric acid, and 46.5 g of EDTA
Tris-acetate (TAE)	Good for analytical gels and preparative gels when the DNA is to be purified by glass beads (*see* Chapter 28)	For 1 L of 50X: 242 g of Tris base, 57.1 mL of glacial acetic acid, and 100 mL of $0.5M$ EDTA, pH 8.0

8. DNA samples (*see* Note 3) are prepared by the addition of 5X loading and loaded into the wells of the gel. All samples are loaded at the same time. It is usual to include a size marker in one of the lanes.
9. The lid to the apparatus is closed and the current applied (*see* Note 4). The gel is usually run between 1 and 3 h, depending on the percentage of the gel and length.
10. After electrophoresis the gel is removed from the apparatus, and the products of the digestion can be viewed on a UV transilluminator.

4. Notes

1. Many workers do not like to include ethidium bromide in their gels and their running buffers, preferring instead to stain their gels after electrophoresis because ethidium intercalation can affect the mobility of the DNA, especially where circular plasmids (either supercoiled or nicked circle) are concerned. However, if the presence or absence of ethidium is kept constant, then no difficulty is encountered.

   A second problem that may be encountered when using ethidium bromide is that it promotes DNA damage under UV illumination (by photonicking). For this reason, if the gel is run with ethidium bromide in the gel and the running buffer, it is best to keep the viewing time to a

minimum to prevent damage to the DNA molecule and subsequent smearing on re-electrophoresis.
2. Loading buffer is a dense solution (usually containing either glycerol or Ficoll) that when mixed with a DNA solution (or restriction digest) gives the sample sufficient density to fall to the bottom of the sample well that has already been filled with running buffer. Loading buffers normally contain either one or two marker dyes that migrate in the electric field in the same direction as the DNA. Two commonly used dyes are bromophenol blue and xylene cyanol. These dyes migrate at different rates from each other and are useful for monitoring the progress of an electrophoretic run, ensuring that the DNA does not pass out of the bottom of the gel. In a 0.8% (w/v) agarose gel, bromophenol blue migrates with DNA of approx 1 kb. Xylene cyanol in the same gel migrates at approx 4 kb.
3. The amount of DNA that can be visualized in a single band with ethidium bromide staining can, in ideal circumstances, be as low as 10 ng. In general circumstances, a single fragment of approx 100 ng can be easily seen. If a restriction enzyme digest produces a large number of bands, then relatively more DNA will have to be loaded so all bands will be seen.
4. When running an analytical gel, the optimal resolution is obtained at about 10 V/cm of gel. When fragments of 5 kb and above are to be analyzed, better resolution is obtained at about 5 V/cm. Fragments smaller than 1 kb are normally resolved better at higher V/cm.

References

1. Sambrook, J., Fritsch, E. F., and Maniatis, T. (1989) *Molecular Cloning: A Laboratory Manual.* Cold Spring Harbor Laboratory, Cold Spring Harbor, NY.
2. Sealey, P. G. and Southern, E. M. (1982) Electrophoresis of DNA, in *Gel Electrophoresis of Nucleic Acids: A Practical Approach* (Rickwood, D. and Hames, B. D., eds.), IRL, Oxford and Washington, pp. 39–76.
3. Boffey, S. A. (1984) Agarose gel electrophoresis of DNA, in *Methods in Molecular Biology Vol. 2: Nucleic Acids* (Walker, J. M., ed.), Humana, Clifton, NJ, pp. 43–50.
4. Fisher, M. P. and Dingman, C. W. (1971) Role of molecular conformation in determining the electrophoretic properties of polynucleotides in agarose-acrylamide composite gels. *Biochemistry* **10**, 895.
5. Aaij, C. and Borst, P. (1972) The gel electrophoresis of DNA. *Biochem. Biophys. Acta* **269**, 192.
6. Helling, R. B., Goodman, H. M., and Boyer, H. W. (1974) Analysis of R.EcoRI fragments of DNA from lambdoid bacteriophages and other viruses by agarose gel electrophoresis. *J. Virol.* **14**, 1235–1244.

CHAPTER 4

Capillary Blotting of Agarose Gels

Duncan R. Smith and David Murphy

1. Introduction

Southern blotting is a well-known technique *(1)*. DNA is cleaved with restriction enzymes (Chapter 2) to produce fragments that are fractionated according to their size in an agarose gel (Chapter 3). The DNA is then partially cleaved by depurination (to facilitate the transfer of larger DNA fragments) and alkali denatured by sequential soaking of the gel in solutions containing HCl and NaOH, respectively. The denatured DNA fragments are then transferred to a solid matrix or filter (usually a nylon membrane) for subsequent hybridization to a specific labeled probe (Chapter 6).

2. Materials

1. Nylon hybridization membrane (e.g., Amersham [UK] Hybond-N; *see* Note 1).
2. Capillary transfer system (*see* Note 2).
3. Depurination buffer: $0.25M$ HCl.
4. Denaturation buffer: $1.5M$ NaCl, $0.5M$ NaOH.
5. Transfer buffer: $1.5M$ NaCl, $0.25M$ NaOH.
6. 20X SCC: $3M$ NaCl, $0.3M$ sodium citrate, pH 7.0.
7. 312-nm UV light transilluminator.

3. Methods

1. Digest DNA and separate by electrophoresis (*see* Chapters 2 and 3; Notes 3 and 4).
2. Remove unused portions of the gel with a clean scalpel blade.
3. Incubate the gel in approx 3 gel volumes of depurination buffer with gentle agitation at room temperature for 30 min, or until the bromophenol blue in the loading dye turns yellow.

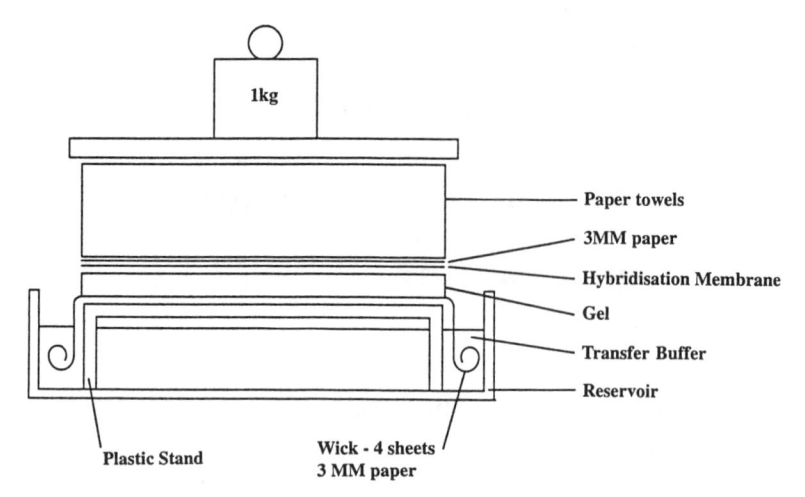

Fig. 1. A typical capillary action Southern transfer system.

4. Decant the depurination buffer, and replace with 3 gel volumes of denaturation buffer. Incubate with gentle agitation at room temperature for 30 min.
5. Decant the denaturation buffer, and replace with 3 gel volumes of transfer buffer. Equilibriate the gel with gentle agitation at room temperature for 30 min.
6. Place the gel on the platform of a capillary transfer system filled with transfer buffer. A diagram of a transfer system is shown in Fig. 1. The system is made up of a platform that sits in a reservoir containing transfer buffer. A wick, made up of four thicknesses of 3MM paper, is placed over the platform, soaked in transfer buffer. All air bubbles must be removed from the wick. The width and length of the platform correspond to the size of the gel, and the wick is cut to the same width. The platform and reservoir are made to the same height.
7. Cut a piece of nylon membrane to the same size as the gel. Wet this by floating it on distilled water and then rinse in transfer buffer. Place the filter on the gel, and smooth out any air bubbles.
8. Cut four pieces of 3MM paper to the same size as the gel. Soak two in transfer buffer, and place them over the filter. Smooth out any bubbles. Place two dry filters onto the sandwich, and then a stack of dry paper towels. Place a 1-kg weight on top, and allow the transfer to proceed for at least 12 h.
9. Disassemble the transfer system. Prior to separating the gel from the filter, the position of the gel slots can be marked. If this is done with a pencil, the marks will appear on the resulting autoradiograph. Rinse the filter in 2X SSC, and bake the filter at 80°C for 20–60 min.

10. Covalently crosslink the DNA to the matrix by exposure to a 312 nm UV light transilluminator. Place the filter, DNA side down, on a piece of Saran Wrap™, and expose for 2–3 min (*see* Note 5).

The filter can be used immediately for hybridization (*see* Chapter 6) or stored dry until required.

4. Notes

1. Note that these methodologies have been developed for neutral nylon membranes (e.g., Amersham Hybond-N) and have not been tested on positively charged membranes (e.g., Amersham Hybond-N+, Bio-Rad [Richmond, CA] Zetaprobe, NEN- Du Pont [Boston, MA] GeneScreen™-*Plus*).
2. The capillary transfer system described here is efficient, but time-consuming. A number of companies now market other systems (e.g., Vacuum blotting: Pharmacia-LKB [Uppsala, Sweden], Hybaid [Twickenham, UK], positive pressure blotting: Stratagene [La Jolla, CA]) that reduce the transfer process to as little as 1 h.
3. The size of a hybridizing band is determined relative to DNA standards of a known size (e.g., bacteriophage λ cut with *Eco*RI and *Hin*dIII, or the convenient 1 kb ladder marketed by Life Technologies [Gaithersburg, MD]). It is best to end–label (*see* Chapter 15) DNA standards radioactively, such that an image of their position is produced on the final autoradiograph.
4. Gene copy number can be determined by Southern blotting by comparing the level of hybridization to copy number standards (prepared by diluting a known quantity of the unlabeled cloned DNA fragments) to the level of hybridization to genomic DNAs. The latter figure should be corrected with respect to the hybridization of a probe to an endogenous standard host gene.
5. Efficient crosslinking of DNA to nylon filters is achieved with an optimal amount of exposure to UV light. After a certain value, the efficiency decreases with increasing exposure. For this reason, it is best to calibrate a UV source before usage. This can be done by taking filters with an identical amount of DNA on each, exposing them to UV for different lengths of time, and then hybridizing them to the same probe. The strongest signal will establish the optimal time for exposure. Note that the energy output of a standard UV transilluminator varies with the age of the bulb, thus necessitating regular recalibration. Some manufacturers (e.g., Stratagene) produce UV crosslinkers that automatically expose the filter to the radiation for the optimal time, delivering a fixed dose of energy.

References

1. Southern, E. M. (1975) Detection of specific sequences among DNA fragments separated by gel electrophoresis. *J. Mol. Biol.* **98**, 503–517.

CHAPTER 5

Random Primed ^{32}P-Labeling of DNA

Duncan R. Smith

1. Introduction

Random primed labeling of DNA has now almost superseded the method of nick translation of DNA. Random primed labeling, based on the method of Feinberg and Vogelstein *(1),* is a method of incorporating radioactive nucleotides along the length of a fragment of DNA. Random primed labeling can give specific activities of between 2×10^9 and 5×10^9 dpm/µg (*see* Note 1). The method below is essentially that described by Feinberg and Vogelstein *(2)* in which a DNA fragment is denatured by heating in a boiling water bath. Then, random sequence oligonucleotides are annealed to both strands. Klenow fragment polymerase is then used to extend the oligonucleotides, using three cold nucleotides and one radioactively labeled nucleotide provided in the reaction mixture to produce a uniformly labeled double-stranded probe. Each batch of random oligonucleotides contains all possible sequences (for hexamers, which are most commonly employed, this would be 4096 different oligonucleotides), so any DNA template can be used with this method.

2. Materials (*see* Note 2)

1. DNA fragment to be labeled in water or 1X TE (10 mM Tris-HCl, 1 mM EDTA, pH 8.0, *see* Note 3).
2. OLB buffer. Make up the following solutions:
   Solution O: 1.25M Tris-HCl, pH 8.0, 0.125M MgCl$_2$ (store at 4°C).

Solution A: 1 mL solution O, 18-µL 2β-mercaptoethanol, 5 µL of dATP,* 5 µL of dTTP,* 5 µL of dGTP.*

Solution B: 2M HEPES, titrated to pH 6.6 with 4M NaOH (store at 4°C).

Solution C: Hexadeoxyribonucleotides evenly suspended (this does not completely dissolve) in TE, pH 8.0, at 90 OD_{260} U/mL (store at −20°C).

To make OLB buffer, mix solutions A:B:C in a ratio of 100:250:150. Store OLB at −20°C.

3. A nucleotide labeled at the α position with phosphorous-32, e.g., ^{32}P-dCTP; SA 3000 Ci/mM. Store at −20°C.
4. Klenow: The large fragment of DNA polymerase 1 (1 U/µL). Store at −20°C.
5. Bovine serum albumin (BSA) at 10 mg/mL in water.

3. Method (*see* Note 4)

1. Take approx 30 ng of DNA to be labeled (the probe), and make the volume up to 31 µL with sterile distilled water.
2. Boil the DNA for approx 3 min, and then immediately place the tube on ice.
3. Add, in the following order: 10 µL of OLB buffer, 2 µL of 10 mg/mL BSA, 5 µL of labeled nucleotide, and 2 µL of Klenow fragment. Mix all the components together by gentle pipeting.
4. Incubate at room temperature for 4–16 h (*see* Note 5).

The probe is ready to use. Remember to denature before use in hybridization (*see* Notes 6 and 7).

4. Notes

1. The specific activity of the probe can be increased by using more than one radiolabeled nucleotide. It is possible to use all four nucleotides as labeled nucleotides, but with the already high specific activities obtainable by this method, there are very few circumstances where this could be justified.
2. Many manufacturers now produce kits (Amersham [Amersham, UK], NEN-Dupont [Boston, MA]) for use in random primed labeling of DNA. These kits are simple and efficient.
3. This protocol is for purified DNA. In most cases, the DNA can be used without purifying the DNA after preparative gel electrophoresis. In this case, the gel slice is diluted with sterile distilled water at a ratio of 3 mL of water/g of gel slice, the DNA denatured, and the gel melted by boiling for 7 min, and then aliquoted into several tubes for either storage at −20°C (in which case the sample is boiled for 3 min before using) or immediate use.

*Where each nucleotide has been previously dissolved in 3 mM Tris-HCl, 0.2 mM EDTA, pH 7.0, at a concentration of 0.1M.

4. When making probes to detect highly abundant nucleic acid sequences, the random primed reaction can be scaled down to half the amounts given above.
5. Incubation times are optimal after about 4 h, whereby >70% of the radioactivity has been incorporated, although it is often convenient to leave the reaction overnight (12–16 h).
6. It is usually not necessary to purify the probe. If the reaction has proceeded correctly, approx 70% of the label will be incorporated. The unincorporated label does not interfere with subsequent usage, although probes can be purified by Sephadex spin columns (*see* Chapter 49).
7. Incorporation of the activity can be checked by diluting down an aliquot of the multiprime reaction to give something on the order of 10^4–10^5 dpm in 1–10 µL (50 µCi = 1.1×10^8 dpm). An aliquot of the diluted radioactivity is then spotted onto two Whatman DE81 disks. One of these is washed five times in $0.5M$ Na_2HPO_4 followed by two washings in water and one in 95% ethanol. Both filters are then dried and counted in a liquid scintillation counter in an aqueous scintillation fluid. The unwashed filter gives the total radioactivity in the sample (and so can be used to correct for counting efficiency), whereas the washed filter measures the radioactivity incorporated into the nucleic acid.

References

1. Feinberg, A. P. and Vogelstein, B. (1983) A technique for radiolabeling DNA restriction endonuclease fragments to high specific activity. *Anal. Biochem.* **132**, 6–13.
2. Feinberg, A. P. and Vogelstein, B. (1984) A technique for radiolabeling DNA restriction endonuclease fragments to high specific activity (Addendum). *Anal. Biochem.* **137**, 266,267.

CHAPTER 6

Hybridization and Competition Hybridization of Southern Blots

Rosemary E. Kelsell

1. Introduction

The separation of DNA restriction enzyme digestion products by gel electrophoresis and immobilization of the fragments onto a solid support (or filter) has been described in the preceding chapters (*see* Chapters 3 and 4). This chapter describes the detection of specific DNA sequences by hybridization to a labeled probe of complementary sequence. This method is suitable for the detection of a wide range of DNA concentrations down to single-copy genes within mammalian genomic DNA (little more than 1 pg of hybridizing DNA in a total of 10 µg).

In principle, hybridization consists of the annealing of a single-stranded labeled nucleic acid probe to denatured DNA fixed to the filter. In practice, it requires a balance between maximizing the specific signal and minimizing the nonspecific background. A number of factors determine the signal strength. Both the length of the probe and the ionic strength of the hybridization solution are important for determining the annealing rate, with probes of approx 200 bp in length and high-ionic-strength buffers giving the best results. The most important factors are the DNA concentration (C_o) and annealing time (t), in accordance with the $C_o t$ relationship, although in the case of filter hybridization, where one of the partners in the annealing reaction is immobilized, it is difficult to predict the exact hybridization kinetics. In theory, the maximum signal strength is therefore best achieved with high probe concentrations

From: *Methods in Molecular Biology, Vol. 58: Basic DNA and RNA Protocols*
Edited by: A. Harwood Humana Press Inc., Totowa, NJ

and long hybridization times, but unfortunately these factors also lead to increased background and a compromise must be reached. The hybridization volume should be kept as small as possible while keeping the filter(s) covered at all times. A further increase in effective probe concentration can be achieved by the inclusion of 10% dextran sulfate, which increases the hybridization rate 10-fold *(1)*. However, it is essential that the dextran sulfate be properly dissolved. Otherwise, its addition can lead to higher background. The hybridization time should be kept as short as possible. A good working time is approximately three times the $C_o t$: between 12 and 16 h for single-copy gene detection on a genomic blot. There is little to be gained from extended hybridization times with probes made from double-stranded DNA fragments, since self-annealing in solution will mean that there is negligible free probe left after this time. A number of blocking agents are added to the hybridization solution to suppress the nonspecific background. These take three forms and are usually used in combination:

1. A wetting agent, usually SDS;
2. An agent to block nonspecific binding to the surface filter, most often Denhardt's solution *(2)*; and
3. Agents to suppress nonspecific annealing to DNA, usually denatured, fragmented DNA.

For further background suppression, the filter is prehybridized in these agents prior to addition of the probe.

Background can also be of a more specific nature. This may occur if the probe is contaminated with other sequences, but more significant is a problem in hybridization of genomic DNA when the probe contains "repetitive elements." There is a wide range of these elements dispersed throughout the genome of all eukaryotic organisms, and they vary in size and frequency. For example, on average, there is one *Alu* repeat every 5 kb in the human genome *(3)*. This means that long probes prepared from genomic DNA, such as whole λ or cosmid clones, may hybridize to many genomic locations and produce a smear that masks the signal from single-copy sequences. One way to overcome this problem is to subclone the DNA fragments used to make the probe and ensure that they hybridize to unique DNA sequences. Subcloning, however, can be both difficult and time-consuming. Competition hybridization offers a more direct way to remove repetitive elements. This technique requires prereassociation of the labeled probe in solution to sheared genomic DNA. As a consequence, all of the repetitive element sequences in the probe are annealed to the competitor DNA, leaving only unique sequences free to hybridize to DNA on the Southern blot.

Competition hybridization has facilitated studies that require the analysis of long stretches of genomic DNA. For example, the use of a large probe may speed up the identification of RFLP markers associated with a genetic disease locus *(4,5)*. The more rapid identification of correct cosmid and λ clones has also been useful in the generation of long-range physical maps *(6,7)* both to analyze the wild-type genomic structure and to identify the junction fragments generated by large genetic rearrangements, such as deletions *(5,8,9)*. Finally, the technique has also been used to study mechanisms of gene amplification at the chromosomal level by utilizing fluorescence *in situ* hybridization with biotinylated cosmid clones, demonstrating that the technique is clearly sensitive enough to allow detection at the single-copy level on metaphase spreads *(10)*.

The conditions of the prereassociation step will vary according to temperature, time, ionic strength, and complexity of the DNA. Various conditions for prereassociation have been developed *(4,11,12)*. This chapter describes a method based on the technique described by Sealey et al. *(13)*. The protocol presented in Section 2. has been used successfully to compete repeats from a variety of λ clones for probing blots prepared from both pulsed-field, as well as ordinary agarose gels.

2. Materials

All general molecular-biology-grade reagents may be obtained from BDH Chemicals Ltd. (UK) or Fisons (UK). All solutions are made up using sterile distilled water and procedures required for molecular biology.

2.1. Hybridization (see Note 1)

1. 20X SSC: 3M NaCl, 0.3M sodium citrate, pH 7.0. Store at room temperature.
2. 100X Denhardt's solution: 2% (w/v) BSA (Fraction V, Sigma [UK]), 2% (w/v) Ficoll (Sigma), 2% (w/v) Polyvinylpyrrolidone (Sigma). Store at −20°C.
3. 10% SDS: Store at room temperature.
4. 10 mg/mL herring sperm DNA (Sigma): Dissolve in water, shear by passing 12 times through a 17-gage syringe needle or by sonication, and then denature by boiling for 15 min. Store at −20°C.
5. Prehybridization solution: 3X SSC, 10X Denhardt's solution; 0.1% (w/v) SDS and 50 µg/mL herring sperm DNA. Prehybridization solution can be made as a stock and stored at −20°C.

6. Labeled probe (*see* Chapters 5, 7, 8, 10, and 12): 100 ng of labeled probe (*see* Notes 2–4). Purify in a Sephadex G50 spin column equilibrated in 3X SSC (*see* Note 5).
7. Hybridization solution: Prehybridization solution plus 10% (w/v) dextran sulfate (Pharmacia, UK). This is made by dissolving dextran sulfate in prehybridization at 65°C, with occasional vigorous agitation. Make the hybridization solution several hours before use to allow adequate time for the dextran sulfate to dissolve; undissolved particles of dextran sulfate result in background spots. Hybridization solution can be made as a stock and stored at –20°C.
8. Hybridization oven and tubes: This author finds tubes the most convenient for hybridization. In this system, the filters are spread around the inner wall of a glass tube, which is then placed on a rotating drum in an oven. Using this method, it is easy to change solutions and the hybridization volume can be kept very small. These are commercially available (e.g., Hybaid Ltd., UK and Techne, UK) or can be homemade. Alternative systems work just as well, but can be harder to set up. A common alternative is to hybridize in sealed bags immersed in a shaking water bath (*see* Note 6).
9. Wash solutions: Make 3 L of 3X SSC, 0.1% (w/v) SDS, and 3 L of 0.1X SSC, 0.1% (w/v) SDS by dilution from stocks.
10. Suitable autoradiographic film: for example, Kodak X-OMAT AR.
11. Stripping solutions:
    a. 500 mL of $0.4 M$ NaOH.
    b. 0.1X SSC, 0.5% SDS, $0.2 M$ Tris-HCl, pH 7.5.

2.2. Competition Hybridization

1. Oligolabeling stop buffer: 20 mM NaCl, 20 mM Tris-HCl, pH 7.5, 2 mM EDTA, pH 8.0, 0.25% (w/v) SDS, 1 µM dCTP.
2. Competitor DNA: Sheared genomic DNA (of the appropriate species or DNA type) at 20 mg/mL in TE. This can be prepared by sonicating total genomic DNA in a suitable volume with 20–30 pulses of 5-s duration. The DNA should be of an average size of 500 bp, and an aliquot is checked on an agarose gel. The DNA is then phenol/chloroform-extracted, chloroform-extracted, and concentrated by ethanol precipitation. The concentration is checked by a spectrophotometer and adjusted to 20 mg/mL.

3. Methods

3.1. Hybridization

1. Rinse the filter in 2X SSC and place in a suitable container (e.g., a hybridization tube) with ~10 mL of prehybridization solution (*see* Note 7). Incubate at 65°C for 3–4 h while rotating or shaking (*see* Note 8).

2. Just prior to the hybridization step, denature the probe by boiling for 5–10 min (*see* Note 4) and add it directly to ~10 mL of hybridization solution, prewarmed to 65°C (*see* Note 9).
3. Remove the prehybridization solution from the filter, and add the completed hybridization solution. If using hybridization tubes, this is easily done by pouring off the prehybridization solution and pouring in the hybridization solution. Incubate with rotating or shaking overnight at 65°C.
4. At the end of the hybridization, remove the filter (*see* Note 10) and place in 500 mL of 3X SSC, 0.1% SDS at room temperature for a minute to wash off unbound probe. Repeat wash twice.
5. Wash the filter in 500 mL of preheated 3X SSC, 0.1% SDS at 65°C for 10 min. Repeat wash twice (*see* Note 11).
6. Wash the filter in 500 mL of preheated 0.1X SSC, 0.1% SDS at 65°C for 15 min. Repeat wash once (*see* Note 12).
7. Take the filter from the wash solution, and place it wet in polythene film or a polythene bag, and if possible, seal the edges. Do not let the filter dry out, or it will become extremely difficult to remove the probe for further washing or reprobing.
8. If a ^{32}P-labeled probe is being used, autoradiograph the filter for 2–24 h (*see* Note 13). If a longer exposure is required, the filter can be laid down for up to a further 2 wk.
9. The same filter can be reprobed after the first probe has been removed from it. This is done by placing the filter(s) in 500 mL of $0.4M$ NaOH at 42°C for 30 min and then transferring it to 500 mL of 0.1X SSC, 0.5% SDS, $0.2M$ Tris-HCl, pH 7.5 at 42°C for a further 30 min. The filter may then be re-exposed to X-ray film to ensure that all the probe has been removed. Finally, it is stored in prehybridization solution at 4°C until ready for use. This author usually adds fresh prehybridization solution the next time the filter is used.

3.2. Competition Hybridization of Probe

1. Make the labeled probe up to a volume of 200 µL with oligolabeling stop buffer. The probe is usually recovered in a volume of ~50 µL after centrifugation through a Sephadex G50 column (*see* Note 5), requiring the addition of 150 µL of oligolabeling stop buffer. Add 50 µL of a 20 mg/mL solution of competitor DNA and 50 µL of 20X SSC.
2. Denature the mixture by boiling for 10 min, and then plunge into ice for 1 min.
3. Place the reaction at 65°C for 10 min (*see* Note 14 and Fig. 1).
4. Add this competed probe to the hybridization solution, and proceed as in Section 3.1., step 3 (*see* Notes 15 and 16).

A Probe 8

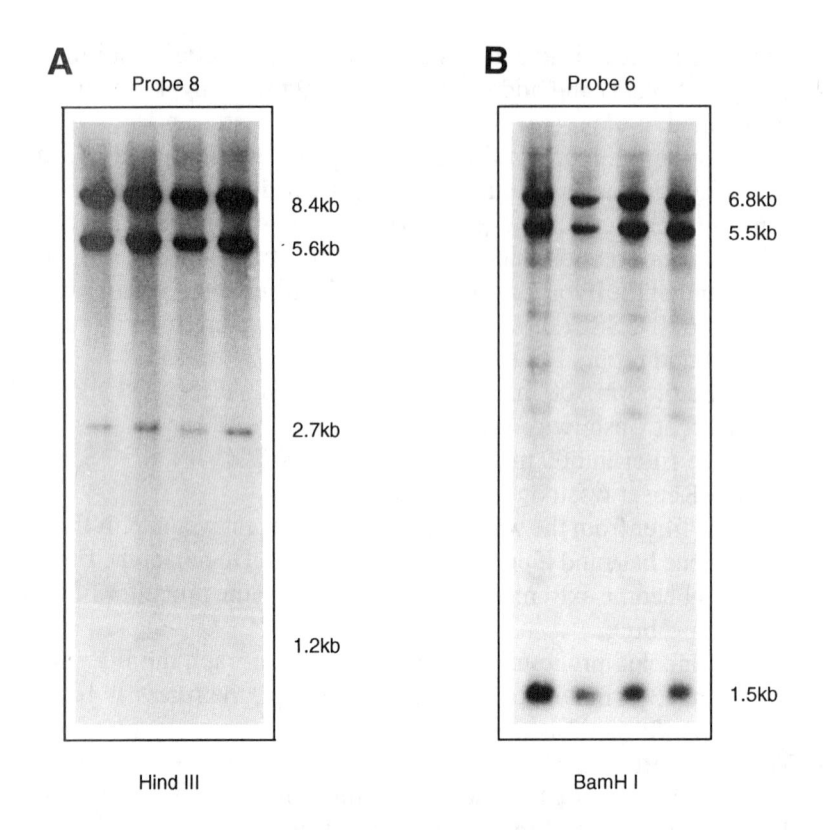

8.4kb
5.6kb
2.7kb
1.2kb

Hind III

B Probe 6

6.8kb
5.5kb
1.5kb

BamH I

Fig. 1. **(A)** Hybridization to *Hin*dIII-digested CHO genomic DNAs using a ^{32}P-labeled λ clone of ~15 kb in length and treated using the competition hybridization protocol described in Section 3.2. There are at least four *Alu*-equivalent repeats situated within the region of DNA covered by this probe (probe 8, Davis and Meuth, ref. 9). **(B)** Hybridization to *Bam*HI-digested CHO genomic DNAs using a different ^{32}P-labeled λ clone (probe 6, Davis and Meuth, ref. 9). In this instance, the repeats have not been competed out of the probe as successfully as in the previous example.

4. Notes

1. Other solutions also work equally well for hybridization (for example, 10% SDS and 7% PEG 6000 for both prehybridization and hybridization of nylon membranes—David Kelsell, unpublished observations). This author's filters are made on good-quality nylon membranes (for example, Genescreen or Genescreen plus, NEN DuPont, UK), allowing the probes to be removed and the filters to be reused with an array of probes.
2. The technique of random priming is the most efficient way of obtaining double-strand probes labeled to a specific activity of >10^9 cpm/μg of DNA.

3. It is usually unnecessary to remove vector sequences from the DNA used for the probe, provided that crosshybridizing marker lanes are cut off the gel before capillary blotting. For example, the arms of λ clones hybridize to λ size markers and plasmid sequences hybridize to some of the marker bands in the 1-kb ladder (Gibco-BRL, UK).
4. Care should be taken if using radioactive labeled probes. This author uses screw-capped 1.5-mL tubes for making radiolabeled probes to avoid the lids popping open during the denaturation steps.
5. Commercial spin columns are available, but in fact Sephadex G50 spin columns are cheap and simple to make. Columns are made as described in Chapter 49, except that in this case, the Sephadex G50 is equilibrated with 3X SSC instead of TE. Columns should not be allowed to dry out and must be made just prior to their use.
6. To set up hybridization in bags, seal the filter in a bag, but make a funnel shape at the top of the bag with the bag sealer. The hybridization mixture containing the probe is then added to the bag through this funnel. It is relatively easy to remove air bubbles by rolling them to the top of the bag with a disposable 10 mL pipet and allowing the air to escape through the funnel. In this way, it is possible to fill bags with the minimum of spillages. This maneuver can be practiced with water in an empty bag!
7. As a general rule, use at least 0.2 mL of prehybridization solution for every cm^2 of filter.
8. Filters can be prehybridized overnight, but this can weaken the filter and may also result in a diminished signal.
9. A probe concentration of 1–10 ng/mL gives the best signal-to-background ratio. If bad background is experienced with a probe at 10 ng/mL, it should be diluted 10-fold and tried again.
10. The probe can be stored and reused. For radiolabeled probes, the storage time is regulated by the isotope half-life, 2 wk for ^{32}P. This is not a problem with nonradioactive probes. Since the probe reanneals during hybridization, the hybridization solution should be boiled for 15 min before reuse.
11. This is a low-stringency wash and will leave probe bound to similar (homologous) DNA sequences, as well as those that are identical. The filter can be autoradiographed at this point in order to detect these related sequences. The stringency of the wash can be increased by sequential washes at lower salt concentrations.
12. This is a high-stringency wash and should only detect identical sequences.
13. The autoradiograph sensitivity depends on the conditions under which the filter is exposed. Preflashing the photographic film with a single flash (<1 ms) of light from a flash gun with an orange filter increases the sensitivity of the film twofold and gives it a linear response to intensity of radioactiv-

ity. Additionally, placing an intensification screen behind the film and putting the assembly at −70°C gives a total 10-fold increase in the sensitivity of the film, but at the expense of the resolution *(14,15)*.
14. Ten minutes are adequate for the prereassociation step. However, if background repetitive sequences are still visible on the autoradiograph (*see* Fig. 1B), check that enough sheared genomic DNA was used. An overestimation of the concentration of the sheared genomic DNA is the most common cause of problems. If this is not the case, extend the prereassociaton time for up to an hour.
15. This author and others *(5)* have found that competed probes appear to generate stronger signals than single-copy probes on Southern blots, possibly because the probes are longer.
16. This author has visualized fragments of between 600 and 7000 bp in size with this method.

References

1. Wahl, G. M., Stern, M., and Stark, G. R. (1979) Efficient transfer of large DNA fragments from agarose gels to diazobenzyloxymethyl-paper and rapid hybridization by dextran sulphate. *Proc. Natl. Acad. Sci. USA* **76**, 3683–3687.
2. Denhardt, D. T. (1966) A membrane filter technique for the detection of complementary DNA. *Biochem. Biophys. Res. Commun.* **23**, 641–646.
3. Deininger, P. L. (1989) SINES: short interspersed repeated DNA elements in higher eukaryotes, in *Mobile DNA* (Berg, D. E. and Howe, M. M., eds.), American Society for Microbiology, Washington, DC, pp. 619–636.
4. Litt, M. and White, R. L. (1985) A highly polymorphic locus in human DNA revealed by cosmid-derived probes. *Proc. Natl. Acad. Sci. USA* **82**, 6206–6210.
5. Blonden, L. A. J., den Dunnen, J. T., van Paassen, H. M. B., Wapenaar, M. C., Grootscholten, P. M., Ginjaar, H. B., Bakker, E., Pearson, P. L., and van Ommen, G. J. B. (1989) High resolution deletion breakpoint mapping in the DMD gene by whole cosmid hybridization. *Nucleic Acids Res.* **17**, 5611–5621.
6. Compton, D. A., Weil, M. M., Jones, C., Riccardi, V. M., Strong, L. C., and Saunders, G. F. (1988) Long-range physical map of the Wilms' tumor-aniridia region on human chromosome 11. *Cell* **55**, 827–836.
7. Compton, D. A., Weil, M. M., Bonetta, L., Huang, A., Jones, C., Yeger, H., Williams, B. R. G., Strong, L. C., and Saunders, G. F. (1990) Definition of the limits of the Wilms tumor locus on human chromosome 11p13. *Genomics* **6**, 309–315.
8. Blonden, L. A. J., et al. (1991) 242 breakpoints in the 200-kb deletion-prone p20 region of the DMD gene are widely spread. *Genomics* **10**, 631–639.
9. Davis, R. E. and Meuth, M. (1994) Molecular characterization of multilocus deletions at a diploid locus in CHO cells: association with an intracisternal-A particle gene. *Somat. Cell. Mol. Genet.* **20**, 287–300.
10. Smith, K. A., Gorman, P. A., Stark, M. B., Groves, R. P., and Stark, G. R. (1990) Distinctive chromosomal structures are formed very early in the amplification of CAD genes in Syrian hamster cells. *Cell* **63**, 1219–1227.

11. Ardeshir, F., Giulotto, E., Zieg, J., Brison, O., Liao, W. S. L., and Stark, G. R. (1983) Structure of amplified DNA in different Syrian hamster cell lines resistant to N(phosphonacetyl)-L-aspartate. *Mol. Cell. Biol.* **3,** 2076–2088.
12. Djabali, M., Nguyen, C., Roux, D., Demengeot, J., Yang, H. M., and Jordan, B. R. (1990) A simple method for the direct use of total cosmid clones as hybridization probes. *Nucleic Acids Res.* **18,** 6166.
13. Sealey, P. G., Whittaker, P. A., and Southern, E. M. (1985) Removal of repeated sequences from hybridization probes. *Nucleic Acids Res.* **13,** 1905–1922.
14. Laskey, R. A. and Mills, A. D. (1975) Quantitative film detection of ^{3}H and ^{14}C in polyacrylamide gels by fluorography. *Eur. J. Biochem.* **56,** 335–341.
15. Laskey, R. A. and Mills, A. D. (1977) Enhanced autoradiographic detection of ^{32}P and ^{125}I using intensifying screens and hypersensitive film. *FEBS Lett.* **82,** 314–316.

CHAPTER 7

Utilization of DNA Probes with Digoxigenin-Modified Nucleotides in Southern Hybridizations

Tim Helentjaris and Tom McCreery

1. Introduction

Probes prepared with either digoxigenin- or biotin-modified nucleotides can be hybridized to Southern blots to detect target nucleic acid sequences. These methods offer an attractive alternative to "radioactively tagged" probes in terms of safety, cost, and efficiency. Most previous nonradioactive strategies utilized the detection of the modified base by the use of a coupled antibody- or avidin-alkaline phosphatase with subsequent exposure to Nitro Blue Tetrazolium (NBT). NBT is converted to an insoluble, colored compound at the site of hybridization. Replacement of this colorimetric reaction with a chemiluminescent process provides better sensitivity, as well as easier reusability of membranes. The development of compounds, such as adamantyl 1,2 dioxetane phosphate (AMPPD) *(1)* provides a convenient alternative to previous detection schemes, since this compound is destabilized by alkaline phosphatase, resulting in the production of light, which will expose a standard X-ray film. Probes produced by methods analogous to those used for NBT detection are also usable in this process. The membranes with the bound alkaline phosphatase are soaked in a dilute solution of AMPPD and then exposed to film at room temperature or 37°C instead of −70°C. Exposure times of 45 min to several hours are all that is necessary to detect single-copy sequences, even in genomic DNA preparations from organisms with very large genomes. Recently, the substrates CSPD, CDP, and CDP-Star

From: *Methods in Molecular Biology, Vol. 58: Basic DNA and RNA Protocols*
Edited by: A. Harwood Humana Press Inc., Totowa, NJ

(Tropix, Bedford, MA), which offer even greater sensitivity and a prolonged signal output, have become available. These can be used in exactly the same way as AMPPD.

The labeled probes required for this process can be conveniently produced by one of two alternative methods. Oligolabeling using random hexanucleotide primers *(2)* is an effective method for producing DNA hybridization probes of high specific activity, although most previous variations have utilized radionuclides as the mechanism for modifying the incorporated nucleotides. Earlier studies reported that DNA polymerase I could utilize digoxigenin-11-dUTP as a substrate and incorporate it into double-stranded DNA *(3)*. We have similarly found that only simple modifications of our earlier protocols are necessary to permit the production of digoxigenin-modified probes that are capable of detection of very low amounts of target DNAs (1 pg or less) in a Southern hybridization. In particular, this technique is effective for labeling DNA fragments isolated from low-melting-point agarose gels (Note 1) and for labeling DNA fragments that are resistant to PCR amplification.

The polymerase chain reaction *(4)* is an efficient method for copying a fragment of DNA using flanking primers. We have found that since most cloning vectors utilize the *lacZ* gene with a synthetic multiple cloning site *(5)*, a single pair of primers will amplify many sequences when inserted into a number of common phage and plasmid vectors, such as the pUC plasmid and λ phage series, and their derivatives. We have experienced little difficulty in amplifying inserts up to 3000 bp in length and now consider this as an alternative to growth of bacterial cultures with subsequent nucleic acid purification to produce DNA fragments of this length for both mapping and sequencing. Since the Taq polymerase also has little difficulty incorporating digoxigenin-11-dUTP into its products, we have found that PCR is a preferred method for producing hybridization probes. It has the following advantages:

1. It utilizes very little input target DNA compared to the final yield of the product;
2. It can utilize either purified DNA (both supercoiled and linear molecules), a broth culture of a plasmid-infected bacteria, or the lysate of a phage-infected bacteria culture as the source of the target sequences; and
3. The progress of the labeling reaction can be easily followed by checking for the production of double-stranded product by standard agarose gel electrophoresis.

PCR labeling also has the three following disadvantages:

1. The target sequences must be cloned into a vector for which flanking primers are available or for which gene-specific primers are available;
2. We have difficulty amplifying most sequences larger than 3000 bp in length; and
3. Even some sequences <3000 bp in length are recalcitrant to PCR amplification, presumably because of high G-C content or the presence of snapback sequences.

Nevertheless, we find this technique to be preferable to oligolabeling and usually attempt to use it for every target sequence until it is determined to be intractable to this approach. Since the modified base itself is relatively cheap when compared to radionuclide-modified nucleotides and the products are stable for much longer periods than ^{32}P-labeled probes, both methods offer an economic and safe mechanism for producing hybridization probes that are quite capable of detecting single-copy sequences within very complex genomes.

2. Materials
2.1. Oligolabeling

All oligolabeling reagents are stored at –20°C and thawed on ice immediately prior to use.

1. Solution O: Dissolve 44.52 g of Tris base and 7.47 g of MgCl$_2$ in 150 mL of H$_2$O, adjust the pH to 8.0 with HCl, bring the volume up to 250 mL.
2. Solution A: 18 µL of β-mercaptoethanol, 25 µL of 100 mM dATP, 25 µL of 100 mM dCTP, 25 µL of 100 mM dGTP, 4 µL of 100 mM dTTP, 899 µL of Solution O (*see* Note 1).
3. Solution B (2M HEPES): Dissolve 119.15 g of HEPES in 100 mL H$_2$O, adjust the pH to 6.6 with 4M NaOH, and bring the volume up to 250 mL.
4. Solution C: Add 555 µL of H$_2$O to 50 U of random hexanucleotides (Pharmacia, Uppsala, Sweden).
5. 5X oligo buffer: Combine solutions A, B, and C at a ratio of 1.0:2.5:1.5.
6. BSA: 10 mg/mL, store at –20°C.
7. Digoxigenin-11-dUTP: Dilute to 400 µM by addition of 37.5 µL of water to 25 µL of 1 mM stock (Boehringer Mannheim, Mannheim, Germany).
8. Klenow fragment of DNA polymerase I from *E. coli*: Dilute the stock to 1 U/µL with 7 mM Tris-HCl, pH 7.5, 7 mM MgCl$_2$, 50 mM NaCl, 50% glycerol.

2.2. PCR-Labeling

1. 10X PCR buffer: Combine 3.7 mL of sterile dH$_2$O, 5.0 mL of 1M 500 mM KCl, 1.0 mL of 1M Tris, pH 8.2, 0.2 mL of 1M MgCl$_2$, and 0.1 mL of 1% gelatin.
2. dXTPs: Make up a single solution containing 2.5 mM of dATP, dGTP, dCTP, and dTTP.
3. Primers: Primers can be specific to each individual application (i.e., usually this simply reflects the cloning vector; see Note 2); however, we use a single pair of primers capable of amplifying inserts in pUC-based plasmids, as well as any other vector with a *lacZ*-multiple cloning site.
    100 µM PCRFSEQ: 5' TTGTA AAACG ACGGC CAGTG 3'
    100 µM PCRRSEQ: 5' GGAAA CAGCT ATGAC CATGA T 3'
    Combine these primers 1:1 to obtain a 50-µM solution of each primer.
4. Taq DNA polymerase: Supplied by the manufacturer at a concentration of 5 U/µL.
5. 1X labeling reaction: 24 µL of this mix are required for each sample. To prepare, combine 19.4 µL of sterile dH$_2$O, 2.5 µL of 10X PCR buffer, 0.5 µL of 2.5 mM dXTP, 1.0 µL of diluted digoxigenin-11-dUTP (*see* Note 3), 0.5 µL of mixed primers, and 0.1 µL of Taq polymerase.
6. Mineral oil: We use USP-grade heavy mineral oil available in drugstores.

2.3. Chemiluminescent Detection

1. Transfer membrane: These should be prepared as they would be for radioactive detection; however, it is very important only to handle the gels and membranes with gloved hands or forceps, since fingerprints cannot be removed and will be made very obvious by the detection protocol. Only nylon membranes should be used, because the AMPPD is stabilized in nylon and will not provide the same level of sensitivity when nitrocellulose or PVDF membranes are used.
2. Prehybridization/hybridization buffer: 5X SSC, 0.1% *n*-lauroylsarcosine (powdered form), 0.02% sodium lauryl sulfate (SDS), 0.1% blocking agent (Tropix or Boehringer Mannheim), 100 µg/mL salmon sperm DNA.
3. Washing solution: 0.1X SSC, 0.1% SDS.
4. Buffer 1: 0.05M Tris-HCl, pH 7.5, 0.15M NaCl. A 10X solution of this buffer can be prepared ahead of time, autoclaved, and diluted immediately before use.
5. Blocking buffer (buffer 2): buffer 1 with 0.1% blocking agent added. It is convenient to prepare buffer 2 the previous day and incubate a portion in a 65°C incubator overnight, which insures that the blocking agent is completely dissolved.

6. Antidigoxigenin solution: This solution consists of 50 mL of blocking buffer with 2.5 µL of antidigoxigenin antibody/alkaline phosphatase conjugate (Boehringer Mannheim #1093 274). This solution should be prepared immediately before use.
7. $1M$ Tris-HCl, pH 9.5, autoclaved.
8. $1M$ MgCl$_2$, autoclaved.
9. $5M$ NaCl; autoclaved.
10. Buffer 3: $0.1M$ diethanolamine (2.4 mL/250 mL final solution), 1 mM MgCl$_2$, 0.02% sodium azide, pH adjusted to 10.0 with concentrated HCl.
11. AMPPD solution: Add 2 µL of Tropix AMPPD to each mL of buffer 3, figuring that approx 2.5 mL of this solution is generally sufficient for a 10 × 10 cm blot. AMPPD does have a limited shelf-life (approx 3 mo) in its undiluted form.
12. Stripping solution: Same as washing solution.
13. Labeled markers (*see* Note 4).

3. Methods
3.1. Oligolabeling

1. Aliquot 50–500 ng of DNA (excised from the vector and isolated by suitable methods on an agarose gel; *see* Note 5) into a 0.65 µL microcentrifuge tube containing 18 µL of H$_2$O.
2. Heat-denature the DNA for 10 min at 95°C.
3. Immediately place the tube containing the DNA on ice.
4. Add the following reagents as quickly as possible and mix thoroughly, 5 µL of 5X Oligo buffer, 0.5 µL of BSA, 0.5 µL of digoxigenin-11-dUTP, 1 µL of Klenow.
5. Incubate at 37°C for 90 min to 18 h (3 h seems to provide efficient labeling).

The probe may be used immediately or stored either at 4°C if use is anticipated within weeks or –20°C for longer term storage (*see* Notes 6 and 7).

3.2. PCR Labeling

PCR conditions will vary depending on the primer pairs utilized. Since we have found that use of higher annealing temperatures can facilitate the amplification of sequences with high G-C content, the development of primers with an annealing temperature of at least 60°C and higher is recommended. A PCR cycle consists of three steps. The first is a denaturation step, which separates the double-stranded DNA into its two single-stranded components. The second is an annealing step where the primers and single-stranded target DNAs are allowed to anneal to each other. The third step is the actual amplification of the target sequence where the Taq polymerase copies the target DNA. Our amplification program typically contains a first cycle of 95°C for 2 min, 60°C for 30 s, and 72°C for 3 min. This cycle has a longer period at 95°C to provide for the efficient initial denaturation of the target DNA. This program is then followed by 18 cycles of 95°C for 30 s, 60°C for 30 s, and 72°C for 3 min. Finally, a last cycle of 95°C for 30 s, 60°C for 30 s, and 72°C for 10 min is used to provide for a longer amplification step, which may aid in "finishing" many previously initiated products. The thermocycler is then also programmed to hold at 6°C for 18 h where overnight runs are most convenient.

1. One microliter of an overnight broth culture or λ phage lysate (*see* Note 8) or 1–10 ng of DNA template (*see* Note 9) is added either to a .65-µL microcentrifuge tube or a single well in a 96-well microtiter plate (for those thermocyclers capable of utilizing them).
2. Prepare the reaction mix in bulk for all of the reactions to be performed, and then aliquot 24 µL of the reaction mix into each tube. Mix by pipeting up and down.
3. After the mixture is added to the tubes, add two drops of mineral oil to prevent evaporation during amplification, where required by thermocycler design.
4. Place the tubes or the microtiter plate into the thermocycler, and start the program.
5. After the last 72°C extension step is complete, the samples can be removed at any time, or they can be allowed to remain overnight at 6°C. After removing the tubes from the thermocycler, draw off the lower aqueous layer with a micropipet, and transfer to a fresh tube.

The expected yield for this reaction is approx 500–1000 ng depending on the size of the target fragment. A reasonably accurate measure of the yield of this reaction can be obtained by analyzing 4 µL of the reaction products on an agarose gel. The reaction can also be scaled up quite easily to obtain more product, if desired. The probes produced by this method are stable for several months or more at –20°C.

3.3. Chemiluminescent Detection

1. Prehybridize the blots at 60–65°C in either plastic bags, plastic boxes, or a hybridization oven for 60 min or longer using 3 mL of prehybridization solution/100 cm^2 of blot.

2. Denature the probes by heating to 95°C for 10 min. Add the probes to the hybridization solution at a concentration of 5–20 ng/mL of hybridization solution. Generally, 2 µL of probe/mL of hybridization solution will produce good results.
3. Replace the prehybridization solution with fresh hybridization solution containing the denatured probe. Incubate from 6 h to overnight at 65°C. Mixing during hybridization is not necessary, unless multiple blots are hybridized in the same container.
4. Remove the hybridization solution (this can be stored at 4°C and reused several times with little effect on signal intensity).
5. Wash the membranes twice for 3 min with at least 200 mL of washing solution with good agitation at room temperature.
6. Wash the membranes for 30 min at 65°C with at least 200 mL of washing solution and with good agitation. This step removes any probe that is not bound to the membrane.
7. Wash the membranes with buffer 1 for 5 min at room temperature (this and all subsequent washes are performed at room temperature and with vigorous agitation). This step removes the washing solution from the membranes and prepares the surface for the blocking agent.
8. Incubate the blots with 200 mL of blocking buffer for 30 min. This blocks the surface of the membrane to prevent nonspecific antibody binding.
9. Incubate the blots in approx 200 mL of antidigoxigenin solution for 30 min. At this point in the process, the antibody:alkaline phosphatase conjugate binds to the digoxigenin-labeled probe that is bound to the membrane.
10. Wash the blots three times for 5 min each with 200 mL of buffer 1. These washes remove the unbound antibody:alkaline phosphatase conjugate.
11. Wash the blots once for 10 min with buffer 3. This wash ensures that the pH of the blot is appropriate for exposure to AMPPD.
12. Remove each blot from buffer 3 and lay out on a Plexiglass™ sheet cut to conform to a the size of X-ray cassette (no intensifying screens needed). Carefully apply approx 2.5 mL of the AMPPD solution to each blot. When all have been laid out on a single sheet, seal with either Saran Wrap™ or a mylar sheet to prevent drying. Blots can also be sealed individually in Saran Wrap™ or mylar page covers, whichever is most convenient.
13. Expose the film for either 3 h at 37°C or overnight at room temperature. The exposure time depends on the level of target sequence, the percentage of modified base in the probe, and the amount of probe added.

The result should be sharp black bands on a clean background (*see* Notes 10 and 11). To reuse the blots, remove the probe by washing for 10 min in boiling stripping solution. Replace the stripping solution with the TE and store for short periods at 4°C, or air-dry and store at 4°C for several months (*see* Note 12).

4. Notes

1. Background may be reduced by decreasing the percentage of modified base in the reaction from 20% of the total dTTP concentration to 5%. This may also increase the efficiency of the reaction, as well as utility of the final product. We have found that digoxigenin-modified probes are slightly different chemically when compared to radioactively labeled probes probably because of the long side chains introduced with the modified base. They form more stable hybrids at lower hybridization temperatures and are more difficult to remove from membranes, once hybridized. Lowering of the dUTP concentration to <10% has a dramatic effect on these qualities.
2. By using an extended primer, you can raise the annealing temperature in the PCR reaction even higher. We have experimented with other primers for the *lac*Z-MCS vectors and have found that the following pair of primers amplifies efficiently and allows the use of even higher annealing temperatures (to 68°C). This can further increase the percentage of sequences that are successfully amplified by PCR.
   FUPSTRM: 5' ATGTG CTGCA AGGCG ATTAA GTTGG G 3'
   RDNSTRM: 5' CACAC AGGAA ACAGC TATGA CCATG 3'
3. The protocol described above calls for 5% labeled base. This provides for a strong signal with minimal background on most autoradiograms. Concentrations as low as 2.5% and as high as 20% are usable.
4. One difference with this type of system is that it is convenient to prepare and utilize a digoxigenin-labeled mol-wt marker in the agarose gel. We find that use of the Klenow fragment of DNA polymerase I to "fill in" the ends of double-stranded DNA fragments created by restriction digestion is an efficient method for creating mol-wt standards that will be visible every time the blot is treated with alkaline phosphatase–antidigoxigenin conjugate. We prefer the 1-kb mol-wt ladder from Gibco-BRL (Gaithersburg, MD), which already possesses accessible ends and yields a set of mol-wt standards every 1000 bp.
5. Fragments may be isolated from low-melting-point agarose by the following method *(6)*. Pour a low-melting-point agarose gel at 4°C, load the DNA samples, and electrophorese at 4°C to prevent the gel from melting. Stain the gel with ethidium bromide. Use a handheld long wavelength UV source to locate your fragment (this reduces the radiation damage to the DNA). Excise the band with a razor blade or drinking straw. Add 2–3 vol of TE,

and incubate for 5 min at 65°C to melt the gel. This solution can then be cleaned up further with phenol: chloroform; however, we usually do not find it to be necessary.
6. We have not really found it necessary to purify the labeled probes, but clean-up may slightly reduce the background seen with Southern blots. We have found ethanol precipitation to be an effective way to purify digoxigenin-modified probes. Add $3M$ NaOAc to a final concentration of $0.2M$ and 3 vol of ethanol (100%), mix well and place at −70°C for 15 min. Microcentrifuge for 5 min and discard the supernatant. Wash the pellet in 700 µL of 70% ethanol, and microfuge for 2 min. Air-dry the pellet thoroughly and resuspend it in 100 µL of TE.
7. Unfortunately, there is no quantitative method to measure digoxigenin-incorporation into the final probe, short of spotting the probe onto a membrane and testing it with antidigoxigenin and alkaline phosphatase conjugate.
8. λ Phage lysates can be labeled directly by PCR by using the following protocol. Prepare plaques in the normal manner, except that both layers in the plate are made with agarose, not agar, since impurities in the agar will inhibit the PCR reaction. Pick individual plaques and transfer them either to .65-mL microcentrifuge tubes or 96-well plates. Add 100 µL of LB broth to each plaque and grow overnight at 37°C with shaking. Add 1 µL of this lysate to PCR tubes or plates, and continue with the standard protocol.
9. The concentration of DNA template is crucial to obtaining a good labeling reaction, and this is also a case where more is definitely not better. Keeping the concentration of target DNA close to 1 ng will maximize the yield of labeled probe with higher specific activity.
10. If there is no signal or a very weak signal on the resulting film:
    a. The probe was not properly prepared or added at too low a level. Check the probe on a gel to determine whether labeling was successful and the yield obtained.
    b. The genomic DNA on the blot is at too low a concentration or is degraded. This is a fairly common problem; try adding more DNA/lane. Check the quality of the DNA by analyzing it undigested on a low-percentage agarose gel.
    c. The AMPPD was too old or degraded. If labeled markers are easily observable, this is not the problem, and you should consider other possibilities.
    d. Antibody was not added or was defective. Again, if the markers are visible, this is not the problem. If the markers are not visible, and then you should check both your conjugate and AMPPD.

11. High background is a more common problem with this technique and can take various forms:
    a. Smudges result from fingerprints and cannot be removed from the blot.
    b. An even gray to very dark background across the entire blot is usually caused by the AMPPD breaking down prematurely. If buffer 3 is improperly prepared or if the AMPPD stock solution has degraded, an even background will begin to develop over the entire blot with a concomitant decreasing signal.
    c. Spots have a number of possible sources and are often noted in the absence of added probe, indicating problems with the antibody conjugate binding nonspecifically to the blots. We have observed that using "dirty" holders for the blots during film exposure has caused the AMPPD to break down presumably because of exogenous alkaline phosphatase or other contaminants. Alternatively, this problem is often associated with buffer 2, if the blocking agent is not completely dissolved.
    d. When the rest of the blot seems relatively clean, but there is high lane background overshadowing any bands, this is a characteristic of the probe used and usually does not reflect on the conditions. If you want to decrease this problem, the only practical alternative is to use higher stringency washing conditions (+68°C and 0.1X SSC).
    e. When processing the blots, you must be careful not to scratch the surface with forceps. This is usually not a problem with radioactive probes, but any rough handling of these membranes can cause higher background.
12. Ghost bands left over from previous uses: This problem may be the result of improper stripping; try stripping again. If this is ineffective, and then it is probably owing to allowing the blots to dry with probes hybridized to them. Do not allow the blots to dry while probes are still bound to them; they cannot be removed and will remain on for nearly the life of the blot. Be careful to seal them during film exposure and strip them immediately after taking the blots off film

Acknowledgments

This work was supported in part by a USDA Hatch project (ARZT-136440-H-25-042) and by a USDA NRI Competitive Grants Program award (91–37300–6453). Many of these protocols were adapted from suggestions from Tropix and Boehringer Mannheim but we are also indebted to D. Hoisington among others for many suggestions.

References

1. Voyta, J. C. and Bronstein, I. (1988) Ultra sensitive detection of alkaline phosphatase activity. *Clin. Chem.* **34**, 1157.

2. Feinberg, A. P. and Vogelstein, B. (1983) A technique for radiolabeling DNA restriction fragments to high specific activity. *Anal. Biochem.* **132**, 6–13.
3. Kreike, C. M., de Koning, J. R. A., and Krens, F. A. (1990) Nonradioactive detection of single-copy DNA-DNA hybrids. *Plant Mol. Biol. Rep.* **8**, 172–179.
4. Saiki, R. K., Scharf, S., Faloona, F. A., Mullis, K. B., Horn, G. T., Erlich, H. A., and Arnheim, N. (1985) Enzymatic amplification of β-globin genomic sequences and restriction sites analysis for diagnosis of sickle cell anemia. *Science* **230**, 1350–1354.
5. Vieira, J. and Messing, J. (1982) The pUC plasmids, an M13mp7-derived system for insertion mutagenesis and sequencing with synthetic universal primers. *Gene* **19**, 259–268.
6. Sambrook, J., Fritsch, E. F., and Maniatis, T. (1989) *Molecular Cloning: A Laboratory Manual.* Cold Spring Harbor Laboratory, Cold Spring Harbor, NY.

CHAPTER 8

The Preparation of Fluorescein-Labeled Nucleic Acid Probes and Their Detection Using Alkaline Phosphatase-Catalyzed Dioxetane Chemiluminescence

Martin Cunningham and Martin Harris

1. Introduction

1.1. Probe Labeling with Fluorescein Hapten

A commonly used method for labeling nucleic acid probes nonradioactively is to introduce a small, relatively inert molecule, such as a hapten or biotin. Such molecules are generally coupled to a nucleoside triphosphate, such as dUTP, via a linker group. The modified nucleotide is then incorporated into the probe by an enzyme-mediated reaction, such as random prime labeling for long probes or terminal transferase-catalyzed 3'-end labeling for oligonucleotides. Following hybridization, detection takes place by means of an antibody (or, in the case of biotin, streptavidin) linked to an enzyme, such as alkaline phosphatase or horseradish peroxidase, which is used to catalyze the detection reaction. A variety of haptens have been employed, most commonly digoxygenin, biotin, and fluorescein. This approach involves a longer detection protocol than a direct enzyme labeling methodology (*see* Chapter 10) because of the additional steps involved in membrane blocking, antibody binding, and subsequent membrane washes. However, it can be more readily adapted to high-sensitivity chemiluminescent detection with alkaline phosphatase-catalyzed dioxetane substrates. It can also be used with oligonucleotide probes, in this case coupled to horseradish peroxidase-catalyzed enhanced chemiluminescence detection, to take advantage of the rapid light output and low backgrounds of this reaction. This approach is described in Chapter 11.

Fluorescein is used as the hapten in the following protocols because it has several advantages for use in nucleic acid systems. These can be listed as follows.

1. It is stable during hybridizations even at elevated temperatures.
2. Fluorescein nucleotide conjugates are accepted by DNA polymerases for incorporation into probes.
3. High-affinity antibodies can be prepared against the fluorescein moiety.
4. The resulting probes have a low affinity for hybridization membranes when suitable blocking agents are used, so there is no necessity for probe purification.
5. There is little chance of sample interference from endogenous agents that would be recognized by the antibody, as may occur with biotin, for example, in *in situ* hybridization.
6. A final, significant advantage of fluorescein as a hapten is that its natural fluorescence can be exploited in a rapid (20-min) transilluminator-based assay to monitor the success of a probe-labeling reaction before embarking on the hybridization stage. This is analogous to the rapid check on incorporation that can be carried out with ^{32}P-labeled probes using β-monitors. A protocol for performing this assay is given in Chapter 9.

The Fluorescein Gene Images system (Amersham International, Bucks, UK) uses fluorescein as a hapten with detection using alkaline phosphatase-catalyzed dioxetane chemiluminescence. Fluorescein is incorporated into the probe by a random prime labeling reaction *(1)* catalyzed by the Klenow fragment of *E. coli* DNA polymerase in the presence of the modified nucleotide fluorescein-11-dUTP. Klenow is particularly efficient in this reaction, and there can be considerable net synthesis of probe during labeling. For example, starting with 50 ng of probe, it is possible to synthesize a further 300 ng of probe in a 1-h reaction. The labeled probe can be used directly in the hybridization (following denaturation) without probe purification, whereas the stringency of the hybridization itself can be controlled by either salt concentration or temperature during the hybridization and/or the stringency washes. Recommended probe concentrations are the same as for the ECL direct system (Chapter 10), i.e., 10 ng/mL for low target applications and 2–5 ng/mL for high levels of target.

Both the hybridization stage and the subsequent membrane blocking step prior to addition of antibody conjugate use a blocking agent that, if used as a solid, can be difficult to dissolve fully, leading to variable levels of background. However, this component is available as a concentrated solution, thus helping to ensure consistent, low backgrounds. Backgrounds have been further minimized by careful preparation of the antibody conjugate used with the system. This has involved optimization of the methods used for preparation of the conjugate, and the use of a monoclonal antifluorescein antibody, which gives a high signal-to-noise ratio in this system.

1.2. Alkaline Phosphatase-Catalyzed Dioxetane Chemiluminescence

Chemiluminescence from the alkaline phosphatase-catalyzed decomposition of a substituted dioxetane provides the basis for the high-sensitivity detection method used in this system. 1,2-Dioxetanes are cyclic peroxides that undergo decomposition into two carbonyl-containing fragments. One of these fragments is initially formed in an electronically excited state, undergoing relaxation to its ground state with the emission of light *(2)*. A phenylphosphate-substituted dioxetane containing a single adamantyl group (4-[phenyl-3-phosphate]-4-methoxyspiro[1,2-dioxetane-3,2'-adamantane]) has proven to be of particular value in a variety of assay formats *(3)*. It is highly stable, but in the presence of calf intestinal alkaline phosphatase, undergoes rapid conversion to the luminescent form via an unstable aryloxide dioxetane.

The maximum light output from this reaction is at 470–480 nm. Under some conditions, for example, in solution assays, the light output can be greatly increased by the inclusion of surfactant or polymers that contain fluorophores, such as fluorescein, that can be activated by energy transfer from the luminescent intermediate. Using fluorescent micelles formed from cetyltrimethylammonium bromide (CTAB) and 5-*(N)*-tetradecanoyl-aminofluorescein, chemiluminescent efficiency can be increased at least 400-fold *(4)*, with maximum light output now at 520–530 nm. It has more recently been found that certain blotting membranes, including nylon membranes, can themselves enhance light output.

Dioxetane-based chemiluminescence allows particularly high levels of sensitivity to be obtained in blotting applications. Low background can also be achieved, although it is necessary to ensure that a number of the reagents used during detection, in particular the antibody conjugate and membrane blocking agent, are optimized for this purpose. Additionally, solutions can easily become contaminated with bacterial alkaline phosphatases, which can cause extensive spotting on the final results on film. It is therefore necessary to use sterile solutions and clean apparatus whenever possible.

In contrast to the HRP-catalyzed chemiluminescence reaction, light output from a standard dioxetane luminescence reaction is initially slow, reaching maximum output only after several hours. This lag phase is caused by the relatively slow accumulation of aryloxide intermediate, which itself decays with a finite half-life. However, light output can continue at a high level for several days with only a slow decline. This allows accumulation of maximum signal during exposure to film, with sensitivity generally being limited by the level of background. The prolonged light output also gives the opportunity for multiple film exposures if required.

The greater sensitivity and prolonged signal of dioxetane chemiluminescence allows lower gel loadings to be used than with HRP-catalyzed enhanced chemiluminescence and provides a greater degree of robustness in the detection of low levels of nucleic acid. Using this system, a sensitivity of at least 100 fg can be achieved in genomic hybridization, using a 0.2-μg loading of human genomic DNA and a 1-kb probe. It is also possible to use this system in many Northern blot applications. As a rough guide, exposures of <48 h with ^{32}P-nucleotides should allow acceptable results to be obtained. The system is, of course, also compatible with lower sensitivity applications.

2. Materials

Components for labeling and for dioxetane chemiluminescent detection are available in kit format as the Fluorescein Gene Images labeling and detection system (Amersham International, plc). These components, together with additional solutions and reagents required, are described below.

2.1. Preparation of Labeled Probe

1. Nucleotide mix: 5X stock solution of fluorescein-11-dUTP (Fl-dUTP), dATP, dCTP, dGTP, and dTTP in 300 mM Tris-HCl, pH 7.8, 50 mM 2-mercaptoethanol, and 50 mM MgCl$_2$. This mixture is supplied in the Fluorescein Gene Images system. It has been optimized to give both a high yield of probe and a good degree of substitution with fluorescein.

Fluorescein-Labeled Nucleic Acid Probes

2. Primers: random nonamers. These primers, supplied with the Fluorescein Gene Images system, have been found to give a greater yield of probe than random hexamer primers.
3. Enzyme solution: 5 U/µL Klenow. This is also supplied in the Fluorescein Gene Images system and gives up to a twofold higher yield of probe than standard Klenow enzyme.
4. TE buffer: 10 mM Tris-HCl, pH 8.0, 1 mM EDTA.
5. 0.2 or 0.5M EDTA, pH 8.0.

2.2. Hybridization and Stringency Washes

1. Liquid block: supplied with the Fluorescein Gene Images system for use in the recommended hybridization buffer and for membrane blocking prior to addition of antibody conjugate.
2. 20X SSC: 3M NaCl, 0.3M sodium citrate, pH 7.0.
3. 10 or 20% (w/v) SDS.
4. Dextran sulfate: mol wt 500,000, for example, Sigma Product no. D-6001. This rate enhancer is required for maximum sensitivity during hybridization.
5. Hybridization buffer: 5X SSC, 1 in 20 dilution of liquid block, 0.1% (w/v) sodium dodecyl sulfate (SDS), 5% (w/v) dextran sulfate, 100 µg/mL sheared, denatured heterologous DNA, e.g., herring sperm DNA (optional).

2.3. Blocking, Antibody Incubation, and Washes

1. Antifluorescein alkaline phosphatase conjugate: 5000X stock supplied in the Fluorescein Gene Images system. The conjugate has been prepared and formulated to give low background in membrane hybridizations.
2. Buffer A: 100 mM Tris-HCl, pH 7.5, 300 mM NaCl. Autoclave in 500-mL aliquots in 1-L bottles for 15 min at 105 kPa (15 psi). Once opened after autoclaving, do not use for more than 1 d.
3. Bovine serum albumin (BSA) Fraction V: for example, Sigma Product no. A-2153 or BDH Product no. 44155.
4. Tween-20: Polyoxyethylene sorbitan monolaurate; for example, Sigma Product no. P-1379 or BDH Product no. 66368.

2.4. Signal Generation and Detection

1. Dioxetane detection reagent in a spray bottle: supplied in the Fluorescein Gene Images system in a form suitable for applying directly to the membrane. It is advisable that the spray be used in a fume cupboard.
2. Detection bags: optically clear plastic bags supplied with the Fluorescein Gene Images system for use during application of the detection reagent and subsequent exposure to film. Alternatively, Saran Wrap™ can be used.
3. Autoradiography film: most standard autoradiography films, such as Hyperfilm-MP™ (Amersham International, plc) or Kodak X-AR, are suitable for detection of the light generated by the dioxetane chemiluminescence reaction.

3. Methods
3.1. Preparation of Labeled Probe

This protocol allows the labeling of 50 ng of template DNA. The standard reaction can be used to label 25 ng to 2 µg of template DNA. Synthesis of labeled probe is most efficient at 50 ng of template, although net synthesis tends to increase with the amount of template.

1. Dilute the DNA to be labeled to a concentration of 2–25 ng/µL in either distilled water or TE buffer (*see* Note 1).
2. Place the nucleotide mix, primers, and water in an ice bath to thaw. Leave the enzyme at –20°C until required, and return it to the freezer immediately after use.
3. Denature the DNA sample by heating for 5 min in a boiling water bath, and then chill on ice. It is advisable to denature in a volume of at least 20 µL.
4. To a 1.5-mL microcentrifuge tube placed in an ice bath, add the appropriate volume of each reagent in the following order: water to a final reaction volume of 50 µL, nucleotide mix (10 µL), primers (5 µL), denatured DNA (50 ng), and enzyme solution (Klenow) at 5 U/µL (1 µL).
5. Mix gently by pipeting up and down and cap the tube. Spin briefly in a microcentrifuge to collect the contents at the bottom of the tube.
6. Incubate the reaction mix at 37°C for 1 h (*see* Note 2).
7. If any of the probe is to be stored, rather than being used in a hybridization immediately, the reaction should be terminated by the addition of EDTA to a final concentration of 20 mM. Probes can then be stored in a freezer at –20°C in the dark for at least 6 mo (*see* Note 3). Do not use a frost-free freezer.

The yield of probe after labeling 25 or 50 ng of template will typically be between 6 and 8 ng/µL. With higher template levels, although the amount of labeled template produced will increase, the overall reaction will be less efficient. At this stage, the rapid labeling assay (*see* Chapter 9) can be used to ensure that the probe has been successfully labeled.

3.2. Hybridization and Stringency Washes

The hybridization and wash conditions given in the following protocol are appropriate for a majority of probes, allowing detection of single-copy mammalian genes without significant crosshybridization to nonhomologous sequences. However, if these conditions are found to be insufficiently stringent for particular probes, then hybridized filters can be washed in 0.2 or 0.1X SSC at 60°C. Alternatively, stringency can be

Fluorescein-Labeled Nucleic Acid Probes

increased by raising the hybridization or wash temperature to 65°C. Such alterations may lead to some decrease in specific signal, although an overall improvement in signal-to-noise should result.

1. Prepare the hybridization buffer as follows: 5X SSC, 0.1% (w/v) SDS, 5% (w/v) dextran sulfate, 20-fold dilution of liquid block (supplied in the Fluorescein Gene Images system), and 100 µg/mL denatured sheared heterologous DNA (optional). Combine all the components and make up to the required volume. Gentle heating with continuous stirring will be required to dissolve the dextran sulfate completely (*see* Note 4).
2. Preheat the required volume of hybridization buffer (0.3 mL/cm^2 for small blots and 0.125 mL/cm^2 or less for large blots) to 60°C, and then place the blots in the buffer, and prehybridize for at least 30 min at 60°C with constant, gentle agitation (*see* Note 5).
3. Remove the required amount of probe to a clean microcentrifuge tube (*see* Note 6). If the volume is <20 µL, make up to this volume with water or TE buffer. Denature the probe by boiling for 5 min, and snap cool on ice.
4. Centrifuge the denatured probe briefly, and then add to the prehybridization buffer. Avoid placing it directly on the membrane and mix gently (*see* Note 6). Hybridize overnight at 60°C with gentle agitation.
5. Prepare two stringency wash solutions; one of 1X SSC, 0.1% (w/v) SDS, and the other of 0.5X SSC, and 0.1% (w/v) SDS, and preheat both solutions to 60°C. Carefully transfer the blots to the 1X SSC, and 0.1% (w/v) SDS and wash for 15 min at 60°C with gentle agitation. Carry out a further wash in preheated 0.5X SSC, 0.1% (w/v) SDS at 60°C for 15 min. These wash solutions should be used in excess volumes of approx 2–5 mL/cm^2 of membrane (*see* Note 7).

3.3. Blocking, Antibody Incubation, and Washes

The following steps are performed at room temperature, and all the incubations require constant agitation of the blots.

1. Following the hybridization washes, briefly rinse the blots in an excess (2 mL/cm^2) of buffer A at room temperature (*see* Note 8).
2. With gentle agitation, incubate blots for 1 h at room temperature in approx 0.75–1.0 mL/cm^2 of a 1 in 10 dilution of liquid blocking agent in buffer A (*see* Note 9).
3. Briefly rinse blots in buffer A as in step 1.
4. Dilute the antifluorescein-AP conjugate 5000-fold in freshly prepared 0.5% (w/v) BSA in buffer A. Incubate the blots in diluted conjugate (0.3 mL/cm^2 of membrane) with gentle agitation at room temperature for 1 h (*see* Note 10).

5. Remove unbound conjugate by washing for 3 × 10 min in 0.3% (v/v) Tween-20 in diluent buffer at room temperature with agitation. Use an excess volume (2–5 mL/cm^2).
6. Rinse blots in buffer A as in step 1.

3.4. Signal Generation and Detection

Read through this whole section before proceeding. If possible, wear powder-free gloves. The dioxetane detection reagent should be applied in a fume cupboard.

1. Cut a section from one of the detection bags supplied with the Fluorescein Gene Images system large enough to cover the blots plus a 1-cm border. Leave one side uncut and open out the bag.
2. After completion of the antibody incubation stage and subsequent washes, drain off any excess wash buffer from the blots by touching the corner of the blot against the box used for washing the blots or other convenient clean surface, and arrange them sample side up on one half of the opened bag. Do not allow the blots to dry out (*see* Note 11).
3. Place the blots in a fume cupboard with the bag opened to expose the blot surface. Hold the dioxetane spray applicator 2–3 cm above the blot and apply the reagent. Approximately 1 spray/14 cm^2 should be sufficient, e.g., a 50-cm^2 blot will require 4 sprays.
4. After spraying, fold the plastic over the top of the blots, and immediately spread the reagent evenly over the blot(s). This can be done either by rolling a 5 mL pipet over the surface or wiping the surface with a gloved hand.
5. For best results, transfer the blots to a fresh bag and place in a film cassette (*see* Note 12). Ensure the outside of the bag is dry before exposing to film. Place the bag containing the blots (sample side up) in a film cassette and take to a darkroom.
6. Switch off the lights and place a sheet of autoradiography film on top of the blots. Close the cassette and expose for 1–2 h for high-target/low-sensitivity applications. Longer exposures (4–16 h) should be used for higher sensitivity applications. Remove film and develop (*see* Note 13).

4. Notes

1. If the solution of DNA to be labeled is too dilute to be used directly, it should be concentrated by ethanol precipitation followed by redissolution in an appropriate volume of water or TE buffer. If an intact plasmid probe is to be used, it may be necessary to correct for the proportion of plasmid sequence present during the hybridization. Closed circular double-stranded DNA can be linearized to avoid more rapid renaturation following denaturation, although this is not usually necessary.

2. In the labeling reaction, a volume of DNA solution up to 25 µL may be used for fragments in low-melting-point agarose. The temperature of incubation and the reaction time can be chosen for convenience, since the reaction reaches a plateau and does not decline significantly overnight at room temperature. However, the rate of reaction does depend to some extent on DNA purity, so for samples of lower purity, such as miniprep DNA or DNA in agarose, a longer incubation period may be required. With purified DNA samples, an increase in probe yield can be obtained with longer reaction times (up to 4 h at 37°C).
3. Labeled probe can be used directly in a hybridization, following denaturation, without removal of unincorporated nucleotide. For long-term storage, it is advisable to keep probes in the dark, since the fluorescein molecule is light-sensitive.
4. The level of ribonuclease activity in the liquid block is low enough for its use in the hybridization buffer for Northern blots. However, if required, liquid block can be autoclaved, but must be subaliquoted into a suitable autoclavable container prior to sterilization. The hybridization buffer can be stored in suitable aliquots at −20°C for at least 12 mo. The addition of sheared, denatured heterologous DNA to the hybridization buffer can reduce nonspecific hybridization, although some reduction in sensitivity may also be observed.
5. Hybridization can be carried out in either boxes, bags, or tubes provided there is sufficient buffer to allow adequate access of probe to the blot. With larger blots (>50 cm^2), or if using minimal hybridization volumes in bags or tubes, the blots should be prewetted in 5X SSC. It is also possible to hybridize several blots in the same solution, again providing there is adequate volume and circulation of buffer. A maximum of two blots should be hybridized together in a single bag, and when using more than one blot in a tube, it is important that the blots do not overlap one another. High backgrounds can result if there is insufficient hybridization buffer or if a blot is not totally immersed. Agitation improves access of buffer to the blot, particularly if several blots are to be hybridized in the same solution.
6. Recommended probe concentrations for hybridization are 10–20 ng/mL for low target applications (for a 50 ng template reaction, this is approx 1–2 µL of reaction product/mL of hybridization buffer) and 1–5 ng/mL for high target applications. To avoid addition of probe directly onto the membrane, a small aliquot of the hybridization buffer may be removed and mixed with the probe before returning the mixture to the bulk of the buffer. For high target applications, it is possible to use shorter hybridization times. Some loss of sensitivity will result, but if necessary, such loss can be offset by the use of higher probe concentrations.

7. During the second stringency wash, the SSC concentration can be varied in the range 0.1–1X SSC to achieve the desired stringency *(5)*. The temperature of these washes may also be increased to achieve the desired stringency. Several blots can be washed in the same solution provided that they can move freely.
8. Once opened after autoclaving, unused buffer A should be discarded at the end of that working day to avoid contamination with exogenous alkaline phosphatases.
9. Diluted block can be stored frozen (in aliquots) for several weeks and is easily thawed when required. However, it is recommended that no aliquot be subjected to repeated thawing and refreezing. Also, it may be convenient to separate any remaining undiluted liquid block into appropriate aliquots for refreezing.
10. Diluted conjugate should be used immediately. Loss of sensitivity occurs with diluted conjugate that has been stored or been through a freeze–thaw cycle. Several blots can be incubated together, but it is important that there is free access of solution to the blot. With larger blots, it is possible to carry out this stage in hybridization bags or tubes to help minimize volume. For a 20 × 20 cm blot, a volume of 50 mL can be used. It may be found that the dilution of conjugate can be optimized further. Greater dilutions will give lower signal, but may reduce background, whereas more concentrated conjugate will give the opposite effect. BSA fraction V (for example, Sigma Product no. A-2153) should be used in the buffer.
11. The transfer of blots to the detection bags can either be carried out directly in the fume cupboard, or else the blots can be arranged while working on the bench and the plastic temporarily closed over the top of the blots while they are taken to the fume cupboard. A sheet of SaranWrap (Dow Chemical Company) can be used in place of the detection bags, although we have found bags to give the most satisfactory results.
12. Changing the bag after application of the detection reagent will avoid the transfer of excess liquid into the cassette. Alternatively, squeezing out all excess liquid from the original bag may suffice. If using SaranWrap, ensure that no liquid can escape.
13. If required, following an initial detection, expose a second film for an appropriate length of time. Depending on the application, the high-sensitivity of this system may mean that overnight exposures of the blot to film result in too dark an image. By the second day, however, the light output will have stabilized at a significantly higher level than that emitted during the first few hours after application of the detection reagent, allowing significantly shorter optimum exposure times. Once the light output has stabilized, it is easier to judge the optimum exposure time.

References

1. Feinberg, A. P. and Vogelstein, B. (1983) A technique for radiolabeling DNA restriction endonuclease fragments to high specific activity. *Anal. Biochem.* **132**, 6–13.
2. Reguero, M., Bernardi, F., Bottoni, A., Olivucci, M., and Robb, M. (1991) Chemiluminescent decomposition of 1,2-dioxetanes: an MC-SCF/MP2 study with VB analysis. *J. Am. Chem. Soc.* **113**, 1566–1571.
3. Schaap, A. P. (1988) Chemical and enzymatic triggering of 1,2-dioxetanes. *Photochem. Photobiol.* **47**, 50.
4. Schaap, A. P., Akhavan-Tafti, H., and Romano, L. J. (1989) Chemiluminescent substrates for alkaline phosphatase: application to ultrasensitive enzyme-linked immunoassays and DNA probes. *Clin. Chem.* **35**, 1863,1864.
5. Meinkoth, J. and Wahl, G. (1984) Hybridization of nucleic acids immobilized on solid supports. *Anal. Biochem.* **138**, 267–284.

CHAPTER 9

Monitoring Incorporation of Fluorescein into Nucleic Acid Probes Using a Rapid Labeling Assay

Martin Cunningham

1. Introduction

The efficiency of incorporation of fluorescein-11-dUTP with both random prime labeling of long probes (Chapter 8) and 3'-tailing of oligonucleotide probes (Chapter 11) can be estimated using the following rapid labeling assay. The basic protocol shown is for use with random prime labeled long probes, but modifications to the protocol for use with oligonucleotides are detailed in the notes. If sample identification is required, a pencil should be used to mark the membrane since some inks can interfere with the assay.

2. Materials

1. 20X SSC: 3M NaCl, 0.3M sodium citrate. Dilute as required.
2. 10% (w/v) Sodium dodecyl sulfate (SDS) stock solution. Dilute as required.
3. TE buffer: 10 mM Tris, pH 8.0, 1 mM EDTA.
4. Nylon membrane (e.g., Hybond-N+, Amersham International, Amersham, UK) or DE 81 paper (Whatman, Maidstone, UK).
5. 3MM paper (Whatman).

3. Methods

1. Prepare 1/5, 1/25, 1/50, 1/100, 1/250, 1/500, and 1/1000 dilutions of the 5X nucleotide mix from the Fluorescein Gene Images system (*see* Chapter 8) in TE buffer (*see* Note 1).
2. Dot out 5 µL of labeled probe and 5 µL of the 1/5 dilution of nucleotide mix (negative control) onto a strip of Hybond-N+, placed on a nonabsorbent backing. Allow the liquid to absorb (but **do not** allow to dry), and

wash the strip with gentle agitation in excess preheated 2X SSC at 60°C for 15 min (*see* Notes 2 and 3).
3. Prepare a reference strip by dotting 5 µL of each nucleotide mix dilution, except the 1/5, onto a separate strip of Hybond-N+. This reference strip can be reused and can be stored, wrapped in Saran Wrap™, for several weeks at –20°C in the dark.
4. Place both the reference and the washed strips on a piece of Whatman 3MM paper lightly moistened with TE buffer, and take to the darkroom (*see* Note 4). Visualize both strips (sample side down) on a UV transilluminator. Optimum contrast is obtained using a short wavelength (254-nm) transilluminator. The labeled probe should be visible as a fluorescent spot with an intensity between the 1/25 and 1/250 dilutions on the reference strip (*see* Note 5). Such a result indicates that the probe has successfully incorporated the fluorescein label. The closer the intensity of the labeled DNA is to the lower dilution (1/25), the more efficient the labeling reaction is. The washed negative control should retain little or no fluorescence, indicating that the fluorescence of the probe is the result of only incorporated fluorescein and not unincorporated Fl-dUTP. If significant fluorescein remains in the negative control, wash the filter for a further 15 min. This is possible only if the strip has not dried.
5. For a hard copy of the results, the strips can be photographed. A Kodak Wratten No. 9 filter (or a similar yellow filter) should be used in conjunction with Polaroid 667 black and white film.

4. Notes

1. For oligonucleotide probes prepared using the ECL 3'-oligolabeling system (*see* Chapter 11), the recommended dilutions are 1/16, 1/100, 1/250, 1/500, 1/1000, and 1/2000 of the fluorescein-11-dUTP (Fl-dUTP) in TE buffer.
2. When using the 3'-oligolabeling system, dot out 5 µL of each labeled oligonucleotide and 5 µL of the 1/16 dilution of Fl-dUTP (negative control).
3. Whatman DE81 paper can be used as an alternative to Hybond-N+ and is recommended for use with oligonucleotides. In this case, 2X SSC, 0.1% (w/v) SDS should be used to wash the strip. A 1-cm edge should be allowed around the edge of the filter to avoid damage during washing. Following the wash stage, rinse the strip successively with water and then ethanol before handling (DE81 paper is fragile in aqueous solutions).
4. If using DE81 paper, the reference strip can be wetted (after applying the dots) successively in water and then ethanol to reduce distortion of the filter, and when using DE81 paper, dry Whatman 3MM paper should be used to transport the strips to the darkroom.
5. An efficiently labeled oligonucleotide should be visible as a fluorescent spot with an intensity between the 1/500 and 1/2000 dilutions.

CHAPTER 10

The Preparation of Horseradish Peroxidase-Labeled Nucleic Acid Probes and Their Detection Using Enhanced Chemiluminescence

Bronwen Harvey, Ian Durrant, and Martin Cunningham

1. Introduction

A range of labeling and detection systems for nucleic acids, many of which show specific advantages for particular applications is currently available to the researcher. The labels generally fall into one of two categories: primary labels, in which a detectable signal, such as a radioisotope or a fluorophore, is introduced directly into the probe, and enzymatic labels, in which an enzyme is used to catalyze a signal-generating reaction. In the latter case, the signal may be a fluorophore, a visibly colored compound (chromophore), or light generated by a chemiluminescent reaction. The enzyme may itself be directly coupled to the probe (a direct label) or may be introduced by means of an antibody–enzyme or streptavidin–enzyme complex, which detects hapten- or biotin-labeled probe (an indirect label). This type of indirect approach has been widely used with probes labeled with molecules, such as fluorescein, biotin, and digoxygenin. An example of such a system using fluorescein-labeled probes detected by an antifluorescein alkaline phosphatase conjugate, which in turn catalyzes highly sensitive chemiluminescent detection, is described in Chapter 8.

From: *Methods in Molecular Biology, Vol. 58: Basic DNA and RNA Protocols*
Edited by: A. Harwood Humana Press Inc., Totowa, NJ

However, the direct labeling of nucleic acid probes with an enzyme can allow more rapid nonradioactive detection, since the necessity for additional antibody-related steps is avoided. Such direct labeling with an enzyme, horseradish peroxidase (HRP), has been described by Renz and Kurz *(1)*. Following hybridization, the peroxidase can be used to catalyze an enhanced chemiluminescent reaction that leads to very rapid generation of signal *(2)*. This approach has been made available as the ECL™ direct system (Amersham International, Amersham, UK), which provides a convenient method for rapid, nonradioactive detection of moderate to high levels of target.

The labeling step involves the reaction of a single-stranded nucleic acid probe with a positively charged HRP–parabenzoquinone–polyethyleneimine complex that is supplied as the labeling reagent. This complex initially undergoes a charge interaction with negatively charged nucleic acid and is subsequently covalently crosslinked to the nucleic acid probe by the addition of glutaraldehyde. This labeling reaction is rapid, taking only 20 min at room temperature, and has been used to label probes from 50 bases to 50 kb in length. Once labeled, the probe may be stored at –20°C for several months in the presence of 50% glycerol. Denatured DNA and RNA probes can be labeled using this method, and the reaction can be scaled up or down to label between 100 ng and 1.5 µg.

The maximum rate of hybridization occurs at 25°C below the T_m of the probe, in the range 60–75°C under nondenaturing conditions. HRP-labeled probes are sensitive to high temperatures, so it is necessary to employ a T_m modifier to allow hybridization at a lower temperature. Urea (6*M*) is included in the Gold buffer supplied with the system, which allows hybridization to occur at 42°C *(3,4)*. These conditions are equivalent to hybridizing a radiolabeled probe at 65 or at 42°C in 50% (v/v) formamide. Peroxidase is stable under these conditions for at least 24 h. The Gold buffer also contains a novel rate enhancer that both maximizes signal and reduces backgrounds. Stringency control is achieved by varying salt concentration either during hybridization, during the stringency washes, or both.

Since the probe is directly labeled with enzyme, it is not necessary to carry out time-consuming antibody-blocking, incubation, and wash steps following the stringency washes. Directly following the stringency washes, the enhanced chemiluminescence reaction is used to generate light, which is captured on autoradiography film, producing a hard copy result. This reaction involves the oxidation of luminol, a cyclic diacylhydrazide, by peroxidase in the presence of peroxide *(5)*, which is supplied as a stable peracid salt. This salt and the enhancer/luminol mixture must be kept separate until immediately before use to avoid gradual chemical degradation. Luminol is converted via an endoperoxide into an electronically excited 3-aminophthalate that, on falling back to the ground state, emits light. Enhancer molecules modulate the reaction between the enzyme, peroxide, and luminol leading to a stimulation of light emission by more than 1000-fold over an unenhanced reaction *(2)*. The maximum emission of light in this reaction is at 428 nm, appropriate for film sensitive to blue light.

Light output from the enhanced chemiluminescence reaction occurs rapidly with no detectable lag phase. There is a peak of light output at 5–10 min, followed by a decay with a half-life of approx 1 h. This decay is thought to be owing to a gradual loss of enzyme activity resulting from free radical damage. These kinetics, in addition to allowing very rapid detection, also have an advantage for reprobing because, since enzyme is inactivated during the detection step, bands resulting from the presence of the probe will not be visible after subsequent hybridization and detection. Rehybridization of the membrane with a new probe is therefore possible without removal of the previous probe.

The combination of rapid labeling, the absence of antibody-blocking, incubation, and wash stages, and the fast kinetics of light generation make this system ideally suited to high-throughput, large-scale screening applications *(6–8)*. Levels of background on membrane hybridizations are generally low with peroxidase-based enhanced chemiluminescence, and the lower prevalence of peroxidase compared with alkaline phosphatase also means that results are less liable to the "spotting" sometimes seen with alkaline phosphatase-based systems when solutions are not scrupulously clean. Good levels of signal are also achievable. For example, using 10 ng/mL probe concentration with an overnight hybridization, it is possible to detect levels of target as low as 500 fg, equivalent to single-copy mammalian gene detection in a sample of 1 µg mammalian total DNA with a 1-kb probe. For the higher levels of target present in applications, such as plaque or colony screens and in PCR product detection, probe concentrations as low as 2 ng/mL can be used with 2 h of hybridization, allowing a particularly high level of throughput.

2. Materials

Components for labeling, hybridization, and enhanced chemiluminescent detection are available in kit format as the ECL direct nucleic acid labeling and detection system (RPN3000, RPN3001, RPN3005 Amersham International, plc). Detailed protocols are included with each kit. The system is stable for at least 3 mo when stored under appropriate conditions, but the labeling reagent is light-sensitive and should be protected from intense light or protracted periods of exposure to light. These components, together with additional solutions and reagents required, are described below.

2.1. Probe Labeling

1. Labeling reagent: This comprises HRP crosslinked to polyethylenimine (PEI). Its synthesis is described in ref. *1*. It can be purchased ready-made (Amersham), and its use is described in the following protocol.
2. Glutaraldehyde solution (1.5% [v/v] in water): 1.5% (v/v) glutaraldehyde solution in deionized water. Although glutaraldehyde is classified as harmful, for ECL, it is present at a concentration of only 1.5% (v/v), at which level it is not classified as harmful. However, like all chemicals, it should be handled within the principles of good laboratory practice.

2.2. Hybridization

1. Gold hybridization buffer: The buffer can be obtained ready-made (Amersham). In addition to $6M$ urea, it contains a detergent, a pH-stabilizing compound, and a novel rate enhancement system. The new rate enhancer is directly responsible for a significant increase in sensitivity over early versions of the ECL Direct system. The new hybridization buffer ensures a higher level of target coverage and promotes an increase in the stability of the enzyme label during the hybridization process.
2. Blocking agent.
3. Sodium chloride (analytical grade).

2.3. Stringency Washes

1. 20X SSC: $0.3M$ trisodium citrate, $3.0M$ sodium chloride, pH 7.0.
2. Primary wash buffer with urea: 360 g of urea, 4 g of sodium dodecyl sulfate (SDS), 25 mL 20X SSC. Make up to 1 L.
3. Primary wash buffer without urea: 4 g of SDS, 25 mL of 20X SSC. Make up to 1 L.
4. Secondary wash buffer: 2X SSC.

2.4. Detection

1. ECL detection reagent 1 contains specially purified luminol and the ECL enhancer in a borate buffer.
2. ECL detection reagent 2 contains a peracid salt in borate buffer.
3. Saran Wrap™ (Dow Chemical Company) is recommended at various stages of the protocol and is particularly suitable for molecular biology applications, since it does not absorb UV or blue light. Saran Wrap is often available from local suppliers.
4. Blue-sensitive autoradiography film is a film optimized for use with ECL systems that is available (Hyperfilm ECL, Amersham). However, other X-ray films are suitable, e.g., Hyperfilm-MP, Kodak-AR. It is also possible to detect the light on a charge-coupled device camera.

3. Methods
3.1. Probe Labeling

1. Dilute the nucleic acid to be labeled to a concentration of 10 ng/μL in a 1.5-mL microcentrifuge tube (*see* Note 1). Close the tube and incubate for 5 min in a boiling water bath to denature the nucleic acid (*see* Notes 2 and 3).
2. Immediately cool the sample on ice for 5 min, and then centrifuge the tube briefly (10 s) to collect the contents at the bottom.
3. Add a volume of labeling reagent equal to that of the sample, and mix briefly by taking up the solution in the tip of a micropipet and gently expelling (*see* Note 4).
4. Add a volume of glutaraldehyde solution equal to that of the labeling reagent. Again mix gently by pipet.
5. Incubate at 37°C for 10 min. The probe is now labeled with HRP and can be stored on ice for up to 15 min before use (*see* Note 5).

3.2. Hybridization

1. Prepare the hybridization buffer as follows. Take the required volume of Gold hybridization buffer and add solid sodium chloride to a final concentration of $0.5M$ (*see* Note 6). This may vary for different probes, although $0.5M$ NaCl has been found to give satisfactory results in most cases. Add blocking agent to a final concentration of 5% (w/v). Rapidly mix the blocking agent into the buffer to avoid the formation of clumps. Continue to mix thoroughly using a magnetic stirrer at 42°C for 30–60 min. A few undissolved particles of blocking agent will not affect the hybridization.
2. Place the blot containing the target nucleic acid sequence onto the surface of the fully prepared hybridization buffer prewarmed to 42°C (*see* Note 7). A volume of buffer equivalent to 0.25 mL/cm^2 of membrane is satisfactory for most applications (*see* Notes 8 and 9). Allow the blot to saturate fully

before submerging in the buffer. Prehybridize at 42°C for 15–60 min in a shaking water bath with gentle agitation.
3. Add the labeled probe from Section 3.1., step 5 to the prehybridization buffer at a concentration of either 10 or 2–5 ng/mL depending on application (*see* Note 10). Avoid placing the probe directly on the membrane and mix gently. Continue incubation with gentle agitation for between 2 h and overnight at 42°C.

3.3. Stringency Washes

1. Preheat enough primary wash buffer to 42°C to allow two washes of 2 mL buffer/cm^2 membrane (*see* Note 11).
2. Carefully transfer the blot from the hybridization container to half of the prewarmed primary wash buffer in a clean container. Incubate at 42°C in a shaking water bath for 20 min. Discard the wash buffer, replace with the remaining buffer, and incubate with agitation for a further 20 min at 42°C.
3. Discard the wash buffer, place the blot in a fresh container, and add an excess of secondary wash buffer (at least 2 mL buffer/cm^2 membrane). Incubate for 5 min at room temperature with gentle agitation. Discard the wash buffer and repeat this step with fresh secondary wash buffer (*see* Note 12). The blot may be temporarily stored at this stage if required (*see* Note 13).

3.4. Detection

1. Combine equal volumes of detection reagents 1 and 2 so that there is a sufficient amount for 0.125 mL combined reagents/cm^2 membrane (*see* Note 14).
2. Using forceps, carefully remove the blot from the final secondary wash buffer. Drain off the excess wash buffer by touching the bottom edge of the blot onto a paper towel. Place the blot, sample side uppermost, onto a sheet of Saran Wrap on the bench.
3. Cover the blot evenly with the freshly mixed detection reagents and leave for 1 min. From this stage, it is advisable to work as quickly as possible to minimize the delay between addition of substrates and exposure to film.
4. Using forceps to hold the blot, drain off the excess detection reagents by touching the bottom edge of the blot onto a paper towel. Place the blot nucleic acid side down onto a sheet of Saran Wrap in a film cassette.
5. Fold over the Saran Wrap to wrap the blots completely, but ensure that only a single layer covers the sample side. Turn the blots so that the sample side is uppermost.
6. In a darkroom with a red safelight, place a piece of blue-light-sensitive film on top of the blot, shut the cassette and expose at room temperature for the required length of time (*see* Note 15). Develop film under normal conditions.

4. Notes

1. Between 10 (100 ng) and 150 µL (1.5 µg) probe may be labeled in each tube. Labeling efficiency will be reduced if the volume of nucleic acid exceeds 150 µL/tube.
2. Complete denaturation of double-stranded probes is essential for maximal labeling. Extending the length of this step should be considered when using heating blocks, since these can be less efficient at denaturation than boiling water baths. The denaturation step is not strictly necessary for single-stranded probes, although in general, denaturation of RNA is advisable to avoid secondary structure.
3. Since labeling is achieved in part through an ionic interaction between the negatively charged nucleic acid and the positively charged labeling reagent, the concentration of salt in the sample to be labeled should not exceed 10 mM.
4. It is important to use equivalent volumes of nucleic acid and labeling reagent for maximum labeling efficiency. Accurate assessment of probe concentration (10 ng/µL) is essential to achieve maximum sensitivity.
5. Probes once labeled are single-stranded and may be used in any membrane hybridization application. If they are to be stored for more than 15 min, sterile glycerol should be added to a final concentration of 50% (v/v), and the probe stored at −20°C until required. Probes stored for up to 12 mo have been used in hybridization with no appreciable loss of sensitivity.
6. A concentration of 0.5M NaCl has been found to be suitable for a variety of probes, although in some cases, it may be found necessary to vary this concentration for optimal stringency.
7. Blots >100 cm^2 should be prewet in 5X SSC before addition to the hybridization buffer. The highest sensitivity detection is obtained with charged nylon hybridization membranes, such as Hybond-N+ (Amersham International). However, for many applications, satisfactory results can be obtained with uncharged nylon (e.g., Hybond-N) or nitrocellulose (e.g., Hybond-ECL).
8. For hybridization carried out in boxes, a volume of buffer equivalent to 0.25 mL/cm^2 membrane is recommended. If the box is significantly larger than the blot, then the volume used should correspond to the area of the bottom of the box. Several blots may be hybridized together if adequate circulation of the buffer is maintained, so that the blots can move freely. Increased backgrounds will otherwise result.
9. Hybridizations may also be carried out in bags or tubes. In these cases, a buffer volume of 0.125 mL/cm^2 may be sufficient. When using tubes, it is important to exclude any air bubbles trapped between the hybridization tube and the membrane, since this can lead to patches of high background following detection. Nylon mesh may be used if there is significant overlap of the blot to ensure good access of probe.

10. Final probe concentration (10 ng/mL) and overnight hybridization will give maximum sensitivity and are necessary for the detection of low levels of target, such as single-copy genes on mammalian genomic Southerns. However, for low-sensitivity applications, such as many plaque or colony screens or for detection of PCR products or bacterial genes, it may be possible to use a lower probe concentration (2–5 ng/mL) in conjunction with a 2-h hybridization.
11. Stringency is controlled in the primary washes by the concentration of SSC. The primary wash contains 5X SSC, but higher stringency and hence more specific hybridization signal can be obtained with lower concentrations of SSC, such as 0.1X SSC. Generally, this wash will contain $6M$ urea and is performed at 42°C. However, it is possible to achieve similar results using primary wash without urea with incubation at 55°C. In this case, the total wash time should not exceed 20 min in order to conserve peroxidase activity.
12. The secondary washes are included to ensure removal of SDS, which may interfere with the enhanced chemiluminescence reaction.
13. Blots may be stored for up to 24 h at 4°C, wetted in secondary wash buffer, and wrapped in Saran Wrap. After storage, the blots should be rinsed briefly in fresh secondary wash buffer just before detection.
14. The detection reagents should ideally be kept separate until immediately before detection to avoid gradual chemical decomposition of the substrates. However, if not used immediately, it is possible to store the prepared solution on ice for up to 30 min.
15. For initial experiments, an exposure of 30–60 min is recommended for low target applications. An alternative approach is to carry out a short 1–2 min initial exposure and use this to judge the length of the subsequent exposure. For plaque and colony screens in contrast, it is important to use the minimum film exposure time that enables positive colonies/plaques to be unambiguously determined in order to avoid false positives.

References

1. Renz, M. and Kurz, C. (1984) A colorimetric method for DNA hybridization. *Nucleic Acids Res.* **12**, 3435–3444.
2. Whitehead, T. P., Thorpe, G. H. G., Carter, T. J. N., Groucutt, C., and Kricka, L. J. (1983) Enhanced chemiluminescence procedure for sensitive determination of peroxidase-labelled conjugates in immunoassay. *Nature* **305**, 158,159.
3. Hutton, J. R. (1977) Renaturation kinetics and thermal stability of DNA in aqueous solutions of formamide and urea. *Nucleic Acids Res.* **4**, 3537–3555.
4. Stone, T. (1992) ECL direct system: an analysis of filter hybridization kinetics. *Life Sci. (Amersham International)* **7**, 8–10.
5. Roswell, D. F. and White, E. H. (1978) The chemiluminescence of luminol and related hydrazides. *Methods Enzymol.* **57**, 409–423.

6. Durrant, I., Benge, L. A. C., Sturrock, C., Devenish, A. T., Howe, R., Roe, S., Moore, M., Scozzafava, G., Proudfoot, L. M. F., Richardson, T. C., and McFarthing, K. G. (1990) The application of enhanced chemiluminescence to membrane-based nucleic acid detection. *Biotechniques* **8,** 564–570.
7. Cohn, D., Chumakov, I., and Weissenbach, J. (1993) A first-generation physical map of the human genome. *Nature* **366,** 698–701.
8. Sorg, R., Enczmann, J., Sorg, U., Kogler, G., Schneider, E. M., and Wermet, I. (1990) Specific non-radioactive detection of PCR-amplified sequences with enhanced chemiluminescence labeling. *Life Sci.* **2,** 3,4.

CHAPTER 11

3'-End Labeling of Oligonucleotides with Fluorescein-11-dUTP and Enhanced Chemiluminescent Detection

Martin Cunningham, Ian Durrant, and Bronwen Harvey

1. Introduction

In this reaction, the enzyme terminal transferase is used to introduce a short tail of fluorescein nucleotides to the 3'-end of an oligonucleotide *(1)* followed by horseradish peroxidase-catalyzed enhanced chemiluminescent detection of the labeled hybrids. This approach is used in the ECL 3'-oligolabeling and detection system (Amersham International, Amersham, UK) for hybridizations on membranes. In general, oligonucleotide probes are used for lower sensitivity applications, and the combination of fluorescein-labeling and enhanced chemiluminescent detection allows as little as 20×10^{-18} mol of homologous target to be detected within a 1-h exposure.

2. Materials

Components for labeling and enhanced chemiluminescent detection are available in kit format as the ECL 3'-oligolabeling and detection system (Amersham International, plc). These components, together with additional solutions and reagents required, are described below.

2.1. Oligonucleotide Labeling

1. Fluorescein-11-dUTP: This nucleotide is supplied in the ECL 3'-oligolabeling and detection system.
2. Terminal transferase: 2 U/µL as supplied in the ECL 3'-oligolabeling and detection system.

From: *Methods in Molecular Biology, Vol. 58: Basic DNA and RNA Protocols*
Edited by: A. Harwood Humana Press Inc., Totowa, NJ

3. Cacodylate buffer: 10X concentrated buffer, pH 7.2, as supplied in the ECL 3'-oligolabeling and detection system. **Warning:** Cacodylate is a highly toxic compound; follow all safety warnings and precautions.
4. TE buffer: 10 mM Tris-HCl, pH 8.0, 1 mM EDTA.

2.2. Hybridization

1. Liquid block: supplied in the ECL 3'-oligolabeling and detection system for use in hybridization buffer and membrane blocking solution.
2. Hybridization buffer component: supplied in the ECL 3'-oligolabeling and detection system to maximize signal to noise during hybridization.
3. 20% (w/v) SDS stock solution.
4. 20X SSC: 0.3M sodium citrate, 3M NaCl.

2.3. Washing the Membrane

1. 20% (w/v) SDS stock solution.
2. 20X SSC: 0.3M sodium citrate, 3M NaCl.

2.4. Membrane Blocking, Antibody Incubations, and Washes

1. Antifluorescein HRP conjugate: 1000X stock supplied in the ECL 3'-oligolabeling and detection system. The conjugate has been prepared and formulated to give low background during membrane hybridizations.
2. Buffer 1: 0.15M NaCl, 0.1M Tris, pH 7.5.
3. Buffer 2: 0.4M NaCl, 0.1M Tris, pH 7.5.
4. Bovine serum albumin (BSA) fraction V: for example, Sigma (St. Louis, MO) product code A-2153.

2.5. Signal Generation and Detection

1. ECL detection reagent 1: This contains specially purified luminol and the ECL enhancer in a borate buffer.
2. ECL detection reagent 2: This contains a peracid salt in borate buffer.
3. Saran Wrap™ (Dow Chemical Company): This is used at various stages during the procedure and does not absorb UV or blue light.
4. Blue-sensitive autoradiography film: a film optimized for the light detection is available (Hyperfilm™-ECL, Amersham). However, other X-ray films are suitable, e.g., Hyperfilm-MP, Kodak-AR. It is also possible to detect the light output on a charge-coupled device camera.

3. Methods

3.1. Oligonucleotide Labeling

This protocol allows the labeling of single-stranded oligonucleotides bearing a 3'-hydroxyl group.

3'-End Labeling

1. Place the required tubes from the labeling components, excluding the enzyme, on ice to thaw (*see* Note 1).
2. To a 1.5-mL conical polypropylene tube on ice, add the labeling reaction components in the following order: 100 pmol oligonucleotide, 10 µL fluorescein-11-dUTP, 16 µL cacodylate buffer, water to 144 µL, 16 µL terminal transferase, giving a total of 160 µL (*see* Note 2).
3. Mix gently by pipeting up and down in the pipet tip.
4. Incubate the reaction mixture at 37°C for 60–90 min.
5. If desired, the reaction can be checked using the rapid labeling assay (*see* Chapter 9).
6. Store labeled probe on ice for immediate use or place at −20°C in the dark for long-term storage (*see* Note 3).

3.2. Hybridization

Hybridization temperature is dependent on the particular probe sequence to be used. Ideally, stringency is controlled by the posthybridization washes so that, for many systems, a hybridization temperature of 42°C will suffice. Generally, the hybridization temperature should be 5–10°C below the melting temperature (T_m) of the oligonucleotide *(2)* (the presence of the 3'-tail does not significantly alter the T_m value).

1. The recommended hybridization buffer consists of the following components: 5X SSC, 0.1% (w/v) hybridization buffer component, 0.02% (w/v) SDS, and a 20-fold dilution of liquid block. Combine all the components, make up to the required volume, and mix well by stirring (*see* Notes 4 and 5).
2. Place blot(s) into the hybridization buffer and prehybridize at the required temperature, for a minimum of 30 min, in a shaking water bath.
3. Add the labeled oligonucleotide probe to the buffer used for the prehybridization step at a final concentration of 5–10 ng/mL (*see* Note 6).
4. Hybridize at the required temperature for 1–2 h in a shaking water bath (*see* Note 7).

3.3. Washing the Membrane

The stringency of the hybridization is controlled at this stage. The presence of the 3'-tail does not affect the stringency conditions that would be required for an unlabeled sequence. Stringency can be altered by a combination of temperature and SSC concentration during the wash steps (*see* Note 8).

1. Remove the blot(s) from the hybridization solution and place in a clear container. Cover with an excess of 5X SSC, 0.1% (w/v) SDS at a volume equivalent to approx 2 mL/cm^2 of membrane. The blot(s) should move freely.

2. Incubate at room temperature for 5 min with constant agitation.
3. Replace the wash buffer with fresh 5X SSC, 0.1% (w/v) SDS, and incubate for a further 5 min.
4. Discard the wash solution. Place blot(s) in a clean container, and cover with an excess of appropriate prewarmed stringency wash buffer. For many systems, a suitable wash buffer would contain 1X SSC, 0.1% (w/v) SDS.
5. Incubate at the desired temperature (typically 42–50°C) for 15 min in a shaking water bath.
6. Replace the stringency wash buffer with fresh solution (as in step 4), and incubate for a further 15 min at the same temperature as in step 5.

3.4. Membrane Blocking, Antibody Incubations, and Washes

The following steps are performed at room temperature and all the incubations require constant agitation of the blot(s).

1. Place the filters in a clean container, and rinse with buffer 1 for 1 min. Use 2 mL of buffer for each cm^2 of membrane.
2. Discard buffer 1 and replace with block solution (a 20-fold dilution of liquid block in buffer 1) equivalent to 0.25 mL/cm^2 of membrane. Incubate for at least 30 min.
3. Rinse blot(s) briefly (1 min) in buffer 1 using a volume as in step 1 above.
4. Dilute the antifluorescein HRP conjugate 1000-fold in buffer 2 containing 0.5% (w/v) BSA (fraction V). The amount of diluted antibody-conjugate solution prepared should be equivalent to 0.25 mL/cm^2 of membrane.
5. Incubate the blot(s) in the diluted antibody conjugate solution for 30 min.
6. Place filters in a clean container, and rinse with an excess of buffer 2 for 5 min using a volume as in step 1 above.
7. Repeat step 6 a further three times to ensure complete removal of nonspecifically bound antibody (*see* Note 9).

3.5. Signal Generation and Detection

This section is carried out as for ECL direct detection (*see* Chapter 10). The exposure time will vary depending on the level of target material. Typical exposure times will vary from 1 to 10 min for high target applications to 10–60 min for low target applications. With high target systems, it is possible to perform multiple exposures.

4. Notes

1. Place the terminal transferase enzyme on ice immediately prior to use. Return the enzyme to a freezer at –20°C as soon as the required amount has been removed.

3'-End Labeling

2. It is recommended that the oligonucleotide for labeling be dissolved (or diluted) in sterile distilled water. The weight corresponding to 100 pmol of an oligonucleotide is dependent on the length of the sequence:

Number of bases	Amount, ng
15	500
20	660
30	1000
50	1660

    If alternative levels of probe are to be labeled, then the volumes of the reactants and the final reaction volume should be altered in the same ratio as the change in the amount of probe. Reactions ranging from 25–250 pmol have been undertaken successfully.
3. Purification of the probe after labeling is not necessary.
4. For convenience, it is recommended that the hybridization buffer be made up in a large volume and stored in suitable aliquots at –20°C. Under these conditions, the buffer is stable for at least 3 mo.
5. The volume of hybridization buffer should be equivalent to 0.25 mL/cm^2 of membrane. This may be reduced to 0.125 mL/cm^2 for large blots hybridized in plastic bags or hybridizations in ovens. It is possible to alter the volume of buffer depending on the size of the container and the number of blots to be hybridized. It is essential that the blots should move freely within the buffer.
6. Avoid placing the probe directly onto the blot when initiating hybridization. Alternatively, a small aliquot of the buffer may be removed and mixed with the probe before returning the mixture to the bulk of the hybridization buffer.
7. Hybridization time can be increased up to 17 h without any significant change in sensitivity.
8. If the stringency wash solution shows any precipitation during storage at room temperature, it should be warmed slightly before use, until all material has dissolved. The SSC concentration can be varied in the range 0.1–1X SSC to achieve the desired stringency, whereas the SDS concentration should remain constant at 0.1% (w/v). At a temperature of 3–5°C below the T_m of the probe, these conditions should allow the discrimination of perfectly matched sequences from those containing mismatches (2). The stringency wash temperature can be as high as 65°C, although when combined with changes in the SSC concentration, it may not be necessary to go higher than 50°C to achieve the desired stringency.
9. It is possible to store the blots in the final wash buffer for up to 30 min prior to proceeding to the detection stage. If longer-term storage is required,

remove the filter from the wash buffer, wrap in Saran Wrap, and store in a refrigerator at 2–8°C. Do not allow the filter to dry out.

References

1. Bollum, F. J. (1974) Terminal deoxynucleotidyl transferase, in *The Enzymes*, vol. 10 (Boyer, P. D., ed.), Academic, New York, pp. 145–171.
2. Wallace, R. B. and Miyada, C. G. (1990) Oligonucleotide probes for the screening of recombinant DNA libraries. *Methods Enzymol* **152,** 432–442.

CHAPTER 12

The Preparation of Riboprobes

Dominique Belin

1. Introduction

The isolation and characterization of RNA polymerases from the *Salmonella* phage SP6 and the *E. coli* phages T7 and T3 have revolutionized all aspects of the study of RNA metabolism *(1–6)*. Indeed, it is now possible to generate unlimited quantities of virtually any RNA molecule in a chemically pure form. This technology is based on a number of properties of the viral transcription units. First, and in contrast to their cellular counterparts, the enzymes are single-chain proteins that are easily purified from phage-infected cells and are now produced by recombinant DNA technology. Second, they very specifically recognize their own promoters, which are contiguous 17–20 bp long sequences rarely encountered in bacterial, plasmid, or eukaryotic sequences. Third, the enzymes are highly processive, allowing the efficient synthesis of very long transcripts from DNA templates. In this chapter, the preparation of the DNA templates and the transcription from the template of ^{32}P-labeled synthetic RNA molecules, commonly called riboprobes, will be discussed.

Riboprobes can be used in Southern and Northern hybridizations by following the same general principles as for DNA probes. They avoid the self-annealing problem of double-stranded DNA probes that leads to decreased availability of the probe for hybridization to the immobilized target; this is particularly critical with heterologous probes, where the reannealed probe may displace incompletely matched hybrids. Self-annealing, of course, do not occur with single-stranded RNA probes, but most difficulties encountered with riboprobes stem from the increased thermal stability of RNA:RNA hybrids. Thus, crosshybridization of GC-rich probes to rRNAs can generate unacceptable backgrounds. This problem is often solved by increasing the stringency of hybridization, as illustrated in Fig. 1. Alternatively, the template may have to be shortened to remove GC-rich regions from the probe.

2. Materials

These protocols require the use of standard molecular biological materials and methods for carrying out subcloning, polymerase chain reaction (PCR), and gel electrophoresis, in addition to those listed below. More rigorous precautions are required for working with RNA than commonly used for most DNA studies, since it is important to eliminate RNase contamination. The two major sources of unwanted RNases are the skin of investigators and microbial contamination of solutions. Most RNases do not require divalent cations and are not irreversibly denatured by autoclaving. Gloves should be worn and frequently changed. Sterile plasticware should be used, although some mechanically manufactured tubes and pipet tips have been used successfully without sterilization. Glassware should be incubated at 180°C in a dried baking oven for several hours.

Restriction enzyme digestion and PCR are described in Chapters 2 and 34, respectively.

1. Water: Although a number of protocols recommend treatment of the water used for all the solutions with diethylpyrocarbonate, I find this to be unnecessary. Double-distilled water is used for the preparation of stock solutions that can be autoclaved (121°C, 15–30 min), and sterilized water is used otherwise.
2. TE: 10 mM Tris-HCl, pH 8.1, 1 mM EDTA.
3. 10X TB: 0.4M Tris-HCl, pH 7.4, 0.2M NaCl, 60 mM MgCl$_2$, 20 mM spermidine. If ribonucleotides are used at a concentration >0.5 mM each, the MgCl$_2$ concentration should be increased to provide a free magnesium concentration of 4 mM.
4. 0.2M DTT: The solution is stored in small aliquots at –20°C. Aliquots are used only once. EDTA can be included at 0.5 mM to stabilize DTT solutions.
5. Ribonucleotides: Neutralized solutions of ribonucleotides are commercially available or can be made up from dry powder (*see* Note 1). Ribonucleotide solutions can be stored at –20°C for a few months.

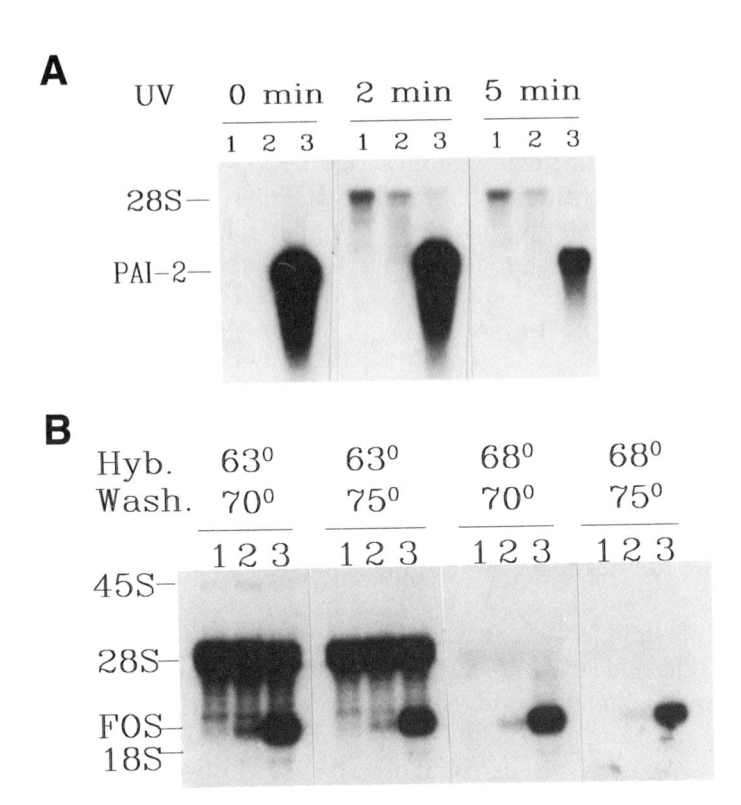

Fig. 1. **(A)** Effect of UV crosslinking. Northern blot hybridization of PAI–2 mRNA in murine total cellular RNA with homologous cRNA probe. Lanes 1 and 2: 5 μg of placental RNA (15.5 and 18.5 d of gestation), which do not contain detectable levels of PAI-2 mRNA. Lane 3: 1 μg of LPS-induced macrophage RNA, an abundant source of PAI-2 mRNA *(15)*. All samples were electrophoresed and transferred together. After cutting the membrane, each filter was UV treated as described. The filters were hybridized together at 58°C, washed at 70°C, and exposed together. Crosshybridization of the probe to 28S rRNA is more pronounced after UV irradiation, and specific hybridization is decreased after 5 min of UV exposure. **(B)** Effect of hybridization temperature. Northern blot hybridization of c-*fos* mRNA in rat total cellular RNA with a murine v-*fos* cRNA probe. Lane 1: uninduced cells. Lane 2: partially induced cells. Lane 3: fully induced cells (M. Prentki and D. B., unpublished). All samples were electrophoresed and transferred together. The filters, which were not UV crosslinked, were hybridized and washed in parallel at the indicated temperatures. The four filters were exposed together. Crosshybridization of the probe to 28S rRNA was essentially abolished by hybridizing at 68°C. Some specific signal was lost with the 75°C stringency wash.

6. BSA: bovine serum albumin at 2 mg/mL. Store at –20°C.
7. RNase inhibitor: placental RNase inhibitor at U/µL. Store at –20°C.
8. RNA polymerase stocks: the three RNA polymerases, T3, T7, and SP6, are available commercially. Store at –20°C.
9. RNA polymerase dilution buffer: 50 m*M* Tris-HCl, pH 8.1, 1 m*M* DTT, 0.1 m*M* EDTA, 500 µg/mL BSA, 5% glycerol. Diluted enzyme is unstable and should be stored on ice for no more than a few hours.
10. RNase-free DNase.
11. Stop mix: 1% SDS, 10 m*M* EDTA, 1 mg/mL tRNA. The tRNA may be omitted.
12. TEN: 10 m*M* Tris-HCl, pH 8.1, 1 m*M* EDTA, and 100 m*M* NaCl.

3. Methods
3.1. Linearized Plasmid Templates for Runoff Transcription

1. Subclone the desired gene fragment into an appropriate transcription vector (*see* Note 2).
2. Isolate plasmid DNA by alkaline lysis from a 30–100 mL of saturated culture. Plasmid DNAs are purified by CsCl/ethidium bromide centrifugation or by precipitation with polyethyleneglycol (*see* Note 3).
3. Linearize 2–20 µg of plasmid DNA with an appropriate restriction enzyme (*see* Note 4). Verify the extent of digestion by electrophoresing an aliquot (0.2–0.5 µg of DNA) in agarose minigels in the presence of ethidium bromide (0.5 µg/mL) (*see* Note 5).
4. Purify the restricted DNA by two extractions with phenol/chloroform, and remove residual phenol by one extraction with chloroform. Precipitate the DNA with ethanol. If little DNA is present (below 5 µg), add 10 µg of glycogen as a carrier; this carrier has no adverse effect in the transcription reactions. After washing the ethanol pellet, the DNA is dried in air and resuspended in TE at 1 µg/µL.

3.2. Synthetic and PCR-Derived Templates

The major limitation in using restriction enzymes to clone inserts and to linearize plasmid templates is that appropriate sites are not always available. Furthermore, the transcripts will almost always contain 5'–and 3'-portions, which differ from those of the endogenous RNA. One possibility to circumvent these difficulties is based on the transcription of small DNA fragments obtained by annealing of synthetic oligodeoxynucleotides *(6,7)*. An alternative and more general approach is presented that generates the transcription templates via PCR amplification of plasmid DNA. In theory, such templates could direct the synthesis of virtually any RNA sequence.

1. Design the 5'-primer, which has a composite sequence: Its 5'-portion is constituted by a minimal T7 promoter, and its 3'-portion corresponds to the beginning of the transcript (see Note 6). Six to ten nucleotides are usually sufficient to prime DNA synthesis on the plasmid template.
2. Design the 3'-primer, which is usually 17–20 nucleotides long and defines the 3'-end of the transcript (see Note 7).
3. PCR-amplify 2–50 ng of plasmid DNA in a total volume of 100 µL. Verify that a DNA fragment of the expected size has been amplified, and estimate the amount of DNA by comparison with known standards.
4. Purify the DNA as described in Section 3.1., step 4 and resuspend in TE at an appropriate concentration. The PCR-derived templates are transcribed at lower DNA concentrations than plasmids to maintain the molar ratio of enzyme to promoter; I use 3 µg/mL for a 100-bp fragment.

3.3. Basic Transcription Protocol for Radioactive Probes

1. Assemble the transcription mixture to a total volume of 20 µL by adding in the following order: water (as required), 2 µL of 10X TB, 1 µL of BSA, 1 µL of 0.2M DTT, 0.5 µL of RNase inhibitor, 2 µL of a 5 mM solution of each ribonucleotide (i.e., ATP, GTP, CTP), 10 µL of $\alpha[^{32}P]$-UTP (400 Ci/mmol, 10 mCi/mL), 1 µL of restricted plasmid DNA template, and 0.25–0.75 µL of RNA polymerase (see Notes 8 and 9).
2. Incubate for 40 min at 37°C for T7 and T3 RNA polymerases or 40°C for SP6 RNA polymerase. After addition of the same number of units of enzyme (see Note 9), incubate for a further 40 min.
3. Degrade the template with RNase-free DNase (1 U/µg of DNA) for 20 min at 37°C. Stop the reaction by adding 40 µL of Stop mix.
4. After two extractions with phenol/chloroform, in which the organic phases are back-extracted with 50 µL of TEN, the combined aqueous phases are purified from unincorporated ribonucleotides by spun-column centrifugation. The spun-column is prepared by filling disposable columns (QS-Q, Isolab) with a sterile 50% slurry of G-50 Sephadex (Pharmacia) in TEN and followed by centrifugation for 5 min at 200g. The samples are carefully deposited on top of the dried resin, and the column is placed in a sterile conical centrifuge tube. After centrifugation for a further 5 min at 200g, 200 µL of TEN are deposited on top of the resin and the column recentrifuged. The eluted RNA, approx 200–400 µL, is ethanol-precipitated and resuspended in water (see Note 10).
5. Measure the incorporation efficiency by counting an aliquot (see Note 11). The size of the transcript may be verified by electrophoresis on polyacrylamide/urea gels (see Chapter 47 and Note 12).

4. Notes

1. Powdered ribonucleotides should be resuspended in water, neutralized to pH 6–8 with 1M NaOH or HCl, and adjusted to the desired concentration by measuring the UV absorbance of appropriate dilutions:
    100 mM ATP: 1540 absorbance units at 259 nm.
    100 mM GTP: 1370 absorbance units at 253 nm.
    100 mM CTP: 910 absorbance units at 271 nm.
    100 mM UTP: 1000 absorbance units at 262 nm.
    The integrity of ribonucleotide solutions can be verified by thin-layer chromatography on PEI-cellulose (PEI-CEL300). The resin is first washed with water by ascending chromatography to remove residual UV-absorbing material and dried. Ten to thirty nanomoles of ribonucleotides are deposited on the resin, which is then resolved by ascending chromatography with 0.5M KH$_2$PO$_4$, adjusted to pH 3.5 with H$_3$PO$_4$. After drying, the ribonucleotides are detected by UV shadowing at 254 nm.
2. All the vectors that are commercially available consist of high-copy-number *E. coli* plasmids derived from ColE1. The original plasmids (pSP64 and pSP65, Promega) contained only one SP6 promoter located upstream of multiple cloning sites (MCS) *(2)*. In the second generation of plasmids, two promoters in opposite orientation flank the MCS, allowing transcription of both strands of inserted DNA fragments. The pGEM series (Promega) contain SP6 and T7 promoters *(5)*, whereas the pBluescript™ series (Stratagene, La Jolla, CA) contain T7 and T3 promoters.
    The choice of plasmid is mostly a matter of personal preference, although it can be influenced by the properties of individual sequences. For instance, premature termination with SP6 transcripts has frequently been observed. The nature of the termination signals is not completely understood *(4,8)*, and their efficiency can be more pronounced when one ribonucleotide is present at suboptimal concentration. The problem is often solved by recloning the inserts in front of a T7 or T3 promoter. The partial recognition of T7 promoters by T3 polymerases may result, however, in the transcription of both strands when the ratio of enzyme to promoter is not carefully controlled. This can be a source of artifacts, particularly when the templates are linearized within the cloned inserts.
3. It is possible to use plasmid DNA from "minipreps," although transcription efficiency can be reduced, particularly with SP6 polymerase. The RNA present in the "minipreps" is digested with pancreatic RNase (20 µg/mL), which is then removed during purification of the linearized templates. Spun-column centrifugation of the digested "minipreps" can improve transcription efficiency.

4. Since RNA polymerases can initiate transcription unspecifically from 3'-protruding ends, restriction enzymes that generate 5'-protruding or blunt ends are usually preferred *(4,9)*. If the only available site generates 3' protruding ends, the DNA can be blunt-ended by exonucleolytic digestion using T4 DNA polymerase or the Klenow fragment of DNA polymerase I. Restriction with enzymes that cut the plasmids more than once may also be used, provided that the promoter is not separated from the insert.
5. The plasmids must be linearized as extensively as possible. Since circular plasmids are efficient templates, their transcription may yield RNA molecules that can be up to 20 kb long and thus incorporate a significant portion of the available ribonucleotide.
6. There are additional constraints on the 5' sequence of transcripts, since it has been shown that the sequence immediately downstream of the start site may effect the efficiency of elongation mode. The first 6 nucleotides have a strong influence on promoter efficiency. In particular, the presence of uracil residues is usually detrimental *(6,7)*. It may be necessary, therefore, to include in the 5'-end of the transcripts 5–6 nucleotides that differ from those present in natural RNA. A similar case has been reported for SP6 polymerase *(10,11)*. This author has used a number of composite T7 promoters whose efficiency is summarized below. In addition to the promoter sequence (5'-TAATACGACTCACTATA) at the 5'-end, the next 6 nucleotides of the templates are:

Efficient promoters	Inefficient promoters
GGGAGA (T7 consensus)	GTTGGG (5% efficiency)
GGGCGA (pBS plasmids)	GCTTTG (1% efficiency)
GCCGAA	

7. The 3'-end of the transcripts should be exactly defined by the 5'-end of the downstream primer. However, during transcription, template-independent addition of 1–2 nucleotides usually generates populations of RNA with differing 3'-ends. The proportion of each residue at the 3'-end may depend on context and is influenced by the relative concentration of each ribonucleotide, a limiting ribonucleotide being less frequently incorporated *(2,6,7,12)*.
8. The order of addition of components may be changed. Remember that dilutions of RNase inhibitor are very unstable in the absence of DTT, and that DNA should not be added to undiluted 10X TB. The temperature of all components must be at least at 25°C to avoid precipitation of the DNA:spermidine complex. Since the radioactive nucleotides are provided in well-insulated vials, they can take up to 10–15 min after thawing to reach an acceptable temperature. I routinely incubate all components, except the RNase inhibitor and the polymerase, at 30°C for 15–20 min.

When more than one probe is to be synthesized, a reaction mixture with all components is added to the DNA template. It has also been possible to reduce the total volume to as small as 8 µL in order to conserve materials. However, since the enzymes are sensitive to surface denaturation, these incubations are done in 400-µL vials.

9. For SP6 polymerase, use 4–6 U/µg of plasmid DNA (size: 3–4 kbp). For T7 and T3 polymerases, use 10–20 U/µg of plasmid DNA.
10. The purification step (Section 3.3., step 4) may not be required, and some investigators use the transcription mixtures directly in hybridization assays. However, low backgrounds, consistency, and quantitation of the newly synthesized RNAs may justify the additional time and effort.
11. More than 50% of the labeled ribonucleotide is routinely incorporated and often this is >80%. An input of 50 µCi may therefore yield up to 4×10^7 Cerenkov-cpm of RNA. This represents 125 pmol of UMP and, assuming no sequence bias (i.e., 25% of residues are uracil residues), 130 ng of RNA (SA: 3×10^8 Cerenkov-cpm/µg).

    Similar results are obtained with labeled CTP and GTP, although GTP rapidly loses its incorporation efficiency on storage. ATP is not routinely used because of its higher K_m (2). UTP is particularly stable and can be used even after several weeks, once radioactive decay is taken into account. With each labeled ribonucleotide, the initial concentration must be >12.5 µM to ensure efficient incorporation and to prevent polymerase pausing. ^{3}H-labeled probes may also be made for use in *in situ* hybridization (*see* ref. *13* for a detailed protocol).

    The T4 gene *32* protein can increase transcription efficiency when added at 10 µg/µg of template DNA (D. Caput, personal communication). It has been recently reported that a reduction of premature termination and an increase synthesis of large full-length transcripts (size >1 kb) can be obtained by performing the transcription at 30°C or at room temperature *(4,14)*.
12. Large transcripts that do not enter the polyacrylamide gel are diagnostic of incompletely linearized templates. Small transcripts are indicative of extensive pausing or premature termination.

Acknowledgments

I thank P. Vassalli for his early encouragment to use riboprobes for detecting rare mRNAs. Over the last few years, many collegues, students, and technicians have contributed to the methods outlined in this chapter, including M. Collart, N. Busso, J.-D. Vassalli, H. Krisch, S. Clarkson, J. Huarte, S. Strickland, P. Sappino, M. Pepper, A. Stutz, G. Moreau, D. Caput, M. Prentki, P. Gubler, F. Silva, V. Monney, D. Gay-Ducrest, and N. Sappino. Research is supported by grants from the Swiss National Science Foundation.

References

1. Butler, E. T. and Chamberlin, M. J. (1984) Bacteriophage SP6-specific RNA polymerase. *J. Biol. Chem.* **257,** 5772–5788.
2. Melton, D. A., Krieg, P. A., Rebagliati, M. R., Maniatis, T., Zinn, K., and Green, M. R. (1984) Efficient in vitro synthesis of biologically active RNA and RNA hybridisation probes from plasmids containing a bacteriophage SP6 promoter. *Nucleic Acids Res.* **12,** 7035–7056.
3. Davanloo, P., Rosenberg, A. H., Dunn, J. J., and Studier, F. W. (1984) Cloning and expression of the gene for bacteriophage T7 RNA polymerase. *Proc. Natl. Acad. Sci. USA* **81,** 2035–2039.
4. Krieg, P. A. and Melton, D. A. (1987) In vitro RNA synthesis with SP6 RNA polymerase. *Methods Enzymol.* **155,** 397–415.
5. Yisraeli, J. K. and Melton, D. A. (1989) Synthesis of long, capped transcripts in vitro by SP6 and T7 RNA polymerases. *Methods Enzymol.* **180,** 42–50.
6. Milligan, J. F. and Uhlenbeck, O. C. (1989) Synthesis of small RNAs using T7 RNA polymerase. *Methods Enzymol.* **180,** 51–62.
7. Milligan, J. F., Groebe, D. R., Witherell, G. W., and Uhlenbeck, O. C. (1987) Oligoribonucleotide synthesis using T7 RNA polymerase and synthetic DNA templates. *Nucleic Acids Res.* **15,** 8783–8798.
8. Roitsch, T. and Lehle, L. (1989) Requirements for efficient in vitro transcription and translation: a study using yeast invertase as a probe. *Biochim. Biophys. Acta* **1009,** 19–26.
9. Schenbon, E. T. and Mierendorf, R. C. (1985) A novel transcription property of SP6 and T7 RNA polymerases: dependence on template structure. *Nucleic Acids Res.* **13,** 6223–6234.
10. Nam, S. C. and Kang, C. (1988) Transcription initiation site selection and abortive initiation cycling of phage SP6 RNA polymerase. *J. Biol. Chem.* **263,** 18,123–18,127.
11. Solazzo, M., Spinelli, L., and Cesareni, G. (1987) SP6 RNA polymerase: sequence requirements downstream from the transcription start site. *Focus* **10,** 11,12.
12. Moreau, G. (1991) RNA binding properties of the *Xenopus* La proteins. Ph.D. thesis, University of Geneva.
13. Sappino, A.-P., Huarte, J., Belin, D., and Vassalli, J.-D. (1989) Plasminogen activators in tissue remodeling and invasion: mRNA localisation in mouse ovaries and implanting embryos. *J. Cell Biol.* **109,** 2471–2479.
14. Krieg, P. A. (1991) Improved synthesis of full length RNA probes at reduced incubation temperatures. *Nucleic Acids Res.* **18,** 6463.
15. Belin, D., Wohlwend, A., Schleuning, W.-D., Kruithof, E. K. O., and Vassalli, J.-D. (1989) Facultative polypeptide translocation allows a single mRNA to encode the secreted and cytosolic forms of plasminogen activators inhibitor 2. *EMBO J.* **8,** 3287–3294.

CHAPTER 13

Native Polyacrylamide Gel Electrophoresis

Adrian J. Harwood

1. Introduction

Agarose gel electrophoresis is generally adequate for resolving nucleic acid fragments in the size range of 100 nucleotides to around 10–15 kb. Below this range, fragments are both difficult to separate and hard to visualize because of diffusion within the gel matrix. These problems are solved by native polyacrylamide gel electrophoresis (PAGE). Using native PAGE, fragments as small as 10 bp and up to 1 kb can be separated with a resolution of as little as 1 in 500 bp.

Native PAGE also has a number of other advantages. It has a high loading capacity; up to 10 µg of DNA can be loaded into a single well (1 cm × 1 mm) without significant loss of resolution. Polyacrylamide contains few inhibitors of enzymatic reactions, so PAGE is an ideal gel system from which to isolate DNA fragments for subcloning and other molecular biological techniques. It has two disadvantages. The mobility of the fragments can be affected by base composition, making accurate sizing of bands a problem. In addition, polyacrylamide quenches fluorescence, making bands containing <25 ng difficult to visualize by ethidium bromide staining; alternative means of visualizing DNA fragments are discussed in Chapters 14 and 15 and Note 1.

In this chapter I describe the preparation and use of native PAGE gels. I also describe a method for gel purification from polyacrylamide gel slices.

2. Materials
2.1. Separation of DNA Fragments

1. Gel apparatus: Many designs of apparatus are commercially available. The gel is poured between two vertical plates held apart by spacers (*see* Note 2).
2. Deionized H_2O: Autoclaved water is not necessary for the gel mix or running buffer, but should be used for diluting samples and purification from gel slices.
3. 10X TBE: 108 g of Trizma Base (Tris), 55 g of boric acid, and 9.3 g of EDTA (disodium salt). Make up to 1 L with deionized H_2O, but it should be discarded when a precipitate forms.
4. Acrylamide stock: 30% acrylamide, 1% *N,N'*-methylene bisacrylamide. Store at +4°C. This is commercially available, or alternatively can be made by dissolving acrylamide and bisacrylamide in water; this should be filtered. Acrylamide is a neurotoxin and therefore must be handled carefully. Gloves and a mask must be worn when weighing out.
5. APS: 10% ammonium persulfate (w/v). Can be stored at +4°C for several months.
6. TEMED: *N,N,N',N'*-Tetramethyl–1,2-diaminoethane. Store at +4°C.
7. 5X sample buffer: 15% Ficoll, 2.5X TBE, 0.25% (w/v) xylene cyanol, and 0.025% (w/v) bromophenol blue.
8. Ethidium bromide: A 10 mg/mL solution of ethidium bromide. Ethidium bromide is a potent mutagen and should be handled with care. Store at 4°C in the dark.

2.2. Purification of DNA Fragments

9. Elution buffer: 0.5*M* Ammonium acetate, 1 m*M* EDTA pH 8.0.
10. TE: 10 m*M* Tris-HCl, pH 7.5, 1 m*M* EDTA. Sterilize by autoclaving.
11. Sodium acetate: 3*M* sodium acetate, pH 5.2.

3. Methods
3.1. Separation of DNA Fragments

1. For 50 mL, enough for a 18 × 14 × 0.15 cm gel, mix 10X TBE, acrylamide, H_2O, and APS as described in Table 1.
2. Just prior to pouring add 50 µL of TEMED, and mix by swirling.
3. Immediately pour the gel mix between the gel plates and insert the gel comb. Leave to set; this takes about 30 min.
4. Fill the gel apparatus with 0.5X TBE and remove the comb. Use a syringe to wash out the wells; this may take multiple washes. It is important to remove as much unpolymerized acrylamide as possible because this impairs the running of the samples (*see* Note 3).

Table 1
Gel Mixes for 3.5, 5, and 12% Polyacrylamide Gels

Acrylamide concentration	3.5%	5%	12%
10X TBE	2.5 mL	2.5 mL	2.5 mL
Acrylamide stock	5.8 mL	8.3 mL	20.0 mL
dH$_2$O	41.3 mL	38.8 mL	27.1 mL
APS	350.0 µL	350.0 µL	350.0 µL
Effective range of separation	100–2000 bp	80–500 bp	40–200 bp

6. Add 0.2 vol of 5X sample buffer to each sample, usually in 10–20 µL of TE, water, or enzyme buffer. Mix and spin the contents to the bottom of the tube (*see* Note 4).
7. Load the samples on the gel and run at 200–300 V (approx 10 V/cm) until the bromophenol blue band is two-thirds of the way down the gel; this takes about 2.5 h (*see* Note 5).
8. Disassemble the gel apparatus and place the gel to stain in 1 mg/mL of ethidium bromide for approx 30 min. View the stained gel on a transilluminator. Alternative visualization methods are described in Notes 1 and 6 and Chapter 14.

3.2. Purification of DNA Fragments

A number of methods are available to extract DNA from polyacrylamide gel slices. I find the "crush and soak" method *(1)* a simple and effective way of purifying DNA from both native and denaturing polyacrylamide gels.

1. Cut the DNA band from the gel using a new scalpel blade, and place in a 1.5-mL tube.
2. Break up the polyacrylamide gel slice with a yellow pipet tip, and add between 1 and 2 vol of elution buffer.
3. Seal the tube with Para-film™, and incubate at 4°C on a rotating wheel for 3–4 h for fragments smaller than 500 bp or overnight for larger fragments.
4. Spin out the lumps of polyacrylamide by centrifugation at 10,000*g* for 10 min, and carefully remove the supernatant. Add an additional 0.5 vol of elution buffer to the pellet, vortex briefly, and recentrifuge. Combine the two supernatants.
5. Add 2 vol of ethanol and place on ice for 10 min. Centrifuge at 12,000*g* for 15 min.
6. Redissolve the pellet in 200 µL of TE and add 25 µL of 3*M* sodium acetate. Add 400 µL of ethanol, and precipitate as in step 5.

7. Rinse the pellet once with 70% ethanol, and air-dry.
8. Resuspend in an appropriate volume of TE, and check the yield by gel electrophoresis.

4. Notes

1. High concentrations of DNA, e.g., as experienced when purifying oligonucleotides, can be visualized without ethidium bromide staining by UV shadowing. This is carried out by wrapping the gel in a UV-transparent plastic film, such as Saran Wrap™, and then placing it onto a thin-layer chromatography plate that contains an ultraviolet (UV) fluorescent indicator (Merck 60F254, cat. no. 5554). Long-wave UV light is shone through the gel onto the chromatography plate, causing it to glow. Regions of high DNA concentration leave a "shadow" on the plate as the transmitted UV is absorbed by the DNA. The position of the DNA can be marked on the Saran Wrap with a fiber-tip pen and then cut from the gel.
2. Grease and dirt on the plates can cause bubbles to form while pouring the gel; the plates therefore should be cleaned thoroughly and then wiped with ethanol. To help ensure that the gel only sticks to one plate when the apparatus is disassembled, siliconize one of the gel plates. This is easily done by wiping the plate with a tissue soaked in dimethyl dichlorosilane solution and then washing in distilled water followed by ethanol. If the plates are baked at 100°C for 30 min, the siliconization will last 4–5 gel runs.
3. If separating very small fragments, <50 bp, the gel should be prerun for 30 min, because this elevates the resolution problem experienced with fragments running close to the electrophoresis front.
4. High salt buffers (above 50 mM NaCl) will affect sample mobility and tend to make bands collapse. In this case, salt should be removed by ethanol precipitation.
5. Do not run the gel faster than 10 V/cm, because this will cause the gel to overheat, affecting the resolution. The gel can be run more slowly, e.g., 75 V will run overnight.
6. If the samples are radiolabeled (see Chapter 15), the gel should be fixed in 10% acetic acid, transferred to 3MM paper (Whatman, Maidstone, UK), and dried. The dried gel is autoradiographed. If a band is to be isolated from the gel, it can be wrapped in plastic film and autoradiographed wet. The autoradiograph is then aligned with the gel and the band cut out and purified as in Section 3.2.

Reference

1. Maxam, A. M. and Gilbert, W. (1977) A new method for sequencing DNA. *Proc. Natl. Acad. Sci. USA* **73**, 668–671.

CHAPTER 14

Use of Silver Staining to Detect Nucleic Acids

Lloyd G. Mitchell, Angelika Bodenteich, and Carl R. Merril

1. Introduction

Silver stains are useful for the detection of nanogram amounts of proteins or nucleic acids in acrylamide gels or on various membranes. They have been shown to be more sensitive than organic stains in detecting proteins and DNA. They are capable of detecting as little as 0.03 ng/mm^2 of DNA *(1)*. In addition, silver stains avoid the mutagenic hazards presented by both ethidium bromide and radioactive detection methods.

The thirteenth century alchemist, Count Albert von Bollstadt, was the first to record that silver nitrate would stain organic material, including human skin. However, silver was not employed in scientific studies until 1844, when Krause used silver to stain histological tissues. Histological silver stains were further developed by the anatomists Golgi and Cajal. The first silver stains used on polyacrylamide gels were adapted from these histological stains. They were quite tedious, requiring numerous steps and solutions. They took 3 h to perform *(2)*. Continued research on the use of silver staining to detect proteins and nucleic acids separated in polyacrylamide has resulted in a number of simplified protocols. It has also extended our knowledge of the mechanisms underlying these stains.

Silver images of protein or nucleic acid patterns are produced by a difference in the oxidation-reduction potential in regions occupied by nucleic acids or proteins compared to the surrounding gel or membrane *(3,4)*. This redox potential catalyzes the reduction of ionic to metallic silver. A positive (dark) image will be produced if the region occupied by nucleic acid or protein has a higher redox potential than the surrounding region, whereas a negative image will result if the redox potential is higher in the surrounding matrix. The redox potentials may be altered by manipulating the chemistry of the staining solutions. Experiments with nucleic acids and their components have implicated the purines as the active subunits in the silver staining reaction (5).

There are three general chemistries utilized in silver staining: diamine or ammoniacal stains, nondiamine chemical reduction stains, and photochemical stains. The diamine silver stains, which were initially developed for the visualization of nerve fibers (6), were the first type of silver stain employed to detect biopolymers separated on polyacrylamide gels. The diamine stains rely on the formation of silver diamine complexes in the presence of ammonium hydroxide. To produce an image, the solution containing the silver diamine complexes must be acidified, most commonly with citric acid, in the presence of formaldehyde. The resulting decrease in ammonium ions liberates silver ions from the silver diamine complexes, thereby facilitating the reduction of the silver ions with the formaldehyde to metallic silver. The concentration of citric acid in the image-developing solution must be precisely controlled to reduce background deposition of silver.

The photodevelopment stains rely on light to reduce ionic to metallic silver. These stains are simple and quick, requiring fewer solutions, but they lack the sensitivity of the other methods (7).

Most nondiamine chemical reduction stains were developed by adapting photochemical protocols (3). They employ silver nitrate, which reacts with biopolymers under acidic conditions. This is followed by the selective reduction of ionic silver by formaldehyde under alkaline conditions. Sodium carbonate is often utilized in these stains to maintain alkalinity during image development since formic acid is generated by the oxidation of formaldehyde. The nondiamine chemical stains are fast and easy to perform. They also work well for gels that are under 1 mm thick. A nondiamine silver staining protocol, such as the one described in this chapter, can be performed in 40 min using solutions that are stable for months.

2. Materials

All reagents must be prepared with ultrapure deionized water with a conductivity of <1 μmho. The water should also be free of both ionic and

Silver Staining to Detect Nucleic Acids

organic contaminates. Such contaminants are often responsible for background staining. Care must also be taken to prevent proteins from skin or other potential contaminating sources from contact with the gels or the reagents, since this silver stain is also an excellent protein detector.

Silver can be used to stain native and denaturing acrylamide gels, as well as dehydratable gels using discontinuous buffer systems and varying amounts of crosslinkers *(8)*. We recommend the use of gels that are 0.4 mm thick (*see* Note 1).

The solutions are easy to prepare and all but the developer may be kept for at least 6 mo. They can also be made during the procedure.

1. Fixative: 10% ethanol (v/v) (*see* Note 2).
2. Oxidizer: $0.0034M$ potassium dichromate (1 g/L) and $0.0032M$ nitric acid (or 0.2 mL of concentrated nitric acid, $16M$/L) (*see* Note 3). Avoid direct contact with the skin since this solution is an irritant. It is a strong oxidizer and should be kept away from reducing agents. It is also photosensitive and should be stored in an amber or dark bottle.
3. Silver solution: 2% silver nitrate (2 g/L) (*see* Note 3). Silver nitrate solution is caustic to eyes and mucous membranes, and it will stain both skin and clothing. Store in an amber or dark bottle.
4. Developer: $0.28M$ sodium carbonate (30 g/L) plus 0.5 mL/L of 37% formalin (*see* Note 4), added just prior to use. This solution should be used at 4°C and it may be stored at that temperature. Since formaldehyde vapor may be a carcinogen, this solution should be made and used in a fume hood.
5. Stop solution: 5% acetic acid (v/v).

3. Methods

Following the electrophoretic separation of nucleic acids, each polyacrylamide gel is placed in a separate clean glass or plastic tray. The use of clean vinyl or latex gloves, with the powder washed off, and a lab coat will reduce the risk of contaminating the gel and the staining of the investigator. All solutions should be at room temperature except the developer, which is precooled at 4°C. Subdued laboratory lighting will help to decrease background staining. Development should not be done on a light box. The gel should be handled gently to avoid crushing that may produce artifacts. Solutions should not be poured directly on the gel; instead, tilt the tray and pour solutions into a corner or use individual trays for each solution and move the gels from tray to tray. Solution volumes should be sufficient to completely immerse the gels. The trays should be gently agitated during staining.

1. Following electrophoretic separation soak the gel in fixative for 10 min (*see* Note 5). Use a volume of fixative double that needed to cover the gel (*see* Note 6). The gel may be stored indefinitely in the fixative at this point if development needs to be delayed.
2. The gel is then placed in the dichromate oxidizer soak for 3–5 min (*see* Note 7). Then decant the oxidizer. Wash the gel three times with deionized water, for 3–5 min each (*see* Notes 7 and 8).
3. Place the gel in the silver nitrate solution for 15–20 min (*see* Note 9). Briefly wash the gel (for 20–30 s) in deionized water to remove silver nitrate from the gel surface (*see* Note 10).
4. Form the image by washing the gel with precooled developer for approx 30 s with constant agitation (*see* Note 11). Observe the gel closely during this process and discard the solution as soon as it loses clarity (*see* Note 12). Replace with fresh developer and again discard as soon as it loses clarity. Repeat this process until the gel image has formed or until the background staining becomes objectionable.
5. When the DNA bands reach the desired intensity relative to the background, stop development by placing the gel in 5% acetic acid (*see* Notes 13 and 14). The DNA bands normally are dark brown or black on a faint yellow/brown background (*see* Fig. 1).

Stained gels cast on polyester backing may be dried in a microwave oven or left at room temperature to dry. Nonbacked gels can be dried on cellophane or filter paper under vacuum with mild heat. For a permanent record, the gel should be photographed on a light box since gels silver-stained with this procedure may darken over time.

4. Notes

1. Thinner gels can be stained faster, but they are more fragile and difficult to work with whereas gels thicker than 0.7 mm require longer to stain. To increase the durability of gels that are thinner than 0.7 mm, we recommend casting them on a polyester backing (such as GelBond PAG, FMC, Rockland, ME or Gel-Fix, Serva, Heidelberg, Germany). Alternatively, the gel may be cast and bonded to silanized glass (Bind-Silane, Pharmacia-LKB, Piscataway, NJ). Gels that are attached to polyester sheets offer the advantage that after electrophoresis they can be trimmed to remove the regions that do not contain DNA, thereby decreasing the amount of staining solutions required.

    If gels bound to polyester sheets are utilized, the polyester sheets must be flattened and held tightly by capillary attraction against the backing glass plate of the gel apparatus prior to casting. To accomplish this, a few milliliters of distilled water or 5% glycerol is pipeted onto the backing gel plate, then the polyester sheet is placed on the plate. One must be careful

Silver Staining to Detect Nucleic Acids

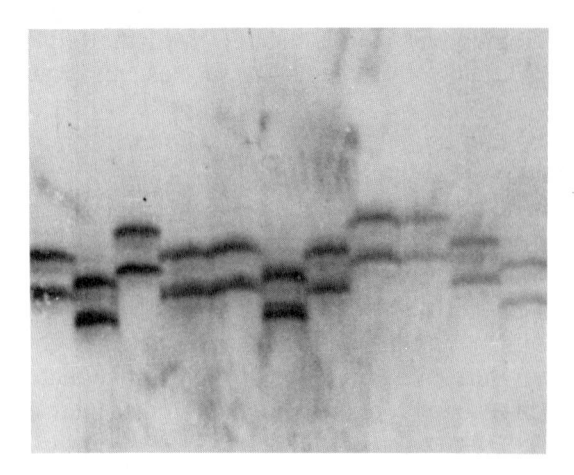

Fig. 1. A typical silver-stained gel. This gel contains a region of the mitochondrial DNA, from several unrelated individuals, that is rich in AC repeat sequences. The DNA was PCR amplified and 1 µL (8–35 ng/µL) from each reaction was denatured by adding 3 µL of a loading buffer containing formamide followed by heating at 96°C for 1 min. The DNA was electrophoresed at 55°C on 6% polyacrylamide gel containing 6M urea. This gel was cast on GelBond PAG (FMC, Rockland, ME). This example also illustrates some of the surface staining artifacts that may occur with this technique.

to place the nonbinding surface against the backing plate. The sheet is then flattened with a rubber print roller or pipet. It is important that all bubbles and irregularities be removed so that the gel will have uniform thickness. To prevent the polyester sheet from moving while rolling, a thin film of silicon grease can be applied between the sheet and plate on the lower end of the gel. The polyester sheet should be large enough to cover and include the gel, loading wells, and the spacers that are placed on top of the sheet. Once the polyester sheet is in place with the spacers, the gel is cast as usual.
2. Methanol may be substituted, but it is toxic and may be absorbed by contact through the skin or by respiration. Many fixative solutions for DNA and proteins separated on polyacrylamide gels contain acetic acid. However, this protocol does not include acetic acid in the fixative, since darker bands were obtained in its absence.
3. This solution may also be made as a 10× stock.
4. Concentrated formalin stock is generally 37%.
5. Fix for 20 min with gels thicker than 0.7 mm. This time may be reduced by using fixative preheated to between 40–50°C.

6. Besides minimizing the diffusion of DNA in the gel, fixation also removes the buffer ions and other chemicals (such as urea) that could interfere with staining.
7. Soak for 10 min for gels thicker than 0.7 mm.
8. It is important to wash the gel until the concentration of the dichromate oxidizer is reduced such that the gel is no longer visibly yellow. If the gel is still yellow or turns red in the dichromate oxidizer, an acid wash of 0.01M nitric acid may be used to remove the coloration.
9. Soak gels thicker than 0.7 mm for 25–30 min. If any contaminants are present at this point in the procedure, particularly chlorides, a silver precipitate may form. The presence of such a precipitate usually results in a reduction of sensitivity of the silver stain and an increase in the background staining.
10. A wash that is too long at this step will allow silver to diffuse out of the gel and will diminish the sensitivity of the staining.
11. Cooled developer is used to permit controlled image development.
12. A brownish silver carbonate precipitate forms as silver ions escape from the gel. This precipitate will deposit on the gel surface if the solutions are not changed frequently.
13. Gels that are overdeveloped or that have a mirror-like appearance owing to surface deposition of silver can be destained with the following procedure: Dissolve 37 g of anhydrous cupric sulfate and 37 g of sodium chloride in 850 mL of deionized water. With constant stirring add concentrated ammonium hydroxide until the solution becomes deep blue and there is no precipitate, then bring the volume to 1 L. A second solution containing 436 g sodium thiosulfate pentahydrate in 1 L of deionized water is also prepared. Equal parts of these two solutions are combined just prior to use. The gel is soaked in this solution until almost completely clear. The destaining is stopped by 10% acetic acid for 15 min, followed by a minimum of 1 h washing with several changes of deionized water. The gel may then be restained by beginning the procedure at the silver nitrate step (step 3). The dichromate step is not required for restaining. This destaining procedure is also useful for the removal of silver stains from clothing.
14. Some artifacts and precipitate on the surface of the gel can be removed by rubbing the dried gel gently with a damp piece of absorbant cotton. This technique should be tried in a noncritical region. The procedure should be stopped when the surface becomes tacky. However, it can be repeated after the gel is allowed to dry.

References

1. Goldman, D. and Merril, C. R. (1982) Silver staining of DNA in polyacrylamide gels: linearity and effect of fragment size. *Electrophoresis* **3**, 24–32.

2. Merril, C. R., Switzer, R. C., and Van Keuren, M. L. (1979) Trace polypeptides in cellular extracts and human body fluids detected by two dimensional electrophoresis and a highly sensitive silver stain. *Proc. Natl. Acad. Sci. USA* **76,** 4335–4339.
3. Merril, C. R. (1987) Development and mechanisms of silver stains for electrophoresis. *Acta Histochem. Cytochem.* **19,** 655–667.
4. Merril, C. R. (1990) Silver staining of proteins and DNA. *Nature* **343,** 779–780.
5. Merril, C. R. and Pratt, M. E. (1986) A silver stain for the rapid quantitative detection of proteins nucleic acids on membranes or thin layer plates. *Anal. Biochem.* **156,** 96–110.
6. Bielschowsky, M. (1904) Die silberimpragnation der neurofibrillen. *J. Psychol. Neurol.* **3,** 169–189.
7. Merril, C. R., Harrington, M., and Alley, V. (1984) A photodevelopment silver stain for the rapid visualization of proteins separated on polyacrylamide gels. *Electrophoresis* **5,** 289–297.
8. Allen, R. C., Graves, G., and Budowle, B. (1989) Polymerase chain reaction amplification products separated on rehydratable polyacrylamide gels and stained with silver. *Biotechniques* **7(7),** 736–744.

CHAPTER 15

End-Labeling of DNA Fragments

Adrian J. Harwood

1. Introduction

End-labeling is a rapid and sensitive method for radioactively, or nonisotopically, labeling DNA fragments and is useful where there is too little DNA to visualize by ethidium bromide staining. This can be a serious problem when separating small fragments by polyacrylamide gel electrophoresis (*see* Chapter 13). End-labeling has the second advantage that it can be used to produce fragments labeled at one end. Because all of the enzymes employed are specific to either the 5' or 3' termini of DNA molecules, it will only incorporate the label once per DNA strand. When double-stranded DNA restriction enzyme fragments are used, the label is incorporated at both ends, but can be removed from one by a second restriction enzyme digestion. This works well with DNA fragments cloned into polylinkers because one end can be removed as a tiny piece of DNA, making subsequent purification easier. These single labeled molecules can help to order restriction enzyme fragments and are a prerequisite for Maxam-Gilbert DNA sequencing (*see* Chapter 53). End-labeled synthetic oligonucleotides have numerous applications, such as short DNA sequence-specific probes for mutation screening *(1)*; probes for gel retardation and Southwestern assays *(2)*; and for sequencing PCR products where the samples are contaminated with other oligonucleotides (*see* Chapter 49).

There are two commonly used methods of end-labeling, the "fill-in" and the kinase reaction. It is also possible to end-label at the 3' end by using terminal transferase *(3)*, but this method is now less commonly used. The "fill-in" reaction uses the Klenow fragment of *E. coli* DNA

A

      Klenow
      + dGTP, dATP, dTTP
      + dCTP*

ATATG-3' ⟶ ATATGGATC*-3'
TATACCTAG-5'   TATACCTAG-5'

Restriction enzyme Incubation with
digested DNA   labelled dCTP

B

i) Removal of 5' terminal phosphate.

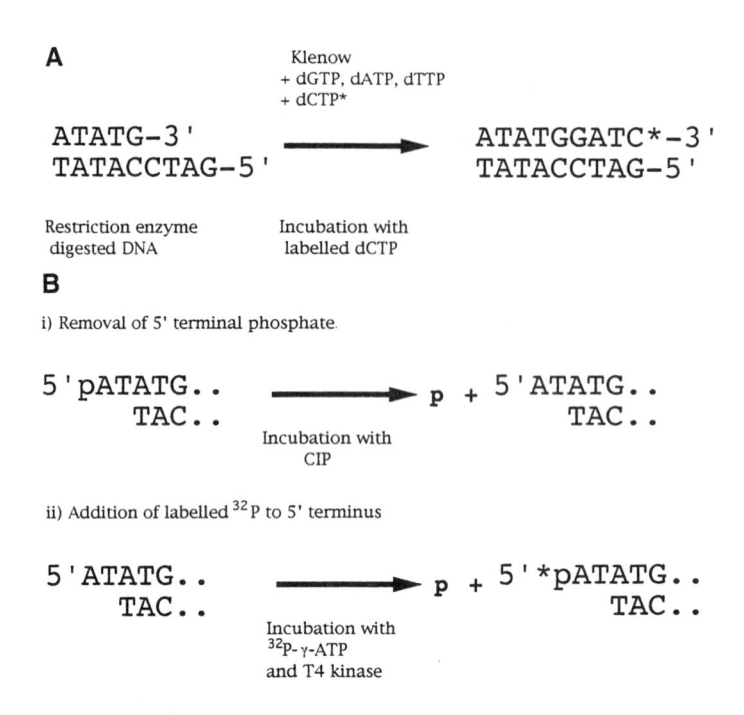

5'pATATG.. ⟶ p + 5'ATATG..
  TAC..   TAC..
    Incubation with
    CIP

ii) Addition of labelled ^{32}P to 5' terminus

5'ATATG.. ⟶ p + 5'*pATATG..
  TAC..   TAC..
    Incubation with
    ^{32}P-γ-ATP
    and T4 kinase

Fig. 1. **(A)** The fill-in reaction. **(B)** The kinase reaction.

polymerase *(4)*. This fragment is generated by an enzymatic digestion of the polymerase that retains both polymerase and 3'-5' exonuclease activity but removes the 5'-3' exonuclease activity. To end-label a piece of DNA it must be digested first with a restriction enzyme that generates a 5' overhang. The polymerase activity of Klenow is then used to extend the 3' recessed end of one DNA strand by using the 5' overhang of the other strand as a template (Fig. 1A). Because of its ease, this is the method of choice for double-stranded DNA fragments. When suitable termini are not available, or when the substrate is single-stranded, the kinase reaction must be used. The kinase reaction uses T4 polynucleotide kinase (T4 kinase) to transfer a labeled phosphate to the 5' end of the DNA molecule (Fig 1B). This method is ideal for labeling oligonucleotides, which are synthesized without a 5' phosphate. To label restriction-enzyme-digested DNA fragments, the terminal phosphate must be removed prior to the kinase reaction using a phosphatase, such as Calf Intestinal Alkaline Phosphatase (CIP). All of these reactions can be used without labeled nucleotides to modify the DNA fragments for further recombinant DNA manipulations.

2. Materials

Molecular biology grade reagents should be utilized whenever possible. Manipulations are performed in 1.5 mL disposable, sterile polypropylene tubes, with screw tops to prevent leakage of radioactivity. Local safety regulations must be obeyed when using radioactivity.

2.1. End-Labeling with Klenow

1. 10X Klenow buffer: 200 mM Tris-HCl, pH 7.6, 100 mM MgCl$_2$, 15 mM β-mercaptoethanol, 25 mM dithiothreitol.
2. Labeled nucleotide: ^{32}P-α-dNTP, most commonly supplied as dATP or dCTP, but dGTP and dTTP are available. It is also possible to substitute nonisotopic label, such as fluoroscein-11-dUTP and dioxigenin-11-dUTP.
3. Unlabeled dNTPs:
   a. dNTP mix; a mixture of 0.25 mM of each unlabeled dNTP, excluding that which corresponds to the labeled nucleotide (*see* Note 1).
   b. Cold dNTP: 0.25 mM dNTP corresponding to the labeled nucleotide (*see* Note 1).
4. Klenow: The Klenow (large) fragment of DNA polymerase I at 1 U/µL. Store at –20°C.
5. TE: 10 mM Tris-HCl, pH 7.5, 1 mM EDTA. Autoclave and store at room temperature.
6. Phenol: Tris-HCl-equilibrated phenol containing 0.1% hydroxyquinoline (as an antioxidant). Use Ultrapure, redistilled phenol. Extract repeatedly with 0.5M Tris-HCl, pH 8.0, until the aqueous phase is 8.0 and then extract once with 0.1M Tris-HCl, pH 8.0. This can be stored at 4°C for at least 2 mo. Phenol is both caustic and toxic and should be handled with care.
7. Chloroform.
8. Phenol:chloroform mixture: A 1:1 mixture was made by adding an equal volume of chloroform to 0.1M Tris-HCl, pH 8.0, equilibrated phenol. This can be stored at 4°C for at least 2 mo.
9. Ethanol and 70% ethanol (v/v in water).
10. 5M Ammonium acetate, pH 7.5: Store at room temperature.

2.2. End-Labeling with T4 Kinase

11. 10X CIP buffer: 10 mM ZnCl$_2$, 10 mM MgCl$_2$, 100 mM Tris-HCl, pH 8.3.
12. CIP: Calf intestinal alkaline phosphatase (Boehringer Mannhiem Gmbh, Mannheim, Germany) at 1 U/µL. Store at 4°C.
13. 10X kinase buffer: 700 mM Tris-HCl, pH 7.6, 100 mM MgCl$_2$, 50 mM dithiothreitol.
14. ^{32}P-γ-ATP: Specific activity > 3000 Ci/mmol.
15. T4 kinase: T4 polynucleotide kinase at 1 U/µL. Store at –20°C.
16. Cold ATP: 1.0 mM ATP (freshly made from 20 mM stock).

3. Methods
3.1. End-Labeling with Klenow

1. Resuspend 1–1000 ng of DNA in 42 µL of dH$_2$O (*see* Note 2). Add 5 µL of 10X Klenow buffer, 1 µL of ^{32}P-α-dNTP, 1 µL of dNTP mix, and 1 µL of Klenow. Incubate at room temperature for 15 min (*see* Note 3).
2. Chase with 1 µL of cold dNTP. Incubate at room temperature for a further 15 min (*see* Note 1).
3. Add 50 µL of TE followed by 100 µL of phenol:chloroform. Vortex briefly and separate by centrifugation at 12,000g in a microfuge (*see* Notes 4 and 5).
4. Remove the aqueous (top) phase to a fresh tube and add 100 µL of chloroform. Separate the layers as in step 3 and remove the aqueous phase to a fresh tube. Care must be taken because the discarded reagents are contaminated with unincorporated ^{32}P-α-dNTP.
5. Add 60 µL (0.6 vol) of 5M ammonium acetate and 200 µL (2 vol) of ethanol (*see* Note 6) and place on ice for 5 min. Centrifuge at 12,000g for 15 min. Carefully remove the supernatant (remember that it is radioactive) and wash the pellet in 70% ethanol.
6. Air-dry the pellet for 10 min and resuspend in the required amount of TE (10–100 µL).

The labeled DNA can be either immediately separated by gel electrophoresis and detected by autoradiography (*see* Note 7), or digested further with a second restriction enzyme. In either case, it is a good idea to count a 1 µL sample in a scintillation counter. Between 5000 and 10,000 cpm are required to detect the fragment by autoradiography. Possible causes of poor labeling and possible solutions are discussed in Notes 8–10.

3.2. End-Labeling with T4 Kinase

1. Dissolve 1–2 µg of restriction-enzyme-digested DNA in 44 µL of dH$_2$O. Add 5 µL of 10X CIP buffer and 0.05–1 U of CIP (*see* Note 11). Incubate for 30 min at 37°C (*see* Notes 12 and 13).
2. Heat inactivate at 60°C for 10 min. Phenol extract, and precipitate as in Section 3.1., steps 3–5 (*see* Notes 14 and 15).
3. Resuspend the DNA in 17.5 µL of dH$_2$O. Add 2.5 µL of 10X kinase buffer, 5 µL of ^{32}P-γ-ATP, and 1 µL of T4 kinase. Incubate at 37°C for 30 min.
4. Add 1 µL of cold ATP and incubate for a further 30 min (*see* Note 16).
5. Phenol extract and precipitate as in steps Section 3.1., steps 4–6 (*see* Note 17).

4. Notes

1. Unlabeled dNTPs are required for two reasons. First, the labeled nucleotide may not correspond to the first nucleotide to be filled within the restriction enzyme site. In the example shown in Fig. 1A, which is a *Bam*HI site, the labeled nucleotide, dCTP*, corresponds to the fourth nucleotide, therefore the other three nucleotides must be filled with cold dNTPs before the label is incorporated. For convenience, a general 7.5-mM mix of all the unlabeled dNTPs can be used regardless of the actual composition of the restriction enzyme site. Second, a "chase" is required to generate molecules with flush ends because the polymerase stalls in the limited concentrations of labeled nucleotide. This step may be omitted in cases where the heterogeneous sized termini are not a problem, for example, when labeling large DNA fragments for separation by agarose gel electrophoresis.
2. The fill-in reaction is very robust and provided Mg^{2+} is present can be carried out in almost any buffer. This means that it is possible to carry out the reaction by simply adding the labeled dNTP, unlabeled dNTPs, and Klenow directly to restriction enzyme mix at the end of digestion.
3. Because only a small region of DNA is labeled in this reaction, it proceeds very quickly. Incubation at room temperature is sufficient, unless ^{35}S-labeled dNTP is used, when labeling should be carried out at 37°C. Prolonged incubation can result in degradation of the DNA ends.
4. The labeled DNA may be used for gel electrophoresis at this point, but it must be remembered that unincorporated ^{32}P-α-dNTP will be present in the DNA solution. This may increase the exposure of the operator and increase the risk of contamination when carrying out gel electrophoresis.
5. An alternative purification is to pass the DNA through a Sephadex G50 spin column (*see* Chapter 49).
6. If only very small amounts of DNA are present, it may be necessary to add carrier, such as 10 µg of tRNA or glycogen.
7. The gel should be fixed in 10% acetic acid or trichloroacetic acid (TCA) before drying to prevent contamination of the gel dryer.
8. Klenow is rarely affected by inhibitors, but it rapidly loses its activity if it is warmed in the absence of substrate. It can be one of the first enzymes to be lost from the general enzyme stock. If the activity of the enzyme is in doubt, carry out a test reaction by labeling control DNA. Generally, DNA markers are good for this, but check the structure of the ends before proceeding.
9. The structure of the end is important, because the enzyme can only "fill-in" those bases present in the site. Recheck the sequence of the single-strand end produced by restriction enzyme digestion. It may be possible to exchange the ^{32}P-α-dNTP for another that has higher specific activity.

10. The Klenow "fill-in" reaction only incorporates a small number of ^{32}P-labeled nucleotides per DNA molecule. If higher levels of incorporation are required, T4 DNA polymerase may be used. T4 DNA polymerase has 200-fold higher 3'-5' exonuclease activity than Klenow. If the DNA fragments are incubated in the absence of dNTPs, this enzyme will produce a region of single-stranded DNA that can be subsequently labeled with higher incorporation by addition of ^{32}P-α-dNTP and cold dNTPs to the mix *(5)*.
11. One unit of CIP dephosphorylates 50 pmol of ends in 1 h (for a 5 kb fragment, 1 pmol of ends is approx 2 µg).
12. The efficiency of dephosphorylation of blunt and 5' recessed ends is improved by incubating the reaction at 55°C.
13. The phosphatase reaction can be carried out in restriction enzyme buffer, by addition of 0.1 vol of 500 m*M* Tris-HCl, pH 8.9, 1 m*M* EDTA, and the required amount of enzyme.
14. It is important to remove all phosphatase in order to prevent removal of the newly incorporated labeled phosphate.
15. The T4 kinase reaction is very sensitive to inhibitors, such as those found in agarose. Care should be taken to ensure that the DNA is inhibitor-free. In addition, T4 kinase will readily phosphorylate RNA molecules, therefore the presence of RNA should be avoided because this will severely reduce the incorporation of labeled ^{32}P into the DNA.
16. The labeling reaction is only approx 10% efficient. To get all molecules phosphorylated it is necessary to chase the reaction with excess cold ATP.
17. This is a poor way to purify oligonulceotides; instead I recommend a Sephadex G25 spin column *(see* Chapter 19).

References

1. Wallace, R. B., Shaffer, J., Murphy, R. F., Bonner, J., Hirose, T., and Itakura, K. (1979) Hybridisation of synthetic oligodeoxyribonucleotides to φχ174 DNA: the effect of single base pair mismatch. *Nucliec Acid Res.* **6**, 3543–.
2. Harwood, A. J., ed. (1994) *Protocols for Gene Analysis. Methods in Molecular Biology,* vol. 31. Humana, Totowa, NJ.
3. Walker, J. M., ed. (1984) *Nucleic Acids. Methods in Molecular Biology,* vol. 2. Humana, Clifton, NJ.
4. Klenow, H., Overgaard-Hansen, K., and Patkar, S. A. (1971) Proteolytic cleavage of native DNA polymerase into two different catalytic fragments. *Eur J. Biochem.* **22**, 371–381.
5. Challberg, M. D. and Englund, P. T. (1980) Specific labelling of 3' termini with T4 DNA polymerase. *Methods Enzymol.* **65**, 39–43.

PART II
RNA Analysis

CHAPTER 16

Northern Blot Analysis

Robb Krumlauf

1. Introduction

In the analysis of gene expression, the steady-state level of RNA transcripts is one of the most convenient parameters used to monitor the gene activity in cell lines and tissues. A variety of methods, such as S1 hybridization, RNase protection, and Northern blotting, can be used to measure RNA levels, and the choice of the best assay system depends largely on the type of information required, levels of sensitivity, and limitations of the particular in vivo system being examined. This chapter details the method for analyzing RNA by Northern blotting, which basically involves the isolation of RNA, its size fractionation by electrophoresis, transfer to membrane, and detection by nucleic acid hybridization and autoradiography.

There are several advantages to Northern blotting analysis. Transcription patterns of genes are often complex, and multiple RNA species can be generated from the same gene. Northern analysis provides information on the relative number, size, and abundance of RNAs derived from a gene, which cannot be obtained by the alternative methods. This technique also generates a record of the RNA that is stored on a membrane that can be used many times. Therefore, the expression of several genes can be analyzed on the same RNA samples, by using multiple probes to rehybridize the filter. There are two disadvantages often associated with this technique. The RNA isolated from cells or tissues must be of high quality and not degraded, which can be difficult in some tissues or for inexperienced workers. This is not a problem for the alternative S1 or

From: *Methods in Molecular Biology, Vol. 58: Basic DNA and RNA Protocols*
Edited by: A. Harwood Humana Press Inc., Totowa, NJ

RNase protection methods, since these assays use enzymes to degrade the RNA as a part of the procedure. Northern analysis is also generally considered to be less sensitive and involves more steps than the alternatives. It therefore requires larger amounts of RNA or starting material, and more time to obtain the results.

Many of these problems associated with the Northern blotting method can be eliminated by optimizing the transfer and coupling of RNA to membrane supports and utilizing more sensitive single-stranded probes. Nylon membranes, such as GeneScreen™ supplied by Dupont (UK), are flexible, tear-resistant supports to which the RNA can be permanently coupled by UV crosslinking. One of the main advantages of performing filter hybridization to UV crosslinked RNA is that very high stringency conditions can be used, which improves sensitivity, reduces background, and enables convenient multiple reuse of the filters. In many cases where the concentration of the desired RNA species is high or large amounts of sample are available, these may not be critical parameters. However, RNA samples frequently need to be isolated from small amounts of tissues or cells that are very hard to obtain (transgenic mice, clinical samples, or embryonic material). In transfection experiments, it is desirable to minimize the amount of RNA needed for analysis to enable more DNA constructs to be examined and reduce the amount of cell culturing required. In these cases, optimal conditions are essential. Sensitivity is greatly enhanced by using single-stranded RNA probes, but these often generate very high backgrounds owing to nonspecific hybridization to rRNA species, especially when total RNA is being used. It is therefore beneficial to isolate polyA$^+$ RNA, even if the amount of total RNA in the sample is very small. Analyzing only polyA$^+$ RNA will improve the signal-to-noise ratio and avoid RNA overloading, resulting in incomplete coupling of the RNA to the filter.

RNA:RNA hybrids are very stable, and often the proper stringency conditions for hybridization are not used, resulting in high background or crosshybridization to other RNA species. By increasing the stringency of both the hybridization and washing conditions, the nonspecific signals are greatly reduced or eliminated, but if the RNA has not been UV crosslinked to the filter, these conditions can remove much of the sample resulting in loss of signal. The conditions can also be so severe that nitrocellulose membrane supports become brittle and break apart, which is why nylon membranes are essential in obtaining optimal reproducible results. The choice of hybridization buffers is not critical, but minimizing the temperature of hybridization and washing can reduce the degree of RNA degradation and extend the reusable life of the filter. This can be accomplished by using formamide in the buffers to reduce the melting temperatures of the hybrids.

Optimizing the Northern blotting method in these ways gives an assay that is as sensitive as RNase protection or S1 analysis, and provides a convenient long-term record of important samples that may be repeatedly analyzed. This chapter describes in detail the protocols for carrying out high-sensitivity Northern blots that have been used successfully in the laboratory to examine genes that are expressed at very low levels in small amounts of early embryonic tissues and in transgenic mice *(1–3)*. Generally, the protocol is not drastically different from existing Northern methods and requires no specialized equipment except a UV-light source. The method depends on an easy means of isolating high quality polyA$^+$ mRNA, UV crosslinking of the RNA to nylon membranes, and hybridization to single-stranded RNA probes. However, it is equally applicable to Northern analysis of all types and difficulties, and alternative RNA isolation procedures or hybridization methods can be substituted into the protocol with little change in efficiency or results. DNA probes may also be used in most cases, but they will be less sensitive that the single-stranded RNA probes.

2. Materials

Methods for RNA isolation require pure reagents and care in preparation to avoid RNA degradation. It is advisable to wear gloves at all times. Heavy metals or other contaminants can result in the chemical cleavage of RNA. Thus, deionized distilled water and high-grade analytical reagents should be used. Autoclaving buffers alone are not sufficient to inactivate contaminating RNases, since renaturation may follow heat denaturation. The chemical diethylpyrocarbonate (DEPC) binds to RNase, irreversibly inactivating it, and is the most convenient means of producing RNase-free solutions (**Caution:** DEPC is a carcinogen; use gloves).

2.1. Isolation of RNA
2.1.1. LiCl-Urea Method

1. DEPC-treated water: Add 1–2 drops of DEPC/L of deionized water and mix by shaking. Autoclave to sterilize. This removes the DEPC, which breaks down to ethanol, CO_2, and water.

2. Homogenization buffer: $3M$ LiCL, $6M$ urea. Dissolve 126 g of LiCl and 360 g of urea in 1 L of sterile DEPC water and filter through a 0.2-mm filter.
3. $2M$ Tris-HCl, pH 7.6: Make buffer in DEPC-treated water and adjust to pH 7.6. Place in a glass bottle, add 1–2 drops of DEPC, mix by shaking, and autoclave immediately. DEPC can destroy Tris, so check the pH of the buffer after autoclaving, and never autoclave dilute Tris buffers.
4. $0.5M$ EDTA: Make the EDTA stock, add 1–2 drops of DEPC, and autoclave.
5. 10% SDS: Make SDS stock, add 1–2 drops of DEPC, and incubate at 68°C overnight. Do not autoclave.
6. $3M$ sodium acetate, pH 5.2: Adjust the acetate stock to pH 5.2, and add 1–2 drops of DEPC. Autoclave.
7. 95% Ethanol.
8. Redistilled phenol (Gibco-BRL): Melt at 65°C, saturate with equal volume of TES, and let set overnight. Add 5 mL of m-Cresol, 0.2 mL of 2-mercaptoethanol, and 0.1 g of 8-hydroxy quinolate/100 mL of phenol. Mix until dissolved, and then remove as much of the aqueous layer as possible.
9. Phenol:chloroform:isoamyl alcohol (24:24:1): Use treated phenol (step 8), AR-grade chloroform, and isoamyl alcohol. Keep in the dark, and do not store for long periods.
10. TES: 10 mM Tris-HCl, pH 7.6, 1 mM EDTA, 0.5% SDS. Make in DEPC water, and use DEPC-treated $2M$ Tris-HCl, pH 7.6, and $0.5M$ EDTA stocks.
11. TE: 10 mM Tris-HCl, pH 7.6, 1 mM EDTA. Make in DEPC water, and use DEPC-treated $2M$ Tris-HCl and $0.5M$ EDTA stocks.
12. Sterile disposable RNase-free centrifuge tubes: Standard 1.5-mL Eppendorf microfuge tubes work for small preparations. They can be used directly from the box if handled with gloves. Falcon also supplies 5- (#2063), 15- (#2059), and 50-mL (#2070) sterile polypropylene tubes.

2.1.2. Hot Phenol Method

1. 50 mM sodium acetate, pH 5.2: Add 1–2 drops of DEPC, and autoclave.
2. $5M$ NaCl: Add 1–2 drops of DEPC, and autoclave.
3. Saturated phenol: Melt redistilled phenol (Gibco-BRL), add 0.25 vol of 50 mM sodium acetate, pH 5.2, and mix. Let stand until phases separate, and remove as much of the aqueous layer as possible. Store the phenol in small aliquots at −20°C until needed. Refreeze any unused phenol after RNA isolation.
4. NaCl/ice slurry: Add NaCl to ice, and mix until temperature reaches −10 to −13°C.

2.2. Isolation of PolyA+ RNA

1. OligodT cellulose: Use T3 grade (Collaborative Research Labs, USA).
2. $0.1N$ NaOH: Dilute Analar-grade $10N$ NaOH stock with DEPC-treated water.

3. 6M LiCl: Dissolve 252 g of LiCl in a liter of water, add 1–2 drops of DEPC, and autoclave.
4. Binding buffer: 20 mM Tris-HCl, pH 7.6, 0.6M LiCl, 1 mM EDTA, and 0.2% SDS made from RNase-free stocks and DEPC-treated water.
5. Low-salt-binding buffer: 20 mM Tris-HCl, pH 7.6, 0.1M LiCl, 1mM EDTA, and 0.2% SDS made from RNase-free stocks and DEPC-treated water.
6. Elution buffer: 10 mM Tris-HCl, pH 7.6, 1 mM EDTA, and 0.1% SDS made from RNase-free stocks and DEPC-treated water.
7. Plastic oligodT columns, top and bottom closures, and rack for 20 columns: from Isolabs (Cleveland, OH) (Quicksep #QS-P) or Advanced Labs Techniques (Tunbridge Wells, Kent, UK).

2.3. Electrophoresis, Transfer, and Crosslinking

1. 20X MOPS stock: 0.4M MOPS, 0.1M sodium acetate, 20 mM EDTA. For 1 L, dissolve 83.6 g of MOPS, 8.2 g of sodium acetate, and 40 mL of 0.5M EDTA stock in DEPC-treated water.
2. Agarose: SeaKeem.
3. Formamide: There is no need for deionization.
4. 37% Formaldehyde: Supplied as a liquid stock.
5. Blue juice dye: 0.1% xylene cyanol, 0.1% bromphenol blue, 1X MOPS, 50% glycerol. Make with DEPC water and autoclaved glycerol.
6. 0.1M ammonium acetate.
7. 20X SSC: 3M NaCl, 0.3M sodium citrate. Make 6X SSC and 2X SSC by dilution with autoclaved distilled water.
8. Denaturing buffer: 50 mM NaOH, 0.1M NaCl.
9. Neutralizing buffer: 100 mM Tris-HCl, pH 7.6., diluted in DEPC-treated water.
10. Nylon membrane: 200 × 200 mm of Nylon membrane, e.g., GeneScreen (*see* Note 1).
11. Whatmann 3MM filter paper, flat paper towels, and plastic wrap (Saran Wrap™).

2.4. Hybridization

1. 50X Denhardt's solution: 0.05% (w/v) BSA, 0.05% (w/v) polyvinyl pyrolodone, and 0.05% (w/v) Ficoll 400.
2. Prehybridization buffer: 50–60% formamide, 5X SSC, 5X Denhardt's, 50 mM sodium phosphate buffer, pH 6.8, 250 µg/mL of sheared denatured salmon sperm DNA, 100 µg/mL of yeast tRNA, 1% SDS. Make with DEPC-treated water.
3. Hybridization buffer: 50–60% formamide, 5X SSC, 5X Denhardt's, 50 mM sodium phosphate buffer, pH 6.8, 250 µg/mL of sheared denatured salmon sperm DNA, 100 µg/mL of yeast tRNA, 1% SDS (v/v), 10% dextran sulfate (w/v), and appropriately labeled probe (*see* Chapters 5, 7, 8, and 10). Make with DEPC-treated water.

3. Methods
3.1. Isolation of Total RNA
3.1.1. LiCl-Urea Method

This method works for large and small amounts of fresh or frozen tissue, and may also be used for cultured cells. It is efficient at inhibiting nucleases in order to produce high-quality biologically active RNA and is based on the method of Auffray and Rougenon *(4)*. No ultracentrifugation steps are required, and the entire method can be performed using a refrigerated benchtop centrifuge with a swing-out rotor capable of handling standard sterile disposable plastic centrifuge tubes (1.5–50 mL vol). This eliminates the need for specially treated glassware, and the method can be used to isolate RNA from large numbers of samples simultaneously.

1. Weigh the fresh or frozen tissue, and place in a sterile plastic disposable centrifuge tube of appropriate size.
2. Add homogenization buffer to a ratio of 5–10 mL of buffer/g of starting material.
3. Homogenize the sample in the plastic tubes for 2 min at 0°C (on ice) using a Polyton, Ultra-turrax or similar motor-driven homogenizer (*see* Note 2).
4. To complete the DNA shearing, sonicate the sample on ice for 1–2 min. A microtip should be used on small samples placed in Eppendorf 1.5-mL microcentrifuge tubes.
5. Spin the sample at low speed (1000 rpm) for 2–5 min at 0°C in a benchtop centrifuge to pellet any nonhomogenized material and membrane debris.
6. Pour the supernatant into a new centrifuge tube, and store overnight at 0–4°C (*see* Note 3).
7. The DNA should remain in solution, and the RNA is pelleted by centrifugation. Centrifugation is carried out in either in a Sorvall HB4 rotor at 9000 rpm for 10 min at 0°C or in a refrigerated benchtop centrifuge at 2000–4000 rpm (use maximum rated speed) for 30 min at 0°C, depending on the sample size and type of tube.
8. Discard the supernatant, and add 0.1 vol of LiCl-urea homogenization buffer, precooled to 4°C. Vortex and leave on ice for 30 min.
9. Centrifuge and discard the supernatant (*see* Note 4). Dissolve the pellet in either TES at 5 mL/g of the original tissue weight or 0.5 vol of the original homogenzation buffer. The sample should be mixed well to break up and dissolve the RNA pellet.
10. Add an equal volume of phenol:chloroform:isoamyl alcohol (24:24:1), and extract the sample for 5–10 min with vigorous shaking (*see* Note 5).

Northern Blot Analysis

11. Separate the phases by spinning at 3000–4000 rpm in a benchtop centrifuge at room temperature. Carefully remove the top aqueous phase and place in a new tube. If the interface is very thick, re-extract the organic phase by adding 0.5 vol of TES and recentrifuging.
12. Combine the aqueous phases, and re-extract a second time with an equal volume of phenol:chloroform:isoamyl alcohol. Spin at 3000–4000 rpm in a benchtop centrifuge to separate the phases and carefully remove the top aqueous layer to a new tube. Repeat the extraction procedure until the interface is clear.
13. Add 0.1 vol of 3M sodium acetate, pH 5.2, and 2 vol of 95% ethanol. Mix and place at –20°C. The RNA is stable indefinitely at this stage.
14. Pellet the RNA by centrifugation (as in step 7), and discard the supernatant. Rinse the pellet in an equal volume of 70% ethanol and recentrifuge.
15. Dissolve the RNA in TE, and read absorbance at 260 nm (40 μg/mL = 1 A_{260} U). Adjust the volume to desired concentration, and store at –20°C. (*See* Notes 6 and 7.)

This is a pure preparation of total RNA and can be used directly in many assays (S1, RNase protection, and Northerns) or for the preparation of polyA$^+$ RNA.

3.1.2. Hot Phenol Method

This method is a fast protocol for the isolation of RNA from tissue-culture cells and is especially useful for handling large numbers of different samples. The DNA is removed, since it is soluble in phenol at 60°C and pH 5.2.

1. Pellet cells by centrifugation in 50-mL plastic tubes. Resuspend in PBS and repellet. Repeat the PBS wash a second time. This removes all traces of medium and serum.
2. Resuspend the cell pellet in 1 mL of 50 mM sodium acetate, pH 5.2, by vortexing. Ensure that all of the cell clumps are dispersed.
3. Add 10 mL of 50 mM sodium acetate, pH 5.2, containing 1% SDS. Mix by inversion to lyse the cells.
4. Add an equal volume (11 mL) of phenol saturated with 50 mM sodium acetate, pH 5.2, which has been prewarmed to 60°C. Mix by vortexing, and place in a 60°C bath for 15 min. During the incubation, frequently remove the tube and vortex for a few seconds.
5. Remove the tubes from the bath, and immerse in an NaCl/ice slurry (–10°C) for 5 min.
6. Spin at 3000 rpm for 10 min at 2°C in a refrigerated benchtop centrifuge. Insert a pipet, and remove the bottom phenol layer leaving the interface.

7. Add 10 mL of saturated phenol at 60°C and vortex. Incubate the sample for a further 15 min at 60°C again with repeated vortexing.
8. Remove from bath, and incubate in NaCl/ice slurry for 5 min. Spin at 3000 rpm for 10 min at 2°C.
9. Carefully remove the top aqueous layer, and add 0.1 vol of 5M NaCl and 2 vol of 95% ethanol. Place at –20°C.
10. Pellet total RNA at 9000 rpm for 1 min at 4°C in a Sorval HB-4 swing-out rotor, and pour off the supernatant. Rinse with 70% ethanol and repellet.
11. Redissolve the RNA in TE, and read the A_{260} to measure the concentration. Store frozen at –20°C.

3.2. Isolation of PolyA+RNA

The isolation of polyA+RNA can be done by chromatography on oligodT cellulose in batch operation using centrifuge tubes or in sterile RNase-free plastic columns, such as those used for radioimmune assays. The columns fit in racks, and 20 samples can be processed simultaneously. It works for both large or small amounts of total RNA provided the binding capacity of the oligodT cellulose is not exceeded. In cases where the amount of total RNA is very small, carrier RNA is added to the sample to reduce background losses. In this way, polyA+RNA can be easily isolated from as little as 10–100 µg of total RNA *(5)*. This method is based on the method of Aviv and Leder *(6)*.

1. OligodT cellulose is suspended in water and allowed to swell for several hours.
2. Gently mix the cellulose, and allow the material to settle for 20 min. Decant the fines, add sterile water, and resuspend the cellulose. Repeat to remove fine particles that may clog the column.
3. Resuspend the oligodT cellulose in 0.1N NaOH, and pour the slurry into the column. Allow the liquid to flow through until a packed volume of 5–7 mm of cellulose forms at the bottom.
4. Rinse the cellulose by carefully pipeting 5 mL of binding buffer into each column, and allow it to flow through. Repeat three times to remove all traces of NaOH. The eluant should be checked with pH indicator paper. The columns are ready for use (*see* Note 8).
5. Dissolve the total RNA in TES, usually 1–2 mL for small samples and 5–10 mL for large samples (5–10 mg of RNA). In the case of very small samples (i.e., <100 µg total RNA) add 5 µL of a 100 µg/mL polyA+RNA carrier.
6. Heat the RNA at 60°C for 10 min in a water bath. Remove and adjust the salt to 0.6M LiCl by adding 0.1 vol of 6M stock.
7. Vortex and immediately pour the samples carefully onto the prepared oligodT columns. Minimize the disturbance to the cellulose layer.

Northern Blot Analysis

8. Collect the run-through and reapply to the column to ensure complete binding of the polyA$^+$RNA. Collect the run-through again, and save as polyA$^-$RNA.
9. Add 5 mL of binding buffer, being careful to minimize disturbance to the cellulose layer, and allow the binding buffer rinse to flow through. Repeat three more times with 5 mL of binding buffer.
10. Rinse three times with 5 mL of low-salt binding buffer, which removes traces of ribosomal RNA from the column.
11. Allow the column to run dry, and add elution buffer.
    a. If large amounts of RNA are bound, use 5–10 successive 1-mL rinses of elution buffer, collected in sterile 1.5-mL microfuge tubes (*see* Note 9). Determine which fractions contain the RNA by reading the A$_{260}$ of each fraction blanked against elution. Usually this is fraction 3–4.
    b. For small amounts of RNA, use 75 µL of elution buffer, and allow it to follow through. Collect the RNA in a single 1.5-mL microfuge tube using four 150-µL washes of elution buffer. Allow 1–2 min between each wash to enable the RNA to elute. Collect another set of four 150-µL washes of elution buffer in a second tube. Usually all of the polyA$^+$RNA is in the first tube.
12. Precipitate the RNA with 0.1 vol of 3*M* sodium acetate, pH 5.2, and 2 vol of 95% ethanol. Place at –20°C. Spin for 15–20 min in a microfuge at 4°C, or in a Sorval HB-4 rotor at 9000 rpm at 4°C for 10 min. Discard the supernatant, and wash the sample by adding an equal volume of 70% ethanol and respin. Discard the supernatant and dissolve the RNA in TE. Store at –20°C.

3.3. Gel Electrophoresis, Transfer, and UV Crosslinking

The methods for electrophoresis and transfer of nucleic acids by capillary action have been described in detail in many methods books; the protocols below represent slightly modified versions of the standard techniques *(6–8)*. The UV crosslinking method is based on tests designed to optimize the permanent binding of RNA to GeneScreen membranes *(9)*.

3.3.1. Electrophoresis

1. Carefully clean gel box, casting tray (plates), and combs with mild detergent, and rinse with clean distilled water. Give a final rinse with sterile distilled water. Use gloves in all steps to avoid contamination with RNase.
2. Set up the casting tray and comb in a fumehood, since formaldehyde vapors given off during the pouring and setting of the gel are irritating to the eyes.
3. For 200 mL of a 1.4% gel, add 2.89 g of agarose to 156 mL of sterile distilled water in a 500-mL flask. Boil until the agarose is melted, and allow to cool to 60°C (*see* Note 10).

4. Add 10 mL of 20X MOPS stock and 34 mL of 37% formaldehyde. Mix briefly by gentle swirling and immediately pour the gel. Avoid air bubbles on the surface of the gel. If they form, they can be removed by touching with a sterile syringe needle before the gel cools.
5. Allow the gel to set, and gently remove the comb. Gels with formaldehyde have a lower tensile strength than normal agarose gels and tear easily. Take care when handling.
6. RNA samples are placed in a 1.5-mL microfuge tube with a dye-denaturation cocktail as follows: 1 µL of 20X MOPS, 10 µL of formamide, 3.5 µL 37% formaldehyde, 2 µL of blue juice dye, and 5 µL of RNA in TE. Usually 1–2 µg of polyA$^+$ is sufficient. If RNA is too dilute, it can be dried down and resuspended in 5 µL of TE. Samples can be prepared and stored at 4°C for a few hours or overnight.
7. Heat-denature the RNA samples at 60°C for 10–15 min in a water bath, and cool on ice. Load the gel immediately.
8. Run the gel for 5–7 h at 60–70 V in 1X MOPS with no formaldehyde (*see* Notes 11–13).

3.3.2. Transfer

1. Cut the nylon membrane to approx the gel size and completely wet by first floating and then immersing in the water. After 10 min, transfer to 2X SSC and store until use.
2. Soak the gel in denaturation buffer for 20 min with gentle shaking (*see* Note 14).
3. Neutralize the gel in neutralization buffer for 20 min, and then place in 2X SSC for 20 min with gentle shaking (*see* Note 15).
4. Set up the blotting apparatus by placing sheets of Whatman 3MM paper that are soaked in 20X SSC on a glass plate over a resevoir of 20X SSC. The ends are immersed in the reservoir to act as a wick for the SSC buffer.
5. The gel is placed on the wet 3MM with the comb-side down to provide a smooth, even surface for contact with the membrane.
6. Surround the gel with strips of parafilm on the exposed areas of the 3MM wick to allow flow of 20X SSC only through the gel.
7. Wet the surface of the gel with 20X SSC, and carefully lay the wet membrane on top of the gel. Be careful to avoid trapping air bubbles between the gel and membrane.
8. Place three layers of 3MM cut to the exact size of gel, and prewet with 20X SSC on top of the membrane. Again, be careful to avoid trapping air bubbles.
9. Place several inches of paper towels on the 3MM filter papers, and cover with a glass plate and a lead weight. Transfer overnight, and replace the towels if they become completely wet.

Northern Blot Analysis

10. Following transfer, remove the paper towels, and mark the position of the lanes and gel orientation on the membrane with a pen.
11. Trim the edges of the membrane with a scapel, and peel the filter from the gel. Immerse the membrane in 6X SSC for 1–2 min.

3.3.3. UV Crosslinking

1. Pipet 2–4 mL of 6X SSC onto the middle of a large piece of Saran Wrap placed on a benchtop.
2. Remove the membrane from the 6X SSC, and carefully place it onto the plastic with the side that was in direct contact with the gel during transfer facing down. Avoid air bubbles between membrane and the plastic.
3. Fold over the excess plastic wrap to enclose the membrane completely, and place on a glass plate. The side of the membrane in contact with the gel should face up.
4. Place the filter on the plate under a UV source that delivers 600 µW/cm^2 at a wavelength of 254 nm (see Notes 16–20). Expose for 5 min.
5. Remove from the source, and take filter out of plastic wrap. Place the filter between two sheets of 3MM paper, and bake at 80°C for 60 min.

The filter is now ready for hybridization.

3.4. Filter Hybridization

A large variety of hybridization buffers or systems are available and can be used with equal success in the filter hybridizations. This method is based on Wahal et al. (10). Regardless of protocols, high backgrounds can be a problem using nylon membranes. This can usually be avoided by higher concentrations of SDS in both the hybridization and washing steps. The UV crosslinking prevents the high levels of SDS from eluting the RNA from the membrane. A typical example of a final Northern result is shown in Fig. 1. The figure also illustrates how special RNase wash treatments can be useful.

1. Place the membrane in a plastic bag sealed on three sides. Pour in the prehybridization buffer, and seal the remaining side with a bag sealer.
2. Place the bag in a water bath at the hybridization temperature determined by the type of probe being used (see Table 1). For single-stranded RNA probes, this temperature is 60–65°C.
3. Prehybridize for 2–4 h, and then remove the bag, cut a corner, and pour out the prehybridization buffer (see Note 21).
4. Place the probe in 10–15 mL of hybridization buffer and prewarm at the hybridization temperature for 20 min.

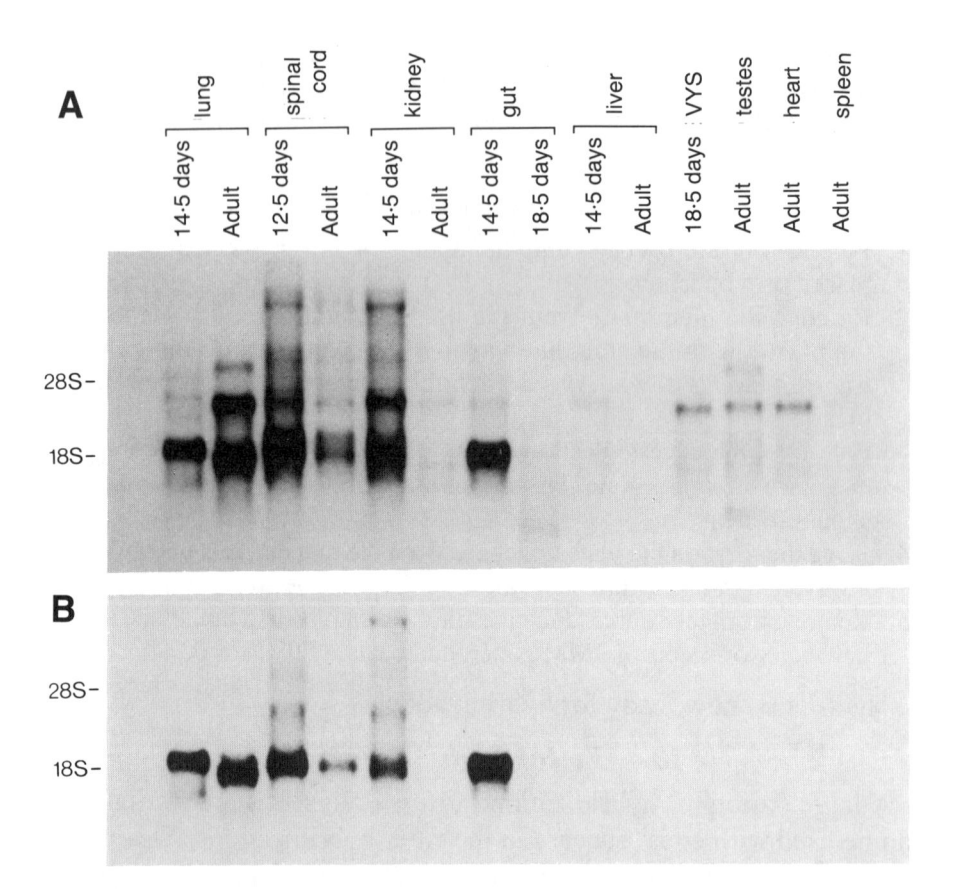

Fig. 1. Example of a Northern blot using a Homeobox probe on mouse embryonic tissues. Lanes contain 1–2 µg of polyA⁺RNA from the tissues indicated at the top of the lanes. **(A)** Results with washing at 80°C in 0.1X SSC, 0.5% SDS. Some crosshybridization is seen with other members of the homeobox family, and this is removed when the same filter is washed using the RNase treatment method **(B)**. Note many bands remain unchanged, and these represent transcripts of a single gene. However, bands resulting from closely related genes are removed in this high-stringency treatment. Probe: ^{32}P-single-stranded Hox-2.1 RNA generated by SP6 in in vitro transcription from a plasmid template *(1)*.

5. Carefully pour the hybridization buffer with probe into the plastic bag, and reseal avoiding as many air bubbles as possible. Hybridize the filter overnight (12–24 h) (*see* Note 22).
6. Cut open the bag, and remove the hybridization buffer. Remove the filter, and rinse three times in 500 mL of 2X SSC at room temperature.

Table 1
Hybridization and Washing Conditions for DNA and RNA

Probe	Filter	Hybridization temperature, °C%	Formamide in hybridization	Wash conditions
RNA	RNA	60–65°C	60	0.1X SSC, 0.5% SDS, 75°C
DNA	RNA	50–55°C	50	0.1X SSC, 0.5% SDS, 65°C
RNA	DNA	50–55°C	50	1X SSC, 0.5% SDS, 65°C
DNA	DNA	42°C	50	1X SSC, 0.5% SDS, 55°C

7. Wash the filter in 700 mL of 0.1X SSC, and 0.5% SDS at 70–80°C for 1 h. Change the buffer, and perform a second identical wash. It is helpful to use a hand-held monitor to check background. A third wash may be necessary.
8. Remove the filter from the wash buffer, and wrap in plastic film while moist. Expose to autoradiography at −70°C with an intensifying screen and Kodak XAR–5 film (*see* Notes 23–25).
9. After the results are obtained, the filters are stripped for reuse by washing in 70% formamide at 80–90°C for 10–15 min. This can be repeated if necessary. The filter can be autoradiographed to ensure probe removal.

4. Notes

1. Many types of nylon membranes have been tested, and they all work to varying extents. Charged membranes, such as GeneScreen Plus, or Hybond-N+, do not work as efficiently. In our hands, the GeneScreen membrane reproducibly produced the highest degree of RNA binding and low backgrounds.
2. Avoid foaming as much as possible. It is important that the sample be well homogenized to shear the nuclear DNA. More homogenization buffer can be added if the sample is extremely viscous.
3. It is important when removing the supernatant to be sure that the sample is not viscous and that there are no clumps of nonhomogenized material. Samples can be stored for several months at this stage.
4. Wipe the sides of the tube dry with a sterile tissue to remove as much of the supernatant as possible. This reduces DNA contamination of the sample. In the case of large samples, it may be advisable to perform a second LiCl-urea wash to completely remove DNA.
5. If the RNA pellet is slow to dissolve, add phenol-chloroform, and vortex or place on a shaker until it dissolves.
6. If a large number of samples are to be read, it is convenient to use sterile disposable UV cuvets, which minimizes crosscontamination between RNA samples and reduces the handling time of the RNA.

7. When handling large samples, the DNA may not be completely sheared or removed by the LiCl-urea washes. It is important to have as much of the DNA removed as possible, since contamination drastically reduces the flow rates of the oligodT columns used in the isolation of polyA$^+$ RNA. If after the final ethanol precipitation DNA is still present, it may be removed by redissolving the RNA in TE and adjusting the sample to 2M LiCl. Overnight incubation at 4°C reprecipitates the RNA, but leaves the DNA in solution. The RNA is then pelleted by centrifugation, dissolved in TE, and reprecipitated with ethanol as in Section 3.1.1., step 13.
8. If small amounts of RNA (i.e., <100 µg) are to be loaded onto the column, add 2 mL of binding buffer containing 5 µg/mL of yeast tRNA. Allow this to run through, and then rinse with 5 mL of normal binding buffer.
9. The oligodT columns can be reused after successive rinses with 5 mL of TES, followed by 5 mL of 0.1N NaOH. The column is left at room temperature for 10–15 min to destroy any remaining traces of RNA. To store the column, refill it with TES or 0.1N NaOH and cap the top and bottom. The columns can be stored at 4°C indefinitely and reused repeatedly without future RNA loss or contamination. Storage in NaOH can result in some degradation of the glass-fiber filter at the bottom of the column after long periods, in which case the oligodT should be transferred to a new column.
10. The percentage of the gel may vary with need, but for separation of 1.5–8 kb, RNA, 1.4% is ideal.
11. Best results are obtained using high voltage and, hence, short running times.
12. The pH of the buffer can change during the run, so it is advisable to recirculate the buffer by either using a pump or by mixing the buffer several times during the run. Higher voltages can be used to shorten the length of the run if a recirculating pump is used. The pH can easily be checked with pH indicator paper to monitor any changes.
13. Markers may be used on the gels, in which case they must be stained at the end of elctrophoresis. To do this, cut off marker lanes to be stained, rinse twice in distilled water for 20 min, and twice in 0.1M ammonium acetate for 20 min. These rinses are necessary to remove formaldehyde, since this binds ethidiumbromide. Incubate the slices for 15 min in 0.1M ammonium acetate with 0.2 µg/mL of ethidum bromide. Destain the gel in two washes of 25 min in 0.1M ammonium acetate. If background is still high, the gel can be destained several hours or overnight at 4°C in 0.1M ammonium acetate without the RNA bands diffusing. A photograph can be taken on standard UV transilluminator setup.

Staining can prevent efficient transfer of the RNA to the membrane, so do not stain part of gel to be used for hybridization. Useful markers are the ribosomal RNAs found in poly⁻RNA or commercial RNA size ladders (supplied by Gibco-BRL).

14. This partially cleaves the RNA and aids transfer of large species. Do not leave longer than 20 min, and use DEPC-treated water at this step if transfering RNA <800 bp in size.
15. These steps remove formaldehyde, which can inhibit efficient transfer.
16. It is important to use a 254-nm wavelength bulb, without the polarizing filter. Attempts to crosslink the samples on a standard transilluminator (normally set at 305 nm) usually do not work. The polarizing filters usually exclude too much energy for effective crosslinking.
17. Many hand-held UV lamp units will work if the protective cover is removed. We recommend a standard four-bulb transilluminator without the polarizing filter mounted on a stand upside down in a protective box or cabinet to prevent UV exposure to user.
18. The dose of energy is important. Often specific distances from the source are described in protocols; however, it is important to determine the proper distance for your own UV source. Use a standard UV safety measuring device that measures in units of $\mu W/cm^2$ at 254 nm. These devices are not expensive and easily obtained from a scientific supply. Find the distance from the source that registers 600 $\mu W/cm^2$, and then fix the height of the UV source to that height.
19. A commerical UV crosslinker is now available from Stratagene (the StrataLinker™) and crosslinking to GeneScreen is very efficent using the manufacturer's instructions.
20. Several workers describe UV crosslinking with a dry filter. This can be successful and has the advantage that much shorter times are required, usually 10–30 s. However it is possible to overlink the RNA to the filter and reduce the usefulness of the RNA on the membrane. The optimum crosslinking times for the wet or dry method can be tested easily and directly. Prepare a test gel with a single large-slot preparative comb, and transfer the RNA to the filter. Set up the crosslinking, but cover most of the filter with a black paper. At various time-points, gradually expose more of the filter to the UV source. This will provide a time course of crosslinking. After hybridization of the filter, the optimum linking times and length of the plateau for effective linking are read directly from the filter. This plateau is generally very broad when the filter is linked wet, and the conditions are therefore very reproducible. On dry filters the plateau may be only a few seconds and difficult to reproduce, which is why the wet-linking method is preferred.
21. The prehybridization and hybridization buffers can be made up and stored for long periods at 4°C.
22. All of these transfer, UV crosslinking, and hybridization methods can be applied to Southern as well as Northern analysis. The only difference is the selection of hybridization temperatures depending on the nucleic acid involved and the probe. *See* Table 1 for alternative conditions for DNA.

23. It is important that the filter never becomes completely dried out, or it will not be possible to remove all of the bound probe.
24. If the background is high or a higher degree of stringency is required owing to crosshybridization, wash the filter in 100 mL of 0.1X SSC, 10–30% formamide at 65°C, followed by 0.1X SSC, 0.5% SDS at 70°C for 30 min. Re-expose the filter.
25. Alternatively, background can be eliminated by RNase treatment of the filter *(1)*, but it may no longer be reusable. To RNase treat, place the filter in 2X SSC, 0.1 µg/mL RNase A at room temperature for 20 min. Then wash in 500 mL 2X SSC, 0.5% SDS at 40–50°C for 30 min, and re-expose the membrane. This treatment is analogous to an RNase protection assay. *See* Fig. 1 for an example of this treatment.

References

1. Krumlauf, R., Holland, P. W. H., McVey, J. H., and Hogan, B. L. M. (1987) Developmental and spatial patterns of expression of the mouse homeobox gene, Hox-2.1. *Development* **99,** 603–618.
2. Graham, A., Papalopulu, N., Lorimer, J., McVey, J. H., Tuddenham, E. G. D., and Krumlauf, R. (1988) Characterization of a murine homeobox gene, Hox-2.6, related to the *Drosophila deformed* gene. *Genes Dev.* **2,** 1424–1438.
3. Graham, A., Papalopulu, N., and Krumlauf, R. (1989) The murine and *Drosophila* homeobox gene complexes have common features of organization and expression. *Cell* **57,** 367–378.
4. Auffray, C. and Rougeon, F. (1980) Purification of mouse immuno:globulin heavy-chain RNAs from total myeloma tumor RNA. *Eur. J. Biochem.* **107,** 303–324.
5. Krumlauf, R., Hammer, R., Tilghman, S. M., and Brinster, R. (1985) Developmental regulation of alpha-fetoprotein in transgenic mice. *Mol. Cell Biol.* **5,** 163–168.
6. Aviv, H. and Leder, P. (1972) Purification of biologically active globin mRNA by chromatography on oligothymidylic acid-cellulose. *Proc. Natl. Acad. Sci. USA* **69,** 1408–1412.
7. Sambrook, J., Fritsch, E. F., and Maniatis, T. (1989) *Molecular Cloning: A Laboratory Manual.* Cold Spring Harbor Laboratory, Cold Spring Harbor, NY.
8. Southern, E. (1975) Detection of specific sequences among DNA fragments separated by gel electrophoresis. *Mol. Biol.* **98,** 503–512.
9. Krumlauf, R. (1989) GeneScreen hybridization transfer membrane: UV-crosslinking protocols. Dupont Manufacturers Instruction Booklet, Du Pont, Wilmington, DE.
10. Wahal, G. M., Stern, M., and Stark, G. R. (1979) Efficient transfer of large DNA fragments from agarose gels to diazobenzloxymethal-paper and rapid hybridization by using dextran sulfate. *Proc. Natl. Acad. Sci. USA* **76,** 3683–3688.

CHAPTER 17

RNA Slot Blotting

David Murphy

1. Introduction

In a slot blot, RNA is applied, unfractionated, to a solid matrix, and therefore, slot blotting is a rapid and sensitive method for analyzing changes in RNA quantity following developmental or physiological changes. Its disadvantage is that it cannot be used when the RNA sample contains additional homologous sequences, which will hybridize to the probe and confuse the results.

2. Materials

1. 3:1 Loading buffer: 5 mL of formamide (Fluka [Buchs, Switzerland]; see Note 1), 1.7 mL of formaldehyde (Fluka), DEPC-treated water, and 0.5 mL of 20X MAE (20X MAE: $0.4M$ MOPS, pH 7, $0.1M$ Na acetate, and $0.02M$ EDTA).
2. DEPC-treated 20X SSPE: $3.6M$ NaCl, 200 mM sodium phosphate, pH 6.8, and 20 mM EDTA.
3. Slot blot applicator linked to vacuum line (e.g., Schleicher and Schuell, Dassel, Germany).
4. Nylon hybridization matrix (e.g., Amersham [UK] Hybond-N); see Note 2.
5. A UV crosslinker (see Chapter 16).
6. RNA samples: see Notes 3–5.

3. Method (See Note 4)

1. Add 3 vol of 3:1 loading buffer to the RNA sample (up to 50 µg total cellular RNA). Incubate at 65°C for 15 min to denature.
2. Add SSPE to 2X to a final vol of up to 500 µL.
3. Apply to the nylon hybridization matrix using the slot-blot applicator.

4. Disassemble the slot-blot system, and rinse the filter in DEPC-treated 2X SSPE.
5. Bake the filter at 80°C for 1 h. This step dries the filter and assists with the fixing of the RNA to the matrix.
6. Covalently crosslink the RNA to the matrix in a UV crosslinker (see Chapter 16).
7. Proceed with hybridization as described in Chapter 16.

4. Notes

1. Formamide should be deionized before use. To do this, formamide is mixed and stirred for 1 h with Dowex XG8 mixed-bed resin, followed by filtration through 3MM paper. Fluka formamide (catalog number 47670) does not require deionization.
2. These methods have been developed for neutral nylon membranes (e.g., Amersham Hybond-N). We have not tested positively charged membranes (e.g., Amersham [UK] Hybond-N+, Bio-Rad [Richmond, CA] Zeta-Probe, NEN-Du Pont [Boston, MA] Genescreen™ Plus).
3. Total cellular RNA prepared using rapid techniques can be contaminated with genomic DNA and therefore make accurate quantitation by spectrophotometry difficult. In addition, it is important to control for loading artifacts. For semiquantitative analysis, each sample should be slot blotted twice, one filter hybridized to a probe for the RNA under test, and the other with a probe for an RNA that does not change in expression between samples. Alternatively, one slot blot can be made per sample and the same filter reprobed.
4. Slot blotting of RNA can be used to quantitate changes in the steady-state levels of specific RNA species rapidly following developmental or physiological change. The hybridization of the probe to dilutions of the RNA samples under test is compared to the signal generated by hybridization of the same probe to dilutions of specific RNA standards generated by in vitro transcription of cloned cDNAs by bacteriophage RNA polymerases (see Chapter 12).
5. Slot blotting is notoriously prone to giving false-positive signals. This can best be avoided by:
    a. Rigorous testing of the probe using Northern blots to ensure that it is specific;
    b. Inclusion in all assays of adequate negative controls; and
    c. Assiduous preparation of all probes, solutions, RNAs, and so forth, in order to avoid contamination.

CHAPTER 18

The RNase Protection Assay

Dominique Belin

1. Introduction

The RNase protection assay is based on the resistance of RNA:RNA hybrids to single-strand specific RNases, after annealing to a complementary ^{32}P-labeled probe in solution. It can be used to map the ends of RNA molecules or exon-intron boundaries. It also provides an attractive and highly sensitive alternative to Northern blot hybridization for the quantitative determination of mRNA abundance. Hybridization is carried out with an excess concentration of probe so that all complementary sequences are driven into the labeled hybrid. Unhybridized probe or any single-stranded regions of the hybridized probe are then removed by RNase digestion. The "protected" probe is detected and quantitated on a denaturing polyacrylamide gel. Originally, single-stranded DNA probes were used for the assay [1]. However, these have the disadvantage of being lengthy to prepare. The ease with which labeled RNA molelcules (riboprobes; see Chapter 12) can now be made makes an assay based on RNA:RNA hybridization much more favorable [2,3].

RNase protection has a number of advantages. First, solution hybridization tolerates high RNA input (up to 60 µg of total RNA), and is not affected by the efficiency of transfer on membranes or by the availability of membrane-bound RNAs. Second, the signal-to-noise ratio is much more favorable, since crosshybridizing RNAs yield only short protected fragments. Third, a significant fraction of mRNAs is often partially degraded during RNA isolation; in Northern blots, this generates a trail of shorter hybridizing species, which reduces the sensitivity of detection.

Finally, the detection of hybridized probes on sequencing gels is much more sensitive, because the width of the bands can be reduced to a tenth of those used in agarose gels. Only two features of Northern blots are lost in RNase protection assays: complete size determination of target RNAs and multiple use of each sample.

2. Materials

1. ^{32}P-labeled probe: A riboprobe prepared as in Chapter 12 (*see* Notes 1–3). Resuspend the probe in water at 1–2 ng/µL.
2. Hybridization mixture: 80% deionized formamide (*see* Note 4), 0.4M NaCl, 40 mM Na-PIPES, pH 6.8, and 1 mM EDTA.
3. RNase digestion buffer: 300 mM NaCl, 10 mM Tris-HCl, pH 7.4, and 4 mM EDTA.
4. Pancreatic RNase: A 10 mg/mL solution in TE containing 10 mM NaCl. Vials containing lyophilized RNases should be carefully open in a ventilated hood to avoid contamination. Boil for 15 min and slowly cool to room temperature. Store in aliquots at –20°C.
5. T1 RNase: A 1 mg/mL solution in TE. Adjust the pH to 7. Store in aliquots at –20°C.
6. Proteinase K: Dissolve the enzyme at 20 mg/mL in water, and store in aliquots at –20°C.

3. Method

1. For each sample, add 1 µL of probe to 29 µL of hybridization mixture. The probe is resuspended in water and constitutes 3% of the final volume; the exact amount of probe is not critical, since it is in excess over its specific target (*see* Note 5).
2. Lyophilize or ethanol-precipitate 10 µg of the sample RNA. Resuspend in 30 µL of complete hybridization mixture (including probe). Heat for 2 min at 90°C, then incubate overnight, usually at 45°C (*see* Notes 6 and 7).
3. Cool the samples on ice, and add 300 µL of RNase digestion buffer. Digest for 1 h at 25°C with pancreatic RNase, which cleaves after uracil and cytosine residues, and T1 RNase, which cleaves after guanine residues, or with both RNases (*see* Notes 8 and 9).
4. Add 20 µL of 10% SDS, and remove the enzyme(s) with 0.5 µL (10 µg) of proteinase K for 10–20 min at 37°C. Extract twice with phenol/chloroform, and precipitate the RNAs with ethanol with 10 µg of carrier tRNA.
5. Resuspend the RNAs in sample buffer, denature the hybrids for 2 min at 90°C, and electrophorese on a suitable percentage polyacrylamide/urea sequencing gel (*see* Chapter 47). Fix the gels with 20% ethanol and 10% acetic acid to remove the urea, dry, and autoradiograph (Note 10).

RNase Protection Assay

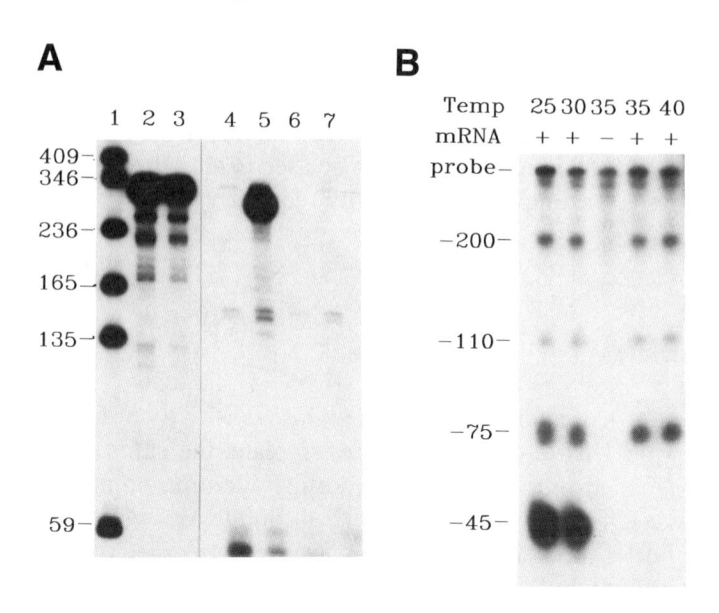

Fig. 1. RNase protection assay. **(A)** Discrimination between target-specific signal and complete probe protection by residual DNA. Detection of PN-I mRNA in total RNA from murine tissues *(6)*. The probe (310 nucleotides long) was gel purified. Lane 1: size markers. Lane 2: purified probe. Lane 3: probe hybridized and processed without RNase digestion. Lane 4: control hybridization with 10 µg of tRNA; traces of fully protected probe are visible. Lane 5: 10 µg of RNA from seminal vesicles, an abundant source of PN-I mRNA; the specific protected fragment is 260 nucleotides long. Lane 6: 10 µg of liver RNA, which does not contain detectable levels of PN-I mRNA. Lane 7: 10 µg of testis RNA, which contains trace levels of PN-I mRNA. **(B)** Effect of hybridization temperature on the detection of short complementary RNAs. The 5'-ends of phage T4 gene 32 transcripts in total RNA from bacteria carrying a gene 32 expression cassette were mapped by hybridization to a cRNA probe containing 400 nucleotides of gene 32 upstream sequences *(4)*. The probe was not gel-purified, and hybridization was performed at the indicated temperatures. Traces of fully protected probe result from incomplete DNase digestion of the template; they are also visible in the control hybridization without target RNA. The 44-nucleotide protected fragment is no longer detected above 30°C.

4. Notes

1. The probes should be 100–400 nucleotides long and should include at least 10 nucleotides that are not complementary to the target RNA. Residual template DNA generally produces a trace of full-length protected probe, which must be distinguishable from the fragment protected by the target RNA (Fig. 1).

2. It is often useful to decrease the specific activity of the probe, since more RNA is synthesized as a consequence of the higher ribonucleotide concentration, the probes are less susceptible to radiolysis, and less radioactivity is used. The following guideline can be used to alter the specific activity of the probes according to the sensitivity required:

RNA target abundance	Unlabeled UTP	^{32}P-UTP	Probe/sample*
High	100 µM	2.5 µM, 10 µCi	6–12 kcpm
Moderate	10 µM	2.5 µM, 10 µCi	60–120 kcpm
Low	—	12.5 µM, 50 µCi	300–600 kcpm

3. Purification is often necessary for maximal sensitivity or for mapping purposes. After DNase digestion of the template, the entire reaction can be directly loaded onto a 5–6% polyacrylamide/urea preparative gel (gel thickness: 0.4–1.5 mm), provided that enough EDTA is present in the sample buffer to chelate all the magnesium. After electrophoresis, cover the wet gel within Saran Wrap™ and expose for 0.5–5 min at room temperature to localize the full-length transcript. Cut the exposed band on the film with a razor blade. After aligning the cut film on the gel, excise the gel band with a sterile blade. The cut gel should be re-exposed to verify that the correct band has been excized. Elute the RNA from the gel either by (a) electroelution or (b) soaking.
   a. Electroelute for 1–2 h at 30 V/cm in 0.1X Tris/borate/EDTA in a sterile dialysis bag, after which, invert the polarity for 30 s to detach the eluted RNA from the membrane. Purify the eluate by two phenol/chloroform extractions and ethanol precipitation with a known amount of tRNA carrier (*see* Note 6). This procedure is very sensitive to RNase degradation.
   b. Incubate the gel fragment in an Eppendorf tube in 500 µL of 0.5M ammonium acetate, 1% SDS, and 20 µg/mL tRNA for 1–3 h at 37°C. The eluate and residual gel can be counted to ensure that more than 60% of the RNA is eluted. After two extractions with phenol/chloroform, recover the eluted RNA by ethanol precipitation.
4. To deionize formamide, incubate at –80°C until 75–90% of the solution has crystallized. Discard the liquid phase, thaw, and incubate at 4°C for several hours with a mixed-bed resin (AG501-X8, Bio-Rad [Richmond, CA]). Use a Teflon™-covered magnet that has been freed of RNase by treatment with 0.1M NaOH for 10 min, and rinsed with water and with crude formamide. Check the conductivity, which should be below 20 µS. Filter the solution onto sterile paper (Whatman, LS-14) over a sterile funnel. The absorbance at 270 nm should be below 0.2.

*The amount of probe required per sample for the detection of the target (*see* Note 5).

5. The amount of probe must be in excess of the target RNA. Since initially it is not possible to know the abundance of the target, this may have to be calculated empirically. The figures presented in Note 2 may be used as a guide. In addition, for very low abundant target RNA, the amount of sample RNA may be increased up to 60 µg.

   An input of 1–2 ng of a 300 nucleotide probe in a total volume of 30 µL will drive the hybridization of target RNA to completion (4–8 $T_{1/2}$) in approx 16 h. Shorter hybridizations can be performed, but require higher probe input (R_o) to achieve the same extent of saturation, i.e., to maintain the $R_o \times T_{1/2}$ value.

6. To facilitate the RNase digestion step, each sample should contain the same amount of total RNA. Inequalities may be eliminated by addition of tRNA. A negative control sample, containing only tRNA, is always included (Fig. 1A, lane 4, Fig. 1B, lane 3).

7. The temperature of hybridization must be reduced to detect small or very AU-rich protected fragments. For instance, a 44 nucleotide fragment of a phage T4 gene 32 transcript (containing 35 A/U and 9 C/G) was only protected by performing the hybridization at 25–30°C (*4;* Fig. 1B).

8. The amount of RNase is determined by the total amount of RNA present in the samples, including that contributed by the probe. This author usually adds 0.5–1.0 µg of pancreatic RNase and/or 0.25 µg of T1 RNase/µg of RNA. In most cases, digestion with pancreatic RNase alone is sufficient. When the probe and the target RNAs are from different species, the extent of homology can be sufficient to generate discrete protected fragments, particularly if digestion is performed with RNase T1 only. The temperature of digestion can be increased to 30–37°C, although this often leads to partial cleavages within the RNA:RNA hybrids.

9. To ensure that the probe remains intact during the hybridization, it may be useful to include a parallel control that is hybridized and processed without RNase treatment (Fig. 1A, lanes 2 and 3).

10. Minor shorter protected fragments are often detected, which may complicate the interpretation of mapping assays. To distinguish between digestion artifacts and rare target RNAs that are only partially complementary to the probe, it may be useful to use as a control target RNA a synthetic sense transcript fully complementary to the probe *(5)*.

Acknowledgments

I thank P. Vassalli for his early encouragment to use riboprobes for detecting rare mRNAs. Over the last few years, many collegues, students, and technicians have contributed to the methods outlined in this chapter, including M. Collart, N. Busso, J.-D. Vassalli, H. Krisch, S. Clarkson,

J. Huarte, S. Strickland, P. Sappino, M. Pepper, A. Stutz, G. Moreau, D. Caput, M. Prentki, P. Gubler, F. Silva, V. Monney, D. Gay-Ducrest, and N. Sappino. Research is supported by grants from the Swiss National Science Foundation.

References

1. Williams, D. L., Newman, T. C., Shelnes, G. S., and Gordon, D. A. (1986) Measurements of apolipoprotein mRNA by DNA-excess solution hybridization with single-stranded probes. *Methods Enzymol.* **128,** 671–689.
2. Belin, D. (1994) The use of riboprobes for the analysis of gene expression. *Methods Mol. Biol.* **31,** 257–272.
3. Srivastava, R. A. K. and Schonfeld, G. (1994) Quantification of absolute amounts of cellular messenger RNA by RNA-excess solution hybridization. *Methods Mol. Biol.* **31,** 273–281.
4. Belin, D., Mudd, E. A., Prentki, P., Yi-Yi, Y., and Krisch, H. M. (1987) Sense and antisense transcription of bacteriophage T4 gene 32. *J. Mol. Biol.* **194,** 231–243.
5. Belin, D., Wohlwend, A., Schleuning, W.-D., Kruithof, E. K. O., and Vassalli, J.-D. (1989) Facultative polypeptide translocation allows a single mRNA to encode the secreted and cytosolic forms of plasminogen activators inhibitor 2. *EMBO J.* **8,** 3287–3294.
6. Vassalli, J.-D., Huarte, J., Bosco, D., Sappino, A.-P., Sappino, N., Velardi, A., Wohlwend, A., Erno, H., Monard, D., and Belin, D. (1993) Protease-nexin I as an androgen-dependent secretory product of the murine seminal vesicle. *EMBO J.* **12,** 1871–1878.

CHAPTER 19

Primer Extension Analysis of mRNA

Mark W. Leonard and Roger K. Patient

1. Introduction

Primer extension is a relatively quick and convenient means by which gene transcription can be monitored. The technique can be used to accurately determine the site of transcription initiation or to quantify the amount of cap-site-specific message produced.

The principle of this technique is shown in Fig. 1. In brief, a radiolabeled probe fragment (preferably single-stranded) is hybridized to its complementary sequence near the mRNA 5' terminus. This primer is then extended by the enzyme reverse transcriptase back to the initiation point of the message. The products of the reaction are run on a denaturing polyacrylamide gel and exposed to autoradiography.

The major advantage of primer extension for mRNA analysis is its convenience (compared to S1 mapping or RNase mapping). The technique enables precise determination of the start point of transcription. Also, because very clean results can be obtained, more than one mRNA can be quantitatively analyzed in a single reaction. For example, the transcripts from a transfected marked gene can be distinguished from the cell's endogenous gene products, or, by judicious choice of primers, transcription from a test gene and a cotransfected (internal standard) control gene can be monitored in the same reaction (*see* Fig. 2). Finally, in genes with multiple cap sites, the amount of transcription from each site can be distinguished in a single reaction.

Many of the above considerations apply equally to the analysis of mRNA by S1 or RNase mapping (*see* Chapters 18 and 20). Although these

From: *Methods in Molecular Biology, Vol. 58: Basic DNA and RNA Protocols*
Edited by: A. Harwood Humana Press Inc., Totowa, NJ

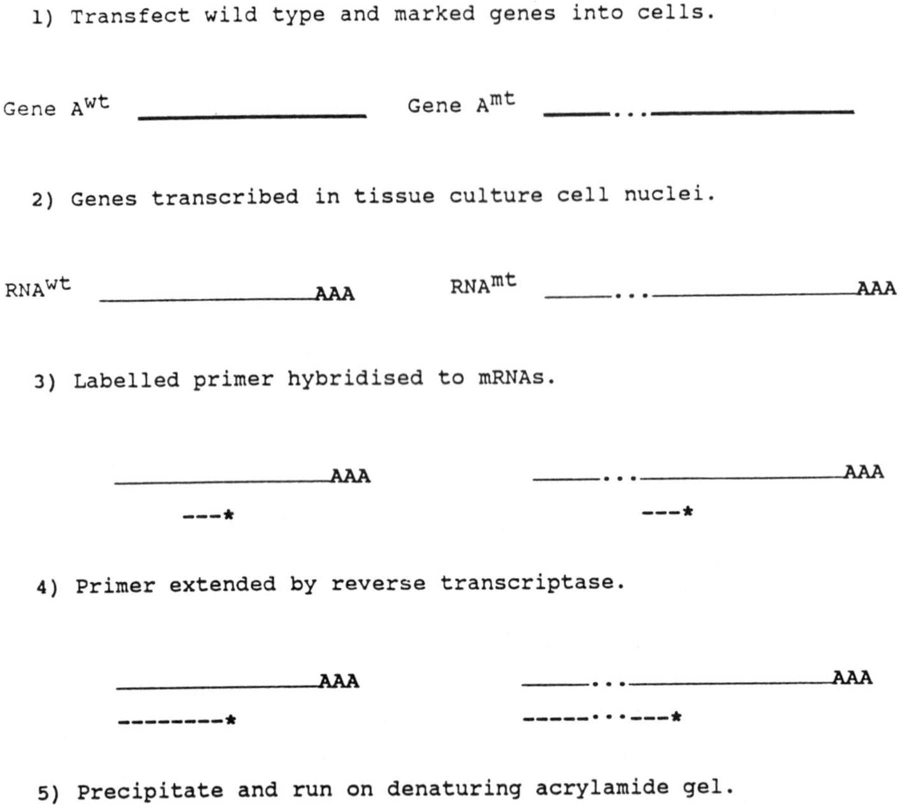

1) Transfect wild type and marked genes into cells.

Gene A^{wt} ─────────────   Gene A^{mt} ─────...─────────

2) Genes transcribed in tissue culture cell nuclei.

RNA^{wt} ──────────AAA   RNA^{mt} ──────...──────────AAA

3) Labelled primer hybridised to mRNAs.

```
         _____AAA              ____..._____AAA
         ---*                         ---*
```

4) Primer extended by reverse transcriptase.

```
         _____AAA              _____..._____AAA
         ---------*                   ---------...---*
```

5) Precipitate and run on denaturing acrylamide gel.

Fig. 1. Distinguishing the different sized transcripts from wild-type and marked copies of the *X. laevis* β-globin gene by primer extension: schematic representation of the primer extension reaction.

latter techniques are usually more sensitive, the convenience of primer extension means that it is often the method of choice for RNA analysis.

2. Materials

As with all procedures involving RNA, extreme care must be taken to avoid RNase contamination and degradation of samples *(1)*. Work surfaces should be clean, and gloves should be worn at all times. Stock solutions and glassware should be treated with the RNase inhibitor diethylpyrocarbonate (DEPC) (0.1% for 30 min or longer at 37°C). DEPC cannot be used to treat solutions containing Tris, since it is unstable in the presence of this buffer. Tris buffers should be made with

Primer Extension Assay

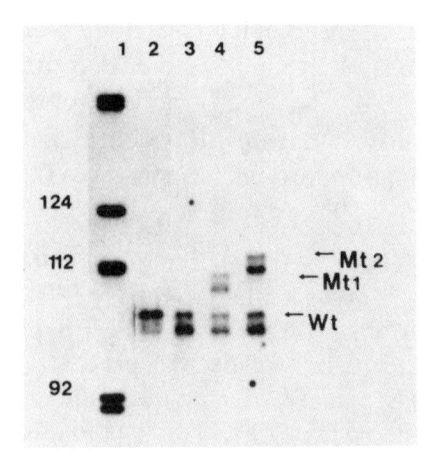

Fig. 2. An example of a primer extension reaction: total cytoplasmic RNA from cultured cells containing *Xenopus* β-globin gene constructs, analyzed using a single 5'-labeled oligo primer. Lane 3, wild-type, β-globin gene introduced alone; Lanes 4 and 5, wild-type gene plus genes marked by insertion of different-sized oligonucleotides into the first exon. Samples are run alongside a labeled DNA size marker (Lane 1) and total cytoplasmic RNA (50 ng) from *X. laevis* erythroblasts (Lane 2) to show the cap site used in vivo.

DEPC-treated autoclaved water and then reautoclaved. It is important to remove all traces of DEPC (by autoclaving) prior to use to prevent carboxymethylation of RNA and proteins. Sterile disposable plasticware can be treated in the same way, but is usually sufficiently RNase-free that pretreatment is not required. Separate stocks of plasticware and reagents should be maintained solely for RNA work.

Glass capillaries for the hybridization reaction should be siliconized (optional for plasticware). Items for siliconization are rinsed in a 5% solution of dichlorodimethyl silane in chloroform in a fume hood (this solution is toxic and highly volatile). Siliconized items should be rinsed several times with DEPC-treated water and baked (100°C, 2 h) before being DEPC-treated themselves.

The most convenient source of single-strand primer is a chemically synthesized short oligonucleotide (*see* Note 1). This should be 18 nucleotides or longer, and should ideally be located within 100 bases of the mRNA cap site. Care should be taken to ensure that the chosen oligomer does not contain repeats capable of inter- or intraprobe hybridization, since this will reduce the efficiency of annealing to mRNA.

Usually, the primer is 5' end-labeled *(2–4)* (*see* Note 2), particularly if the sequence of the extension products is to be determined by the method of Maxam and Gilbert *(3,5)*. The specific activity of the primer should be determined by Cerenkov counting after removal of unincorporated nucleotide (by centrifugation through a Sephadex G25 column). Primers should have a SA of >5 × 10^5 dpm/pmol.

1. 10X Kinase buffer: 500 m*M* Tris-HCl, pH 7.6, 100 m*M* $MgCl_2$, 50 m*M* dithiothreitol (DTT), 1 m*M* spermidine, 1 m*M* EDTA.
2. 5X Hybridization buffer: 2*M* NaCl, 50 m*M* PIPES, pH 6.4.
3. 1X Extension buffer: 5 µL of 1*M* Tris-HCl, pH 8.3, 5 µL of 200 m*M* DTT, 5 µL of 120 m*M* $MgCl_2$, 2.5 µL of 1 mg/mL actinomycin D, 5 µL each of 10 m*M* dATP, dCTP, dGTP, dTTP, 1 µL (approx 40 U) of RNasin, and DEPC water to 89 µL.
4. Formamide loading dye: 80% deionized formamide, 45 m*M* Tris-HCl borate, 45 m*M* boric acid, 1.25 m*M* EDTA, 0.02% xylene cyanole.
5. G25 Sephadex: Suspend Sephadex in approx 50 vol of TE (10 m*M* Tris-HCl, pH 7.5; 1 m*M* EDTA). Autoclave for 15 min (Sephadex will swell) and allow to cool to room temperature. Store at 4°C in a capped bottle.
6. Carrier (yeast) tRNA: Make up to 25 mg/mL with DEPC H_2O. Use the maker's bottle to avoid RNase contamination during weighing. Store in sterile plastic tubes at –20°C.
7. 3*M* sodium acetate (NaOAc), brought to pH 5.4 with glacial acetic acid, DEP-treated, and autoclaved for 15 min.
8. 70% EtOH/DEPC H_2O: 70 mL absolute ethanol, 30 mL DEPC-treated water.

The NaCl, $MgCl_2$, EDTA, and PIPES should be DEPC-treated as described above. The remaining buffer components should be made up in DEPC-treated autoclaved water, but not treated this way themselves. Actinomycin D is light-sensitive and toxic, and should not be autoclaved. DTT is thermally unstable and cannot be autoclaved (stock 1*M* DTT should be made up in DEPC-treated 10 m*M* NaOAc, filter-sterilized, and aliquoted at –20°C). Stocks of dNTPs should be filter-sterilized.

These buffers can be aliquoted and stored at –20°C. If the RNase inhibitor RNasin is to be present, it should be added fresh to the extension buffer on the day it is to be used. Reverse transcriptase (stored in aliquots at –70°C) should, similarly, be added fresh to the extension mix immediately prior to use.

3. Methods

Protocols for the labeling, hybridization, and extension reactions are given below. The hybridization reaction is frequently carried out in sealed

Primer Extension Assay

capillaries as described; however, the protocol is greatly simplified by using the modification given in Note 3.

3.1. Kinasing Primer

1. In a plastic microfuge tube, mix 1–50 pmol of oligonucleotide 5' (unphosphorylated) ends, 2 µL of 10X kinase buffer, 70 µCi of aqueous ^{32}P-γ ATP (3000 Ci/mmol), 10–20 U of T4 polynucleotide kinase, and DEPC water to 20 µL.
2. Incubate at 37°C for 30 min.
3. Add 4 µL of 250 mM EDTA.
4. Remove unincorporated labeled nucleotide by centrifugation down a Sephadex G25 column (see Note 4).
5. Extract with phenol:chloroform:isoamyl alcohol (25:24:1) (v/v).
6. Add 10 µg of tRNA (carrier for the precipitation).
7. Add 1/10 vol of 3M NaOAc; ethanol-precipitate by adding 2 vol of 100% ethanol and pelleting in a microfuge for 15 min. Wash the pellet in 70% ethanol and dry under vacuum.
8. Resuspend (0.25–2.5 fmol/µL) in DEPC water.

3.2. Hybridization Reaction

1. In a plastic microcentrifuge tube (siliconization is not essential), mix 2 µL of (0.5–5 fmol) labeled primer (see Note 5), 2 µL of 5X hybridization buffer, 0.1–1 fmol of target mRNA (up to 20 µg total RNA), and DEPC water to 10 µL.
2. Take the mixture up in a siliconized glass capillary (see Note 3).
3. Seal both ends of the capillary in a Bunsen flame and label the sample with indelible marker (on water-proof tape if necessary).
4. Heat the sample to 70°C by submerging in a water bath for 3 min (see Note 6).
5. Hybridize for 3–6 h, submerged in a water bath at the optimum temperature (see Notes 7–11).

3.3. Extension Reaction

1. Dry the capillaries and remove the ends with a glass cutter.
2. Remove both ends of the capillary, connect to a pipet, and expel the hybridization mix into 89 µL of extension buffer in a microfuge tube (including 1 µL [approx 40 U] of RNasin if desired; see Note 12).
3. Add 1 µL (approx 20 U) of AMV reverse transcriptase per sample.
4. Incubate the extension mix at 42°C for 1 h (see Note 13).
5. Add 11 µL of 3M NaOAc, ethanol-precipitate total nucleic acid as in step 7, Section 3.1., wash with 70% ethanol, and vacuum dry.
6. Take the pellet up in 5–10 µL of formamide loading dye.
7. Heat-denature at 90°C for 3 min; chill on ice.

8. Electrophorese on an appropriate percentage of denaturing polyacrylamide gel, along with suitable size markers or DNA sequencing ladders.

The labeled extension products are run on a standard denaturing urea–acrylamide gel of suitable percentage (dependent on the distance of the primer from the 5' end of the message) to enable good resolution of cap site length transcripts. This is particularly important when the site of transcription initiation is to be determined, but is also important for quantification experiments if multiple start sites exist in the gene of interest (*see* Notes 14–16).

4. Notes

1. A single-stranded (5' unphosphorylated) oligonucleotide is the most convenient source of primer. Alternatively, the primer can be generated by restriction enzyme digestion of the cloned gene under study. A pair of infrequently cut restriction sites with cohesive ends of different lengths, separated by approx 20–100 bp, are cleaved so as to produce coding and noncoding strands of different lengths. A radiolabeling step between the first and second cleavages yields a uniquely end-labeled fragment that can be separated from the noncoding strand by denaturing acrylamide gel electrophoresis *(3)* and autoradiography *(6)*. The greater the difference in size between the two strands, and the smaller the restriction fragment, the better the probe separation. It is possible to use a uniquely end-labeled double-stranded primer, since RNA–DNA hybrids are more stable than the corresponding DNA–DNA hybrids (by approx 5°C). However, the use of a double- stranded primer requires precise determination of the hybridization temperature and reduces the efficiency of the technique.
2. Ideally, the primer should be 5' end-labeled. The 5' terminal phosphate groups of restriction fragments must be removed by treatment with alkaline phosphatase *(4)* prior to kinasing; oligomers should be obtained without a terminal phosphate to preclude this step. Such 5' end-labeling is particularly important if the sequence of the extension products is to be determined by the method of Maxam and Gilbert. However, if the objective is merely to quantitate the amount of a particular message, then a 3' labeled primer can be used *(4,7)*. Practically, this may be simpler in some instances in which restriction-enzyme-generated probes are to be used (for example, labeling may be important in generating strands of different sizes). The sensitivity of the primer extension analysis can be improved by using a uniformly labeled (M13 vector-generated) probe *(4,7)*.
3. The hybridization reaction is frequently performed in siliconized capillaries but the protocol can be greatly simplified by carrying out this step directly in plastic microfuge tubes. It is important to completely submerge

the microfuge tubes in the water bath to prevent evaporation, since this will alter the salt concentration of the reaction and affect the hybridization. Care must be taken to ensure the tubes are adequately sealed; screw-capped microfuge tubes are best for this purpose. The extension reaction can then be carried out in the same tubes.

4. Prepare G25 Sephadex columns as follows: Plug the nipple of a 2-mL sterile plastic syringe with polymer wool and fill the syringe with a concentrated slurry of Sephadex. Suspend the syringe over a sterile plastic centrifuge tube and spin at 1500 rpm in a swing-out rotor for 5 min. Discard the centrifuge tube and buffer contents, and transfer the syringe to a fresh tube. Load the labeled primer on to the top of the column, rinsing out the tube with 50 µL of TE to ensure transfer of all the label to the column. Spin at 1500 rpm for 5 min. Labeled primer will be found in the eluate, unincorporated label will be located at the top of the column. A successfully labeled probe should register full-scale deflection on a Geiger counter. If this does not happen, rewash the column with a further 50–100 µL of TE. Remove 1 µL of the probe and determine the SA by Cerenkov counting. Probes should not be used with SA <2–5 × 10^6 dpm/pmol.

5. It may be more convenient to mix the labeled primer with the RNA that is to be analyzed while preparing the RNA. The primer and the RNA can then be coprecipitated, after which the pellet can be taken up in DEPC-treated water and the remaining components of the hybridization reaction added directly to the resuspended pellet.

6. The primer and RNA for analysis are mixed with hybridization buffer, briefly heated to 70°C to remove any RNA (or primer) secondary structure, and then allowed to anneal.

7. The hybridization is carried out with the primer in approx 10-fold molar excess over the target complementary RNA; too great an excess (particularly at suboptimal hybridization stringency) can result in nonspecific priming.

8. The optimum temperature for the hybridization will depend on the length of primer and its base composition. Formulae for the estimation of RNA–DNA hybrid melting temperatures *(8,9)* are not accurate for short DNA primers. For most probes, the value is in the range 45–65°C in the buffer given, but pilot experiments should be carried out over this range of temperatures to determine the optimum hybridization temperature for the specific primer–mRNA combination being used.

9. A simple way to monitor the hybridization efficiency is to perform a modified S1 nuclease mapping reaction on the RNA–primer hybrids (*see* Chapter 20). Briefly, the oligomer and RNA under analysis are hybridized (3–6 h) at a range of temperatures, the hybrids digested with S1 nuclease, and the amount of protected (hybridized) oligomer determined by denaturing acrylamide electrophoresis. The temperature at which the maximum

amount of S1 nuclease resistant labeled primer occurs should be used for subsequent hybridization reactions.

10. It may also be necessary to optimize the amount of target mRNA present in a typical reaction by carrying out hybridizations with varying amounts of the RNA preparation. Total cytoplasmic RNA can be used in the reaction, although greater sensitivity (and cleaner results) may be achieved using poly A^+ RNA.

11. The hybridization is carried out for 3–6 h. RNA is susceptible to thermal degradation at elevated temperatures (although this effect is reduced at the acid pH used here), so primers with high optimum hybridization temperatures should be annealed for as short a time as possible.

12. The ribonuclease inhibitor RNasin can be included in the extension reaction (1 µL, approx 40 U) but is less effective in the hybridization reaction. This is because the 70°C incubation to remove secondary structure and the high temperatures required in the annealing reaction are likely to cause denaturation of the RNasin protein. Additionally, the activity of RNasin is critically dependent on the presence of a minimum amount of 1 mM dithiotreitol, which is also unstable at elevated temperatures. Ribonucleoside–vanadyl complex RNase inhibitors can reduce the efficiency of extension under certain conditions (>2 mM inhibitor in the presence of low [1 mM] concentrations of dNTPs).

13. The elongation reaction is carried out at 42°C to reduce the amount of mRNA secondary structure, which can cause premature termination of the reverse transcriptase.

14. The further the primer is located from the cap site, the greater the likelihood of premature termination by the reverse transcriptase at sites of RNA secondary structure or RNA cleavage (e.g., by RNase). The presence of the resulting discrete "drop off" bands will cause a reduced cap-site signal. Changing the position of the primer to a site 5' to strong secondary structure extension barriers will improve the yield of cap-site product.

15. The extension reaction often generates more than one band in the vicinity of the cap-site (see Fig. 2). These may represent genuine multiple start sites of transcription. Additionally, it is thought that methylation of the mRNA at the cap-site, or adjacent nucleotide, causes premature termination of the reverse transcriptase enzyme in a portion of the molecules. Such effects should be taken into account when determining the precise point of transcription initiation. The ratio of the cap-site bands can vary between different transfected cell types, but is constant for a given source of mRNA.

16. Additional bands may be produced as a result of fold-back cDNA synthesis by reverse transcriptase, although the incorporation of actinomycin D in the extension buffer should reduce this activity. Sodium pyro-

phosphate (80 m*M*) appears to be even more effective in this function, but may be inhibitory to reverse transcriptase from some sources. Spurious bands occasionally may arise by cross hybridization of the primer to either endogenous tissue-culture-cell RNA or carrier tRNA. Such bands can be identified by performing a control primer extension reaction on RNA isolated from cells known not to express the gene.

Further Reading

Sambrook, J., Fritsch, E. F., and Maniatis, T. (1989) *Molecular Cloning: A Laboratory Manual* Cold Spring Harbor Laboratory, Cold Spring Harbor, NY.
Williams, J. G. and Mason, P. J. (1985) Hybridization and analysis of RNA, in *Nucleic Acid Hybridization—A Practical Approach* (Hames, B. D. and Higgins, S. J., eds.), IRL, Oxford, Washington, DC, pp. 139–160.
Krug, M. S. and Berger, S. L. (1987) First strand cDNA synthesis primed with oligo dT, in *Methods in Enzymology,* vol. 152 (Berger, S. L. and Kimmel, A. R., eds.), Academic, London, New York, pp. 316–325.
Calzone, F. J., Britten, R., and Davidson, E. H. (1987) Mapping gene transcripts by nuclease protection assays and cDNA primer extension, in *Methods in Enzymology,* vol. 152 (Berger, S. L. and Kimmel, A. R., eds.), Academic, London, New York, pp. 611–632.

References

1. Blumberg, D. (1987) Creating a ribonuclease-free environment, in *Methods in Enzymology,* vol. 152 (Berger, S. L. and Kimmel, A. R., eds.), Academic, London, New York, pp. 20–24.
2. Richardson, C. C. (1965) Phosphorylation of nucleic acid by an enzyme from T4 bacteriophage infected *E. coli. Proc. Natl. Acad. Sci. USA* **54,** 158–165.
3. Maxam, A. M. and Gilbert, W. (1980) Sequencing end-labeled DNA with base specific chemical cleavages, in *Methods in Enzymology,* vol. 65 (Grossman, L. and Moldave, K., eds.), Academic, London, New York, pp. 499–560.
4. Arrand, J. E. (1985) Preparation of nucleic acid probes, in *Nucleic Acid Hybridization—A Practical Approach* (Hames, B. D. and Higgins, S. J.), IRL, Oxford, Washington, DC, pp. 139–160.
5. Maxam, A. M. and Gilbert, W. (1977) A new method for sequencing DNA. *Proc. Natl. Acad. Sci. USA* **74,** 560–564.
6. Boffey, S. (1987) Autoradiography and fluorography, in *Techniques in Molecular Biology,* vol. 2 (Walker, J. M. and Gaastra, W., eds.), Croom and Helm, London, Sydney, pp. 288–295.
7. Tsang, A. S., Mahbubani, H., and Williams, J. G. (1982) Cell-type-specific actin mRNA populations in dictyostelium discoideum. *Cell* **31,** 375–382.
8. Schildkraut, C. and Lifson, S. (1965) Dependence of the melting temperature of DNA upon salt concentration. *Biopolymers* **3,** 195–208.
9. Britten, R. J. and Davidson, E. H. (1985) Hybridization strategy, in *Nucleic Acid Hybridization—A Practical Approach* (Hames, B. D. and Higgins, S. J., eds.), IRL, Oxford, Washington, DC, pp. 3–15.

CHAPTER 20

S1 Mapping Using Single-Stranded DNA Probes

Stéphane Viville and Roberto Mantovani

1. Introduction

The S1 nuclease is an endonuclease isolated from *Aspergillus oryzae* that digests single- but not double-stranded nucleic acid. In addition, it digests partially mismatched double-stranded molecules with such sensitivity that even a single base-pair mismatch can be cut and hence detected. In practice, a probe of end-labeled double-stranded DNA is denatured and hybridized to complementary RNA molecules. S1 is used to recognize and cut mismatches or unannealed regions and the products are analyzed on a denaturing acrylamide gel. A number of different uses of the S1 nuclease have been developed to analyze mRNA taking advantage of this property *(1,2)*. Both qualitative and quantitative information can be obtained in the same experiment *(3)*.

Qualitatively, it is possible to characterize the start site(s) of mRNA, to establish the exact intron/exon map of a given gene (*see* ref. *4* and Fig. 1), and to map the polyadenylation sites. Quantitatively, it can be used to study gene regulation both in vivo and in vitro, for example, in the study of the Eα gene promoter *(5)*.

In this chapter we describe a S1 mapping method based on the preparation and use of a single-stranded DNA probe (*see* Fig. 2). This offers many advantages:

1. Oligonucleotides allow the exact choice of fragment for a probe.
2. Oligonucleotide labeling is easy and efficient, resulting in a high specific activity probe.

From: *Methods in Molecular Biology, Vol. 58: Basic DNA and RNA Protocols*
Edited by: A. Harwood Humana Press Inc., Totowa, NJ

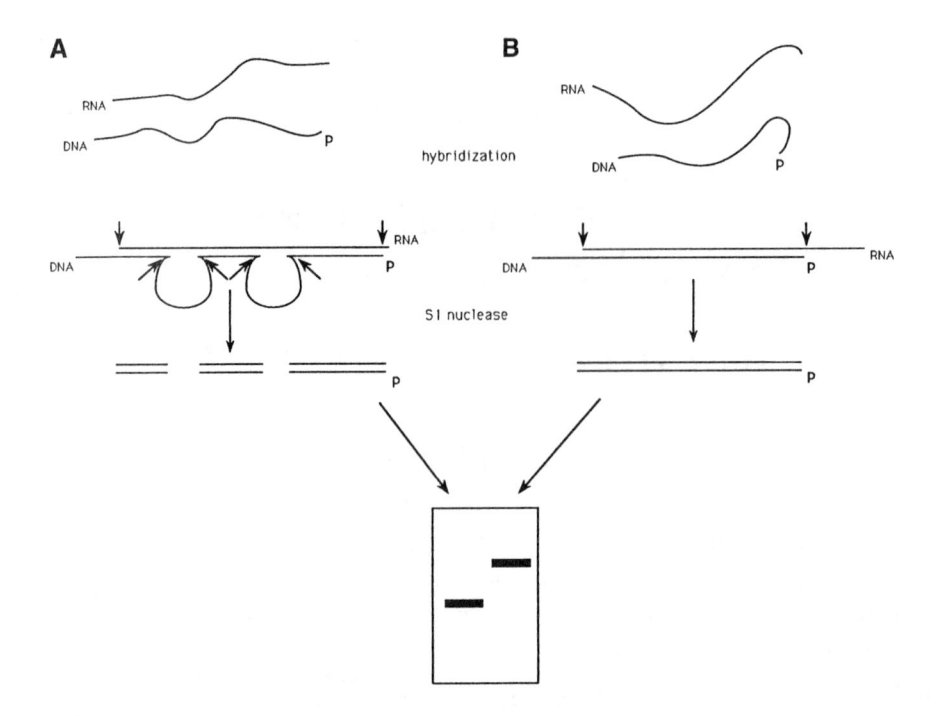

Fig. 1. Illustration of the use of the S1 nuclease. (**A**) Mapping the intron/exon organization of a given gene. The labeled DNA hybridizes to the corresponding mRNA; the introns are cut by the S1 nuclease. (**B**) Mapping of the 5' end of a mRNA. The labeled DNA is used as a probe to map the 5' end of a messenger RNA. In both cases the labeled fragments are visualized on a denaturing acrylamide gel.

3. The probe can be prepared from single-stranded DNA (e.g., M13, BlueScript) as well as a double-stranded template.
4. The single-stranded probe avoids problems often encountered setting up hybridization conditions of double-stranded probes *(6)*.
5. The probe is stable for 3–4 wk.

We illustrate the process with examples from the analysis of the Eα promoter.

2. Materials
2.1. Preparation of Single-Stranded DNA Probe from a Single-Stranded DNA Template

1. 10X Kinase buffer: 400 mM Tris-HCl, pH 7.8, 100 mM MgCl$_2$, 100 mM β-mercaptoethanol, 250 µg/mL bovine serum albumin. Store at –20°C.

Mapping of Single-Stranded DNA Probes 149

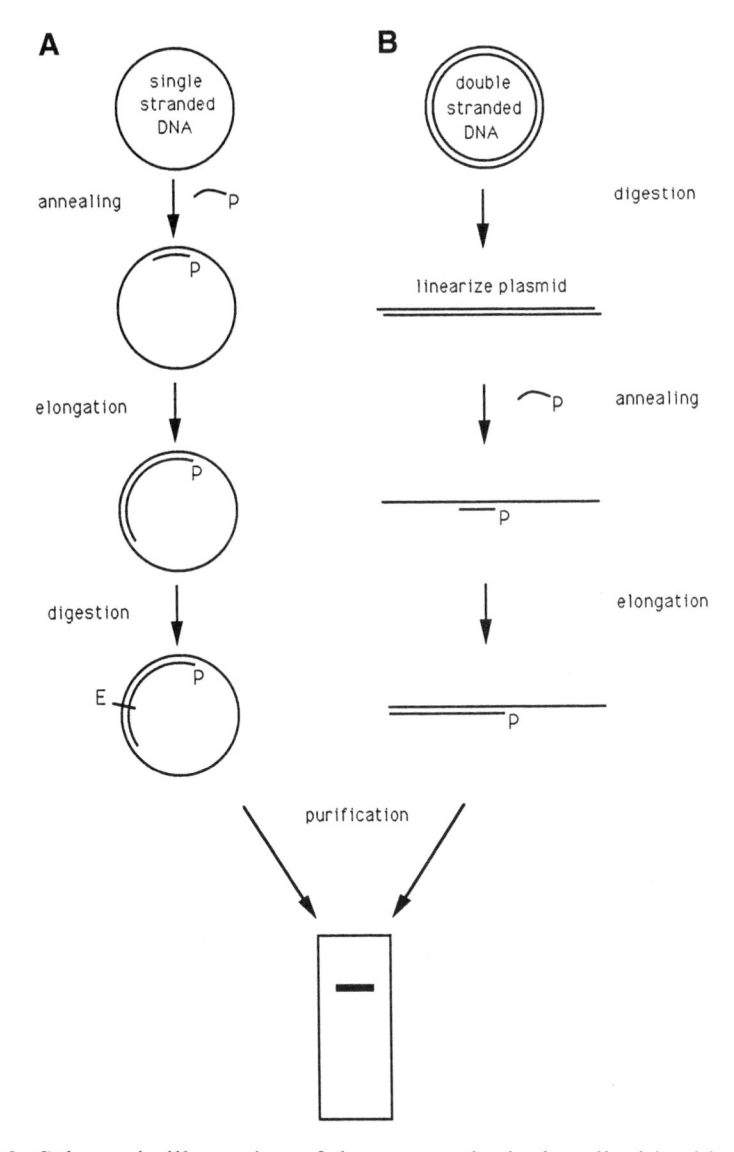

Fig. 2. Schematic illustration of the two methods described in this chapter for the preparation of a single-stranded DNA probe. (**A**) Using a single-stranded plasmid as a template. (**B**) Using a double-stranded DNA plasmid to synthesize the probe.

2. [^{32}P]-γ-ATP: specific activity >3000 Ci/mmol.
3. Suitable oligonucleotide (*see* Note 1). Prepare a 10 pmol/µL solution in distilled water. Store at –20°C.

4. PNK: Polynucleotide kinase at a stock concentration of 10 U/µL. Store at –20°C.
5. 10X Annealing buffer: 100 m*M* Tris-HCl, pH 7.5, 100 m*M* MgCl$_2$, 500 m*M* NaCl, 100 m*M* dithiothreitol.
6. DNA template: 1 mg/mL (*see* Note 2).
7. 10X dNTP mix: 5 m*M* dATP, 5 m*M* dCTP, 5 m*M* dGTP, and 5 m*M* dTTP.
8. Klenow: The large fragment of *E. coli* DNA Polymerase I at a stock concentration of 10 U/µL. Store at –20°C.
9. Suitable restriction enzyme (*see* Note 3).
10. 6% Polyacrylamide/8*M* urea solution in 0.5X TBE and formamide dye (*see* Chapter 48).
11. X-ray film: A film of suitable sensitivity, e.g., Kodak XAR.
12. Elution buffer: 50 m*M* Tris-HCl, pH 7.5, 0.5 m*M* EDTA.
13. 3*M* sodium acetate, pH 7.4.
14. tRNA: Stock solution 10 mg/mL. Store at –20°C.

2.2. Hybridization and S1 Analysis

15. 4X Hybridization buffer: 1.6*M* NaCl, 40 m*M* PIPES, pH 6.4.
16. Deionized formamide.
17. Paraffin oil.
18. S1 buffer: 300 m*M* NaCl, 30 m*M* sodium acetate, pH 4.5, 4.5 m*M* zinc acetate.
19. S1 enzyme: A stock concentration of 10 U/µL. Store at –20°C.
20. S1 stop buffer: 2.5*M* ammonium acetate, 50 m*M* EDTA.
21. Isopropanol.

3. Methods

3.1. Preparation of Single-Stranded DNA Probe from a Single-Stranded DNA Template

1. Label the oligonucleotide by incubation of 1 µL of cold oligonucleotide, 10 µL of [^{32}P]-γ-ATP, 2 µL of 10X kinase buffer, 6 µL of dH$_2$O, and 1 µL of PNK at 37°C for 45 min. Inactivate the PNK at 95°C for 2 min.
2. Add 2 µL of single-stranded DNA template, 4 µL of 10X annealing buffer, and 14 µL of H$_2$O and incubate for 10 min at 65°C, followed by 10 min at 55°C, then 10 min at 37°C.
3. Add 4 µL of dNTP mix and 1 µL of Klenow and leave at room temperature for 10 min. Inactivate the Klenow by placing the tube at 65°C for 15 min.
4. Correct the mix for the restriction enzyme digest by either adding a suitable amount of sodium chloride or by dilution. Add 20 U of restriction

Mapping of Single-Stranded DNA Probes

enzyme and incubate at 37°C for 60 min (*see* Note 3).
5. Precipitate the sample by adding 2 µL of tRNA stock, 0.1 vol of 3M sodium acetate, and 3 vol of ethanol, and place the sample in dry ice for 10 min. Centrifuge at 12,000g for 15 min in a microfuge.
6. Resuspend the pellet in 20 µL of formamide dye, heat at 95°C for 5 min, and load a 6% polyacrylamide/urea gel. Run the gel until the bromophenol blue reaches the bottom.
7. Locate the single-stranded DNA fragment by exposing the gel for 5 min to a X-ray film. Excise the corresponding band and elute it overnight in 500 µL of elution buffer.
8. Add 50 µL of sodium acetate, 20 µg of tRNA, and 1 mL of ethanol; place at –20°C for 2 h. Centrifuge for 20 min at 12,000g in a microfuge.
9. Count the pellet and resuspend it in formamide (10^5 cpm/µL). Store at –20°C. The probe is now ready for hybridization and S1 analysis.

3.2. Preparation of Single-Stranded DNA Probe from Double-Stranded DNA Template

Follow the same protocol described in Section 3.1., but for the following exceptions (*see* Fig. 2B).

1. Initially cut 20 µg of double-stranded plasmid with a restriction enzyme of choice. Precipitate and resuspend in H_2O at a concentration of 1 mg/mL.
2. The annealing step (Section 3.1., step 2) must be carried out extremely quickly: After heat inactivation of PNK, add 2–5 µg of restricted plasmid, 4 µL of 10X annealing buffer, and 14 µL of H_2O, and incubate for 5 min at 95°C. Immediately place the tube in ice-cold water and then proceed to step 3.
3. Omit the restriction enzyme digestion (step 4).

3.3. Hybridization and S1 Analysis

1. In a 0.5 mL Eppendorf tube put 10 µL of probe (10,000 cpm total), 5 µL of 4X hybridization buffer, and 5 µL of RNA (total poly A^+ or synthesized in vitro). Add 20 µL of paraffin oil to prevent evaporation and incubate 4–16 h at 37°C (*see* Note 4).
2. Add 200 µL of S1 buffer containing 100 U of S1 nuclease, mix well, and place at 37°C for 5–30 min (*see* Note 5). Stop the reaction by adding 50 µL of S1 stop buffer, 40 µg of tRNA, and 300 µL of isopropanol. Place in dry ice for 15 min and centrifuge at 12,000g for 15 min.
3. Wash the pellet with 80% ethanol and dry. Resuspend in 3 µL of formamide dye, heat for 5 min at 95°C, and load a 6% polyacrylamide/urea gel. Run until the bromophenol blue reaches the bottom. Expose the gel for 8–48 h (*see* Fig. 3).

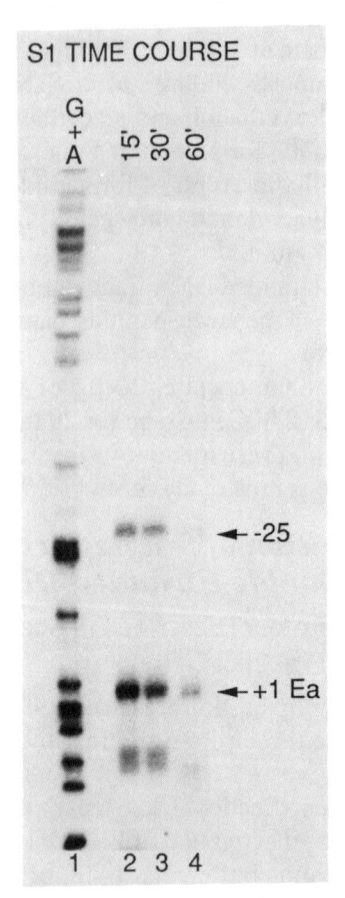

Fig. 3. Time-course of S1 analysis. In vitro synthesized RNA hybridized to Eα probe is incubated at 37°C for the indicated time with 100 U of S1 nuclease.

4. Notes

1. The oligonucleotide should be 20–25 nucleotides long and should be complementary to the DNA strand to be analyzed.
2. Both the single-stranded and double-stranded DNAs are prepared according to standard protocols. The advantage of using a single-stranded DNA template is basically a higher yield of the probe: 1–1.5 × 10^6 cpm vs the 4–6 × 10^5 cpm expected from a double-stranded plasmid template.
3. The restriction enzyme should be used to cut 200–600 nucleotides 3' of the position to which the oligonucleotide hybridizes to increase fragment recovery. Longer fragments are harder to recover from acrylamide gels. It

does not matter if the restriction enzyme cuts the template DNA more than once as long as it is outside the probe sequence.
4. Hybridization time varies with the type of RNA to be analyzed: Using total cytoplasmic or poly A$^+$ RNA, the samples should be left at 37°C at least 12 h; when hybridizing RNA generated from in vitro transcription, 2–4 h are usually sufficient.
5. An S1 time-course should be performed as shown in Fig. 3 for in vitro synthesized RNA. This will establish the conditions for which all single-stranded nucleic acids, including the free probe, are completely digested. The optimal cutting time for total cytoplasmic and poly A$^+$ RNA is usually longer (30 min) than for in vitro synthesized RNA (15 min).

References

1. Berk, A. J. and Sharp, P. A. (1977) Sizing and mapping of early adenovirus mRNAs by gel electrophoresis of S1 endonuclease digested hybrids. *Cell* **12,** 721–732.
2. Favaloro, J., Treisman, R., and Kamen, R. (1980) Transcription maps of polyoma virus-specific RNA: analysis by two-dimensional nuclease S1 gel mapping. *Meth. Enzymol.* **65,** 718–749.
3. Weaver, R. F. and Weissmann, C. (1979) Mapping of RNA by a modification of Berk-Sharp procedure: the 5' termini of 15 S β-globin mRNA precursor and mature 10 S b-globin mRNA have identical map coordinates. *Nucleic Acids Res.* **7,** 1175–1193.
4. Lopata, M. A., Sollner-Webb, B., and Cleveland, D. W. (1985) Surprising S1-resistant trimolecular hybrids: potential complication in interpretation of S1 mapping analysis. *Mol. Cell. Biol.* **5,** 2842–2846.
5. Viville, S., Jongeneel, V., Koch, W., Mantovani, R., Benoist, C., and Mathis, D. (1991) The Eα promoter: a linker-scanning analysis. *J. Immunol.* **146,** 3211–3217.
6. Dean, M. (1987) Determining the hybridization temperature for S1 nuclease mapping. *Nucleic Acids Res.* **15,** 6754.

CHAPTER 21

Nonradioactive *In Situ* Hybridization for Cells and Tissues

Ian Durrant

1. Introduction

In situ hybridization (ISH) is a widely used technique that has great power in many applications, including diagnosis of viral infections *(1)*, chromosome analysis *(2)*, and mRNA analysis *(3)*. Traditionally, researchers have used radioactive labels to prepare probes for ISH *(4)*. Such probes can be labeled DNA, RNA, or oligonucleotides. RNA and oligonucleotide probes are most often used because they have several features of advantage for ISH (*see* Table 1); in particular, they are single stranded, offering high sensitivity (RNA) and high-specificity (oligonucleotide). Recently, nonradioactive probes have become more popular, and the same features hold true for these. In this chapter, the ISH process is illustrated by use of a system for the detection of mRNA on tissue sections with RNA probes, but it is equally applicable to work on cell lines and for DNA detection in such systems. In addition, the detection procedures described with the *in situ* process are applicable to RNA, DNA, and oligonucleotide probes. The analysis of chromosomes and nuclei is a distinct procedure that utilizes nonradioactive DNA probes and fluorescent detection *(5)*.

1.1. Probe Labeling

Nonradioactive labels for nucleic acid probes have existed for some time, notably biotin *(6)* and digoxigenin *(7)*, in which the label is attached to the base in dUTP, or UTP, through a spacer arm. However, relatively

From: *Methods in Molecular Biology, Vol. 58: Basic DNA and RNA Protocols*
Edited by: A. Harwood Humana Press Inc., Totowa, NJ

Table 1
Features of Probe Types for *In Situ* Hybridization

Probe type	Positive	Negative
RNA probes	RNA hybrids have a higher stability	Subcloning required
	Single-stranded	RNase degradation possible
	No vector sequences present	Critical hybridzation conditions
	Controls easy to produce	
	Low background	
	High sensitivity	
Oligonucleotides	Single-stranded	Low level of labeling
	Highly specific	Sequence optimization required
	Easy access to target	Cocktails of probes may be needed
	Stable	
	Easy to produce in large amounts	
DNA probes	No subcloning required	Probe denaturation required
	Easy and reliable labeling methods	Probe reanneals in hybridization
	Less critical hybridization conditions	Difficult to remove vector sequences

high backgrounds may occur, particularly with biotin owing to the high level of endogenous biotin and biotin-binding sites present in many biological systems. Recently, fluorescein has been developed as an alternative label *(8)*. This compound is not found in biological systems, incorporates well into nucleic acid probes, and gives low backgrounds in ISH owing to the availability of high-specificity, high-affinity antibodies.

All types of probe can be made, as follows.

1. DNA probes: DNA probes are generated by the incorporation of fluorescein-dUTP by random prime labeling. This reaction is detailed in Chapter 8. Fluorescein-dUTP can also be incorporated by nick translation *(9)* and in the PCR reaction *(10)*.
2. Oligonucleotide probes: Oligonucleotide probes are labeled at the 3'-end by the addition of a short tail of fluorescein-dUTP. The details of this reaction are given in Chapter 11 and elsewhere *(11)*. The reaction yields a short

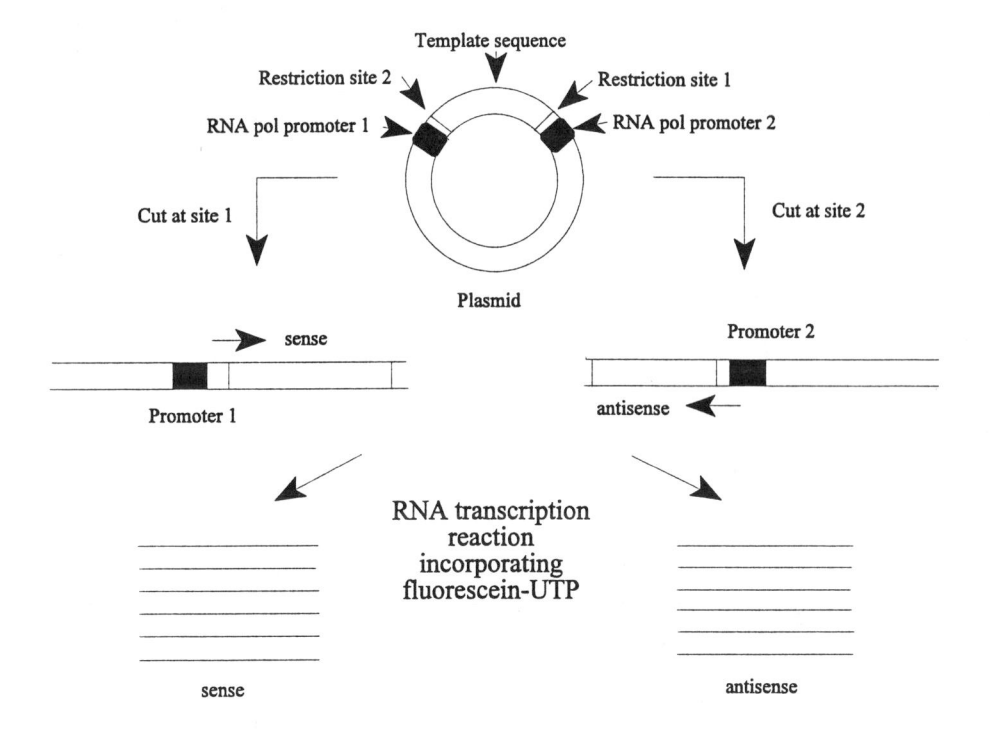

Fig. 1. Diagram of the RNA transcript probe labeling reaction using a dual promoter vector to produce sense (negative) and antisense (positive) probes.

tail that gives good sensitivity without compromising the stringency of the hybridization process.

3. RNA probes: RNA probes are generated by use of RNA polymerase enzymes and plasmid (or other appropriate) constructs containing a promoter element specific for a particular RNA polymerase. The plasmid acts as a template and is supplied in a double-stranded form; the polymerase produces single-stranded RNA transcribed from only one of the strands (see Fig. 1). By using a plasmid carrying two promoters on either side of the cloned template sequence, an antisense probe is made that will hybridize to mRNA and a control, sense probe can also be produced that should show no hybridization signal, thus confirming the specificity of the hybridization process.

There are three commonly used RNA polymerases that have been isolated from bacteriophage T3, T7, and SP6 *(12)*. Each has a distinct promoter sequence, so that the transcription reaction is specific for each enzyme. The labeling reaction is simple to perform and reliable, pro-

Table 2
Yield of Fluorescein-Labeled RNA with Different Polymerases

Vector/transcript	Size, kb	Yield, µg with SP6	T7	T3
pAM19/–	3.5	4.5		
pAM19/lysozyme	0.27	4.5	2.5	
pAM18/Nras	0.5	3.7	1.2	
pSP65/Nras	1.5	3.0		
pAM18/Actin	1.8		1.8	
pSP64/POMC	1.4	4.0		
pSP65/POMC	1.4	4.0		
pGEM/–	4.0	4.5	2.7	6.2
Average yield		3.9	2.1	6.2

vided that precautions are taken to reduce exposure of the reaction mixture to exogenous RNase to a minimum. The use of sterile pipet tips, reaction tubes, and solutions, and the wearing of protective gloves are recommended.

The labeling reaction was originally developed for the introduction of radioactive labels and has been adapted for the production of nonradioactive probes. The modified reactions are efficient processes, leading to high levels of incorporation of fluorescein-labeled nucleotide and relatively large amounts of synthesized probe (*see* Table 2).

1.2. In Situ *Hybridization*

The process of ISH is not technically very different from that performed on membranes; the kinetics and theory of hybridization that have been presented for membranes *(13)* can generally be applied to ISH. The important features of ISH that distinguish it from membrane studies are the processes of sample preparation and pretreatments that are required before hybridization can begin *(14,15)*. Each system needs to be optimized, since these treatments are dependent on the fixation, fixative, tissue type, probe type, and possibly even the label (radioactive or nonradioactive) that is used. Time spent getting this process working well is essential, and it is probably even more important for nonradioactive methods. However, for ISH, it must be noted that the sensitivity achieved with nonradioactive probes may not match that of radioactive probes, although greater resolution is generally obtained with nonradioactive systems. In addition, the nonradioactive ISH system as a whole may be less robust in that it is likely to require precise optimization and accurate operation at all stages. Sensitivity of ISH depends on a number of factors, including the fixation and pretreatments, the labeling and detection, and also the hybridization. These points are addressed with these systems by operator optimization of the prehybridization events, efficient labeling, high-affinity and specificity antibody detection, and by the use of a carefully formulated hybridization buffer.

2. Materials

Labeling of DNA probes and oligonucleotide probes is described elsewhere. The labeling process set out below is for the generation of fluorescein-labeled RNA probes. All the rest of the procedures regarding the treatment of samples, controls, hybridization, and detection are applicable to all three types of probe. Reagents for labeling DNA, RNA, or oligonucleotide probes with a fluorescein-labeled nucleotide for carrying out the ISH process and subsequent detection are available, as complete systems, from Amersham International (Amersham, UK) as the DNA, RNA, or Oligo color kit (RPN3200, 3300 or 3400, respectively) *(16,17)*.

2.1. Probe Labeling

1. Transcription buffer: 200 m*M* Tris-HCl, pH 7.5, 30 m*M* MgCl$_2$, 10 m*M* spermidine, 0.05% (w/v) BSA.
2. Human placental ribonuclease inhibitor (HPRI): 20 U/µL.
3. 0.2*M* DTT: freshly prepared solution, sterilized by filtration.
4. Linearized DNA template: typically 0.5–1 µg/µL.
5. SP6, T7, or T3 RNA polymerase, supplied, for example, by Amersham International (Amersham, UK).
6. Nucleotide solution: 1.25 m*M* ATP, 1.25 m*M* GTP, 1.25 m*M* CTP, 0.312 m*M* UTP, and 0.312 m*M* fluorescein-11-UTP (available from Amersham) in sterile water (*see* Note 1).

2.2. Postlabeling Purification and Processing

1. RNase-free DNase I: (Amersham E2210), freshly diluted to 10 U/80 µL in sterile water.
2. 0.4*M* NaHCO$_3$: sterilized by autoclaving.
3. 0.6*M* Na$_2$CO$_3$: sterilized by autoclaving.
4. Glacial acetic acid.
5. 3*M* sodium acetate, pH5.2: sterilized by autoclaving.
6. 10 mg/mL yeast tRNA: for example, Sigma (St. Louis, MO) R8759.

2.3. Hybridization and Stringency Washes

1. Hybridization buffer: Standard hybridization buffers (*see* Note 2) are usable, but a preformed, optimized buffer, available from Amersham, is recommended that consists of 4X SSC, 600 µg/mL herring testes DNA, 2X Denhardt's solution, and a proprietary rate enhancement compound. This buffer should be diluted 1:1 with deionized formamide before use. Once made, the working buffer can be stored at –20°C.
2. 20X SSC stock: 3M sodium chloride, 0.3M trisodium citrate, pH 7.0.
3. 10% (w/v) SDS.

2.4. Blocking, Antibody Incubations, and Washes

1. TBS: 100 mM Tris–HCl, pH7.5, 400 mM NaCl.
2. Block buffer: 0.5% (w/v) blocking agent (RPN3023, Amersham International) in TBS.
3. Antifluorescein-alkaline phosphatase conjugate: available from Amersham.

2.5. Detection

1. Detection buffer: 100 mM Tris–HCl, pH 9.5, 100 mM NaCl, 50 mM MgCl$_2$.
2. NBT solution: 50 mg/mL nitroblue tetrazolium in dimethyl formamide.
3. BCIP solution: 75 mg/mL bromo-chloro-indolyl phosphate in 70% (v/v) dimethyl formamide.

3. Methods

3.1. Probe Labeling

1. To a 1.5-mL microcentrifuge tube at room temperature (*see* Note 3), add the following reaction components in the order given: Water to ensure a final volume of 20 µL, 4 µL of transcription buffer, 1 µL of 0.2M DTT, 20 U of HPRI, 8 µL of nucleotide solution, 1 µg of linearized DNA template, and 25 U of RNA polymerase. Mix gently by pipeting up and down.
3. Incubate for a minimum of 2 h; for SP6, incubate at 40°C, and for T7 and T3, incubate at 37°C.
4. The resulting labeled RNA probes can be stored at –20°C (*see* Note 4).

3.2. Postlabeling Purification and Processing

1. Add 10 U of RNase-free DNase I to the RNA probe mixture (*see* Note 5), and incubate at 37°C for 10 min.
2. Add 20 µL of 0.4M NaHCO$_3$, 20 µL of 0.6M Na$_2$CO$_3$ and 60 µL of sterile water. Mix gently and incubate at 60°C to allow alkaline hydrolysis to occur for a time based on the transcript length and the probe size required (*see* Note 6).
3. Add 1.3 µL of glacial acetic acid, 20 µL of 3M sodium acetate, pH 5.2, 2 µL of 10 mg/mL yeast tRNA, and 500 µL of ethanol. Allow the RNA to precipitate at –20°C for at least 2 h.

4. Pellet the RNA at maximum speed in a microcentrifuge for 15 min. Remove the supernatant.
5. Rinse the pellet in 70% ethanol, kept at –20°C. Centrifuge for 5 min. Remove and discard the supernatant.
6. Redissolve the RNA probe in sterile water at a concentration of 10–20× that required in the hybridization experiment. Store the RNA probe at –20°C.

3.3. Pretreatment of Cells and Tissues

This stage is vital to the successful outcome of any ISH experiment (*see* Note 7), and each individual system will require its own optimized pretreatment protocol. The pretreatments are designed to balance the requirements for probe and antibody conjugate access to their target molecules with retention of the morphology and the target molecule within the tissue. In addition, the pretreatment schedule may also be used to establish some of the controls that are required for ISH experiments (*see* Note 8 and Fig. 2).

3.4. Hybridization and Stringency Washes

The control of stringency may be achieved in the hybridization step by manipulation of the temperature, salt concentration, and formamide concentration. However, it is more conveniently controlled by use of these parameters in the posthybridization stringency washes.

1. Briefly warm the supplied hybridization buffer at 37°C for 5 min. Dilute 1:1 with deionized formamide, and mix thoroughly.
2. Add the required amount of probe to the hybridization buffer, prewarmed at 55°C. A probe concentration of 300–600 ng/mL will be suitable for most applications (*see* Note 9).
3. Apply a suitable volume of hybridization buffer/probe mix over the section (*see* Note 10).
4. Place slides on a hot plate set at 95°C for 3–6 min depending on the target (*see* Note 11).
5. Hybridize in a humidified chamber at the desired temperature (typically 55°C) for the required length of time (*see* Note 12).
6. Wash the slides at the desired stringency. Typical washes would consist of 1X SSC, 0.1% (v/v) SDS for 2 × 5 min at room temperature followed by 0.2X SSC, 0.1% (v/v) SDS at 55°C for 2 × 10 min.
7. RNA probes can be further treated to reduce background by incubation in RNase A. This removes single-stranded, unhybridized material only. Rinse slides in 2X SSC for 2 min. Incubate slides in prewarmed 10 µg/mL RNase A in 2X SSC at 37°C for 20–30 min. Rinse slides in 2X SSC before proceeding to the next stage.

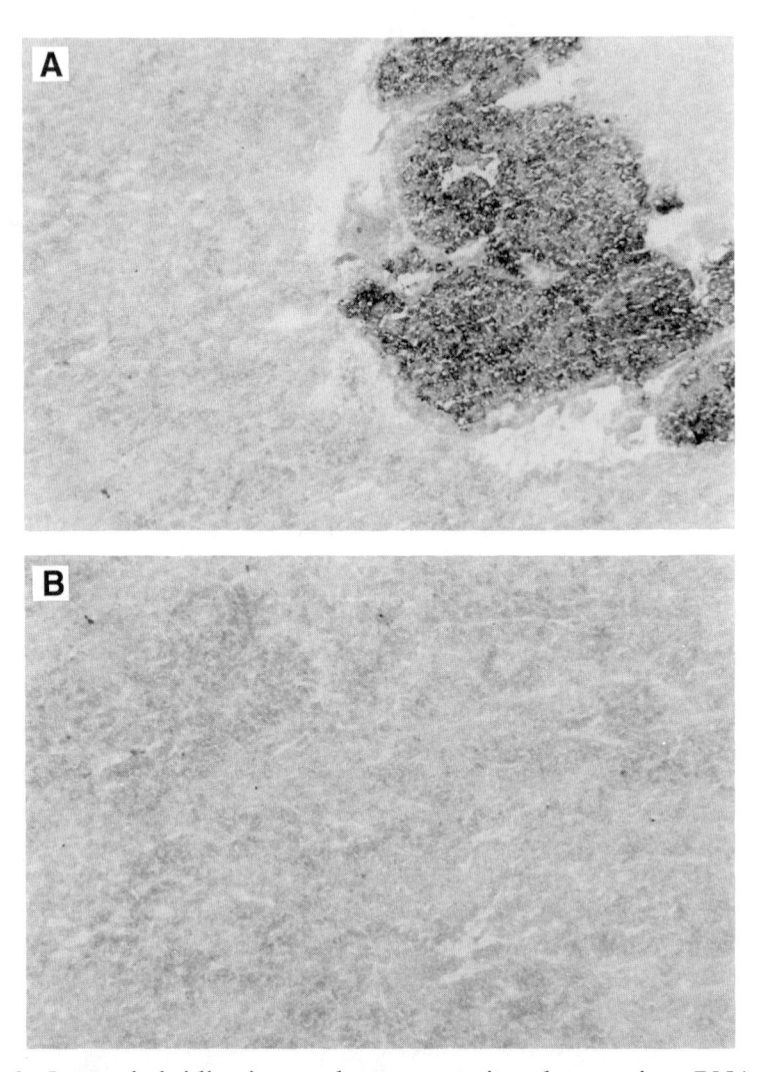

Fig. 2. *In situ* hybridization to detect pro-opiomelanocortin mRNA in rat pituitary. Fresh frozen rat pituitary sectioned on a cryostat onto VectaBond (Vector Labs) coated slides. Probe labeling, hybridization, and detection as detailed in the text. **(A)** Hybridized with antisense probe, **(B)** hybridized with sense probe.

3.5. Blocking, Antibody Incubations, and Washes

The following series of incubations should be performed with gentle shaking and at room temperature, unless otherwise stated. Note: The TBS used in these steps has a higher salt concentration than standard TBS.

1. Wash slides in TBS for 5 min.
2. Incubate the slides in block solution (0.5% [w/v] blocking agent in TBS) for 1 h.
3. Rinse slides in TBS for 1 min. Drain the buffer off the slides, and if necessary, dry the surface of each slide around the section (*see* Note 13).
4. Dilute the antibody 1:1000 in 0.5% (w/v) BSA Fraction V in TBS (*see* Note 14). Cover the section with the antibody solution, typically using 100 µL for each section. Incubate for 1 h without shaking.
5. Wash the slides in TBS for 3 × 5 min.

3.6. Detection

1. Wash the slide in detection buffer for 5 min. Drain the buffer off each slide and, if necessary, dry the surface of the slide around the section (*see* Note 15).
2. Add 45 µL of NBT stock and 35 µL of BCIP stock to 10 mL of detection buffer. Apply 500 µL of detection reagent to each section and leave to develop in the dark for 4–24 h (*see* Note 16).
3. Rinse the slides for 2 × 2 min in distilled water.
4. Counterstain, if required, and apply a mountant solution and a cover slip (*see* Note 17).
5. View in a suitable light microscope (*see* Note 18).

4. Notes

1. Alternative hapten-labeled nucleotides can be used, including biotin-UTP (Sigma B8280) and digoxigenin-UTP (Boehringer 1209 256). Final concentrations of CTP, GTP, and ATP in the reaction mix should be 0.5–1 mM. The combined final concentration of UTP and labeled UTP should be 0.25–1 mM. Depending on the label used, the yield of RNA probe may benefit from optimization of the nucleotide ratios within the reaction mixture.
2. The hybridization buffer supplied by Amersham is a 2X stock and requires 1:1 dilution with deionized formamide. Working solutions of standard buffer formulations would include an optimized combination of the following components *(18):* 2–4X SSC, 1–5X Denhardt's solution, nonspecific DNA, nonspecific RNA, 25–50% deionized formamide, and 5–10% dextran sulfate.
3. The labeling reaction is set up at room temperature to avoid precipitation of the template DNA by the spermidine in the reaction buffer.
4. Although it is generally difficult to monitor probe yields when using nonradioactive labels, it is possible, with fluorescein as the label, to design a simple and rapid assay making use of its fluorescent properties *(19)*. This assay is able to indicate that probe synthesis has occurred and gives an estimate of probe yield (*see* Chapter 9). The test should be performed on an aliquot withdrawn from the in vitro transcription reaction mixture. Fur-

ther processing of the transcript, notably alkaline digestion, may reduce the fluorescence, but this does not appear to affect the antigenicity of the label, which is important since the detection system is based on recognition by antifluorescein antibody.

5. When using fluorescein-labeled RNA probes, there is no absolute requirement to purify the probes further, provided that the transcript is already of a suitable size for ISH. The presence of unincorporated nucleotides does not appear to have any significant effect on the background observed with this system. However, it may still be advantageous to remove the template DNA by enzymatic digestion. If alkaline hydrolysis is not required, then the probe can be ethanol-precipitated after this process.

6. The presence of the fluorescein label does not interfere with alkaline hydrolysis when this is required. As a consequence of this process, the unincorporated nucleotides are also largely removed, but that is not the prime reason for performing this procedure. The time required for the limited alkaline hydrolysis of the primary transcript is derived from the following equation *(20)*:

$$t\ (min) = (L_o - L_f)/(k\ L_o\ L_f) \quad (1)$$

where L_o = primary transcript length (kb), L_f = average probe length required (kb), and the rate constant $k = 0.11$ cuts/kb/min. For ISH the optimal probe size is 200–400 bases.

7. Pretreatments cover all the processes required before hybridization can occur *(21)*, including slide coating, sample preparation, and fixation. A general scheme would include the following steps:
    a. Rapid fixation or rapid freezing of the biological sample.
    b. Sectioning onto coated slides.
    c. Sample pretreatments that include a combination of incubations in acids, detergents, and proteases. The most important step is a Proteinase K treatment; the concentration and incubation time of Proteinase K have to be optimized empirically, and different tissues may require different conditions. Cells may not require any treatment at all.
    d. Treatment to remove endogenous alkaline phosphatase activity.

8. It is essential to perform a number of negative controls alongside the main reactions used for ISH *(22)*. These can take various forms, for example:
    a. Incubation in hybridization buffer alone (no probe)—allows an assessment to be made of the level of signal resulting from the hybridization buffer components and nonspecific binding of the antibody.
    b. DNase or RNase pretreatment of the slide prior to hybridization—destroys the target DNA or RNA, respectively, and should show that the probe binds to the predicted target species.

c. Hybridization with a totally nonspecific probe—indicates that the stringency conditions are sufficient to remove nonspecific probes.
d. With RNA probes, labeled transcripts can be made that are either complementary or identical to an mRNA target. Hybridization with the identical strand probe should show that the stringency conditions are sufficient to remove a probe with the same GC content and same length as the complementary, specific probe.
e. If in doubt, probes should be validated on Northern blots.
9. The hybridization time and probe concentration are linked. Short hybridizations are possible with high probe concentrations (600–1000 ng/mL). Overnight incubations may be possible with a lower probe concentration (200–400 ng/mL).
10. The hybridization buffer volume is dependent on the size of the section and on whether cover slips are used. Small sections can be covered easily with 25–50 µL without using a cover slip; 10 µL may be sufficient with a cover slip.
11. If the target is DNA, then the slide must be denatured by placing onto a hot plate at 95°C for 5–6 min. If the target is RNA, this procedure, although not strictly required, is recommended. An incubation on the hot plate for 3–4 min is thought to remove the secondary structure of the target mRNA, thus increasing the efficiency of hybridization.
12. The chamber should be sealed and kept at a high humidity throughout the hybridization to avoid hybridization buffer evaporation. In this way, it may be possible to avoid the need to seal cover slips onto the slide or even to avoid the use of cover slips entirely. At present, most experiments are performed in sealed boxes placed in incubators, but recently, intelligent heating block technology has been used to design machines that can be used for this purpose (e.g., OmniSlide™, Hybaid). Typical experiments require an overnight hybridization. However, it is possible to perform 2–4 h hybridizations using high probe concentrations with high levels of target.
13. It is important to avoid the presence of excess buffer around the section since this may lead to overdilution of the antibody conjugate.
14. A dilution of 1:1000 of antibody conjugate should be suitable for most applications. The volume of diluted antibody added to each section may also need to be altered to account for the size of the sections and for whether cover slips are used. These alterations will need to be determined empirically.
15. It is important to avoid the presence of excess buffer around the section, since this may lead to overdilution of the substrates.
16. The volume of detection reagent applied to each section must be enough to flood the section. The length of time required to obtain a result will depend on the target level, but typically an overnight incubation is required for

maximum sensitivity. Other substrates for alkaline phosphatase can also be used, for example, Fast Red, although in our laboratory the NBT/BCIP system gives the best signal:noise.

17. A histochemical stain can be used before mounting in order to provide some contrast across the tissue section. However, care must be taken that the stain does not interfere with the colored reaction product already deposited on the slide. With NBT/BCIP substrates, it is possible to counterstain the section utilizing a light green counterstain, such as 0.5% (w/v) Methyl Green, or 0.5% (w/v) Fast Green, in water for 1 min. A number of standard mountant solutions are suitable for use with this system. Aqueous-based mountants include PBS/glycerol (CitiFluor) and glycerol/gelatin (Sigma). However, these cannot be used with counterstains, since the stain does fade and the NBT/BCIP color may fade with time. Nonaqueous mountants include DPX (BDH/Merck) and Histotech (Serotec). It is important to ensure that the sections are washed in graded ethanols, ending in 100% ethanol, and are dried before applying the nonaqueous mountant, since the presence of water in the section will lead to loss of color with time. The Histotech mountant does not require a cover slip, because it is baked onto the section.

18. A good-quality microscope is recommended with a transmitted white light source and 10× and 40× objective lenses. In our laboratory, the microscope is fitted with a 35-mm camera, and Kodak Ektar™ 25 color print film is used to record the results.

References

1. McQuaid, S., Allan, G. M., Taylor, M. J., Todd, D., Smith, J., Cosby, S. L., and Allen, I. V. (1991) Comparison of reporter molecules for viral in situ hybridization. *J. Virol. Methods* **31**, 1–10.
2. Price, C. M. (1993) Fluorescence in situ hybridization. *Blood Rev.* **7**, 127–134.
3. Mitchell, V., Gambiez, A., and Beauvillain, J. C. (1993) Fine structural localization of proenkephalin mRNAs in the hypothalamic magnocellular dorsal nucleus of the guinea pig: a comparison of radioisotopic and enzymatic in situ hybridization methods at the light- and electron-microscope levels. *Cell Tissue Res.* **274**, 219–228.
4. Brady, M. A. W. and Finlan, M. F. (1990) Radioactive labels: autoradiography and choice of emulsions for in situ hybridization, in *In Situ Hybridization: Principles and Practice* (Polak, J. M. and McGee, J. O.'D., eds.), Oxford University Press, Oxford, pp. 31–57.
5. Viegas-Pequignot, E. (1992) In situ hybridization to chromosomes with biotinylated probes, in *In Situ Hybridization: A Practical Approach* (Wilkinson, D. G., ed.), IRL, Oxford, pp. 137–158.
6. Langer, P. R., Waldrop, A. A., and Ward, D. C. (1981) Enzymatic synthesis of biotin-labeled polynucleotides: novel nucleic acid affinity probes. *Proc. Natl. Acad. Sci. USA* **78**, 6633–6637.

7. Holtke, H.-J. and Kessler, C. (1990) Non-radioactive labeling of RNA transcripts in vitro with the hapten digoxigenin (DIG): hybridization and ELISA-based detection. *Nucleic Acids Res.* **18**, 5843–5851.
8. Cunningham, M. W. (1991) Nucleic acid detection with light. *Life Sci.* **6**, 1–5.
9. Wiegant, J., Wiesmeijer, C. C., Hoovers, J. M. N., Schuuring, E., d'Azzo, A., Vrolijk, J., Tanke, H. J., and Raap, A. K. (1993) Multiple and sensitive fluorescence in situ hybridization with rhodamine-, fluorescein-, and coumarin-labeled DNAs. *Cytogenet. Cell Genet.*, **63**, 73–76.
10. Harvey, B. and Ellis, P. (1992) Nonradioactive labelling of DNA probes during PCR. *Life Sci.* **9**, 3,4.
11. Chadwick, P. M. E. and Durrant, I. (1994) Labeling of oligonucleotides with fluorescein, in *Protocols for Nucleic Acid Analysis by Nonradioactive Probes* (Isaac, P. G., ed.), Humana, Totowa, NJ, pp. 101–105.
12. Melton, D., Kreig, P. A., Rebagliati, M. R., Maniatis, T., Zinn, K., and Green, M. R. (1984) Efficient in vitro synthesis of biologically active RNA and RNA hybridization probes from plasmids containing a bacteriophage SP6 promoter. *Nucleic Acids Res.* **12**, 7035–7056.
13. Keller, G. H. and Manak, M. M. (1989) Molecular hybridization technology, in *DNA Probes*, Stockton, New York, pp. 1–27.
14. Wilkinson, D. G. (1992) The theory and practice of in situ hybridization, in *In Situ Hybridization: A Practical Approach* (Wilkinson, D. G., ed.), IRL, Oxford, pp. 1–14.
15. Moorman, A. M., de Boer, P. A. J., Vermeulen, J. L. M., and Lamers, W. H. (1993) Practical aspects of radio-isotopic in situ hybridization on RNA. *Histochem. J.* **25**, 251–266.
16. Durrant, I., Brunning, S., and Eccleston, L. (1993) DNA and RNA colour kits for nonradioactive in situ hybridization. *Life Sci.* **12**, 7,8.
17. Brunning, S. and Durrant, I. (1994) Oligo colour kit: a new probe labelling kit for nonradioactive in situ hybridization. *Life Sci.* **14**, 19.
18. Angerer, L. M. and Angerer, R. C. (1992) In situ hybridization to cellular RNA with radiolabelled RNA probes, in *In Situ Hybridization: A Practical Approach* (Wilkinson, D. G., ed.), IRL, Oxford, pp. 15–32.
19. Cunningham, M. and Harvey, B. (1992) A rapid method for monitoring nonradioactive labelling of DNA with fluorescein-11-dUTP. *Highlights*, **4**, 8,9.
20. Cox, K. H., DeLeon, D. V., Angerer, L. M., and Angerer, R. C. (1984) Detection of mRNAs in sea urchin embryos by in situ hybridization using asymmetric RNA probes. *Develop. Biol.* **101**, 485–502.
21. Brunning, S., Cresswell, L., Durrant, I., and Eccleston, L. (1993) General introduction to in situ hybridization techniques, in *A Guide to Radioactive and Nonradioactive In Situ Hybridization Systems*. Amersham International, Amersham, pp. 5–12.
22. Terenghi, G. and Fallon, R. A. (1990) Techniques and applications of in situ hybridization, in *Curent Topics in Pathology: Pathology of the Nucleus* (Underwood, J. C. E., ed.), Springer-Verlag, Berlin, pp. 290–337.

PART III
GENE CLONING

CHAPTER 22

In Vitro Packaging of DNA

Jeremy W. Dale and Peter J. Greenaway

1. Introduction

In the normal growth cycle of bacteriophage lambda, the proteins that ultimately form the head of the phage particle are assembled into an empty precursor of the head (prehead); the phage DNA is replicated separately and then inserted into the empty head particles—a process known as packaging. A number of phage gene products play an important role in this process. Among these are:

1. The E protein, which is the major component of the phage head; mutants that are defective in this gene are unable to assemble the preheads, and therefore accumulate the other, unassembled components of the phage particle as well as the other proteins involved in packaging.
2. The D and A proteins, which are involved in the packaging process itself. Mutants that are defective in these genes are able to produce the preheads, but will not package DNA. This results in the accumulation of empty preheads.

In vitro packaging involves the use of two bacterial extracts: one from cells infected with a D mutant phage, and a second from cells infected with a phage mutant in gene E. Mixing these two extracts allows in vitro complementation and results in the ability to assemble mature phage particles. However, the packaging reaction will only work for DNA with certain properties. Lambda DNA is replicated as a multiple length linear molecule; the enzymes involved in packaging (in particular, the A protein) will recognize a specific site on this DNA (the *cos* site), cut the DNA at that point, and start packaging it into the phage head. Packaging continues until a second *cos* site is reached, when a second cut is made, giving a single complete lambda DNA molecule inside the phage head. (The cleavage at the *cos* sites is asymmetric, giving rise to the single-stranded cohesive ends of mature linear lambda DNA.) It therefore follows that only DNA molecules that carry a *cos* site will be packaged.

Furthermore, in order to produce phage particles that are capable of infecting a bacterial cell, the length of DNA between two *cos* sites must be within certain limits (packaging limits). If this length is >105% of the size of lambda DNA (i.e., greater than about 53 kb), the head fills up before the second *cos* site is reached and the second cleavage cannot then take place. If the DNA is less than about 80% of the size of lambda DNA, cleavage will occur when the second *cos* site is reached, but the DNA within the phase head will then be insufficient to maintain the structural integrity of the particle.

The existence of packaging limits in one sense reduces the usefulness of lambda phage vectors, since it places constraints on the size of the inserts that can be cloned; however, it can be turned to an advantage by permitting a selection to be applied for inserts within a desired size range.

A further application of in vitro packaging arises in the use of a special type of cloning vector known as a cosmid *(1,2)*, which is a plasmid carrying the *cos* sequence. The packaging reaction depends on the presence of *cos* sites separated by a suitable length of DNA, but is not selective for the nature of the DNA between the these sites. The cosmid vector itself is much too small to be packaged into a viable particle, and thus a large fragment of foreign DNA must be inserted before a molecule in the correct size range for packaging is obtained.

In vitro packaging, therefore, has two advantages over transformation/transfection as a means of introducing DNA into a bacterial cell. First, it is more effective, especially when using large DNA molecules, such as lambda vectors, or cosmids with large inserts. Second, it provides a selection for the presence of these large inserts.

It is important to realize that this chapter is intended as a brief introduction to a complicated subject. It is advisable to consult specialized texts before attempting to use lambda or cosmid vectors or in vitro packaging. The procedure described in this chapter includes the protocols for preparing the packaging mixes (*see* Note 1); these are also available commercially.

In Vitro Packaging

2. Materials

1. Packaging strains:
   BHB2688: lysogenic for CI$_{ts}$ b2 red3 Eam4 Sam7
   BHB2690: lysogenic for CI$_{ts}$ b2 red3 Dam15 Sam7
   The CI$_{ts}$ mutation makes the lysogen heat inducible, because the mutant repressor is thermolabile. The b2 deletion and the red3 mutation reduce the background arising from the presence of endogenous DNA in the packaging extracts. The Sam7 mutation makes the phage lysis defective, thus allowing the phage products to accumulate within the cell after induction of the lysogen. The purpose of the D and E mutations is described in Section 1. The D, E, and S mutations are amber mutations, thus allocating the phage to be grown normally within an amber suppressor host.
2. L broth: 15 g of tryptone, 5 g of yeast extract, and 5 g of NaCl. Dissolve the ingredients in about 600 mL of water, adjust the pH to 7.2 with NaOH, and make the volume up to 1 L. Transfer 500 mL to each of two 5-L flasks and sterilize by autoclaving.
3. Sucrose-Tris buffer: 10% sucrose in 50 mM Tris-HCl, pH 7.5.
4. Lysozyme: 2 mg/mL in 0.25M Tris-HCl, pH 7.5; prepared fresh.
5. Liquid nitrogen.
6. Packaging buffer: 5 mM Tris-HCl, pH 7.5, 30 mM spermidine, 60 mM putrescein, 20 mM MgCl$_2$, 15 mM ATP, 3 mM 2-mercaptoethanol.
7. Sonication buffer: 20 mM Tris-HCl, pH 8.0, 3 mM MgCl$_2$, 5 mM 2-mercaptoethanol, 1 mM EDTA.
8. Phage buffer: 7 g Na$_2$HPO$_4$, 3 g KH$_2$PO$_4$, 5 g NaCl, 0.25 g MgSO$_4$ · 7H$_2$O, 0.015 g CaCl$_2$, and 0.01 g gelatin. Add water to 1 L. Dispense in small volumes and sterilize by autoclaving.
9. Sensitive *E. coli* strain.

3. Methods

3.1. Preparation of Packaging Mix A (Packaging Protein)

1. Set up an overnight culture of BHB2688 in L broth at 30°C.
2. Use the overnight culture to inoculate 500 mL of L broth and incubate, with shaking, at 30°C. When the OD$_{650}$ has reached 0.3, remove the flasks from the shaker and place in a 45°C waterbath for 15 min, without shaking. This is to induce the phage.
3. Transfer the flask to an orbital shaker and incubate at 37°C for 1 h.
4. Recover the cells by centrifugation at 9000g (8000 rpm in a 6 × 250 mL rotor) for 10 min.
5. Remove all the supernatant, invert the tubes, and allow them to drain for at least 5 min.

6. Resuspend the cells in approx 2 mL of sucrose–Tris buffer. Distribute the suspension in 0.5-mL aliquots, and to each tube add 50 µL of a freshly prepared solution of lysozyme. Mix gently by inversion.
7. Freeze this suspension immediately in liquid nitrogen.
8. Remove the tubes from the liquid nitrogen (**NB:** Use forceps!), and allow the cells to thaw at 0°C for at least 15 min.
9. Pool the thawed cells, add 50 µL of packaging buffer, and centrifuge at 48,000g (20,000 rpm in an 8 × 50 rotor) for 30 min. Recover the supernatant and dispense 50-µL aliquots into cooled microcentrifuge tubes. Freeze the tubes in liquid nitrogen and then store at –70°C until required.

3.2. Preparation of Packaging Mix B (Phage Heads)

1. Using strain BHB 2690, carry out steps 1–5 from Section 3.1.
2. Resuspend the cells in 5 mL of cold sonication buffer. Sonicate the cells until the suspension is no longer viscous. Keep the vessel in an icewater bath during the process, use short bursts of sonication (3–5 s), and avoid foaming.
3. Centrifuge the suspension at 15,000g (10,000 rpm in a 6 × 250 mL rotor) for 10 min.
4. Recover the supernatant and dispense 50-µL aliquots into cooled microcentrifuge tubes. Freeze each tube immediately in liquid nitrogen and then store at –70°C until required.

3.3. Packaging Reaction

1. Thaw one tube of each extract (A and B) as prepared above.
2. Add 1–2 µL of the DNA to be packaged (*see* Chapters 23 and 25) to 7 µL of sonication buffer in a microcentrifuge tube.
3. Add 1 µL of packaging buffer, 10 µL of extract A, and 6 µL of extract B. Incubate at 25°C for 60 min.
4. Dilute with 0.5 mL of phage buffer and plate out using an appropriate lambda-sensitive host onto either BBL agar overlay plates (for lambda vectors) or onto antibiotic selection plates (for cosmids).
5. Score the resulting plaques or transformants and screen them for the presence of specific inserted DNA sequences, e.g., by hybridization or immunoscreening (Chapter 26).

4. Notes

1. There are several alternative procedures for the preparation of packaging extracts. Examples of other procedures will be found in refs. *3–6*.

References

1. Collins, J. and Hohn, E. (1978) Cosmids: a type of plasmid gene cloning vector that is packageable *in vitro* in bacteriophage heads. *Proc. Natl. Acad. Sci. USA* **75,** 4242–4246.
2. Hayley, J. D. (1985) Cosmid library construction, in *Methods in Molecular Biology, vol. 4: New Nucleic Acid Techniques* (Walker, J. M., ed.), Humana, Clifton, NJ.
3. Enquist, L. and Sternberg, N. (1979) *In vitro* packaging of lambda Dam vectors and their use in cloning DNA fragments. *Meth. Enzymol.* **68,** 281–298.
4. Hohn, B. (1979) *In vitro* packaging of lambda and cosmid DNA. *Meth. Enzymol.* **68,** 299–309.
5. Hohn, B. and Murray, K. (1977) Packaging recombinant DNA molecules into bacteriophage particles *in vitro*. *Proc. Natl. Acad. Sci. USA* **74,** 3259–3263.
6. Sternberg, N., Tiemeier, D., and Enquist, L. (1977) In vitro packaging of a lambda dam vector containing EcoRI DNA fragments of *Escherichia coli* and phage P1. *Gene* **1,** 255–280.

CHAPTER 23

Construction of Mammalian Genomic Libraries Using λ Replacement Vectors

Alan N. Bateson and Jeffrey W. Pollard

1. Introduction

Genomic libraries are required for cloning genes, especially those mutated in inheritable diseases, and ultimately provide the means to sequence entire genomes. The ideal genomic library should consist of sufficient clones of overlapping sequence to cover the entire genome. Such an ideal state can be approached by randomly fragmenting genomic DNA and cloning the pieces into a suitable vector *(1)*. Bacteriophage λ vectors were the first vectors available for library construction. Genomic libraries made in these vectors are readily introduced into host bacteria and can contain up to 25 kb of genomic DNA. In addition, cosmids, hybrid λ phage/plasmid vectors, have been constructed that will accommodate larger genomic inserts, up to 40 kb, and that, after library plating, can be handled in the same way as plasmids *(2)*. It is also possible to construct genomic libraries in bacterial plasmid vectors. However, the plasmid transformation efficiency limits their use to small genomes, such as viruses, yeasts, and *Dictyostelium*. To clone large regions of genomic DNA, it is necessary to order many genomic clones via their overlapping sequences. This entails rounds of clone isolation followed by library screening with probes derived from their terminii, a process known as "walking." To increase the "walking" speed, P1 bacteriophage, PAC, BAC, and YAC vectors have been developed that clone large DNA fragments, ranging in size from 100–1000 kb *(3)*. These vectors are now those of choice for the initial generation of libraries from large genomes.

From: *Methods in Molecular Biology, Vol. 58: Basic DNA and RNA Protocols*
Edited by: A. Harwood Humana Press Inc., Totowa, NJ

In these cases, however, λ phage and cosmid libraries are still required to generate sublibraries to break up these larger clones into smaller, more manageable pieces and to clone small regions not readily represented in these other forms of library.

This chapter describes the use of λ replacement vectors, but the same principles apply to the construction of genomic libraries in cosmids. λ phage has a linear genome that can be divided into three. The right and left arms are required for lytic growth, but the central portion is dispensable (4). This central fragment, known as the "stuffer" fragment, can therefore be replaced by foreign DNA. Since λ will efficiently package a viral genome of between 78 and 105% of the size of the wild-type, genomic DNA fragments of between 7 and 24.2 kb can be accommodated. These genomic DNA fragments can be generated either by mechanical shearing or by restriction enzyme digestion. Shearing results in truly random DNA fragments, but, because these fragments are blunt ended, they ligate with poor efficiency. A psuedo-random cleavage can be achieved by partial digestion with restriction enzymes that cut at 4-bp recognition sequences; theoretically in an unbiased genome these cut every 256 bases. DNA fragments in the correct size range are then isolated by size fractionation and can then be efficiently ligated to appropriately cut λ arms. The restriction enzyme *Sau*3A and those with the same recognition sequence, such as *Mbo*I and *Dpn*II, are particular useful for these purposes because their cohesive ends are compatible with the *Bam*HI site present in many vectors. Enzymes that cut at 6-bp recognition sequences can also be used, but their reduced frequency of cutting will reduce the potential for obtaining overlapping clones.

Because λ phage has a minimum packaging size (~40 kb), the background of ligated λ arms that do not contain genomic inserts is eliminated. Background packaging, however, does occur owing to religation of the stuffer fragment. This background can be reduced in a number of ways.

1. Most vectors contain synthetic polylinker sequences that flank the stuffer fragment. In many cases double-enzyme digestion can be used to remove the compatible cohesive ends from the stuffer fragment. In addition, some vectors contain a unique restriction enzyme site(s) in their stuffer fragment that can be used to digest it into smaller fragments. For example, EMBL4 contains a *Sal*I and Charon 40 has a synthetic "polystuffer" that can be cleaved by the enzyme *Nae*I into many small pieces.

Lambda Replacement Vectors

2. In practice, even after multiple digestion, physical removal of the stuffer fragment has been found to improve the efficiency of library construction. This is best achieved by separation on a sucrose gradient.
3. Phages that contain the stuffer fragment can be removed by using the Spi selection system *(5–7)*. This is effective because phages that contain the *red* and *gam* genes cannot be plated on bacterial strains containing P2 lysogens. The *red* and *gam* genes are found within the stuffer fragment and so recombinant phages are Spi$^-$, becoming permissive for growth. This selection, however, results in two complications. First, a *chi* site is required within the vector to maintain a high replication rate. In the absence of a *chi* sequence, crytpic sites within some genomic inserts may lead to the presence of recombinant λ phage molecules with higher rates of replication that then become overrepresented within the population *(5,6)*. Second, *gam*$^-$ phages only replicate efficiently in recA$^+$ bacterial hosts. The presence of recombinogenic sequences within the DNA may, however, require a recA$^-$ host to be used. In these cases, *red*$^-$ and *gam*$^+$ vectors are available in which the *gam* gene has been moved onto one of the λ arms *(8)*.

A number of λ replacement vectors are available that have the properties discussed above and are useful for library construction. λDASH, λFIX, λ2001, and the EMBL vectors have a variety polylinker sequences and all allow the Spi selection. λDASH and λFIX also contain T7 and T3 RNA polymerase promoters to allow the generation of riboprobes (*see* Chapter 12) in order to make genomic walking easier. The vectors Charon 32–35 and 40 are *red*$^-$ and *gam*$^+$ and are hence useful for cloning difficult DNA sequences. For simplicity this chapter describes library construction using EMBL4, but the method can be applied to all of the above vectors.

2. Materials

Use good quality chemicals and reagents. Store all solutions at room temperature, unless otherwise stated. Enzymes, such as *Bam*HI, *Sal*I, *Mbo*I, T4 DNA ligase, and DNaseI are supplied with appropriate buffers. Store at –20°C.

2.1. Preparation of Phage Stocks

1. Bacterial strains and bacteriophage: For λEMBL4 *(6)* use the following hosts:
   QR48: a *recA*$^-$ host for growth of the parent vector.
   NM538: for growth of the recombinant phage for secondary screening and titering EMBL4 parent.
   NM539: to detect *spi*$^-$ recombinants for initial plating of the library.
   LE392: for growth and titering of wild-type λ.

2. L-broth: For 1 L, dissolve 10 g of bacto-tryptone, 5 g of yeast extract, and 5 g of NaCl; adjust to pH 7.2 and autoclave. To grow λ add 10 mM MgSO$_4$.
3. CY medium: For 1 L, dissolve 10 g of casamino acids, 5 g of bacto-yeast extract, 3 g of NaCl, 2 g of KCl, and 2.46 g of MgSO$_4$ · 7H$_2$O; adjust to pH 7.5 and autoclave.
4. 10 mM MgSO$_4$.
5. λ Gradient buffer: 10 mM Tris-HCl, 100 mM NaCl, 10 mM MgSO$_4$, pH 7.4.
6. Cesium chloride: Make solution as directed in Section 3.1., step 17.
7. λ Dialysis buffer: 10 mM Tris-HCl, 25 mM NaCl, 10 mM MgSO$_4$, pH 8.0.

2.2. Preparation of Phage DNA

8. 0.5M EDTA, pH 7.0.
9. 10% SDS: A 10% (w/v) solution of sodium dodecyl sulfate (SDS).
10. Proteinase K: 10 mg/mL solution in 10 mM Tris-HCl, 5 mM ETDA, 0.5% (w/v) SDS, pH 7.8.
11. Phenol: Redistilled phenol equilibrated to a final 0.1M Tris-HCl, pH 8.0. Initially, equilibrate with several changes of 0.5M Tris-HCl to ensure pH 8.0 is reached. Phenol is both caustic and toxic and must be handled with care.

2.3. Preparation of λ Arms

12. Sucrose gradient buffers: Prepare a gradient from 15 and 35% (w/v) sucrose solutions in 1M NaCl, 20 mM Tris-HCl, 5 mM EDTA, pH 8.0.
13. Phenol/chloroform: A 1:1 mixture of phenol and chloroform.
14. 3M Sodium acetate, pH 6.0.
15. Isopropanol: propan-2-ol.
16. 0.01M MgCl$_2$.
17. TE buffer: 10 mM Tris-HCl, 1 mM EDTA, pH 7.5.

2.4. Partial Digestion of the Genomic DNA

18. Loading buffer: 4% (w/v) sucrose, 0.25% (w/v) bromophenol blue, and 0.25% (w/v) xylene cyanol in water.
19. Stop solution: 28 mM EDTA, pH 8.0, in 25% (v/v) loading buffer.

2.5. Library Construction

20. Packaging mix: These are commercially available or can be homemade (*see* Chapter 22).
21. SM medium: For 1 L, dissolve 5.8 g of NaCl, 2 g of MgSO$_4$ · 7H$_2$O, and 20 g of gelatin in 950 mL of water and 50 mL of 1M Tris-HCl, pH 7.5. Autoclave.
22. Chloroform.
23. Top agar: 7 g of bacto-agar in 1 L of L-broth supplemented with 10 mM MgSO$_4$. Autoclave.
24. Bottom agar: 15 g of bacto-agar in 1 L of L-broth supplemented with 10 mM MgSO$_4$. Autoclave.

Lambda Replacement Vectors

3. Method
3.1. Preparation of Phage Stocks

1. From a frozen stock of bacteria QR48, seed a 10-mL liquid culture in L-broth at 37°C.
2. Inoculate 1 mL of overnight culture into a conical flask (at least 4× the volume) containing 50 mL of CY medium and incubate it, with shaking (250 rpm) for 2–2.5 h at 37°C until the optical density is 0.4 at 630 nm.
3. Harvest the cells by centrifugation at 3000g for 10 min at 4°C. Pour off the supernatant and resuspend the pellet in 25 mL of 10 mM MgSO$_4$. This suspension may be stored for up to 10 d at 4°C, but the highest plating efficiency is achieved with 0–2 d-old cells.
4. To obtain the maximum yield of phage, it is necessary to calculate the optimum ratio of phage with which to infect the bacteria: the infectivity ratio. This is determined in a small-scale experiment prior to the full-scale preparation of phage (*see* Note 1). The day before the experiment, determine the titer of both the plating bacterial cells, QR48, and the phage lysate (refs. *1,8, see* Note 1).
5. In separate sterile 1.5-mL tubes, inoculate 0.1 mL containing 2×10^8 plating cells with 0.1 mL of phage in the following titers: 10^8, 5×10^7, 10^7, 5×10^6, 10^6, and 5×10^5 pfu. Incubate at room temperature for 15 min to allow the phage particles to absorb.
6. Carefully transfer each of these cultures to separate 100-mL conical flasks containing 10 mL of CY medium, prewarmed to 37°C. Shake (140 rpm) overnight at 37°C on an orbital shaker.
7. Clear the lysates by centrifugation at 12,000g for 10 min at 4°C. Titer the supernatants and determine the optimum infectivity ratio.
8. Continue with the main phage preparation. Inoculate 4×10^9 plating cells with the optimum infectivity ratio of phage in a 1.5-mL sterile plastic tube. Incubate at room temperature for 15 min.
9. Transfer this carefully to a 2-L conical flask containing 200 mL of CY medium prewarmed to 37°C. Shake overnight at 140 rpm in an orbital shaker at 37°C.
10. At the end of this period, cultures should still appear cloudy, but with extensive lysis showing as stringy clumps of bacterial cell debris. If this is not the case, the lysis of the cells can be increased by placing the flasks at 4°C for 1–2 h.
11. Remove the cell debris by centrifugation at 5000g at 4°C for 20 min.
12. Remove the supernatant to 50 mL polycarbonate ultracentrifuge tubes, and centrifuge these at 75,000g for 1 h at 4°C.
13. Pour off the supernatant; the pellet should have a dirty yellow center (debris) surrounded by a faintly blue opalescent ring of phage.

14. To resuspend the pellet, add 0.48 mL of λ gradient buffer at 4°C to each tube, and tilt the tubes so that the pellet is under the buffer. Leave the phage to diffuse out overnight.
15. Transfer gently to a 15-mL siliconized glass centrifuge tube and centrifuge at 12,000g for 5 min at 4°C to remove cell debris.
16. Transfer the supernatant to a fresh 15-mL glass centrifuge tube, add 100 µg of DNaseI, and incubate at room temperature for 1 h.
17. Prepare a three-step cesium chloride gradient in λ gradient buffer with concentrations of 1.7 g/cm^3 (95 g to 75 mL; η, 1.3990), 1.5 g/cm^3 (67 g to 82 mL; η, 1.3815), and 1.30 g/cm^3 (40 g to 90 mL; η, 1.3621) using a peristaltic pump to carefully layer sequentially 2 mL of 1.7, 3 mL of 1.5, and 2 mL of 1.3 into the 12-mL transparent ultracentrifuge tube, marking the interface of each layer.
18. Carefully layer the phage suspension using the pump and top up the centrifuge tube with gradient buffer.
19. Centrifuge at 160,000g for 1 h 40 min at 20°C using the "slow acceleration" and "break off" modes of the ultracentrifuge.
20. The expected banding pattern is shown in Fig. 1 (*see* Note 2). Carefully remove the phage band with a 21-gage needle and a syringe. This is done by placing a small piece of Scotch tape over the band and then piercing the tube immediately below the band, bevel edge of the needle up. Be careful that you do not go straight through the tube and into your finger.
21. Dialyze the phage against λ dialysis buffer at 4°C using two changes of 2 L each.
22. Titer the phage preparation (*see* Note 3).

3.2. Preparation of Phage DNA

1. To the phage preparation add 1/50 (v/v) of 0.5*M* EDTA.
2. Add 1/50 (v/v) of a 10% SDS solution and incubate for 15 min at 65°C.
3. Cool to 45°C, add proteinase K to 50 µg/mL, and incubate at 45°C for 1 h.
4. Extract with phenol by carefully inverting the tube several times.
5. Break the phases by centrifugation at 7800g for 5 min at 4°C.
6. Repeat this procedure by extracting the aqueous (top) phase with phenol three times.
7. Extract the aqueous phase two times with chloroform to remove the phenol, each time breaking the phases at 7800g for 5 min.
8. Extract the aqueous phase with water-saturated ether three times to remove the chloroform, separating the phases each time at 1000g for 2 min.
9. Remove residual ether by placing briefly under vacuum.
10. Run a range of dilutions of the resulting supernatant on a 0.5% (w/v) agarose gel using known concentrations of λ as a marker.

Lambda Replacement Vectors

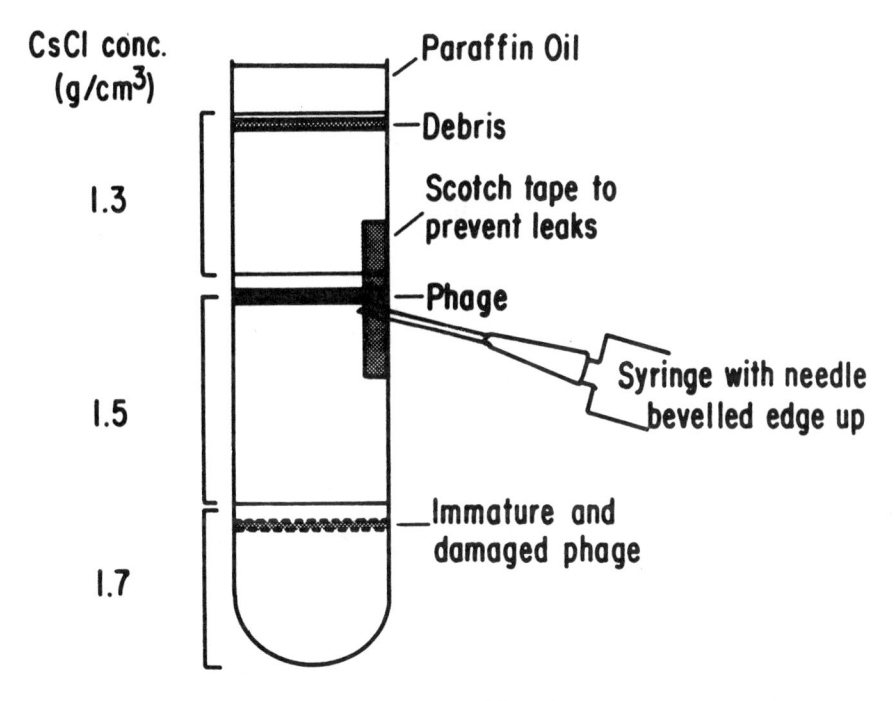

```
CsCl conc.
 (g/cm³)              Paraffin Oil
                      Debris
   1.3                Scotch tape to
                      prevent leaks
                      Phage
                      Syringe with needle
   1.5                bevelled edge up

                      Immature and
                      damaged phage
   1.7
```

Fig. 1. A diagram showing the expected banding pattern of the phage preparation after ultracentrifugation in a cesium chloride step gradient. The intact phage band appearing immediately below the 1.3–1.5 g/cm³ cesium chloride interface is extracted from the gradient with a syringe and needle as shown.

11. Estimate the DNA concentration from the agarose gel, or better, measure the absorbance at 260 and 280 nm. The OD 260/280 ratio should be 1.8; 1 OD = 50 µg/mL (*see* Note 4).

3.3. Preparation of λ Arms

1. It is advisable to ensure that the DNA can be completely digested with *Bam*HI on a small scale before proceeding to the large-scale preparation of arms. Take 0.5 µg of DNA and add 2 U of *Bam*HI in the appropriate buffer (supplied by the manufacturer) and make up to 10 µL and incubate for 1 h at 37°C.
2. Heat the DNA at 65°C for 10 min (to melt *cos* ends), cool, and analyze by 0.5% (w/v) agarose gel electrophoresis.
3. If the above procedure shows, complete digestion, proceed to a large-scale digestion; if not, clean the DNA with further phenol extractions.
4. Digest 150 µg of vector DNA in a 500-µL volume with 300 U of *Bam*HI in the appropriate buffer at 37°C overnight.

5. Check for complete digestion by running 0.5 µg of the DNA after heat shock at 68°C for 10 min on a 0.8% (w/v) agarose gel with appropriate markers. If the reaction is incomplete, add more enzyme, adjust the buffer and incubation volume, and continue the incubation.
6. Once the digestion is complete, add 300 U of *Sal*I with appropriate buffer adjustment, to cut the "stuffer" fragment, and leave this to digest at 37°C overnight (*see* Note 5).
7. Prepare a 12-mL 15–35% sucrose gradient (*see* Note 6).
8. Extract the digested DNA with an equal volume of phenol/chloroform once. Separate the phases by a 5-min centrifugation in a microfuge. Extract with an equal volume of chloroform and again separate the phases. Precipitate the DNA with 300 mM sodium acetate and 0.6 vol of isopropanol (*see* Note 5).
9. Pellet the DNA by centrifugation in a microfuge, dry the pellet in a Speed-Vac, and resuspend it in 200 µL TE buffer. Remove an aliquot containing 0.5 µg to analyze the DNA after allowing the cohesive ends to reanneal at 42°C for 1 h in the presence of 0.01M MgCl$_2$ (*see* Note 7).
10. If the DNA is intact, layer the remainder over the gradient and centrifuge at 140,000g for 19 h at 15°C using the "slow acceleration" and "break-off" modes of the centrifuge.
11. Collect 0.5-mL fractions through a 21-gage needle inserted at the base of the tube. (To stop the gradient from gushing out, the top may be sealed with parafilm and the flow started by introducing a very small hole in the top. Alternatively, a commercial gradient fractionator may be used.)
12. Remove 10 µL of every second fraction, dilute with 35 µL of water, and add 8 µL of loading buffer. Heat to 68°C for 10 min and analyze on a 20 × 20 cm 0.5% (w/v) agarose gel with appropriate restriction digested λ markers adjusted to match the salt and sucrose concentrations in the fractions.
13. Locate the fractions containing the arms (20 and 8.9 kb) and pool.
14. Either dialyze this against TE buffer if the volume is large, or dilute the fractions threefold with TE buffer and precipitate overnight with 2 vol of 95% ethanol without additional salt. Regain the pellet by centrifugation, wash it with 70% ethanol, dry it in a Speed-Vac, and dissolve it in an appropriate volume of TE buffer (to give 300–500 µg/mL).
15. Measure the exact concentration by absorbance at 260 nm and store in 5-µg aliquots at –20°C.

3.4. Partial Digestion of the Genomic DNA *(see Note 8)*
1. Before starting the large-scale digestion, perform a pilot digestion. Take 10 µg of DNA, 3 µL of *Mbo*I buffer (10× stock), and water to give a final volume of 27 µL and place at 37°C to warm.

2. Prepare 13 tubes for time-points ranging from 0–60 min and place on ice. Add 5 µL of stop solution to each tube.
3. Dilute the stock *Mbo*1 enzyme with sterile ice-cold water to give a concentration of 0.5 U/µL and place on ice.
4. Quickly add 3 µL of diluted enzyme to the prewarmed DNA, mix, and start timing.
5. Remove 2 µL immediately and then 2 µL every 5 min for 1 h.
6. Run each time-point on a large (20 × 20 cm) 0.3% (w/v) agarose gel using λ, and λ digested with *Xho*1 or *Hind*III as markers. Stain and photograph the gel.
7. Determine the time of digestion required to give maximum fluorescence in the 15–22 kb range. Ideally this should be approx 30 min of digestion.
8. Scale up the digestion to 150 µg of DNA in a total volume of 450 µL. The digestion requires only half the corresponding enzyme concentration, however, to give the maximum number of molecules in the required size range *(1)*. Determine the exact time by digesting for the optimum time of the pilot experiment ±5 min. At each time, stop the experiment by adding 150 µL of the reaction mixture to an ice-cold microfuge tube containing EDTA, pH 8.0, to give a final concentration of 20 mM.
9. Pool the three fractions and load onto a 15–35% (w/v) sucrose gradient as described for the preparation of vector arms (*see* Note 6).
10. After centrifugation, fractionate the gradient and run 10 µL of every second fraction on a 0.3% (w/v) agarose gel with the appropriate markers.
11. Label the fractions according to the sizes determined from the agarose gel (*see* Note 9).
12. Dilute each fraction threefold and ethanol precipitate with 2 vol of 95% (v/v) ethanol, dry it in a Speed-Vac, and resuspend in 20 µL of sterile TE buffer.
13. Determine the concentration of DNA by electrophoresis on a 0.8% (w/v) agarose gel and comparison with standards of known concentration.

3.5. Library Construction

1. Ligations of 1 µg of DNA total should be tested with ratios of 4:1, 2:1, 1:1, and 0.5:1 arms to insert (*see* Note 10). Carry out reactions at 150 µg DNA/mL with 1 U of T4 DNA ligase overnight at 4°C. Also ligate 0.5 µg of vector arms alone under similar conditions to estimate the background caused by vector contamination with the stuffer fragment (*see* Notes 11 and 12).
2. Package 250 ng of the ligated DNA using a commercial packaging kit or homemade packaging components as described in Chapter 22 *(1,9)*. At this step also package *bona fide* λ to determine the packaging efficiencies of the extract (commercial kits always provide this control).

3. Remove the freeze thaw and sonicated extracts from the −70°C freezer and allow to thaw on ice. The freeze thaw lysate will thaw first. It should be added to the still-frozen sonicated extract. Mix gently, and when they are almost, but not quite, defrosted, add the DNA to be packaged. Mix and incubate for 1 h at room temperature.
4. Add 0.5 mL of SM medium and a drop of chloroform and mix.
5. Remove the debris by centrifugation in a microfuge for 30 s and titer the packaged DNA on the appropriate strain:
    λ—LE392
    arms + insert—NM539 (spi^- selection)
    arms alone—NM538 and NM539
6. Calculate the efficiency of packaging of λ (*see* Note 13) and of the arms + insert. λ should be about 100× more efficient than the religated material.
7. If one of the test ligations gives a suitable efficiency in the packaging reaction, then select this and scale up the reaction to produce a full library (*see* Notes 13–15). Test a small aliquot.
8. Once the library is produced, it should be tested for completeness as outlined in Note 14 and can be stored under chloroform indefinitely. It can thereafter be screened with or without amplification as required *(1,5)* (*see* Chapter 26).
9. The library can be amplified by growing as a plate stock, making sure that the plaques do not overgrow each other, thereby reducing the possibility of recombination between different recombinant phages. To prepare the plate stock, mix aliquots of the packaging reaction containing about 2×10^4 recombinants in a 50-µL volume with 0.2 mL of plating bacteria and incubate it for 20 min at 37°C.
10. Melt 6.5 mL of top agar with each aliquot and spread on a 150-mm plate of bottom agar, previously dried.
11. Incubate for 8–10 h, making sure the plates are absolutely horizontal so that spread of clones is minimized.
12. Overlay the plates with 12 mL of SM medium and store at 4°C overnight.
13. Recover the bacteriophage suspension and transfer it to a sterile polypropylene tube. Rinse the plate with 4 mL of SM medium and pool with original suspension.
14. Add chloroform to 5% (v/v) and incubate for 15 min at room temperature with occasional shaking.
15. Recover the bacteriophage by removal of the cell and agar debris by centrifugation at 4000*g* for 5 min at 4°C.
16. Remove the supernatant and add chloroform to 0.3% (v/v), at which point the libraries may be stored in aliquots at 4°C for many years. They can be concentrated if necessary by centrifugation in a CsCl gradient as described above.

4. Notes

1. To obtain a good yield of phage, it is essential to have very clean glassware. This is achieved by soaking with chromic acid and then thoroughly rinsing with distilled water. It is also necessary to maintain good sterility control because the single-stranded *cos*– ends of lambda are extremely vulnerable and their integrity is essential. It is always worthwhile testing to see if the *cos* ends will anneal by incubation at 42°C for 1 h in the presence of 10 mM MgCl$_2$ and analyzing the product on a 0.3% (w/v) agarose gel. Obviously, it is also necessary to have good housekeeping over the storage and verification of bacterial and phage stocks. Details can be obtained from ref. *1*.
2. If the lowest band on the cesium chloride gradient is larger than the phage λ band, this suggests that mixing and pipeting have been too vigorous. If the top band is too large, it will contaminate the phage preparation. It may, therefore, be necessary to rerun the phage preparation on another gradient. If the yield is low, it may be necessary to titer at each step to ascertain where the losses occur so that you improve that part of the preparation.
3. It is necessary to always maintain the Mg^{2+} concentration; otherwise the phage particles will disintegrate.
4. 10^{10} Plaque-forming units of phage gives approx 0.44 μg of DNA. Therefore a 200-mL cleared lysate with a yield of 6×10^{10} pfu/mL will give 500 μg of DNA. The DNA should be clean by this step. If for any reason the DNA has a low 260/280 ratio, return to Section 3.2., step 4 and start again. The presence of sheared DNA usually means vigorous pipeting or poor quality phenol. Cut pipet tips and redistilled phenol are necessary.
5. Both EMBL 3 and 4 were designed such that by digestion with *Sal*I the stuffer fragment is cut without the arms being destroyed. With EMBL 4 the *Sal*I site lies between the *Bam*HI site and the stuffer fragment (the converse is true with EMBL 3 so *Sal*I digestion will not allow cloning into the *Bam*HI site). Thus, cutting with the two enzymes substantially reduces the probability that there can be religation of complete phage genome. This religation, in fact, is made even further unlikely because the small piece of polylinker produced by *Sal*I digestion is lost during the isopropanol precipitation because it is too small to be precipitated *(4)*. This further improves the quality of the library by preventing false recombinants that may have escaped the genetic selection imposed by the *spi* selection.
6. Both the arms and donor DNA may be fractionated on a low-gelling temperature agarose gel, but these gels have a lower capacity than sucrose gradients and any impurities in the agarose severely inhibit the ligation reaction. In practice, therefore, we have not found this to be a suitable method to fractionate the DNA.

7. The procedure for isolating arms can be further improved if the cohesive ends are allowed to anneal at 42°C for 1 h in the presence of 10 mM MgCl$_2$. This enables the isolation of a single species of DNA from the sucrose gradient and at the same time checks the DNA.
8. The size of the DNA is the most important determinant in the production of a complete library. Although some breakage is inevitable, the DNA should be >100 kb, and the bigger the better; otherwise, when a random collection of fragments is generated by partial restriction endonuclease digestion, there will be a large proportion of noncohesive ends. These fragments substantially lower the efficiency of packaging because they join nonproductively to vector DNA since only one of the two λ arms can join to them. Moreover, since concatemers of phage particles are the usual substrates for packaging, these ragged ends cause termination of this process. Thus if the donor DNA preparation appears smeared down the gel, it is better to go back and prepare it again rather than proceeding with the library. The usual causes for the poor quality of DNA are mechanical shearing (too vigorous pipeting, use of uncut pipet tips, or mixing too roughly) or DNase contamination caused by poor sterility control over solutions and glassware.
9. Do not throw away these tubes. This material is very precious and may be used later for either different vectors or to repeat packaging with a slightly smaller or larger insert size. Store it and the DNA prior to ligation at –20°C.
10. To calculate the amount of vector and donor DNA required, use the following formula: Insert size equals 20 kb and, therefore, has a molecular weight of approx 1.27×10^7. Arms are 30 kb and molecular weight of 1.9×10^7. Therefore, for a 1/1 ratio

$$(\mu g \text{ arms})/(1.9 \times 10^7) = (\mu g \text{ insert})/(1.27 \times 10^7)$$

Therefore, µg arms = $1.496 \times$ µg insert. Thus for a 1-µg reaction, 0.6 µg of arms and 0.4 µg of insert are required. This needs to be adjusted according to the average size of the insert and for different ratios of arms to insert *(1)*. The theoretical ratio of arms to inserts should be 2/1, but some of the molecules may lack a cohesive terminus (*see* Note 8), so the effective concentrations are different from those calculated from the measured DNA concentration. Therefore, it is necessary to do a range of test ratios before the large-scale packaging. Prior to the ligation, it is possible to alkaline phosphatase the donor DNA to prevent self-ligation. This increases the effective concentration of available ends and will increase the efficiency of productive ligations. Care should be exercised over the quality of the alkaline phosphatase since many preparations contain some nuclease activities (*see* Chapter 25).
11. The success of the ligation can be confirmed by agarose gel electrophoresis.

Lambda Replacement Vectors

12. It is also a good idea to anneal the cos ends of the vector by incubating it in the presence of 10 mM MgCl$_2$ at 42°C for 1 h.
13. λ should give at least 10^8 pfu/µg DNA. If not, the packaging extract (or your handling of it) is defective, and precious sample should not be committed to packaging. Remember when packaging not to defrost many aliquots of packaging extracts and leave them on ice; follow the protocol. We have also found it preferable to do several small-scale packagings rather than a single large-scale one, because of the problems of handling the extracts; we only use 2–3 vials at any one time. The vector arms alone should give <10^4 pfu/µg when plated on NM538 and lower when plated on NM539. If, providing the packaging is efficient, the arms + insert give lower than 10^6 pfu/µg, check the activity of the ligase by ligating a commercial λ*Hind*III digest and checking this using agarose gel electrophoresis. If the ligase is efficient, repeat steps 1 and 2 in Section 3.5. using the arms only. If these ligate efficiently as assessed by agarose gel electrophoresis, then the insert DNA is not of suitable quality. As mentioned in Note 8, this is the single most common reason for the failure to prepare a good library.
14. The ideal is to construct as complete a library as possible such that every sequence in a mammalian genome is represented at least once. To calculate the number of plaques to achieve a library of a 99% probability of containing the required sequence, the following formula should be used *(1,4)*.

$$N = [\ln (1 - P)]/[\ln (1 - x/y)]$$

where P is the desired probability, x is the insert size, y is the haploid genome size, and N is the necessary number of recombinants. Therefore, given the 20-kb hypothetical insert in order to get a 99% probability, about 8×10^5 plaques in a single packaging reaction are required. The library should be tested for its validity by picking approx 100 plaques randomly and spotting them as an array onto plating bacteria in top agar. Blot onto nitrocellulose after growth, and probe this with labeled total genomic DNA of the host species. Because of the large numbers of repeat sequences in the mammalian genome, approx 80% of the clones should give a signal after hybridization to this probe.
15. We have always found some loss of efficiency on scale up, which should be compensated for in the final packaging reaction. Remember to add a drop of chloroform and clarify the extract. Packaging extracts can inhibit bacterial growth and lower the titer.

Acknowledgments

We would like to acknowledge that this chapter is based on a previous one first published in *Methods in Molecular Biology,* vol. 4. This paper was prepared while the corresponding author's research was supported by NIH grant DK48960 and the Albert Einstein Cancer Center, grant CA 13330.

References

1. Sambrook, J., Fritsch, E. F., and Maniatis, T. (1989) *Molecular Cloning: A Laboratory Manual,* 2nd ed. Cold Spring Harbor Laboratory, Cold Spring Harbor, NY.
2. Haley, J. D. (1988) Cosmid library construction, in Methods in Molecular Biology, vol. 4: *New Nucleic Acid Techniques* (Walker, J. M., ed.), Humana, Clifton, NJ, pp. 257–283.
3. Schalkwyk, L. C., Francis, F., and Lehrach, H. (1995) Techniques in mammalian genome mapping. *Curr. Opin. Biotechnol.* **6,** 37–43.
4. Hendrix, R. W., Roberts, J. W., Stahl, F., and Weisberg, R. A., eds. (1983) *Lambda 11.* Cold Spring Harbor Laboratory, Cold Spring Harbor, NY.
5. Kaiser, K. and Murray, N. (1985) The use of phage lambda replacement vectors in the construction of representative genomic DNA libraries, in *DNA Cloning: A Practical Approach,* vol. 1 (Glover, D. M., ed.), IRL, Oxford, Washington.
6. Frischauf, A. M., Lehrach, H., Poustka, A., and Murray, N. (1983) Lambda replacement vectors carrying polylinker sequences. *J. Mol. Biol.* **170,** 827–842.
7. Burt, D. and Brammer, W. J. (1985) Bacteriophage λ as a vector, in *Basic Cloning Techniques: A Manual of Experimental Procedures* (Pritchard, R. H. and Holland, I. B., eds.) Blackwell Scientific, Oxford.
8. Loenen, W. A. M. and Blattner, F. R. (1983) Lambda charon vectors (32,33,34, and 35) adapted for DNA cloning in recombination-deficient hosts. *Gene* **26,** 171–179.
9. Dale, J. W. and Greenaway, P. J. (1984) Preparation and assay of phage λ, in *Methods in Molecular Biology, vol. 2. Nucleic Acids* (Walker, J. M., ed.), Humana, Clifton, NJ, pp. 201–209.

CHAPTER 24

The Production of Double-Stranded Complementary DNA for Use in Making Libraries

Steve Mayall and Jane Kirk

1. Introduction

The synthesis of complementary DNA from mRNA templates by the action of reverse transcriptase (RT'ase) is a fundamental technique in molecular cloning. The prime consideration is that a large amount of long cDNA copies should be made. The quality and length of the cDNA product is largely dependent on the quality of the mRNA used as the starting material, and great care must be taken to avoid degradation. The choice of primers and RT'ase used during the synthesis also influences the final product.

A short double-stranded region is generated by the annealing of a primer to the mRNA (*see* Fig. 1). The DNA strand complementary to the RNA is then generated by the action of avian myeloblastosis virus (AMV) or Moloney murine leukemia virus (MoMuLV) RT'ase. Previous methods took advantage of the fact that this first strand synthesis results in the production of a short hairpin loop that provided the primer for the second strand synthesis reaction. Subsequent S1 nuclease digestion was then required to remove this single-stranded hairpin region. This approach has been superseded by a method that uses DNA polymerase I in combination with RNase H to replace the mRNA in the hybrid with small regions of newly synthesized DNA *(1,2)*. The scale of cDNA synthesis is checked by the incorporation of radioactive nucleotides into the first and second strand.

From: *Methods in Molecular Biology, Vol. 58: Basic DNA and RNA Protocols*
Edited by: A. Harwood Humana Press Inc., Totowa, NJ

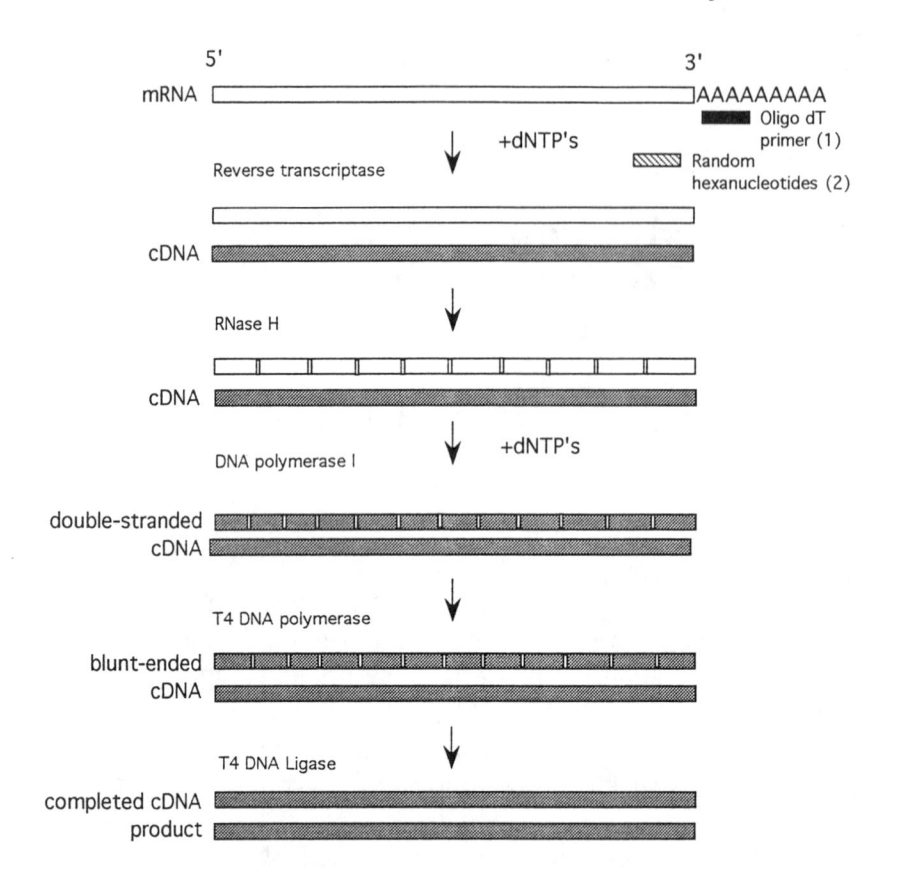

Fig. 1. A schematic representation of the steps employed to synthesize double-stranded cDNA.

The short stretches of double-stranded DNA are ligated into a continuous strand by T4 DNA ligase. At the same time, linkers containing *Eco*RI sites are ligated to the blunt ends of the cDNA and then digested with *Eco*RI to allow cloning into any similarly cut vector. *Eco*RI sites within the cDNA itself are made resistant to cleavage by prior methylation.

A spermine precipitation in the presence of salt removes a large proportion of the unincorporated linkers *(3)*. The remaining linkers are removed by a gel purification step, which also permits size selection of the cDNA. The cDNA is then ready to be ligated to a vector of choice.

2. Materials

All solutions should be autoclaved and where possible treated overnight at 37°C with 0.01% diethyl pyrocarbonate (DEPC) prior to autoclaving. Do not add DEPC to solutions containing Tris-HCl. All glassware should be DEPC-treated and/or baked at 180°C for 3 h. Ideally sterile, disposable plasticware should be used wherever possible. Single-stranded nucleic acids stick to plastic very readily and hence tubes should be siliconized before use. Gloves must be worn at all times to preclude the introduction of contaminating RNases. Great care should also be taken to avoid exogeneous DNA contamination. All enzymes and buffers should be stored at −20°C unless indicated otherwise.

2.1. cDNA Synthesis Reactions

1. 10X First strand synthesis buffer: 500 mM Tris-HCl, pH 8.3, 500 mM KCl, 100 mM MgCl$_2$, 10 mM DTT. Store at −20°C.
2. Sodium pyrophosphate (NaPPi, 150 mM): Available from Sigma (St. Louis, MO).
3. Human placental RNase inhibitor (100 U/µL): Available from Amersham Life Sciences (Amersham, UK). Store at −20°C.
4. 10X dNTP mix: A solution of dATP, dTTP, dCTP, and dGTP (10 mM each) available from Pharmacia (Uppsala, Sweden). Store at −20°C.
5. Oligo dT$_{12-18}$ primer or random hexanucleotide primers (1 mg/mL): Available from Promega (Madison, WI). Store at −20°C.
6. [α-^{32}P]dCTP (3000 Ci/mmol): Available from Amersham Life Sciences.
7. Poly(A)$^+$ RNA (200 µg/mL): Stored in DEPC-treated water at −70°C.
8. AMV RT'ase (20 U/µL): Available from Amersham Life Sciences. Store at −20°C.
9. 2X Second strand synthesis buffer: 50 mM Tris-HCl, pH 8.3, 200 mM KCl, 10 mM MgCl$_2$, 10 mM DTT. Store at −20°C.
10. RNase H (5 U/µL): Available from Amersham Life Sciences. Store at −20°C.
11. DNA polymerase I (5 U/µL): Available from Amersham Life Sciences. Store at −20°C.
12. T4 DNA polymerase (4 U/µL): Available from Amersham Life Sciences. Store at −20°C.
13. Equilibrated phenol: Saturate with 100 mM Tris-HCl, pH 8.0, as described *(4)*. Mix with chloroform (1:1 v/v), and store at 4°C. Phenol is available from Rathbone (Walkerburn, Scotland).
14. Saturated chloroform: Saturate with TE buffer (10 mM Tris-HCl, pH 8.0, 1 mM EDTA), and store at room temperature.
15. DE-81 disks: Available from Whatman (Maidstone, UK).

2.2. Addition of EcoRI Linkers to the cDNA Ends

16. 5X Buffer M: 500 mM Tris-HCl, pH 8.0, 500 mM NaCl, 5 mM EDTA. Store at –20°C.
17. 10X Buffer SAM: 800 μM S-adenosyl methionine, available from Sigma. Store at –20°C.
18. *Eco*RI methylase (20 U/μL): Available from NEB Labs (Beverly, MA) or Amersham Life Sciences. Store at –20°C.
19. 10X Ligase buffer: 200 mM Tris-HCl, pH 7.5, 150 mM MgCl$_2$, 100 mM DTT, 5 mM ATP. Store at –20°C.
20. Phosphorylated *Eco*RI linkers (150 pmol/μL): Available from NEB Labs. Store at –20°C.
21. T4 DNA ligase (400 U/μL): Available from NEB Labs. Store at –20°C.
22. 10X *Eco*RI Buffer: 500 mM NaCl, 100 mM Tris-HCl, pH 7.5, 100 mM MgCl$_2$, 10 mM DTT. Store at –20°C.
23. *Eco*RI (20 U/μL): Available from NEB Labs or Amersham Life Sciences. Store at –20°C.

2.3. Removal of Unincorporated Linkers and Size Selection of cDNA

24. Spermine wash buffer: 70% ethanol, 10 mM MgCl$_2$, 0.3M NaOAc, pH 7.0. Store at 4°C.
25. Low-melting-point agarose: FMC SeaPlaque agarose, available from Flowgen Bioscience (Sittingbourne, UK).
26. 50X TAE buffer: 2M Tris-acetate, 0.05M EDTA.
27. DNA size marker sample: 0.5 μg of DNA ladder in 5% glycerol, 0.04% bromophenol blue, available from NEB Labs.
28. STE: 100 mM NaCl, 10 mM Tris-HCl, pH 8.0, 1 mM EDTA.

3. Methods

3.1. cDNA Synthesis Reactions

1. Mix together the components of first strand synthesis shown in Table 1 in the order given (*see* Note 1).
2. Remove 1 μL for the incorporation assay (*see* step 8). Add 1 μL of AMV RT'ase (*see* Note 4), mix, and incubate at 42°C for 60 min. Place on ice and remove a further 1 μL for the incorporation assay.
3. Mix together the components of second strand synthesis shown in Table 2. Incubate at 12°C for 60 min, then at 22°C for 60 min, and finally at 70°C for 10 min.
4. Spin for a few seconds in a microcentrifuge. Place on ice and add 1 μL of T4 DNA polymerase to blunt the cDNA ends. Mix and incubate at 37°C for 10 min.

Table 1
First Strand Synthesis Mix

10X first strand buffer	2 µL
NaPPi	1 µL
HP RNase inhibitor	1 µL
10X dNTP mix	2 µL
Oligo dT or random primer (see Note 2)	1 µL
[α-^{32}P]dCTP	0.5 µL
poly(A)$^+$ RNA (see Note 3)	5 µL
H$_2$O	7.5 µL

Table 2
Second Strand Synthesis Mix

Completed first strand reaction mix	19 µL
2X second strand buffer	50 µL
[α-^{32}P]dCTP	1 µL
RNase H	1 µL
DNA polymerase I	6 µL
H$_2$O	23 µL

5. Add 2 µL of 0.5M EDTA, pH 8.0. Place on ice and remove 1 µL from the mix for the incorporation assay.
6. Extract twice with an equal volume of equilibrated phenol:chloroform then once with an equal volume of saturated chloroform (see Note 5).
7. To remove unincorporated nucleotides, add 100 µL of 4M ammonium acetate and 200 µL of 100% ethanol. Vortex and leave on dry ice for 15 min then allow the sample to warm to 4°C. Spin for 15 min at 4°C in a microcentrifuge. Remove the supernatant and wash the pellet in ice-cold 70% ethanol. Spin for 5 min at 4°C. Remove the supernatant, air-dry, and redissolve in 10 µL of TE (see Note 6).
8. To measure the incorporation of radioactive nucleotides into the cDNA, spot the retained 1 µL aliquots onto separate DE-81 filters and allow to dry. Wash five times in 200 mL of 0.5M Na$_2$HPO$_4$, pH 7.0, by swirling very gently. Wash the filters two or three times in 200 mL H$_2$O until clear. Rinse in 100 mL of methanol and allow to air-dry. Count the filters in a scintillation counter using a toluene-based scintillant and calculate the level of ^{32}P incorporation.

3.2. Addition of EcoRI Linkers to the cDNA Ends

1. To methylate EcoRI sites in the cDNA (see Note 7) add 4 µL of 5X buffer M, 2 µL of 10X buffer SAM, and 3 µL of H$_2$O to the cDNA. Mix and spin

for a few seconds in a microcentrifuge. Add 1 µL of EcoRI methylase, mix, and incubate at 37°C for 60 min.
2. Inactivate the methylase by heating at 70°C for 10 min and place on ice.
3. To ligate linkers to the cDNA, add 3 µL of 10X ligase buffer, 2 µL of phosphorylated EcoRI linkers, 2 µL of T4 DNA ligase, and 3 µL of H_2O. Mix and incubate at 12°C for 16 h.
4. Heat at 65°C for 10 min to inactivate the ligase and place on ice.
5. To digest the linkered cDNA add 9 µL of 10X EcoRI buffer, 49.5 µL of H_2O, and 1.5 µL of EcoRI. Mix and incubate at 37°C for 5 h, then inactivate at 70°C for 10 min and place on ice.

3.3. Removal of Unincorporated Linkers and Size Selection of cDNA

1. Add 5 µL of 2M KCl and 5 µL of 100 mM spermine to the 90 µL digested, linkered cDNA reaction (*see* Note 8). Vortex and leave on ice for 30 min.
2. Spin for 15 min in a microcentrifuge at 4°C. Remove the supernatant carefully and add 1 mL spermine wash buffer. Leave on ice for 30 min. Carefully remove the supernatant again and then repeat the wash.
3. Wash once in 70% ethanol, air-dry, and dissolve the pellet in 9 µL of TE.
4. Prepare a gel of 1% low-melting-point agarose in 1X TAE containing 0.5 µg/mL ethidium bromide. Leave the gel to set at 4°C for 30 min. Fill the electrophoresis tank with 1X TAE buffer to the height of the gel but do not cover it.
5. Add 1 µL of 50% glycerol to the cDNA sample and load onto the gel. The lack of loading dye makes the sample harder to load but easier to visualize small amounts of DNA under UV light. Load an adjacent lane with a DNA size marker sample containing dye.
6. Electrophorese the gel at constant voltage at 4°C until the bromophenol blue dye has run half the length of the gel. View under long-wave UV light (to minimize UV-induced DNA damage) and cut out a gel slice corresponding to cDNA between 700 and 7000 base pairs in size.
7. If the slice is large it may be beneficial to concentrate the cDNA by inserting the slice in a reverse orientation into a second low-melting-point gel and electrophoresing as before until it is concentrated into a tight band.
8. Add 400 µL of STE/100 mg of gel slice. Incubate at 65°C for 10 min. Extract once with an equal volume of equilibrated phenol, once with equilibrated phenol:chloroform, and once with saturated chloroform. Avoid the interface, which contains white agarose powder.
9. Precipitate with 0.1 vol of 4M ammonium acetate and 2.5 vol of 100% ethanol. Leave on ice for 30 min and spin for 30 min at 4°C. Wash the pellet with 70% ethanol.

10. Resuspend the pellet in 5 µL of TE. Run 0.5 µL on a 1% agarose gel alongside size standards of known concentration to give an indication of cDNA size and concentration. Store the remainder as 1-µL aliquots at −70°C. The stored cDNA should be stable for several months.

4. Notes

1. Reactions can be scaled up to allow the synthesis of larger quantities of cDNA. Components of all reactions should be gently mixed together on ice.
2. Oligo dT or random hexanucleotide primers can be used for cDNA synthesis. The oligo dT primer, which anneals to the poly(A)$^+$ tail of mRNA, is most commonly used. This generates a large proportion of full-length cDNAs although the 5' end of large mRNAs will be underrepresented. Random primers are more useful for certain purposes, such as the generation of expression libraries, enriching the 5' ends of very long transcripts, or solving problems when using the oligo dT primer owing to secondary structure in the mRNA.
3. Sometimes problems with mRNA secondary structure affect cDNA synthesis. These may be averted by heating the poly(A)$^+$ RNA to 70°C for 1 min, then immediately chilling on ice prior to the cDNA synthesis reactions.
4. Both AMV and MoMuLV RT'ase can be used for cDNA synthesis. AMV RT'ase is more commonly used since fewer units of enzyme are required. However, it has been reported that under certain conditions, longer transcripts (up to 10 kb in length) can be synthesized using MoMuLV RT'ase.
5. The recovery of cDNA from this and all subsequent steps can be monitored using a hand-held mini monitor. Expect initially readings of 100–200 cps.
6. Material can be stored at −20°C at this stage. It is advisable to run a small portion (0.5 µL) of the double-stranded cDNA on a 1% agarose gel before proceeding with other steps. This gives an indication of the size range of the cDNA product.
7. The efficiency of the cloning steps should be checked first with a mock cDNA rather than risk losing precious cDNA. The mock should consist of blunt-ended, phosphatased fragments of a similar size to the cDNA.
8. The largest losses of cDNA occur during the spermine precipitation and gel purification steps. These losses are justified because it is imperative to remove excessively short cDNAs and all traces of unincorporated linkers from the reaction. A tiny amount of residual self-ligated linkers in the mix would give a huge molar excess of ends compared to the linkered cDNA when ligated to the vector. This would result in a large proportion of the library containing no detectable inserts, essentially an empty library. The problem can usually be remedied by an additional spermine precipitation of the linkered cDNA after gel purification.

References

1. Gubler, U. and Hoffman, B. (1983) A simple and very efficient method for generating cDNA libraries. *Gene* **25**, 263–269.
2. Okayama, H. and Berg, P. (1982) High-efficiency cloning of full-length cDNA. *Mol. Cell. Biol.* **2**, 161–170.
3. Elledge, S. J., Mulligan, J. T., Ramer, S. W., Spottswood, M., and Davis, R. W. (1991) λYES: a multifunctional cDNA expression vector for the isolation of genes by complementation of yeast and *Escherichia coli* mutations. *Proc. Natl. Acad. Sci. USA* **88**, 1731–1735.
4. Sambrook, J., Fritsch, E. F., and Maniatis, T. (1989) *Molecular Cloning: A Laboratory Manual.* Cold Spring Harbor Laboratory, Cold Spring Harbor, NY.

CHAPTER 25

Construction of cDNA Libraries

Michael M. Burrell

1. Introduction

To obtain a cDNA clone of an mRNA, the mRNA must be copied faithfully into DNA, and the cDNA library must be large enough to represent the abundance class that contains the mRNA of interest. For example, in tobacco, Goldberg *(1)* has shown that it is possible to divide the mRNA population into three classes with most of the mRNAs (11,300) being in the lowest abundance class and making up 39% of the polysomal mRNA. To obtain a cDNA library that contains at least one clone for each mRNA of this class will require about 2×10^5 clones *(2)*. This can be achieved with a few micrograms of mRNA by using the efficient RNase H method of making double-stranded cDNA *(3; see Chapter 24)* and a bacteriophage λ vector that exploits the high efficiency with which in vitro packaged phage can be introduced into *Escherichia coli*. The choice of λ vector is important because if the DNA to be inserted makes the λ genome >105% of the wild-type length, the packaged phage will have a low viability.

The length of mRNA molecules ranges from several hundred bases to a few kilobases (kb). A suitable vector for this length of DNA is λgt10, because it will accept DNA fragments of up to 7.6 kb. If the mRNA of interest is known to approach or exceed this length, then a different vector should be chosen.

It is desirable in a cloning strategy for parent phage to be suppressed when the library is plated out. This is achieved with λgt10 because the parent phage is cI$^+$imm^{434}, and these are efficiently repressed on a

From: *Methods in Molecular Biology, Vol. 58: Basic DNA and RNA Protocols*
Edited by: A. Harwood Humana Press Inc., Totowa, NJ

hflA[150] strain of E. coli (e.g., C600 hfl). Insertion of cDNA into the EcoRI site of this gene produces a cI⁻ phenotype. These phages will form plaques. On a non-Cgl strain (e.g., C600), the cI⁺ phenotype gives a turbid plaque, and the cI⁻, a clear plaque.

If it is desired to screen a cDNA library with antibodies, then another suitable vector is λgt11. λgt11 has several valuable assets as an expression vector. The EcoRI cloning site is toward the 3' end of the *lacZ* gene, which is thought to stabilize the expressed protein. The phage can accept up to 7.2-kb inserts, produces a temperature-sensitive lysis repressor (cI857) that is ineffective at 42°C, and contains the amber mutation S100, which makes it lysis-defective in hosts that do not contain the suppressor SupF. These features may then be combined with host cells, which carry the plasmid pMC9, which actively represses *lacZ* expression. The cells are also *lon* protease deficient, and this minimizes the side effects of any protein coded for by the insert during cloning and amplification. When screening with antisera, however, high levels of expression can still be achieved by using IPTG to inactivate the *lac* repressor.

Although there are many advantages in using these vectors, there are several points that should be considered before they are used. λgt10 is 43.34 kb, and therefore a 1-kb insert is a small proportion of the total DNA. Thus, much more DNA in total must be prepared from a selected λ clone than from a selected plasmid clone to obtain the same amount of insert DNA. High efficiency of cloning requires good packaging extracts, which are time consuming to prepare, although they can be purchased. It is not considered advisable to maintain clones in λgt10 and λgt11 unless it is known that they are stable. Screening with antibodies sounds attractive, but requires specific antibodies that will recognize the peptide of interest as a fusion product and not necessarily folded in its native form. In addition, many of the correct cDNA clones will be inserted into λgt11 in the wrong reading frame and, therefore, not produce the correct protein.

λZAP (produced by Stratagene) is an alternative λ vector designed to simplify subcloning of the cDNA insert after screening *(4)*. This vector contains a Bluescript plasmid so that when cells are superinfected with f1 helper phage the plasmid is excised and packaged as a covalently closed single-stranded circle. The packaged "phagmid" is used to infect a new host where it replicates as a Bluescript plasmid. During library construction cDNA is cloned into the Bluescript polylinker, so λZAP has

Construction of cDNA Libraries

a wide range of sites into which to clone and can be used with blue/white selection as for λgt11. It is also possible to use this vector to make expression libraries. In this chapter I describe construction of λgt10 and λgt11 libraries, but these methods are equally applicable to λZAP, although the reader should follow the modifications suggested by the manufacturer.

There are many steps in making a cDNA library in λgt10 or λgt11. This chapter starts with methods for the growth of λgt10 and λgt11 and with the preparation of vector DNA from the phage. These steps are followed by methods of preparing λ arms from the vector DNA and the ligation of the cDNA to these arms. Therefore, it is assumed that the reader has previously prepared suitable cDNA (*see* Chapter 24). The chapter finishes with methods of packaging the DNA into viable phage and infection of *E. coli* to produce the library. The procedures described below have worked well in our hands. There are probably many other methods and variations that will prove equally successful. For further information, the reader is directed to Glover *(5)* and Sambrook et al. *(2)*.

2. Materials (*see* Note 1)

Enzymes are commercially available. They can be used along with the buffers supplied with them. In some situations, however, it is more appropriate to use homemade buffers, as indicated *(6)*.

2.1. Preparation of Plating Cells for Infection with λgt10 or λgt11

1. L-broth: For 1 L, 10 g bactotryptone, 5 g bacto-yeast extract, 10 g NaCl, adjusted to pH 7.5.
2. Bottom agar: 1.5% agar in L-broth. Autoclave and store.
3. 20% Maltose: Sterilize by autoclaving and store at room temperature.
4. Host bacteria: For λgt10, use C600 (F−, e14−, [McrA−], thr-1, leuB6, thi−, lacY1, supE44, rflbD1, fhuA21) or Bnn93 (F−, e14−, [McrA−], hsdR [$r_k^- m_k^+$], thr-1, leuB6, thi−, lacY1, supE44, fhuA21, mcrB) and C600 hfl. For λgt11, use Y1088 (F−, δlacU169, supE, supF, hsdR [$r_k^- m_k^+$], met, trpR, fhuA21, proC::Tn5[pMC9; tet^r, amp^r]), Y1089 (F−, ΔlacU169, lon-100, araD139, strA, hflA^{150}, proC::Tn10[pMC9; tet^r, amp^r]) or Y1090 (F−, ΔlacU169, lon-100, araD139, rpsl [Str^r] supF, mcrA, trpC22::Tn10[pMC9; tet^r, amp^r]). For λZAP, use XL-1 blue (F− proAB+, lacI^q lacZΔM15, supE, hsdR).
5. 10 m*M* MgSO$_4$: Sterilize by autoclaving and store at room temperature.

2.2. Titering and Plaque Purification of λgt10 Stocks

6. λ Vectors: These are commercially available as whole DNA or pre-prepared λ arms. Sections 3.2.–3.6. describe the production of λ arms.

7. λ diluent: 10 mM Tris-HCl, pH 7.5, 10 mM MgCl$_2$, 0.1 mM EDTA.
8. Top agar: 0.7% agar in L broth. Autoclave and store. Note that for purification of λ DNA substitute agarose for agar.
9. Chloroform: Do not use very old batches chloroform for bacterial lysis because this will inactivate λ phage.

2.3. Preparing λgt10 and λgt11 DNA

10. Formamide: Good quality formamide is required. Ensure that the purity is sufficient or purify by recrystallization.
11. T$_5$E: 10 mM Tris-HCl, pH 7.5, 5 mM EDTA.
12. TE: 10 mM Tris-HCl, pH 7.5, 1 mM EDTA.

2.4. EcoRI Digestion of λ

13. Suitable restriction enzymes and buffers: For λgt10 and λgt11, *Eco*RI is usually used for cloning. Other restriction enzymes may be more appropriate in other λ vectors. For use of enzymes *see* ref. 6.
14. 0.5M EDTA: pH to 8.0 to dissolve. Sterilize by autoclaving.
15. Phenol: Redistilled phenol equilibrated with 100 mM Tris-HCl.
16 Chloroform/butanol: a 50:1 chloroform:butanol mix.
17. 4M Ammonium acetate, pH 7.5.

2.5. Phosphatasing λ

18. 10X CIP buffer: 0.5M Tris-HCl, 1 mM EDTA, pH 8.5.
19. CIP: Calf intestinal alkaline phosphatase (Boehringer) Store at 4°C. CIP is less prone to nuclease contamination than the bacterial equivalent.

2.6. Ligation of cDNA and Vector

20. 10X Ligation buffer: 0.5M Tris-HCl, pH 7.5, 1M MgCl$_2$, 100 mM dithriothreitol. Store at –20°C.
21. 1 mM Spermidine: Heat to 80°C for 10 min. Store at –20°C.
22. 5 mM ATP: Store at –20°C.
23. T4 DNA ligase: Store at –20°C.

2.7. In Vitro Packaging

24. Packaging mix: These are commercially available or can be made in the laboratory (*see* Chapter 22).

2.8. Plating Out λ Libraries for Screening

25. X-gal: 2% X-gal (5-bromo, 4-chloro-3-indolyl-β-D-galactopyranoside) in dimethylformamide. Store at –20°C.
26. 1M isopropylthiogalactoside (IPTG) in sterile distilled water. Store at –20°C.

3. Methods

This section assumes that the experimenter has prepared double-stranded cDNA ready for ligation into an *Eco*RI site.

3.1. Preparation of Plating Cells

1. Streak out on L plates that contain 0.2% maltose the desired strain of *E. coli* (*see* the protocols below for the correct strain). Incubate overnight at 37°C.
2. Place a single colony in a 250-mL conical flask containing 50 mL of L broth + 0.2% maltose and incubate overnight at 37°C with shaking (220 rpm), but without frothing.
3. Centrifuge 40 mL of culture cells for 10 min at 3000*g*.
4. Resuspend the cells in 20 mL of 10 m*M* MgSO$_4$ at 4°C (*see* Note 2).
5. Adjust OD$_{600}$ to 2.0 with 10 m*M* MgSO$_4$.

3.2. Titering and Plaque Purification of λgt10 Stocks

To efficiently prepare good λgt10 DNA for cloning, it is first necessary to determine the titer of the phage stock and check whether it is contaminated with mutant phage-producing clear plaques, which complicate screening the library later.

1. Titer the phage stock by incubating aliquots of phage in 0.1 mL of λ diluent with 0.1 mL of plating cells (C600) at 37°C for 20 min. Gently shake the phage and cells during the incubation.
2. Add 3 mL of top agar, swirl to mix, and pour onto an L plate supplemented with 0.2% maltose (*see* Note 3).
3. Incubate overnight at 37°C.
4. If there are clear plaques, remove a turbid plaque with a sterile fine Pasteur pipet or capillary and place in an Eppendorf tube containing 1 mL of λ diluent. Add one drop (50–100 µL) of CHCl$_3$ to stop bacterial growth and leave at room temperature for 1 h if required immediately or place at 4°C overnight.
5. Assume 10^6 phage, and serially dilute to determine the titer.
6. If there are clear plaques, repeat step 4.
7. When there are no clear plaques, plate at 2000 plaque forming units (pfu)/82-mm plate to screen 15,000 plaques for clear plaques.
8. If there is more than one clear plaque/10^4, repeat the purification until the clone is sufficiently pure and stable.

3.3. Preparing λgt10 DNA

1. Prepare sufficient plating stock to set up between 15 and 30 82-mm plates at 1×10^6 pfu/plate (*see* Note 4).

2. Prepare fresh plating cells (C600). Pour 15–30 L plates + 0.2% glucose (*see* Note 5). Use 0.1 mL of plating cells, 1×10^6 pfu, in 3 mL of top agar per plate. As controls, omit the λgt10 from one plate and both the cells and the λ from another plate. Fewer larger plates can be used, but they require more practice in handling.
3. After 4–6 h of incubation at 37°C, the plates will appear mottled. If in doubt, do not leave the plates too long or the titer will drop. If the plates become evenly turbid, they have gone too far.
4. Place at 4°C or on ice to cool for 30 min. Overlay with 4.5 mL of ice-cold λ diluent and a few drops of chloroform (from a Pasteur pipet). Leave at 4°C overnight for the phage to diffuse into the λ diluent.
5. Carefully remove the λ diluent from the plates without disturbing the top agar and transfer to polypropylene centrifuge bottles.
6. Centrifuge at 10,000g for 20 min to pellet the cells and any agar. Carefully transfer the supernatant to fresh tubes suitable for a swing out rotor.
7. Centrifuge for 90 min at 70,000g.
8. Resuspend the phage pellet in 5 mL of λ diluent, and add 8.5 g of CsCl. Overlay with 14 mL of 1.47 g/mL density CsCl. Centrifuge overnight at 50,000 rpm (184,000g) in a Beckman 70Ti rotor. Accelerate the centrifuge slowly. Decrease the speed to 40,000 rpm for the last 40 min (*see* Note 6).
9. Remove the bluish white phage band in as small a volume as possible. Make the volume to 4 mL with at least an equal volume of λ diluent containing CsCl at a density of 1.5 g/mL. Centrifuge at 60,000 rpm for 4 h in a Kontron 80.4Ti rotor.
10. Remove the band in as small a volume as possible and make to 1 mL with λ diluent.
11. Add 1 mL of formamide, mix, and leave to stand for 2 h at room temperature.
12. Add 1 mL of SDW and 6 mL of absolute ethanol. The DNA should precipitate immediately.
13. Briefly centrifuge in a bench-top centrifuge and pipet off the supernatant.
14. Redissolve the DNA in T_5E and ethanol precipitate twice more (*see* Note 7).
15. Finally, redissolve DNA in TE buffer.

3.4. Preparation of λgt11 DNA

The first steps in preparing λgt11 DNA differ from the above method for λgt10 because it is possible to exploit the temperature-sensitive repression of lysis. Thus, cells are first multiplied at 32°C and then induced to lyse at 44°C.

1. Plate out Y1088 cells with λgt11 at 32 and 42°C to check that lysis only occurs at 42°C.

Construction of cDNA Libraries

2. Pick a single colony and place in a sterile McCartney bottle with 10 mL of L broth + 10 mM MgCl$_2$, swirl, loosen the cap, and incubate overnight at 32°C.
3. Place 10 mL of overnight culture in a 3-L conical flask containing 1 L of L broth at 32°C. Shake at 280 rpm in an orbital shaker until OD$_{600}$ = 0.6 (2–3 h).
4. Place a sterile thermometer in the flask and swirl in a water bath at 60°C until the flask contents reach 44°C. Then place at 44°C with good aeration for 15 min.
5. Place at 37°C with good aeration for 3 h. (The temperature must not drop below 37°C or poor lysis will result.)
6. Check for lysis by mixing about 2 mL of culture with a few drops of chloroform in a small test tube. The culture should clear in 3–5 min.
7. Add 10 mL of chloroform to the culture and shake at 37°C for 10 min.
8. Pellet the bacterial debris at 5000 rpm (300g) in a 6 × 300 mL rotor (MSE.21 or equivalent) for 10 min at 4°C. Use polypropylene screw-cap bottles.
9. Pellet the phage overnight at 10,000 rpm (15,500g) and continue as in Section 3.3., step 8 (*see* Note 8).

3.5. EcoRI Digestion of λ

This and the next step involve preparing the arms of λgt10 and λgt11 so that the cDNA can be ligated between them. Both vectors have a unique *Eco*RI site. In λgt10, this is in the cI gene, and insertion of DNA will produce the cI⁻ phenotype (clear plaques). In λgt11 the restriction site is in the *lacZ* gene. Therefore, when λgt11 phage are plated on media containing X-gal, the phage with inserts will produce clear plaques and the wild-type phage, blue plaques.

1. Mix the following to a total volume of 100 µL: 10 µg of DNA, 10 µL of 10X *Eco*RI buffer, and sterile distilled water, allowing for the addition of the first 30 U of enzyme at step 3.
2. Centrifuge any droplets of solution to the bottom of the tube.
3. Add 30 U of *Eco*RI in <20 µL.
4. Mix and incubate at 37°C for 2 h.
5. Add 30 U of *Eco*RI and incubate for a further 30 min.
6. Add 2 µL of 0.5M EDTA.
7. Remove the protein by adding 50 µL of phenol, and mix gently (Do not vortex; *see* Note 7). Add 50 µL of CHCl$_3$/butan-1-ol (50/1).
8. Centrifuge and remove aqueous layer.
9. Back extract the phenol/CHCl$_3$ layer with an equal volume of 10 mM Tris-HCl, pH 8.0.
10. Chloroform extract (with chloroform/butan-1-ol; 50/1) the combined aqueous phases.

11. Ethanol precipitate at –20°C by adding 2.5 vol of ethanol and 1/40 of the total volume of 4*M* ammonium acetate.

3.6. Phosphatasing λ

Strictly, this step is unnecessary for λgt10, but we prefer to include it since there are no detrimental effects and it makes checking the cDNA clones on C600 easier.

1. Resuspend the digested λ vector in 180 µL of water. Add 20 µL of 10X CIP buffer.
2. Add 0.1 U of CIP and incubate at 37°C for 15 min.
3. Add 2.5 µL of 0.5*M* EDTA, mix, and heat to 56°C for 10 min.
4. Phenol extract with 125 µL of phenol, mix, add 125 µL of CHCl$_3$/butan-1-ol (50/1), mix gently, and incubate at 37°C for 10 min.
5. Re-extract the aqueous phase as above.
6. Back-extract the organic phase with 125 µL of TE buffer, and incubate at 37°C for 10 min.
7. Combine the aqueous phases.
8. Extract twice with CHCl$_3$/butan-1-ol (50/1).
9. Ethanol precipitate at –20°C with 2.5 vol of absolute ethanol and 1/40 of the total volume of 4*M* ammonium acetate. The DNA precipitates in a few minutes.

3.7. Ligation of cDNA and Vector

It is important to be able to assess after packaging and infection which step, if any, has not worked as well as expected. Therefore, the following control ligations are required:

  i. Uncut λ.
  ii. *Eco*RI cut, not phosphatased λ.
  iii. *Eco*RI cut, phosphatased λ.
  iv. *Eco*RI cut, not phosphatased λ, plus test *Eco*RI cut DNA of 2–3 kb.
  v. *Eco*RI cut, not phosphatased λ, plus cDNA (*see* Note 9).

1. Combine the appropriate amount of cDNA with 1 µg of appropriate vector. Also set up the control ligations. Make up to 90 µL with water and add 10 µL of 2*M* NaCl, mix and 2.1 vol of ethanol. Cool.
2. Centrifuge for 20 min at 4°C to precipitate cDNA and vector. Decant ethanol carefully.
3. Wash the pellet with 0.5 mL of ethanol (precooled to –20°C). Spin on a microfuge for 5 min.
4. Take off the ethanol and repeat the ethanol wash (precooled to –20°C). Freeze-dry the pellet for 5–10 min.

Construction of cDNA Libraries

5. Take up the pellet in 3 µL of sterile double-distilled water. Add 1 µL of a mix of 5 µL of 10X ligation buffer + 5 µL of 1 m*M* spermidine.
6. Heat at 65°C for 5 min. Spin on a microfuge for 30 s.
7. Incubate at 46°C for 20 min.
8. Add 0.5 µL of 5 m*M* rATP and 1.0 µL of T4 DNA ligase. Mix well, but do not vortex. Spin for 30.
9. Incubate at 12°C for 24 h.

3.8. In Vitro Packaging

To obtain a large cDNA library from a minimal amount of cDNA, the packaging of the ligated DNA into phage must work well. It often works inefficiently in unskilled hands, however. Therefore, always check the packaging extracts with some control DNA before committing the cDNA. The packaging procedure that we have used differs slightly from that described in Chapter 22.

1. Add 5 µL of sonicated extract to the DNA, and mix with the tip of pipet.
2. Leave on ice for 15 min.
3. Add 10 µL of just-thawed freeze-thaw lysate.
4. Incubate for 1 h at room temperature.
5. Add 180 µL of λ diluent and mix.
6. If not plating immediately, add one drop of $CHCl_3$; store at 4°C.

3.9. Plating out λ Libraries for Screening

Before commencing a large scale library screen, it is necessary to assess the titer of the library (i.e., number of plaques from a given amount of packaged λ library).

1. Dilute 1% of the packaged λ library into 0.1 mL of λ diluent and add 0.1 mL of plating cells (C600 for λgt10, Y1088 for λgt11, or XL-1 blue for λZAP). Incubate at 37°C for 20 min with gentle shaking.
2. During the incubation melt sufficient top agar for plating and cool to 47°C.
3. At the end of the incubation add 3 mL of top agar, swirl to mix, and plate onto a L plate supplemented with 0.2% maltose (*see* Note 3). For λgt11 and λZAP, supplement the top agar with 40 mL of X-gal and 20 mL of IPTG just prior to mixing.
4. Invert the plates an incubate at 37°C overnight.
5. Count the number of plaques to assess the library titer. The percentage of recombinant plaques (i.e., those that contain an insert) is assessed from the number of clear plaques for λgt10 and the number of white plaques for λgt11 and λZAP (*see* Note 10).
6. Scale up and plate sufficient phage to screen for the desired cDNA (*see* Notes 11–14). Remember to plate λgt10 onto C600 hfI cells.

4. Notes

1. All solutions used for nucleic acid work should be sterile and nuclease free. Unless specified, solutions are stored at room temperature. All glassware and plasticware should be siliconized.
2. When resuspending bacteria, always use a small volume of liquid initially until a creamy paste is obtained, then add the rest of the resuspension medium.
3. When plating phage, use well-dried or 2-d cold plates. When adding the top agar + phage, have the top agar at 45–50°C, do each sample in turn, avoid air bubbles, and leave the plates to set on a flat bench for 30 min. Incubate plates upside down.
4. λgt10 can be prepared from liquid lysates, but the presence of cI⁻ phage revertants will not be recognized and will confuse the assessing of phage-carrying inserts.
5. When preparing λgt10 DNA, the presence of glucose in the L plates increases the titer by 10-fold.
6. For the background to this gradient, consult ref. 7.
7. Always treat high-mol-wt DNA gently. Avoid shear forces or it will easily be degraded.
8. If a suitable centrifuge for the overnight run is not available, the phage may be recipitated with polyethylene glycol (PEG). To 250 mL of supernatant at room temperature add 15 g of NCl, 25 g of PEG-6000, 1.5 mL of 10 mg/mL RNase A, and 1 mg/mL DNase I. Stir at 4°C for a least 3 h, preferably overnight. Collect the phage at 10,000 rpm for 30 min, resuspend in 5 mL of l diluent, and add chloroform extract at room temperature to remove the PEG. I find, however, that the recovery of phage by this method seems to be very variable.
9. The amount of cDNA used depends entirely on its length and quality; 1–5 ng has been used to yield large libraries.
10. Do not amplify more than required, because even without IPTG, plaque size varies quite a lot. Remember to use 50 μg/mL ampicillin to maintain the plasmid containing repressor.
11. λgt11 and λZAP plaques that contain β-galactosidase fusion proteins may give a faint blue color.
12. The number of plaques required depends on the likely abundance of the cDNA in the total library. This may range from 10^4 to 5×10^5.
13. The large plaque size of λgt10 means that only about 1000 plaques can be screened on an 82-mm diameter plate. Doubling the plating cell density reduces the plaque size and therefore can help if large libraries are to be screened. With λgt10 lifts can be done after 6 h rather than overnight.
14. Omit the X-gal and IPTG when plating to screen λgt11 and λZAP libraries.

References

1. Goldberg, R. B., Hoschek, G., and Kamalay, J. C. (1978) Sequence complexity of nuclear and polysomal RNA in leaves of the tobacco plant. *Cell* **14,** 123–131.
2. Sambrook, J., Fritsch, E. F., and Maniatis, T. (1982) *Molecular Cloning. A Laboratory Manual,* 2nd ed. Cold Spring Harbor Laboratory, Cold Spring, Harbor, NY.
3. Gubler, E. and Hoffman, B. J. (1983) A simple and very efficient method for generating cDNA libraries. *Gene* **25,** 263–269.
4. Short, J. M., Fernandez, J. M., Sorge, J. A., and Huse, W. D. (1988) Lambda ZAP: a bacteriophage lambda expression vector with in vivo excision properties. *Nucleic Acids Res.* **16,** 7583–7600.
5. Glover, D. M. (1985) *DNA Cloning, vol. 1: A Practical Approach,* IRL, Oxford, UK.
6. Burrell, M. M. (1993) *Enzymes of Molecular Biology. Methods in Molecular Biology,* vol. 16. Humana, Totowa, NJ.
7. Garger, S. J., Griffith, O. M., and Grill, L. K. (1983) Rapid purification of plasmid DNA by a single centrifugation in a two step cesium chloride-ethidium bromide gradient. *Biochem. Biophys. Res. Commun.* **117,** 835–842.

CHAPTER 26

Screening λ Libraries

Janet C. Harwood and Adrian J. Harwood

1. Introduction

The preceding chapters describe the construction of genomic and cDNA λ libraries. In this chapter we describe two methods for screening libraries. The first method can be used on both genomic and cDNA libraries and screens by sequence homology. To do this, the library is plated and then "lifts" are made by briefly laying a membrane, such as nitrocellulose, onto the plate. The recombinant λ phage DNA within each plaque is then denatured and fixed to the membrane. In this way an accurate representation of the plaques on the original plate is made. Specific clones are identified by hybridization with a labeled nucleic acid. The positive clones are then isolated by aligning the agar plate with the resulting autoradiograph from the hybridization. Clones are then purified by sequential rounds of plating and hybridization.

The second method of screening λ libraries is by protein expression. To do this cDNAs are cloned into λ expression vectors, such as λgt11 *(1,2)*. The expression library is plated and protein expression is induced and transferred to the membrane. The most common screening method is to use antibodies raised against the protein of interest (immunoscreening), but it is also possible to screen by protein:DNA, protein:protein, and even protein:ligand interaction *(3,4)*. The simplest method of immunoscreening is to directly label the primary antibody with ^{125}I. Radiolabeled antibodies, however, tend to give higher backgrounds, are more hazardous, and are not necessarily more sensitive than the available enzymatic methods. Most methods employ a second antibody con-

From: *Methods in Molecular Biology, Vol. 58: Basic DNA and RNA Protocols*
Edited by: A. Harwood Humana Press Inc., Totowa, NJ

jugated to an enzyme, such as horseradish peroxidase (HRP) or alkaline phosphatase (AP), for which there are very sensitive detection methods. Enzymatic detection has the added advantage that signals can be produced directly on the filter, making plaque alignment more accurate. The most sensitive immunoscreen method uses avidin-biotin detection systems *(5,6)*. The biological basis for this detection system is the very tight binding of four molecules of biotin to one of avidin, and provides an amplification ability that can be used to increase the sensitivity of the detection system. Maximum amplification is achieved by reacting with a biotin-labeled second antibody with a complex of biotin-labeled HRP and avidin *(7)*.

In this chapter we describe making lifts for both nucleic acid hybridization and immunoscreening. We also describe a simple method of immunoscreening using a secondary antibody. Finally, we describe a method for small-scale λ DNA preparation (a "miniprep") that produces DNA of good enough quality for restriction enzyme digestion.

2. Materials
2.1. Plaque Hybridization

1. Nitrocellulose membrane (*see* Note 1): Two membranes are needed for each plate screened. The nitrocellulose membrane need not be sterile (*see* Note 2).
2. 3MM paper (Whatman, Maidstone, UK): Cut to a size larger than the membrane.
3. Denaturing solution: $1.5M$ NaCl, $0.5M$ NaOH. This solution should be made fresh.
4. Neutralizing solution: $1.5M$ NaCl, $0.5M$ Tris-HCl, pH 7.0.
5. 20X SSC: $3M$ NaCl, $0.3M$ sodium citrate, pH 7.4. This is diluted with deionized water to make a 3X SSC working solution.
6. SM buffer: For 1 L use 5.8 g of NaCl, 2 g of $MgSO_4 \cdot 2H_2O$, 50 mL of $1M$ Tris-HCl, pH 7.5, and 5 mL of 2% gelatin. Sterilize by autoclaving and store at room temperature.

2.2. Screening Expression Libraries

7. IPTG: Make up 50-mL aliquots of 10 mM isopropylthio-β-D-galactosidase (IPTG) and store at –20°C.
8. TBS: 20 mM Tris-HCl, 137 mM NaCl, pH 7.6.
9. TBST: 20 mM Tris-HCl, 137 mM NaCl, pH 7.6, 0.1% Tween 20.
10. TBST-block: 3% BSA (Fraction V: Sigma [St. Louis, MO] A4503) in TBST.
11. Antibodies: Prior to screening antibodies should be titered to give a good signal (i.e., to detect 50 pg of denatured protein) and low background by

Western blotting *(8)*. For the library screen use the highest dilution that gives an acceptable background. Secondary antibodies should be conjugated to the appropriate detection system, e.g., radiolabeled, or coupled to HRP of AP. If the primary antibody is an IgG, it is possible to replace the second antibody with protein A. The primary antibody solution can be stored at 4°C in the presence of 0.02% sodium azide and reused several times.

12. Detection reagents: Use the appropriate detection system. A good chromogenic method that directly stains the membrane is available for AP, using 5-bromo-4-chloro-3-indolyl phosphate (BCIP) and nitro blue tetrazolium (NBT; *8*). For methods that give a film copy *see* Chapters 7, 8, and 10.

2.3. Secondary and Tertiary Screens

13. L-broth: For 1 L, dissolve 10 g of bacto-tryptone, 5 g of yeast extract, and 5 g of NaCl; adjust to pH 7.2 and autoclave. To grow λ add $MgSO_4$ to a final concentration of 10 mM.
14. Top agar: 7 g of bacto-agar in 1 L of L-broth supplemented with 10 mM $MgSO_4$. Autoclave.
15. L-agar: 15 g of bacto-agar in 1 L of L-broth supplemented with 10 mM $MgSO_4$. Autoclave.
16. Plating bacteria: A fresh culture of an appropriate strain of bacteria (*see* Note 3).
17. TM: 10 mM Tris-HCl, pH 8.0, 10 mM $MgSO_4$.

2.4. λ Minipreps

18. RNase A: 10 mg/mL in deionized water. Store at –20°C.
19. DNase A: 1 mg/mL in deionized water. Store at –20°C.
20. 20% PEG/2M NaCl: Make up in TM, autoclave, and store at room temperature.
21. 10% SDS (w/v) in deionized water. Store at room temperature.
22. 0.5M EDTA: pH to 8.0 to dissolve. Autoclave and store at room temperature.
23. Proteinase K: A 10 mg/mL solution. Store at –20°C.
24. Phenol:chloroform: A 1:1 mix of phenol and chloroform: The phenol is buffered with Tris-HCl, pH 8.0.
25. Chloroform.
26. Sodium acetate: 3M sodium acetate, pH 5.2. Autoclave and store at room temperature.
27. Ethanol: both absolute and 70% (v/v) stocks.
28. 10X Gel loading buffer: 50% glycerol, 0.5% SDS, 10 mM EDTA, 0.25% bromophenol blue, 0.25% xylene cyanol FF.

3. Methods

Before making the lift, plate the λ library as described in Chapters 22 and 25.

3.1. Plaque Hybridization

1. Allow the plates to cool at 4°C for at least 1 h before applying the nitrocellulose membranes.
2. Carefully place a single membrane onto each plate. Take care not to form bubbles between the membrane and the agar plate. Leave the membrane on the plate for 3 min.
3. Mark all four corners of the filter by making holes through the filter into the agar with a 19-gage needle (*see* Note 4). Mark the position of the hole in the agar by marking a small cross with a pen on the underside of the plate. Remove each membrane and label (*see* Note 5).
4. Immediately make a second (duplicate) lift by placing a fresh membrane onto the plate and leaving for 5 min. Use a needle to mark this duplicate filter with the same pattern as the first. This allows alignment of duplicate autoradiographs after hybridization (*see* Note 6). Store the plates inverted at 4°C.
5. Transfer each membrane, plaque side up, onto a piece of 3MM paper soaked in denaturing solution. Leave for 5 min.
6. Float the membranes on neutralization solution in a shallow tray for 5 min.
7. Remove the membranes and then float them on 3X SSC for 2 min.
8. Blot the membranes dry on 3MM paper and bake at 80°C for 1 h.
9. Hybridize to a nucleic acid probe as described in Chapter 6. It is important to obtain an accurate alignment of autoradiograph with the membranes to ensure that an adequate number of orientation marks are placed around the membranes before autoradiography.
10. Mark the position of the needle holes on the autoradiograph and align the duplicate. Positive plaques will be present on both autoradiographs (*see* Note 7). Use the needle holes to align the autoradiograph with the plate.
11. Pick positive plaques by removing a plug of agar around the plaque using the wide end of a Pasteur pipet. Expel the plug into 1 mL of SM buffer and leave to soak at room temperature for 2 h or overnight at 4°C before proceeding to the secondary screens (*see* Section 3.3.).

Variations of this technique can be used to screen bacterial colonies, such as from plasmid and cosmid libraries (*see* Note 8) and M13 phage clones (*see* Note 9).

3.2. Screening Expression Libraries

1. Plate the expression library with an appropriate host and incubate the plates at 42°C for 3–4 h (*see* Note 10).
2. Soak the required number of nitrocellulose membranes in 10 mM IPTG. Dry them at room temperature on cling film for a few hours before use.

Screening Libraries

3. When the plaques have grown to an appropriate size (*see* Note 11), apply the first filter and incubate the plates at 37°C for at least 4 h (*see* Note 12).
4. Mark all four corners of the membrane and remove it from the plate (*see* Section 3.1.). Immediately place it in TBS.
5. Apply a second membrane, and incubate the plates at 37°C for 3 h.
6. Wash each membrane for 5 min in TBS twice, and then incubate in TBST-block for a minimum of 1 h at room temperature. This incubation can be done overnight at 4°C.
7. Wash the membranes as follows: 2 rinses in TBST, 1 wash for 15 min in TBST, and 2 washes for 5 min in TBST.
8. Add the primary antibody and incubate on a rocking platform for 1–4 h at room temperature or overnight at 4°C.
9. Wash the membranes in TBST once for 15 min and then twice for 5 min (*see* Note 13). Add the secondary antibody and incubate on a rocking platform at room temperature for 1 h.
10. Remove the membranes from the secondary antibody and wash the membranes as follows: 2 rinses in TBST, 1 wash for 15 min in TBST, and 2 washes for 5 min in TBST.
11. Apply the appropriate detection system and align the processed membranes (or developed films) to the library plates and pick positives as in Section 3.1., step 11.

3.3. Secondary and Tertiary Screens

1. From the SM buffer surrounding each plaque plug, make the following dilutions into SM:
   10 µL into 1 mL of SM (a 10^{-2} dilution);
   100 µL of a 10^{-2} dilution into 1 mL of SM (a 10^{-3} dilution);
   100 µL of a 10^{-3} dilution into 1 mL of SM (a 10^{-4} dilution);
   100 µL of a 10^{-4} dilution into 1 mL of SM (a 10^{-5} dilution).
2. Add 10 µL of 10^{-3}, 10^{-4}, and 10^{-5} dilutions to 0.3 mL of plating bacteria and 3 mL of molten top agar at 50°C. Plate each mix and incubate overnight. This gives plates with 10^{-5}, 10^{-6}, and 10^{-7} dilutions of the original SM phage stock.
3. Rescreen as in Section 3.1. and pick positive plugs into SM.
4. Assume that the titer is 10^7 plaques/mL and make the following dilutions:
   10 µL into 1 mL of SM (a 10^{-2} dilution).
   100 µL into 1 mL of SM (a 10^{-3} dilution).
   100 µL into 1 mL of SM (a 10^{-6} dilution).
   Add 10 µL of the 10^{-2}, 10^{-3}, and 10^{-6} dilutions to 0.3 mL of plating bacteria and 3 mL of molten top agar at 50°C. Plate each mix and incubate overnight. This should give 10, 100, and 1000 plaques, respectively.
5. Rescreen as in Section 3.1. and pick positive plugs into SM.
6. Store the positive plaque as a plate lysate. To do this aim to plate 10^5 plaques on a 9-cm diameter plate (approx 10 µL of a tertiary screen stock).

Plate as before, and incubate at 37°C overnight. The entire bacterial lawn should be lysed.
7. Add 5 mL of TM, and shake slowly at room temperature for 2 h, or overnight at 4°C.
8. Remove the TM with a pipet into a centrifuge tube, and centrifuge at 4000g for 10 min at 4°C. Take the supernatant, add a drop of chloroform, and store at 4°C.

3.4. λ Minipreps

1. Make a plate lysate as in Section 3.3., step 5, except use agarose instead of agar in the plates (*see* Note 14). Overlay the plates with 5 mL of TM, and shake slowly at room temperature for 2 h, or overnight at 4°C.
2. Remove the TM with a pipet and place in a centrifuge tube. Wash the plate with an additional 1.5 mL, pool, and centrifuge at 4000g for 10 min at 4°C.
3. To the supernatant, add 2 µL of RNase and 2 µL of DNase, and leave at 37°C for 30 min.
4. Add an equal volume of 20% PEG/2M NaCl, and incubate on ice for 2 h.
5. Centrifuge at 10,000g for 10 min at 4°C. Decant all of the supernatant, and dry the walls of the tube as much as possible with a tissue.
6. Resuspend the phage pellet in 0.5 mL of TM. Add 5 µL of 10% SDS and 5 µL of 0.5M EDTA. Incubate at 65°C for 5 min.
7. Add 2.5 µL of proteinase K and incubate at 37°C for 30 min.
8. Extract twice with an equal volume of phenol/chloroform and once with chloroform.
9. Precipitate with 0.1 vol of sodium acetate and 2 vol of ethanol.
10. Resuspend the pellet in 50 µL of TE per plate lysate. Heat to 50°C to dissolve if necessary. Run 2 µL on a 0.8% agarose gel with appropriate markers to check the yield.
11. Digest 10 µL of each λ miniprep DNA in 100 µL with 5 µL of RNase and 50–100 U of enzyme. Digest for 2–3 h. Add 0.1 vol of sodium acetate and 2 vol of ethanol, place on ice for 5 min, and spin in a microcentrifuge for 15 min. Remove the supernatant, air dry, and dissolve in 18 µL of TE. Add 2 µL of 10X gel loading buffer. Before electrophoresis heat to 68°C for 3 min and then cool by placing on ice (*see* Note 15).

4. Notes

1. Other hybridization membranes can be used, but we prefer nitrocellulose. Many nylon-based filters are so sticky that they tend to damage the phage plate and have high background.
2. Some workers use a dry membrane for lifts. Alternatively, the membrane can be prewetted in deionized water to improve its flexibility, but excess water must be removed by sandwiching the membrane between two pieces

of 3MM paper. It is important to ensure that the membrane wets evenly. If this is a problem change to a different manufacturer or wash the membrane for 1–2 min in 0.1% SDS and then rinse well with five washes of distilled water.
3. From a glycerol stock, streak an appropriate strain of bacteria, e.g., NM359 for λEMBL, C600 for λgt10, Y1090 for λgt11, and Xl-1 blue for λZAP, onto an L-agar plate. Invert, and incubate at 37°C overnight. If using Y1090 include ampicillin (50 μg/mL) to maintain the *lac* repressor on the plasmid pMC9.

Use the fresh bacterial streak to inoculate 100 mL of L-broth containing 10 mM MgSO$_4$ and 0.2% maltose. Grow at 37°C in a shaking incubator to 0.5 A$_{600}$. Dispense the culture into two 50-mL polypropylene tubes and pellet the bacteria in a bench centrifuge at 2000g for 10 min. Carefully take off the supernatant and resuspend it in 40 mL of 10 mM MgSO$_4$. **Caution:** The pellet may be loose. The plating bacteria can be stored at 4°C for a short period of time until ready to be used.
4. It is important to be able to orientate the final autoradiograph with the agar plate. Because plates are usually square or circular, the pattern of orientation holes must be made as asymmetrical as possible.
5. Be careful not to remove the agar overlay together to the membrane. Drying the plates prior to plating helps the top agar to stick. The use of agarose to replace the agar also helps in this respect. If any top agar does stick to the membrane it must be removed by careful washing.
6. Hybridization often produces a spotty background that can be mistaken for positively hybridizing phage plaques. Therefore it is important to hybridize duplicate lifts to reduce the number of false positive plaques.
7. When the membrane is placed on the plate it often results in a small degree of streaking in the direction in which it was applied. Although excess streaking should be avoided, a small amount is useful because it gives positive plaques a small "comet tail" that can help distinguish true positives from background. The membrane may slightly stretch during hybridization, therefore allow for small amount of leeway when aligning plaques. It is unlikely that the signal on the autoradiograph signal will align with a single plaque, therefore all the plaques in the area are picked as a plug and then are replated for secondary screening. Usually an additional (tertiary) screen is required to unambiguously isolate the positive clone.
8. The same method can be used to screen plasmid and cosmid libraries, or even bacterial transformants while subcloning. The bacteria, however, are lysed in SDS prior to the denaturation step (Section 3.1., step 5) to limit the diffusion of the plasmid DNA. This is done by taking a lift from the surface of the plate and placing it colony side up on 3MM paper soaked in 10% SDS for 3 min. The plate from which the lift has been taken should be

placed at 37°C for a few hours to allow the colonies to regrow before taking a duplicate lift or storing the plate.
9. Screening recombinant M13 bacteriophage is very simple because the DNA is packaged in a single-stranded form. To screen take a lift as described for λ phage and then immediately bake at 80°C.
10. Growth at 42°C prevents lysogeny.
11. Identification of positive plaques is easiest when they are well separated. It is important that the plaques do not merge with each other. To avoid this, do not plate at a density >500 plaques/cm^2 and monitor the plaque growth frequently; they will grow quickly at 42°C.
12. If only one lift is to be made, the membrane can be left on the plate overnight. If duplicates are to be made leave the first membrane for 4 h and the second for 4–6 h.
13. If the primary antibody is prone to give high background signal try the following washes after removing the primary antibody: 1 for 15 min in TBST; 1 for 15 min in 2M urea, 10 mM glycine, 1% NP40; and 2 for 15 min in TBST.
14. It is essential to use top (0.7%) and bottom (1.4%) agarose instead of agar to eliminate the restriction enzyme inhibitors that are present in agar.
15. Cloned cDNA inserts are usually only a small proportion of the λ phage DNA. It is therefore necessary to digest a large amount of the total phage DNA to see the insert after gel electrophoresis. Remember that this will require more enzyme than the usual restriction enzyme digest. Genomic clones will contain a higher proportion of insert and so less DNA needs to be analyzed.

References

1. Hyunh, T. V., Young, R. A., and Davis, R. W. (1985) Constructing and screening cDNA libraries in Lambda gt10 and Lambda gt11, in *DNA Cloning I, A Practical Approach* (Glover, D. M., ed.), IRL, Oxford, UK.
2. Sambrook, J., Fritsch, E. F., and Maniatis, T. (1989) *Molecular Cloning. A Laboratory Manual,* 2nd ed. Cold Spring Harbor Laboratory, Cold Spring Harbor, NY.
3. Cowell, I. G. and Hurst, H. C. (1994) in *Protocols for Gene Analysis. Methods in Molecular Biology,* vol. 31 (Harwood, A. J., ed.), Humana, Totowa, NJ, pp. 363–376.
4. Mutzel, R. (1994) in *Protocols for Gene Analysis. Methods in Molecular Biology,* vol. 31 (Harwood, A. J., ed.), Humana, Totowa, NJ, pp. 397–408.
5. Guesdo, J. I., Ternynck, T., and Avrameas, S. (1979) The use of avidin-biotin interaction for immunoenzymatic techniques. *J. Histochem. Cytochem.* **27,** 1131–1139.
6. Buckland, R. M. (1986) Strong signals from strepavidin-biotin. *Nature* **320,** 557,558.
7. McGookin, R. (1988) *New Nucleic Acid Techniques. Methods in Molecular Biology,* vol. 4 (Walker, J. M., ed.), Humana, Clifton, NJ, pp. 301–306.
8. Harlow, E. and Lane, D. (1988) *Antibodies: A Laboratory Manual.* Cold Spring Harbor Laboratory, Cold Spring Harbor, NY.

PART IV
SUBCLONING METHODS

CHAPTER 27

Subcloning Strategies and Protocols

Danielle Gioioso Taghian and Jac A. Nickoloff

1. Introduction

Subcloning procedures are used to transfer DNA fragments from one vector context (plasmid, cosmid, or phage) to another. They are commonly used to construct expression systems, and to transfer fragments into specialized vectors for the preparation of hybridization probes and single-stranded DNA sequencing templates. Typically, cloning vectors have three essential elements, an antibiotic resistance marker, an origin of replication (to allow selection of transformed *E. coli* host cells), and one or more restriction sites into which foreign DNA may be inserted. Phage vectors, such as those based on M13 *(1)*, do not use antibiotic markers; instead transformants produce virus particles that kill or slow the growth of infected cells, and appear as plaques on lawns of uninfected bacteria. We describe here a variety of general subcloning strategies and protocols; discussions of more specialized strategies may be found elsewhere (e.g., ref. *2*).

Basic subcloning involves four steps. Vector and insert DNAs are first digested with appropriate restriction enzymes, DNAs are then mixed and treated with T4 DNA ligase, transformed into an *E. coli* host, and finally, screening procedures are used to identify transformants containing recombinant plasmids with the desired structure. Restriction enzymes produce termini with 5'- or 3'-extensions, or with blunt ends. Although any two blunt ends will ligate, protruding ends must be cohesive to join. Therefore, vector and insert DNAs are often digested with the same enzyme, and subsequent ligations recreate the original restriction sites.

From: *Methods in Molecular Biology, Vol. 58: Basic DNA and RNA Protocols*
Edited by: A. Harwood Humana Press Inc., Totowa, NJ

However, there are sets of enzymes that recognize different sites, but create ends with cohesive 3'- or 5'-extensions. In these ligations, the original restriction sites are not recreated. Blunt-end ligations offer flexibility for devising subcloning strategies, yet they are relatively inefficient, and as with cohesive ends, the original sites are recreated only if the two ends are produced by the same enzyme. Newer vectors feature clusters of unique restriction sites (a polylinker), which facilitate subcloning by providing many sites with different cohesive and blunt ends. DNA fragments produced by digestions with many enzymes can be inserted into a polylinker, and flanking sites may be used in subsequent manipulations.

Subcloning simply involves the joining of two linear DNA molecules, usually forming a circular recombinant molecule. When vector and insert DNAs have identical cohesive ends, recombinants are formed with the insert DNA oriented in either of two directions at about equal frequencies (Fig. 1A). If two different enzymes that leave noncompatible ends are used to cleave both vector and insert DNAs, the insert will ligate to the vector in only one direction (Fig. 1B). Such "directional" cloning also includes cases in which one end is blunt and the other has a 5'- or 3'-extension. In many subcloning procedures, desired recombinants are produced with very low efficiencies. This is because bimolecular reactions occur at much lower rates than the competing monomolecular reactions that yield the recircularized parent vector and other undesired products. It is possible to promote desired bimolecular reactions by optimizing DNA concentrations *(3)*. However, in our experience, such fine-tuning is unnecessary. We generally use molar ratios of vector to insert ranging from 1:1 to 1:2. Further increases in this ratio tend to reduce the yield of the undesired parent vector, but overall efficiency is not improved since it also increases the yield of products with multiple inserts. Even under optimum conditions, desired products often comprise a relatively small fraction of ligated molecules. In this chapter, we describe a variety of strategies that can be used singly or in various combinations to improve subcloning efficiencies. These strategies include screening procedures, which facilitate the identification of rare products, and selection procedures, which enrich for desired products.

Screening procedures include physical and phenotypic methods. The most direct physical screening strategy is to prepare and map plasmid DNA isolated from small-scale cultures of 10–20 transformants. This fast and efficient strategy is recommended when desired products are

Subcloning

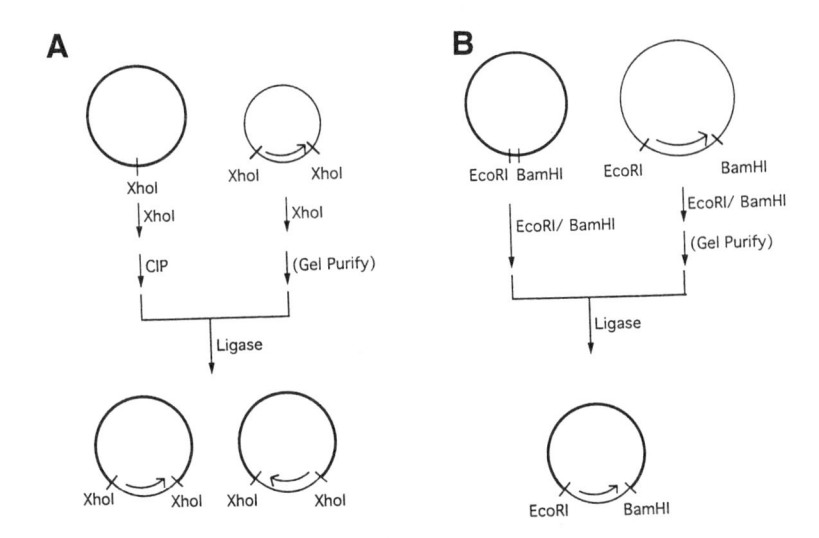

Fig. 1 (**A**) Fragment ligation with identical cohesive ends. *Xho*I digestion linearizes the recipient vector (thick lines) and releases the fragment from its host vector (thin lines). Dephosphorylation of the recipient vector and fragment purification reduce recovery of undesired products. Fragments insert in two orientations. Optional steps in this and subsequent figures are shown in parentheses; such steps may increase the yield of desired products. (**B**) Directional fragment ligation. *Eco*RI and *Bam*HI produce incompatible ends prohibiting self-ligation of either component. Gel purification of the fragment ensures that only the desired ligation product is recovered. Fragments insert in only one orientation.

expected at frequencies of 10% or more, as is often the case when subcloning strategies include steps that enrich for desired products. Several other physical screening strategies enable one to identify likely candidates present at much lower frequencies, but are more time-consuming since a subsequent plasmid DNA preparation is required to produce DNA suitable for definitive identification of the desired clone by restriction mapping. For example, one can prepare and map DNA from cultures containing 6–10 pooled transformants. With this method, it is possible to screen 200 or more transformants. Desired plasmids identified within a pool are then retrieved from a master plate with patches of individual transformants. A fast method for identifying desired products at frequencies as low as 10^{-2} involves electrophoretic analysis of unpurified supercoiled DNA from lysed bacterial colonies ("lid lysates," 4). Lid lysate screening is usually effective only if the inserted fragment increases the molecular weight of the parent vector by 25% or more. PCR-based methods provide a similar level of detection, but at greater cost. Hybridization probes can be prepared from the same DNA fragments used in subcloning, and screening by colony hybridization provides the greatest sensitivity, allowing detection of extremely rare products ($<10^{-6}$). However, screening by hybridization is time-consuming and is rarely justified for routine subcloning.

Phenotypic screens, which are highly effective and easy to perform, are possible when vectors carry cloning sites or polylinkers within a second phenotypic marker. Fragment insertion into these cloning sites usually inactivates the marker. The second marker may be an antibiotic resistance gene, as in pBR322 *(5)*. In this case, transformants containing recombinant plasmids fail to grow when replicated to plates containing the second antibiotic, and are recovered from the first plate. A one-step screening procedure based on marker inactivation uses a portion of the *E. coli lacZ* gene, as in pUC19 *(1)*. These vectors are transformed into specific strains of *E. coli* (i.e., DH5α or JM101) that contain a second defective *lacZ* gene, and the *lacZ* genes are induced by isopropylthio-β-D-galactoside (IPTG). The two gene products complement each other and produce an active β-galactosidase that metabolizes the chromogenic substrate X-GAL, producing blue colonies. Insertion of DNA fragments into *lacZ* inactivates the gene, and transformants appear white. Although insertional inactivation is efficient and convenient, it cannot be used to screen for the insertion of subsequent fragments.

Selection strategies are more powerful than screening strategies. Several selection strategies, based on the 10- to 1000-fold higher transformation efficiency of circular DNA compared to linear DNA *(6,7)*, depend on preventing circularization of unwanted products or linearizing unwanted products. A very effective and general selection involves dephosphorylation of linear vector DNA with calf intestinal alkaline phosphatase (CIP) prior to ligation with insert DNA (Fig. 1A). T4 DNA ligase catalyzes the formation of new phosphodiester bonds between 5'-phosphate and 3'-hydroxyl groups at DNA ends. Normally, this produces two new bonds at each ligated end or four new bonds when a fragment is inserted into a vector. Dephosphorylation of vector DNA prevents vector self-ligation, since T4 DNA ligase only acts on 5'-phosphorylated ends. However, dephosphorylated vectors can ligate to insert DNA with 5'-phosphorylated ends, producing a circular molecule with two new bonds and two single-stranded nicks on opposite strands. Undesired products may also be formed from the vector carrying the fragment to be transferred. To avoid these, desired fragments may be liberated from a plasmid by restriction digestion, separated from other fragments by gel electrophoresis, excised from the gel, and purified before being ligated to a (CIP-treated) vector (Fig. 1A). With directional cloning strategies, CIP treatment of the cleaved vector is unnecessary, since the vector, having two incompatible ends, cannot recircularize. When directional cloning involves cleavage of two closely linked sites (i.e., in a polylinker), it is usually not necessary to separate the large vector fragment from the small polylinker fragment by gel purification. Instead, the small fragment can be removed by passing the DNA through a Sepharose CL-6B spin-column.

A more specialized strategy involves digestions of vector and insert DNAs with two different restriction enzymes that leave noncohesive 4-base 5'-extensions. Each extension is then partially filled using T4 DNA polymerase and only two of the four deoxynucleotide triphosphates (Fig. 2). The resulting 2-base extensions do not support self-ligation, but can ligate to each other *(8)*.

Another simple and efficient strategy employs restriction enzymes that recognize different sites but produce cohesive extensions. In the case shown in Fig. 3A, the recipient vector is digested with *Xho*I, and the fragment to be transferred is liberated by digestion with *Sal*I. *Sal*I fragments ligate efficiently to cleaved *Sal*I or *Xho*I sites. When the desired product is formed, neither site is reformed. Following ligation, the DNA is treated with one (or both) enzymes, which linearizes parental molecules that may have formed. These linearized molecules transform bacteria inefficiently. The desired product is resistant to digestion, remains circular, and transforms bacteria efficiently. Sets of enzymes that create compatible ends are listed in Table 1. This strategy is also effective when DNAs are digested with different enzymes that leave blunt ends. An advantage of this strategy is that CIP treatment is not necessary. Restriction sites at positions other than those used during cloning also may be used to select against undesired products. Thus, any site present in unwanted products, but absent in desired products, can be used to linearize unwanted products in ligation reactions (Fig. 3B). This strategy is often possible in directional cloning since sites between two polylinker sites are deleted.

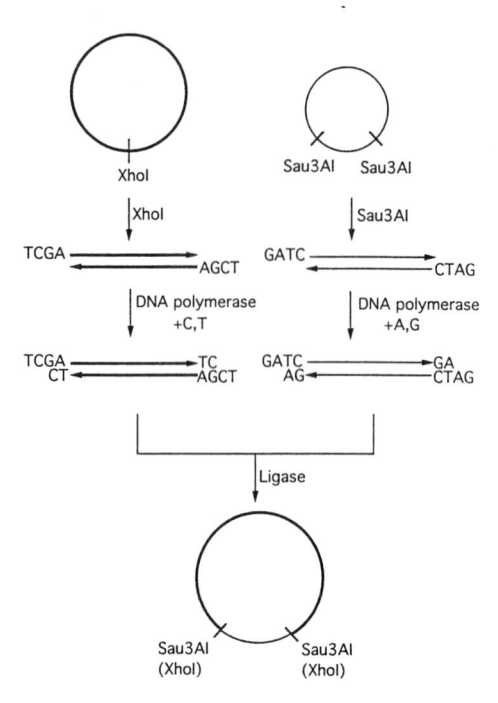

Fig. 2. Fragment ligation with partially filled in ends. *Xho*I and *Sau*3AI digestions produce noncohesive 4-base 5'-extensions. These are partially filled in to create termini that allow ligations between vector and insert molecules, but do not support self-ligation. *Xho*I sites are not reformed.

Highly efficient and very simple subcloning strategies can be designed if donor and recipient plasmids have specific genetic or structural features. Such designs are especially valuable if a particular fragment is to be repeatedly subcloned. One such design employs two starting plasmids with different antibiotic resistance markers. Figure 3C shows a strategy in which a fragment from a kanamycin-resistant (Km^r) plasmid is transferred to an ampicillin-resistant (Ap^r) vector. Desired products are selected on plates containing ampicillin. Treatment of the Ap^r vector with CIP reduces the chance of obtaining the parental Ap^r vector, whereas recircularized donor plasmids are not recovered because they are ampicillin-sensitive. Thus, this strategy eliminates the need for gel purification of the fragment being transferred. As discussed above, it is possible to eliminate the CIP treatment by using directional cloning. Nickoloff and Reynolds *(9)* combined a Km^r–Ap^r strategy with a strat-

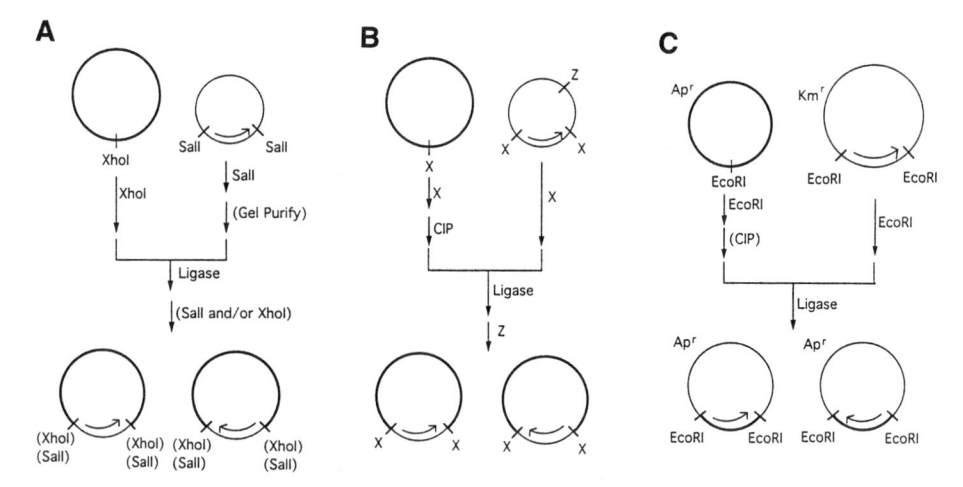

Fig. 3. Three efficient subcloning strategies. **(A)** Digestion with *Sal*I and *Xho*I create nonidentical cohesive ends that reform neither site when ligated to each other. Following ligation, recircularized parent molecules are linearized by digestion with enzymes for either or both of the original recognition sites. **(B)** Digestion of a ligation mixture with an enzyme that recognizes a site (Z), linearizes unwanted products. Site Z must not be present in the desired products. **(C)** A fragment excised from a Km^r plasmid is ligated to an Ap^r vector to create products that are selectable by their unique Ap^r, Km^s phenotype.

Table 1
Sets of Enzymes that Create Cohesive Ends[a]

Set	Enzymes
1	*Eco*RI, *Mun*I
2	*Nco*I, *Bsp*HI (*Afl*III)[b]
3	*Age*I, *Xma*I, *Ngo*MI, *Bsp*EI
4	*Bss*HII, *Mlu*I (*Dsa*I)
5	*Spe*I, *Xba*I, *Avr*II, *Nhe*I
6	*Bgl*II, *Bam*HI, *Bcl*I
7	*Eag*I, *Bsp*120I
8	*Bsi*WI, *Acc*65I, *Bsr*GI
9	*Sal*I, *Xho*I
10	*Ppu*10I, *Apa*L (*Sfc*I)
11	*Pst*I, *Nsi*I (*Bsp*128I)
12	*Psp*1406I (*Bsa*HI)
13	*Cla*I, *Bst*BI

[a]Any pair of enzymes within a set will produce cohesive ends that reform neither site when ligated.

[b]Enzymes in parentheses recognize multiple sites; cleavage of some or all recognized sites produces the proper cohesive ends.

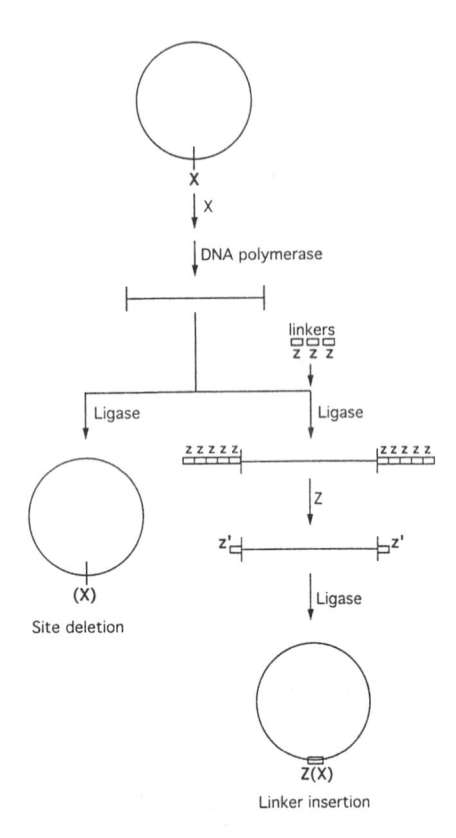

Fig. 4. Site deletion and linker insertion. Blunt ends formed by T4 DNA polymerase may be religated to delete site X **(left)** or linked to create a new site Z **(right)**. Linking sometimes reforms the target site X.

egy involving ligation of cohesive ends produced by different enzymes to generate >90% desired products without CIP or gel purification steps.

Since restriction sites play important roles in efficient subcloning strategies, it is often useful to delete or introduce new restriction sites at specific locations; two procedures that effect such changes are described here. Sites may be deleted simply by self-ligating 5'- or 3'-extensions that have been made blunt by the action of T4 DNA polymerase (Fig. 4). These reactions usually introduce two or four new basepairs of DNA at the deleted site, and may inactivate genes if sites within coding sequences are deleted. Because most restriction sites are palindromic, self-ligation of a filled-in 5'-extension creates a new palindromic sequence that may be recognized by a different restriction enzyme *(10)*. Another way to create a new site is by linker insertion (Fig. 4). Linkers are short (8–10 bp) blunt-ended palindromic DNA sequences that can be ligated to blunt ends to create a new restriction site. Blunt-end ligations with linkers are efficient because they can be used at high molar excess. Linkers usually destroy the insertion site, but they can be designed to recreate these sites. In this case, the linker will be flanked by two copies of the original site. Linkers also can be added to both ends of a fragment before it is inserted into a vector; it is often more efficient to use linkers to clone fragments with blunt ends than to insert blunt-ended fragments directly into a blunt site.

Subcloning involves many sequential enzymatic steps, and usually enzymes must be inactivated or removed between steps. Such purifications previously were performed by phenol:chloroform extraction followed by ethanol precipitation *(2),* a time-consuming (30–60 min) procedure. By using instead a 5-min Sepharose CL-6B spin-column chromatography procedure, multistep procedures are easily performed in 1 d. These columns remove small molecules (e.g., salt, SDS, nucleotides, and DNA fragments smaller than 100 bp) and remove or inactivate most enzymes. DNA is recovered in TE (10 mM Tris–1 mM EDTA) at the same concentration as input DNA *(11)*.

2. Materials
2.1. DNA Digestions

1. Plasmid DNA: Purified DNA is usually used as the starting point for subcloning, however, even quite crude "miniprep" DNA (*see* Chapters 31–33) can be used.
2. Restriction enzymes and buffers: These are available from various manufacturers and are usually supplied with their buffers.
3. TE: 10 mM Tris-HCl, pH 7.5, 1 mM EDTA. Autoclave.

2.2. Sepharose Spin-Column Chromatography

1. 1.5-mL microcentrifuge tubes (with or without caps).
2. 27-gage syringe needle.
3. Siliconized glass beads: 200–300 µm (diameter) (Sigma, St. Louis, MO). Wash the glass beads once with chloroform to remove machine oil, and then soak for 5 min in 5% dimethyldichlorosilane and 95% chloroform. Carry out alternating washes with 95% ethanol and dH$_2$O five times, and finally autoclave under dH$_2$O. Treated glass beads are easy to pipet, and their hydrophobic character prevents DNA loss owing to sticking.

4. Sepharose CL-6B: (Pharmacia, Piscawaty, NJ). Wash Sepharose CL-6B eight times with equal volumes of TE, and then resuspend it as a slurry containing 60% CL-6B and 40% TE. Autoclave for 40 min in 100-mL batches. Care should be taken to maintain sterility, since these steps remove preservatives. CL-6B is stable for at least a year when stored at 4°C.
5. 10X stop mix: 50% glycerol, 0.5% SDS, 10 mM EDTA, 0.25% bromphenol blue, 0.25% xylene cyanol FF (also used as a loading buffer for agarose gels).

2.3. Dephosphorylating Plasmids

1. 10X CIP buffer: 500 mM NaCl, 100 mM Tris-HCl, pH 7.9, 100 mM MgCl$_2$, 10 mM dithiothreitol.
2. CIP: Store at +4°C.
3. Phenol:chloroform:isoamyl alcohol (25:24:1); phenol is buffered with TE, pH 8.0.

2.4. Filling in 5'-Extensions/Removing 3'-Extensions

1. 10X T4 DNA polymerase buffer: 500 mM NaCl, 100 mM Tris-HCl, 100 mM MgCl$_2$, 10 mM dithiothreitol, 500 µg/mL acetylated BSA.
2. 2 mM dNTPs (freshly made from 10-mM stocks).
3. T4 DNA polymerase: 3 U/µL. Store at –20°C.

2.5. Linker Phosphorylation and Ligation

1. Linkers: 10 A$_{260}$/mL in dH$_2$O (*see* Note 1).
2. 10X kinase buffer: 700 mM Tris-HCl, pH 7.6, 100 mM MgCl$_2$, 50 mM dithiothreitol.
3. 1.0 mM ATP: Make a fresh dilution from a 20-mM stock.
4. T4 polynucleotide kinase: 10 U/µL. Store at –20°C.
5. 10X ligation buffer: 500 mM Tris-HCl, pH 7.8, 100 mM MgCl$_2$, 100 mM dithiothreitol, 10 mM ATP, 250 µg/mL bovine serum albumin (BSA).
6. T4 DNA ligase: 400 U/µL (*see* Note 2). New England Biolabs (Beverly, MA) is a good source. Store at –20°C.

2.6. Screening for Recombinants with X-GAL Plates

1. X-GAL: 20 mg/mL X-GAL in dimethylformamide, stored at –20°C.
2. IPTG: 100 mM IPTG in dH$_2$O, stored at –20°C.

3. Methods

3.1. DNA Digestions

The first step for all subcloning procedures is digestion of the participant DNAs with the appropriate restriciton enzymes. Digest 1–2 µg of each DNA in reaction volumes of 20 µL (*see* Note 3). Check for complete digestion by electrophoresing 1/10 of each sample on an agarose gel (see Note 4). Purify the remaining DNA by passing it through a Sepharose CL-6B (see Section 3.2.).

3.2. Sepharose Spin-Column Chromatography

CL-6B columns are used to purify DNAs after each enzymatic step during subcloning.

1. Poke a hole in the bottom of a tube with a 27-gage needle. Then using a Pasteur pipet, add about 50 µL of glass beads.
2. Mix the Sepharose CL-6B vigorously until it is completely resuspended, and then add to the column a volume equal to 10X the volume of the DNA sample to be purified. Gently place the column on a second tube; pressure may build up if a tight seal is formed, which can restrict flow.
3. Centrifuge at 600g for 2 min in a horizontal rotor. Discard the bottom tube, and transfer the column to a new tube.
4. While the column is being prepared, add to the DNA 1/10 vol of 10X stop mix, vortex briefly, and heat to 65°C for 3 min.
5. Apply the DNA sample evenly to the prespun column (within 30 min of column preparation) and centrifuge at 600g for 2 min. The DNA solution collects in the bottom tube (see Notes 5 and 6).

3.3. Dephosphorylation of Plasmids

1. To a 1.5-mL tube, add 2.5 µL of 10X CIP buffer, about 1 µg of linearized vector, and dH$_2$O to 25 µL.
2. Add 0.1–1.0 U of CIP, mix, and incubate for 1 h at 37°C.
3. Add dH$_2$O to 50 µL and stop the reaction by extracting with an equal volume of phenol:chloroform:isoamyl alcohol. Remove the aqueous phase. Purify through Sepharose CL-6B (see Note 7).
4. To assess the efficiency of the reaction, divide the dephosphorylated vector into two equal parts. Use half in a ligation reaction with the fragment to be subcloned (see Section 3.6.) and half in a self-ligation reaction. Both reactions should have equal volumes. After ligation and CL-6B column purification, transform equal amounts of each reaction into *E. coli* to reveal the proportion of background self-ligation products among the potential desired clones.

3.4. Filling in 5'-Extensions/Removing 3'-Extensions

This procedure is used to create blunt DNA termini necessary for various subcloning strategies (see Note 8).

1. To a 1.5-mL tube, add 1.5 µL of 2.0 m*M* dNTPs, 2.5 µL of 10X T4 DNA polymerase buffer, about 1.0 µg linearized plasmid DNA (Section 3.1.), and dH$_2$O to a final volume of 25 µL.

2. Add 2.4 U of T4 DNA polymerase, mix, and incubate at 37°C for 20 min to fill in a 5'-extension and for 5 min to remove a 3'-extension (*see* Note 9).
3. Purify DNA through Sepharose CL-6B.

3.5. Linker Phosphorylation and Ligation

1. To a 1.5-mL tube, add 1 µL of 10X kinase buffer, 1 µL of 1.0 m*M* ATP, 2 µL of linker, and dH$_2$O to 10 µL.
2. Add 10 U of T4 polynucleotide kinase, and incubate for 2 h at 37°C. Do not purify phosphorylated linkers before ligating to DNA (*see* Notes 10 and 11).
3. Add 5 µL of phosphorylated linkers and 3.5 µL of 10X ligation buffer to blunt-ended DNA (*see* Section 3.4.), and add dH$_2$O to 35 µL.
4. Add 400 U of T4 DNA ligase, mix, and incubate for 30 min at 16°C.
5. Purify the DNA through Sepharose CL-6B.
6. Digest the DNA with 50–100 U of restriction enzyme for at least 4 h to remove excess linkers (*see* Note 11). Purify the DNA through Sepharose CL-6B.
7. To ligate into the plasmid vector, mix 4 µL of 10X ligase buffer, 35 µL of the reaction from step 6, and 400 U of T4 DNA ligase. Incubate for 2 h at 16°C. Purify through Sepharose CL-6B, and use to transform *E. coli*.

3.6. Fragment to Vector Ligations

1. If necessary, gel-purify the DNA fragment to be transferred (*see* Chapter 28), and dephosphorylate the recipient DNA (Section 3.3.).
2. To a 1.5-mL tube, add vector and insert DNAs at molar ratios ranging from 1:1 to 1:2 (approx 1 µg each), 3 µL of 10X ligation buffer, and dH$_2$O to 30 µL.
3. Add 400 U of T4 DNA ligase, mix, and incubate at 16°C for 2 h when ligating cohesive ends, and more than 4 h when ligating blunt ends (*see* Notes 12 and 13).
4. Purify DNA through Sepharose CL6B, and use to transform *E. coli*. (*See* Chapters 29 and 30.)

3.7. Screening for Recombinants with X-GAL Plates

Transformants with plasmids containing fragments inserted into *lac*Z appear white; parent vectors with intact *lac*Z genes appear blue.

1. Spread 40 µL each of X-GAL and IPTG onto an LB plate made with an appropriate antibiotic. Let dry completely.
2. Plate transformed cells, and incubate overnight at 37°C.
3. Plates may be refrigerated at 4°C for several hours to enhance color further.

4. Notes

1. Linkers can be purchased in both phosphorylated and nonphosphorylated forms. To ensure maximum efficiency, we recommend phosphorylating linkers just prior to use in linker ligation reactions.
2. Ligase units are defined in at least two ways. Unit recommendations given here are based on the definition used by New England BioLabs. If other sources are used, follow their recommendations.
3. Mixing reactions properly is a seemingly trivial, yet quite important part of all procedures requiring enzymes. Enzymes are always the last component added to a reaction, and should be mixed thoroughly by gentle pipeting or by tapping the tube. Vortexing may inactivate enzymes. When reactions are complete, enzymes are inactivated by adding stop mix, mixing vigorously by vortexing, heating to 65°C, and purifying through Sepharose CL-6B.
4. The initial plasmid DNA digestion is a critical step in subcloning because all subsequent subcloning steps depend on the ends generated by this reaction. Always assay 10% of digestion reactions by agarose gel electrophoresis to check for complete digestion. If any amount of the parent vector remains undigested, it will transform bacteria very efficiently and may produce an unacceptably high background of parental plasmids.
5. The blue indicator dyes serve as internal controls in spin-columns. Any dye evident in eluate suggests that the column failed to retain other small molecules as well. Rerun the sample using slightly more Sepharose CL-6B. When working with multiple columns, the dyes also help to identify columns to which DNA has been applied.
6. Sepharose CL-6B purification is not used when DNA fragments smaller than 100 bp are to be recovered, such as linkers, primers, and small PCR products, since these are retained in the column.
7. CIP must be removed via phenol extraction, because any residual activity may dephosphorylate insert DNA during the subsequent ligation reaction. All other DNA-modifying enzymes are inactivated or removed by Sepharose CL-6B purification, including heat-stable enzymes.
8. Although both the Klenow fragment of *E. coli* DNA polymerase and T4 DNA polymerase are capable of creating blunt ends from 5'- and 3'-extensions, the more active 3'- to 5'-exonuclease activity of T4 DNA polymerase makes this the enzyme of choice for removal of 3'-extensions. We use T4 DNA polymerase for making blunt ends from both 5'- and 3'-extensions.
9. The 3'- to 5'-exonuclease activity of T4 DNA polymerase will rapidly delete large tracts of DNA if nucleotide concentrations are too low. This enzyme repeatedly removes and adds the terminal bases at blunt ends; this cycling can reduce nucleotide pools rapidly. Therefore, do

not exceed the specified reaction times or use excess enzyme in these reactions.
10. If desired, linker phosphorylation can be assessed by setting up a self-ligation reaction and running the products on a 5–10% acrylamide gel. Phosphorylated linkers will ligate to each other and produce a characteristic ladder pattern. In an adjacent lane, run a control sample that has not been treated with ligase.
11. Linker ligations longer than 30 min are not recommended, since this generates products with more linkers that must be removed in the next step. When digesting excess linkers following linker ligation, use as much restriction enzyme as possible (usually comprising 10% of total reaction volume). This is necessary because of the large number of target sites in substrate molecules.
12. Bimolecular ligations between fragments with one blunt and one cohesive end are much more efficient than between fragments with two blunt ends. Monomolecular (recircularization) reactions are efficient with blunt ends or with cohesive ends. We do not assay ligation reactions by gel electrophoresis because the products of these reactions are not usually discrete.
13. DNA concentrations in ligation reactions may be adjusted to ≤50 µg/mL to favor recircularized products instead of concatameric products.

References

1. Yanisch-Perron, C., Vieira, J., and Messing, J. (1985) Improved M13 phage cloning vectors and host strains: nucleotide sequence of the M13mp18 and pUC19 vectors. *Gene* **33**, 103–119.
2. Sambrook, J., Fritsch, E. F., and Maniatis, T. (1989) *Molecular Cloning: A Laboratory Manual,* 2nd ed. Cold Spring Harbor Laboratory, Cold Spring Harbor, NY.
3. Dugaiczyk, A., Boyer, H. W., and Goodman, H. M. (1975) Ligation of *Eco*RI endonuclease-generated DNA fragments into linear and circular structures. *J. Mol. Biol.* **96**, 171–184.
4. Hoekstra, M. F. (1988) Lid lysates: an economical and rapid method for plasmid analysis. *Biotechniques* **6**, 929–932.
5. Bolivar, F., Rodriguez, R. L., Greene, P. J., Betlach, M. C., Heynecker, H. L., Boyer, H. W., Crosa, J. H., and Falkow, S. (1977) Construction and characterization of new cloning vehicles. II. A multipurpose cloning system. *Gene* **2**, 95–113.
6. Cohen, S. N., Chang, A. C. Y., and Hsu, L. (1972) Nonchromosomal antibiotic resistance in bacteria: genetic transformation of Escherichia coli by R-factor DNA. *Proc. Natl. Acad. Sci. USA* **69**, 2110–2114.
7. Conley, E. C. and Saunders, J. R. (1984) Recombination-dependent recircularization of linearized pBR322 plasmid DNA following transformation of *Escherichia coli*. *Mol. Gen. Genet.* **194**, 211–218.

8. Hung, M.-C. and Wensink, P. C. (1984) Different restriction enzyme-generated sticky DNA ends can be joined in vitro. *Nucleic Acids Res.* **12,** 1863–1874.
9. Nickoloff, J. A. and Reynolds, R. J. (1991) Subcloning with new ampicillin and kanamycin resistant analogs of pUC19. *Biotechniques* **10,** 469–472.
10. Nickoloff, J. A. (1992) Converting restriction sites by filling in 5' extensions. *Biotechniques* **12,** 512–514.
11. Nickoloff, J. A. (1994) Sepharose spin-column chromatography: a fast, nontoxic replacement for phenol:chloroform extraction/ethanol precipitation. *Mol. Biotechnol.* **1,** 105–108.

CHAPTER 28

Purification of DNA Fragments from Agarose Gels Using Glass Beads

Etienne Joly

1. Introduction

Size selection of DNA fragments is frequently required before ligation or labeling for the preparation of probes. Many methods are available for purifying DNA fragments following electrophoresis in agarose gels, including the use of low-melting agarose, electrophoresis onto DEAE-cellulose paper, electroelution (with many variations), and usage of various DNA binding matrices. The following method is based on the capacity of sodium iodide (NaI) to dissolve agarose gels and the binding of DNA to glass surfaces at high salt concentration. A very high DNA binding capacity is achieved by using a fine powder of crushed glass. After washing, the DNA can be readily eluted from the glass in water, at a final concentration of up to 100 µg/mL.

The following protocol is derived from the one initially described by Vogelstein and Gillespie *(1)*. It is fast, reliable, cheap, and recovery yields are up to 90%. The recovered DNA is suitable for ligation, cutting with restriction enzymes, and labeling reactions. Some laboratories even use this technique to purify DNA before microinjection into fertilized oocytes for the making of transgenic mice. This method can be used for high-mol wt DNA without much noticeable degradation. It is less efficient, however, for fragments under 400 bp, and recovery yields can fall dramatically when purifying very short fragments (under 200 bp). The only drawback is the requirement to separate the DNA on agarose gels cast using TAE as a buffer.

From: *Methods in Molecular Biology, Vol. 58: Basic DNA and RNA Protocols*
Edited by: A. Harwood Humana Press Inc., Totowa, NJ

2. Materials
2.1. Preparation of Glass Bead Slurry

1. Silica powder, 325 mesh: This is powdered flint glass and can be obtained from ceramic stores. The one manufactured by American Flint Glass Co. is made from ground scintillation vials. Alternatively, one can use Silica Powder obtained from Sigma (St. Louis, MO) (S-5631), but this contains a large proportion of extremely fine particles that need to be eliminated by differential decantation (follow Section 3.2.).
2. Nitric acid: Analar grade. Care must be taken when using the concentrated acid. Wear gloves and only open in a fume hood.
3. Sterile distilled water: Use double-distilled water and autoclave.

2.2. Purification of DNA from Agarose Gels

1. 50X TAE gel running buffer: for 1 L, 242 g of Tris-base, 57.1 mL of glacial acetic acid, 100 mL of $0.5 M$ EDTA, pH 8.0.
2. NaI solution: 90.8 g of NaI (Sigma S-8379), 1.5 g of Na_2SO_3 (Sigma S-0505), and sterile distilled water to a final volume of 100 mL. Store in the dark at 4°C, preferably in a dark or opaque container (*see* Notes 1 and 2).
3. EtOH wash: 50% ethanol, 100 mM NaCl, 10 mM Tris-HCl, pH 7.5, 1 mM EDTA. Store in an air-tight container at room temperature.

3. Methods
3.1. Preparation of Glass Bead Slurry from Crushed Glass

1. Place 250 mL (220–250 g) of powder in a 500 mL beaker and fill to 500 mL with distilled water. Stir for 1 h.
2. Let it settle for 1 h, and then pour the supernatant into a centrifuge bottle. Centrifuge at 4000*g* for 5 min. Resuspend the pellet in 200 mL of water, and transfer to a Pyrex™ beaker.
3. Add 200 mL of concentrated nitric acid (around 70%), and bring close to boiling in a fume hood.
4. Let it cool, and then transfer into a centrifuge bottle. Wash four times with sterile distilled water by resuspension and centrifugation.
5. Resuspend the pellet in an equal volume of sterile distilled water to make a 50% slurry (v/v). Store in aliquots in air-tight containers to prevent evaporation at 4°C (*see* Note 3).

   From 250 mL (220–250 g) of powder, one should obtain around 25 mL of "fines" (50 mL of 50% slurry). Before using, check that the proportion of water in the slurry is around 50%. If water has evaporated, just add some more (distilled and sterile).

3.2. Preparation of Glass Bead Slurry from Silica Powder

This powder contains extremely fine particles that need to be eliminated by a crude sedimentation procedure.

1. Place 100 g of silica into a 500-mL beaker, fill to 500 mL with distilled water, and stir for 1 h.
2. Let settle for 2 min, and eliminate the lumps that have already settled by pouring into a new beaker. Let settle for 2 h.
3. Recover the settled fraction, which should amount to 50–75 mL, by decanting supernatant. This should be done very carefully, since the settled fraction is very fluid and can be poured away very easily. It is therefore a good idea to pour the supernatant into another beaker, at least the first time you prepare this material.
4. Add distilled water to 200 mL, and proceed as in Section 3.1., step 3.

From 100 g of silica, one should obtain 100–150 mL of 50% slurry.

3.3. Purification of DNA from Agarose Gels

1. Separate the DNA fragment on an agarose gel made using TAE gel buffer (*see* Note 4).
2. Visualize the DNA to be purified on a UV transilluminator and cut out as a block of agarose (*see* Notes 5 and 6). Place the block of agarose in a 1.5-mL microfuge tube.
3. Estimate the weight of the agarose block, and add 2–3 µL of NaI solution/mg of agarose gel.
4. Incubate at 55°C until the agarose is dissolved. This should take 5–10 min, and can be accelerated by occasional shaking.
5. Add 2–5 µL of well-resuspended glass bead slurry to the tube and vortex (*see* Note 7). Incubate at room temperature for 10 min. with occasional shaking or vortexing.
6. Spin for 20 s in a microfuge, and discard the supernatant.
7. Wash with 500 µL of EtOH wash by resuspending the pellet by pipeting up and down and then spin as in step 6. Repeat twice.
8. After the last wash, spin once more, and carefully aspirate the last drop of EtOH wash. Let the pellet air-dry for a few minutes.
9. Resuspend the pellet in 10–20 µL of water, and incubate at 55°C for 5 min. Spin as in step 6, and collect the supernatant, which contains the purified DNA.
10. Repeat steps 8 and 9, and pool the two supernatants of eluted DNA.
11. Spin the pooled eluted DNA once more as in step 6, and transfer 90% of supernatant to a fresh tube, making sure to stay away from any visible or invisible pellet of glass beads (*see* Note 8).

The DNA fragment is now ready to use. The protocol can also be used to "clean up" plasmid preparations (*see* Note 8), for example, to use as a template for sequencing (*see* Chapter 45).

4. Notes

1. Not all the sulfite will dissolve. Do not worry, since it is only required as an antioxidizing agent.
2. This solution can turn yellow with time, but this does not noticeably affect activity.
3. For longer-term storage, the slurry can be stored at –20 or –70°C.
4. Agarose should be of "molecular biology" grade, especially if the DNA is to be ligated afterward. It should not be low-melting-temperature agarose, and TBE gels are not suitable.
5. Cut the DNA from the gel in the smallest possible block of agarose.
6. This operation should be carried out as quickly as possible, because UV light is harmful both to the DNA being purified and the experimenter.
7. One microliter of glass beads slurry is sufficient for up to 2 µg of DNA, but it is a good idea to use at least 2 µL of glass beads in order to obtain a good pellet.
8. If a trace of glass beads remains in the purified DNA, it may cause the DNA to rebind to the glass when placed in buffers containing salt, such as most enzyme buffers. To ensure that no traces of glass beads are left, it is a good idea always to spin the tube briefly just before use, pipeting the desired volume of DNA from near the surface of the liquid.
9. To purify plasmid DNA preparations, add 2 vol of 7*M* guanidium HCl to the plasmid solution and 1 µL of glass bead slurry/µg of plasmid DNA, and proceed as from Section 3.3., step 5. Note that plasmids purified in this manner will be in a released form.

Reference

1. Vogelstein, B. and Gillespie, D. (1979) Preparative and analytical purification of DNA from agarose. *Proc. Natl. Acad. Sci. USA* **76,** 615–619.

CHAPTER 29

Transformation of *E. coli*

Fiona M. Tomley

1. Introduction

In recent years several techniques have been described for the introduction of DNA molecules into *Escherichia coli*. These are based on the findings of Mandel and Higa *(1)*, who demonstrated that incubation of cells with naked bacteriophage DNA in cold calcium chloride resulted in uptake of virus and that a transient heat shock of the mixture greatly enhanced the efficiency of transformation. Subsequently, this method was used to introduce a variety of circular and linear DNAs into strains of *E. coli*, and many variations in the technique have been described that are aimed at increasing the yield of transformants (or transfectants in the case of M13 bacteriophage).

The mechanisms involved in DNA transformation are not fully understood, but the central requirements for success remain the presence of multivalent cations, an incubation temperature close to 0°C, and a carefully controlled heat-shock at 42°C. Even with the most efficient methods, however, the proportion of cells that become "competent" for transformation is limited to approx 10% of the total population. Prepared competent cells can be used immediately, or alternatively, frozen aliquots of competent cells can be stored at –70°C and thawed as required. This chapter contains two alternative methods for preparing competent cells, a fast procedure based on the original calcium method and a second based on that of Hanahan that gives a higher efficiency of transformation *(2)*. The procedures for introducing both plasmid and M13 bacteriophage DNA into competent cells are described.

From: *Methods in Molecular Biology, Vol. 58: Basic DNA and RNA Protocols*
Edited by: A. Harwood Humana Press Inc., Totowa, NJ

2. Materials
2.1. Preparation of Competent Cells by the Calcium Method

1. A suitable *E. coli* strain: For example, to use the blue-white β-galactosidase selection, the host must possess a deletion of the 5'-coding region of the chromosomal *LacZ* gene, and for propagating M13 bacteriophage vectors the F'-episome must be present. Recommended strains for M13 propagation include JM101, JM103, JM107, JM109, TG1, and TG2 (*see* Note 1).
2. L-broth: 1% tryptone, 0.5% yeast extract, 200 m*M* NaCl. Sterilize by autoclaving in suitable aliquots.
3. Sterile, detergent-free 1-L conical flasks fitted with porous tops or cotton-wool bungs and an orbital incubator capable of shaking these vigorously (around 250 rpm) at 37°C.
4. 50 mL polyproplyene tubes: For example, disposable Falcon 2070 or reusable Oakridge, sterile and detergent-free, and a centrifuge, preferably refrigerated, capable of spinning these tubes at 4000g.
5. 100 m*M* calcium chloride: Store 1*M* CaCl$_2$ at –20°C in 1.5-mL aliquots. Thaw when needed, dilute to 15 mL, filter, and chill on ice (*see* Note 2).

2.2. Preparation of Competent Cells Using the Hanahan Method

1. SOB: 2% tryptone, 0.5% yeast extract, 10 m*M* NaCl, 2.5 m*M* KCl, pH 7.0. Sterilize by autoclaving in suitable aliquots. Just before use, add Mg^{2+} to 20 m*M* from a sterile filtered stock of 1*M* MgCl$_2$ or 1*M* MgSO$_4$.
2. TFB: 10 m*M* K-MES, 100 m*M* RbCl or KCl, 45 m*M* MnCl$_2$, 10 m*M* CaCl$_2$, 3 m*M* Hexamminecobalt chloride. Make up a 1*M* stock of MES, adjust the pH to 6.3 with 5*M* KOH, and store at –20°C in 10-mL aliquots. To make up 1 L of TFB, use one 10-mL aliquot of 1*M* K-MES, add all the other salts as solids, filter and store in 15-mL aliquots at 4°C where it is stable for over a year. It is important that the final pH of the buffer be 6.15 ± 0.1.
3. FSB: 10 m*M* potassium acetate, 100 m*M* KCl, 45 m*M* MnCl$_2$ · 4H$_2$O, 10 m*M* CaCl$_2$ · 2H$_2$O, 3 m*M* hexamminecobalt chloride, 10% glycerol. Make up a 1*M* stock of potassium acetate, adjust the pH to 7.2 with 2*M* acetic acid and store at –20°C in 10-mL aliquots. To make up 1 L of FSB, use one 10-mL aliquot of 1*M* potassium acetate, add the other salts as solids, glycerol to 10%, adjust the pH to 6.4 with 0.1*N* HCl, filter, and store in 15-mL aliquots at 4°C. During storage, the pH of this buffer drifts down to 6.1–6.2 and then stabilizes. If too much HCl is added, do not attempt to readjust the pH. Instead, discard the batch and start again.
4. DMSO/DTT (*see* Note 3): Take a fresh bottle of highest grade DMSO, divide into 10-mL aliquots and store in sterile, tightly capped tubes at –70°C.

To make up 10 mL of DMSO/DTT use one 100-mL aliquot of 1M potassium acetate (see Section 2.2., step 3), 9 mL of DMSO, 1.53 g of DTT, sterilize through a filter that will withstand organic solvents (e.g., Millex SR, millipore), and store as 300-µL aliquots at –20°C.

2.3. Transformation Procedure

1. X-gal: 20 mg/mL dissolved in dimethylformamide. Store in the dark at –20°C.
2. IPTG: 24 mg/mL dissolved in water. Store in the dark at –20°C.
3. L-agar/antibiotic plates: 1% tryptone, 0.5% yeast extract, 200 mM NaCl, 1.2% bacto-agar. Sterilize by autoclaving in suitable aliquots. To pour plates, melt agar by boiling or microwaving, cool to around 50°C, and add an appropriate antibiotic (see Note 4). Pour the molten agar into petri-dishes placed on a level surface, using 15–20 mL/plate. Once the agar has set, dry plates by inverting at 37°C for several hours before use.

2.4. Introduction of M13 DNA to Competent Cells

1. H-agar: 1% tryptone, 140 mM NaCl, 1.2% bacto-agar. Sterilize by autoclaving in suitable aliquots. To pour plates, melt agar by boiling or microwaving, cool to around 50°C, and pour into Petri-dishes on a level surface (15–20 mL/plate). Once agar has set, dry plates by inverting at 37°C for several hours before use.
2. H-top agar: 1% tryptone, 140 mM NaCl, 0.8% bacto-agar. Sterilize by autoclaving in suitable aliquots. Melt agar as above, and hold at 48°C in a water bath until needed.
3. Plating cells: A freshly grown 50-mL culture of *E. coli* cells, the same strain as the competent cells, which should be set up alongside the competent cell culture.
4. Sterile, disposable 5-mL tubes, e.g., Falcon 2054.

3. Methods

3.1. Preparation of Competent Cells Using Calcium Chloride

1. Pick a single bacterial colony from a freshly streaked plate, or take 500 µL of a fresh overnight culture and transfer into 50 mL of L-broth in a 1-L conical flask.
2. Incubate flask at 37°C with vigorous shaking until the cell concentration reaches around 5 × 10^7 cells/mL. For most *E. coli* strains this is A$_{600}$ 0.3–0.4 and will take approx 3 h of incubation from a single colony and approx 2 h from an overnight culture.
3. Transfer the contents of the flask to a sterile, precooled 50-mL tube, and place on ice for 10 min.

4. Pellet cells by spinning at 4000g for 10 min, and carefully pour off the broth, inverting the tube to drain away the last traces.
5. Suspend cells in 10 mL of cold 0.1M CaCl$_2$, place on ice for 5 min, and then pellet as in step 4.
6. Suspend cells gently in 2.5 mL of cold 0.1M CaCl$_2$, and incubate on ice until ready for use.

At this point, cells can be dispensed in small aliquots into cooled sterile 1.5-mL tubes, snap-frozen in liquid nitrogen, and stored at −70°C until needed. To use frozen, thaw the cells rapidly and immediately place on ice (*see* Note 5).

3.2. Preparation of Competent Cells Using Hanahan Method

1. Pick a single bacterial colony from a freshly streaked plate, or take 500 µL of a fresh overnight culture and transfer into a flask of 50 mL of SOB-broth.
2. Incubate the flask at 37°C with vigorous shaking until the cell concentration reaches approx 5 × 10^7 cells/mL (*see* Section 3.1. step 2).
3. Transfer the contents of the flask to a sterile, precooled 50-mL tube, and place on ice for 10 min.
4. Pellet the cells by spinning at 4000g for 10 min, and carefully pour off the broth, inverting the tube to drain away the last traces.
5. Suspend cells in 10 mL of cold TFB and place on ice for 10 min. Pellet the cells as in step 4.
6. Suspend cells gently in 4 mL of cold TFB, and place on ice.
7. Add 140 µL of DMSO/DTT, and immediately mix by gentle swirling. Place cells on ice for 15 min. Add a further 140 µL of DMSO/DTT, and again ensure immediate mixing with the cells.
8. Place on ice for at least 15 min, and then dispense as small aliquots into cooled 5-mL tubes.

If the cells are to be stored frozen, the same procedure should be followed, except that TSB should be replaced by FSB and DMSO used without DTT. At the end of step 7, aliquot the cells into cooled 5-mL tubes, and then snap-freeze in liquid nitrogen. The cells should be stored at −70°C. To use them, rapidly thaw an aliquot, and then place cells on ice for 15 min.

3.3. Transformation Procedure (see Note 6)

1. Prior to use, store the competent cells in an ice-water bath. For most purposes, 50-µL aliquots of cells will generate sufficient colonies, but more may be used if very high numbers are required (*see* Notes 5 and 7).

Transformation of E. coli

2. Carefully add to each aliquot between 40 and 400 ng of plasmid DNA, or ligation product, in no greater than a 5 µL volume, and swirl gently (*see* Note 8). It is useful to include two transformation controls, one containing 5–10 pg of plasmid DNA and one containing no DNA.
3. Leave in the ice-water bath for 30–45 min.
4. Heat-shock the cells by transferring the tubes to a rack in a preheated water bath at 42°C, and leave for exactly 90 s without shaking (*see* Note 9).
5. Immediately place the tubes of cells back into the ice-water bath, and make up to 200 µL with SOB-broth (without Mg^{2+}). If a large number of colonies are expected, split the competent cell mix to leave 180 µL in the first tube and 20 µL in a second (*see* Note 7).
6. Incubate the tubes of cells for 50 min at 37°C.
7. Add 40 µL of X-gal and 40 µL of IPTG to each sample (200 µL vol), and then plate onto L-agar antibiotic plates. Wait for the plates to dry, and then invert each plate, and incubate overnight at 37°C.

3.4. Introduction of M13 DNA to Competent Cells (see Note 10)

1. Melt sufficient H-top agar (3 mL/plate, plus a few excess milliliters) and keep it in a water bath at 48°C.
2. Carry out transformation as described in Section 3.3., steps 1–3.
3. Just before the heat-shock, make a mix of 200 µL of plating cells, 40 µL of X-gal, and 40 µL of IPTG for each plate.
4. Transfer the tubes of competent cells/ligations to a rack in a preheated water bath at 42°C, and heat shock for 90 s without shaking (*see* Note 9).
5. Place the cells in the ice-water bath, and make up to 200 µL with SOB-broth (minus Mg^{2+}). If you think you are going to get too many plaques, then split the samples at this point (*see* Note 7).
6. Add 280 µL of the plating cell/X-gal/IPTG mix to each sample and then, working one sample at a time, add 3 mL of H-top agar, mix quickly, and pour onto a dry H-agar plate. Be careful not to introduce air bubbles.
7. Allow plates to dry thoroughly, invert, and incubate overnight at 37°C (*see* Note 11).

4. Notes

1. M13 host bacteria can be stored for short periods at 4°C on minimal (M9) agar, which maintains the F'-episome, but should be replated frequently from a master stock stored at −70°C in L-broth plus 15% glycerol.
2. For making up and storing all transformation buffers, use only high-quality pure water, e.g., Milli-Q or equivalent. Use sterile, detergent-free glassware or plasticware; sterilize solutions by filtration through 0.45-µm pore filters, e.g., Nalgene or Acrodisc.

3. In the original Hanahan method, DMSO and DTT solutions were added sequentially. However, the efficiency of transformation was just as high as it would have been if they had been added together *(3)*.
4. It is important that the correct selection is used. The most common selection used is for ampicillin resistance, however, kanamycin and tetracyclin may also be encountered. When using tetracyclin selection, care must be taken since some hosts, such as XL-1 blue cells and Sure™, utilize tetracyclin resistance to maintain the F'-episome.
5. The efficiency of transformation using frozen competent cells is reduced compared to freshly prepared cells, but unless very high numbers of plaques are required, frozen cells are adequate and very convenient.
6. Plasmid selection operates by conferring antibiotic resistance on the host cells, which then form colonies after overnight incubation. An additional selection may be utilized with some plasmids, such as pUC and Bluescript, that express the α-fragment of the *LacZ* gene. Cells transformed with these plasmids form blue colonies when plated with IPTG and X-gal. Subcloning DNA into the plasmid disrupts expression of the α-fragment and results in white colonies.
7. The number of colonies or plaques that are obtained from each transfection varies enormously and is dependent on many factors, such as the amount of DNA added, the ratio of insert to vector in the ligation reaction, the efficiency of ligation, and the efficiency of transformation. As a guide, closed circular plasmid DNA should give over 10^7 colonies/µg using the Hanahan method. Thus, the control transfected with 10 pg of plasmid DNA should yield over 100 plaques. Vector DNA that has been linearized and then religated to an excess of insert with compatible ends will give anything from 10^2–10^4 fewer colonies/µg, depending on the religation efficiency. Thus, each test transfection of 40–400 ng may give between 40 and 40,000. For the calcium method, the numbers of colonies are approx 2- to 10-fold lower. Freshly prepared calcium-treated cells can be stored for up to 48 h in $CaCl_2$ at 4°C, and the efficiency of transfection increases up to sixfold over the first 24 h, but then declines to the original level *(4)*.
8. The volume of ligated DNA added to the cells should not exceed 10% of the total volume.
9. The length of time for heat-shock is calibrated for Falcon 2054 tubes, and for others, the heat-up time may be different.
10. M13 DNA can be introduced into *E. coli* in exactly the same way as for plasmid DNA. In this case, no antibiotic selection is used since the transformed cells are plated with a bacterial lawn, producing plaques as they infect their surrounding cells (*see* Chapter 41). The blue/white selection can be used in the same way as for plasmids. Remember to set up the plating cells in parallel to the transformation.

11. It is important that agar plates be thoroughly dried before top agar is poured onto them.

References

1. Mandel, M. and Higa, A. (1970) Calcium-dependent bacteriophage DNA infection. *J. Mol. Biol.* **53**, 159–162.
2. Hanahan, D. (1983) Studies on transformation of *Escherichia coli* with plasmids. *J. Mol. Biol.* **166**, 557–580.
3. Sambrook, J., Fritsch, E. F., and Maniatis, T. (1989) Preparation and transformation of competent *E. coli,* in *Molecular Cloning, A Laboratory Manual, 2nd ed. Plasmid Vectors*, Cold Spring Harbor Laboratory, Cold Spring Harbor, NY, pp. 1.74–1.84.
4. Dagert, M. and Ehrlich, S. D. (1979) Prolonged incubation in calcium chloride improves the competence of *Escherichia coli* cells. *Gene* **6**, 23–28.

CHAPTER 30

Transformation of Bacteria by Electroporation

Lucy Drury

1. Introduction

The use of an electrical field to permeabilize cells reversibly (electroporation) has become a valuable technique for transference of DNA into both eukaryotic and prokaryotic cells. Many species of bacteria have been successfully electroporated (1) and many strains of *E. coli* are routinely electrotransformed to efficiencies of 10^9 and 10^{10} transformants/µg DNA. Frequencies of transformation can be as high as 80% of the surviving cells and DNA capacities of nearly 10 µg of transforming DNA/mL are possible (2).

The benefit of attaining such high efficiency of transformation is apparent, for example, in the case of plasmid libraries. It is often preferable to construct a library in a plasmid owing to its small size and flexibility. In addition, it is invaluable where the use of a shuttle vector is required for the subsequent transfection of eukaryotic cells. Chemical methods of making cells transformation-competent are unable to produce high enough efficiencies to make this kind of library possible.

Several commercial machines are available that deliver either a square wave pulse or an exponential pulse. Since most of the published data has been obtained using an exponential waveform, this discussion will be confined to that pulse shape. An exponential pulse is generated by the discharge of a capacitor. The voltage decays over time as a function of the time constant τ.

$$\tau = RC \quad (1)$$

R is the resistance in ohms, C is the capacitance in Farads, and τ is the time constant in seconds. The potential applied across a cell suspension will be experienced by any cell as a function of field strength ($E = V/d$, where d is the distance between the electrodes) and the length of the cell. A voltage potential develops across the cell membrane, and when this exceeds a threshold level, the membrane breaks down in localized areas resulting in cell permeability to exogenous molecules. The permeability produced is transient, provided the magnitude and the duration of the electrical field do not exceed some critical limit, otherwise the cell is irreversibly damaged. Since there is an inverse relationship between field strength and cell size, prokaryotes require a higher field strength for permeabilization than do eukaryotic cells. If the voltage and therefore the field strength are reduced, a longer pulse time is required to obtain the maximum efficiency of transformation. However, this range of compensation is limited (2). Increasing the field strength causes a decrease in cell viability, and maximum transformation efficiencies are usually attained when about 30–40% of the cells survive.

This chapter will describe and discuss the methodology of bacterial electroporation with particular reference to *E. coli*.

2. Materials

2.1. Making Electrocompetent Bacteria

1. A suitable strain of *E. coli:* All strains tested by the author attain a higher transformation efficiency than if made competent by chemical means. See ref. *1* for a list of some strains that have been made electrocompetent.
2. L-Broth: 1% bacto-tryptone, 0.5% Bacto-yeast extract, 0.5% NaCl.
3. 0.1 m*M* HEPES, pH 7.0. This may be replaced by distilled H_2O.
4. Distilled H_2O: sterilized by autoclaving.
5. 10% Glycerol (v/v): in sterile distilled H_2O.

2.2. Electroporation of Competent Bacteria

1. Electroporator: Transformation requires a high-voltage electroporation device, such as the Bio-Rad (Richmond, CA) gene pulser apparatus used with the pulse controller, and cuvets with 0.1 or 0.2-cm electrode gap.
2. TE: 10 m*M* Tris-HCl, pH 8.0, 1 m*M* EDTA.
3. SOC: 2% bacto-tryptone, 0.5% bacto-yeast extract, 10 m*M* NaCl, 2.5 m*M* KCl, 10 m*M* $MgSO_4$, 20 m*M* glucose.

3. Methods

3.1. Making Electrocompetent Bacteria

1. Start with a freshly streaked agar plate of the chosen *E. coli* strain.
2. Pick a single colony, and grow an overnight culture in L-broth or any other suitable rich medium.
3. On the next day, inoculate 1 L of L-broth with 10 mL of the overnight culture, and grow at 37°C with good aeration; the best results are obtained with rapidly growing cells (*see* Note 1).
4. When the culture reaches an OD_{600} of between 0.5 and 1.0, place on ice. The optimum cell density may vary for each different strain, but the author has found that usually about 0.5 is the best.
5. Leave on ice for 15–30 min.
6. Centrifuge the bacteria for 6 min at $4000g_{max}$ keeping them at 4°C. Remove the supernatant and discard.
7. Resuspend the cells in an equal volume of 0.1 mM HEPES, previously chilled on ice (*see* Note 2).
8. Spin down cells at 4°C and resuspend in half the volume of HEPES. Care must be taken because the cells form a very loose pellet in these low ionic solutions.
9. Harvest at 4°C once again, and resuspend in 20 mL of ice-cold 10% glycerol.
10. Harvest for the last time, and resuspend in 2–3 mL of 10% glycerol. The final cell concentration should be about 3×10^{10} cells/mL.

The cells may be used fresh or frozen on dry ice and stored at –70°C where they will remain competent for about 6 mo. Cells may be frozen and thawed several times with little loss of activity (*see* Note 3).

3.2. Electroporation of Competent Bacteria

1. Chill the cuvets and the cuvet carriage on ice (*see* Note 4).
2. Set the apparatus to the appropriate settings, i.e., 25 µF capacitance and either 1.8 or 2.5 kV (for 0.1- or 0.2-cm cuvets, respectively). Set the pulse controller unit to 200 Ω.
3. Thaw an aliquot of cells on ice, or use freshly made cells (also kept on ice).
4. To a cold, 1.5-mL polypropylene tube, add 40 µL of the cell suspension (use 45 µL when using the 0.2-cm cuvet; *see* Note 5) and 1–5 µL of DNA in H_2O or a low-ionic-strength buffer, such as TE. Mix well and leave on ice for about 1 min (*see* Notes 6–9). There is no advantage to a longer incubation time (*see* Note 10).
5. Transfer the mixture of cells and DNA to a cold electroporation cuvet, and tap the suspension down to the bottom if necessary.

6. Apply one pulse at the above settings. This should produce a time constant of 4.6–4.8 ms (the field strength will be 18 kV/cm for 0.1-cm cuvets or 12.5 kV/cm for the 0.2 cuvets).
7. Immediately add 1 mL of SOC medium (kept at room temperature) to the cuvet. Resuspend the cells, remove to a 17 × 100-mm polypropylene tube, and incubate the cell suspension at 37°C for 1 h (see Note 11). Shaking the tubes at 225 rpm during this incubation may improve recovery of transformants.
8. Plate out appropriate dilutions on selective agar.

Possible problems that may be encountered are discussed in Notes 12–16.

4. Notes

1. To achieve highly electrocompetent *E. coli*, the cells must be fast-growing and harvested at early to mid-log phase. It is also important to keep the cells at 4°C and work as quickly as possible.
2. Washing and resuspending the bacteria in solutions of low ionic concentration are important to avoid arcing in the cuvet owing to conduction at the high voltages required for electroporation.
3. A 10% glycerol solution provides ideal cryoprotection for *E. coli* cells at –70°C. The cell suspension is frozen by aliquoting into prechilled 1.5-mL polypropylene tubes and placing in dry ice. Quick-freezing in liquid nitrogen may be deleterious (3). Several rounds of careful freeze-thawing on ice does not seem to affect the level of the cell's competence to a great extent.
4. Because of the high field strength necessary, it is best to perform the electroporation at 0–4°C for most species of bacteria. Electroporation of *E. coli* performed at room temperature results in a 100-fold drop in efficiency. This may be related to the state of the cell membranes or may be a result of the additional joule heating that occurs during the pulse (4).
5. When using cuvets with 0.2-cm electrode gap, 40 µL of cell suspension are just adequate to cover the bottom of the cuvet. It has therefore proven better practice to use 45 µL of the cell suspension to circumvent any small pipeting errors. Insufficient liquid in the cuvet results in arcing.
6. Transformation efficiency may be adjusted by changing the cell concentration. Raising cell concentration from 0.8 to 8×10^9/mL increases transformation efficiency by 10–20-fold (4). A steady increase in the number of transformants obtained has been found at cell concentrations of up to 2.8×10^{10}/mL using a fixed concentration of DNA (3).
7. Transforming DNA must be presented to the cells as a solution of low ionic strength. As mentioned in Note 2, high-ionic-strength solutions cause arcing in the cuvet or a very short pulse time with resulting cell death and loss of sample. Salts, such as CsCl and ammonium acetate, must be kept to 10 m*M* or less. It is advisable to have the DNA dissolved in TE or H_2O.

Electroporation

This is particularly relevant after a ligation, since the ligation buffer has an ionic concentration too high for use directly in an electroporation. The DNA must be precipitated in ethanol/sodium acetate (carrier tRNA can be used in the precipitation without affecting the transformation frequency). Alternatively, the ligation can be diluted 1/100 and 5 µL used for electroporation *(5)*.

8. The concentration of transforming DNA present during an electroporation is directly related to the proportion of cells that are transformed. With *E. coli* this relationship holds over several orders of magnitude, and at high DNA concentrations (up to 7.5 µg/mL) nearly 80% of the surviving cells are transformed *(2)*. This is in contrast to chemically treated competent cells where saturation occurs at DNA concentration 100-fold lower and where a much smaller fraction of the cells are competent to become transformed *(6)*. For purposes where a high efficiency, but a low frequency of transformation is required (for library construction where cotransformants are undesirable), a DNA concentration of <10 ng/mL and a cell concentration of $<3 \times 10^{10}$ is appropriate. Alternatively, when a high frequency of transformation is required, use 1–10 µg/mL, which transforms most of the surviving cells *(2)*.

9. The size and topology of the DNA molecules may affect transformation efficiency. It is reported that plasmids of up to 20 kb transform with the same molar efficiency as plasmids of 3 kb and converting these plasmids to a relaxed form does not affect their transforming activity *(2)*. Larger molecules can be taken up but at much lower efficiencies, for example, linear λ DNA (48 kb) has a molar transformation efficiency of 0.1% that of small plasmids *(2)*. No direct comparison between *E. coli* plasmids containing the same origin of replication, promoters, and markers, but differing only in size has been published, and in my hands, different plasmid constructs transform with different molar efficiencies. Powell et al. *(7)* have compared the uptake of related plasmids in *Streptococcus lactis,* and observed no clear relationship between size and molar transformation efficiency.

10. There is no evidence for binding of the DNA to the cell surface during the transformation process, and therefore increasing the preshock incubation time up to 30 min makes very little difference to the number of resulting transformants *(2)*. In support of this observation, experiments by Calvin et al. *(8)* show that when cells are mixed with radioactively labeled plasmid, only a small percentage of the label remains bound after two washes. In addition, certain species of bacteria, such as *Lactobacillus casei*, secrete nucleases, so increasing the preshock incubation time may be detrimental *(9)*.

11. Immediately after the pulse, *E. coli* cells are quite fragile, and rapid addition of the outgrowth medium greatly enhances their viability and transformation

efficiency. Even after 1-min delay, the efficiency drops by three- to five-fold and increases to 20-fold after 10 min *(2)*. Outgrowth is necessary for the cells to express any resistance marker introduced by the transforming plasmid, and is usually for an hour.

12. There are a number of causes of arcing in the cuvet. One reason could be that the ionic strength of the DNA solution or the cell suspension is too high. It is important that the DNA is resuspended in TE or H_2O. If it is a ligation mixture, it must be precipitated with $0.3 M$ Na acetate and 2–3 vol of ethanol or diluted 10- to 100-fold in TE or H_2O. The same problem can be caused by failure to tap the cell/DNA mixture to the bottom of the cuvet or too little solution in the cuvet (*see* Note 5).

    Another likely cause may be owing to the cuvets and the chamber having been chilled on ice and residual H_2O on the surfaces induced an arc. If you are electroporating many samples, it is not necessary to chill the carriage between every pulse, but it is a good idea to dry the carriage between every few samples, since condensation can accumulate and cause arcing.

13. Failure to obtain colonies after transformation could be the result of problems with the cells or the DNA. It is advisable to make a large quantity of an accurate dilution of a supercoiled plasmid, such as pUC18, to use as a positive control in all experiments. Use this routinely to check the cells you make (5 pg supercoiled DNA will give about 10,000 transformants if your cells are at efficiencies of 10^9 transformants/µg DNA.)

    If no colonies are obtained from the positive control, ensure that the growth conditions and harvesting of the cells were correct. The most competent cells are made from fast-growing cells harvested at early to midlog phase. Keep all the wash solutions at 4°C and keep the cells cold while harvesting. When making a new strain competent, it is best to harvest the cells at a range of densities at an OD_{600} of between 0.4 and 1.0. We have found the best density usually to be around 0.5.

    If the electrocompetent cells were previously stored at –70°C, ensure that they are still viable. To do this, plate out an appropriate dilution of the cells on a nonselective plate.

    Should the cells only transform with the control, first check the concentration of your DNA. It may also be possible that the DNA contains toxic contaminants, such as phenol or SDS. The viability of the cells after electroporation can be checked by plating a sample on a nonselective plate. A survival of 30–40% would be expected using the parameters set out in Section 3., but check against an equivalent aliquot of the cells transformed with the control DNA. If the DNA is contaminated, reprecipitate and wash with 70% ethanol, or extract the DNA with glass beads (*see* Chapter 28) to remove unwanted chemicals.

14. If a recently prepared batch of cells already tested for electrocompetence gives a reduced transformation efficiency, it is possibly because of problems with the electroporation. It is important that the cuvets and the carriage are chilled so that the starting temperature of the cells is 0–4°C. It is crucial to add the outgrowth medium (kept at room temperature) as quickly as possible to the cells after electroporation. Alternatively, the storage conditions of the cells may not have remained constant, for example, a freezer not maintaining its temperature.
15. An unexpectedly high apparent transformation efficiency may have a number of explanations. The simplest explanation is that the selective plates have exceeded their shelf-life. DNA contamination can also be a problem because of the high competency of the cells. It is important to maintain good sterile technique and careful use of micropipets to avoid crosscontamination with DNA used in previous experiments. Since electroporation can release plasmid from cells, the effects of contamination with previously transformed bacteria will be greatly heightened, especially if the plasmid is present at a high copy number in the contaminating cells.
16. The particular problems outlined above apply to *E. coli* problems encountered with other bacterial species could be owing to the characteristics of that strain. For example, if the bacterium is encapsulated, the entry of the DNA may be impeded, and some species secrete nucleases that could destroy the DNA. Certain types of bacteria may require a longer recovery time or a longer time to express the selective marker. If the size of the cell is unusual, it may require a different field strength. To establish electroporation conditions for a novel species, it is best to consult references concerning similar bacterial types (*see* ref. *1* for a list of references) for general parameters from which to optimize further.

References

1. Bacterial species that have been transformed by electroporation (1990) Bio-Rad Laboratories, 1414 Harbor Way South, Richmond, CA 94804. Bull. 1631.
2. Dower, W. J., Miller, J. F., and Ragsdale, C. W. (1988) High efficiency transformation of *E. coli* by high voltage electroporation. *Nucleic Acids Res.* **16,** 6127–6145.
3. Dower, W. J. (1990) Electroporation of bacteria: a general approach to genetic transformation, in *Genetic Engineering,* vol. 12 (Setlow, J. K., ed.), Plenum, New York, pp. 275–295.
4. Shigekawa, K. and Dower, W. J. (1988) Electroporation of eukaryotes and prokaryotes: a general approach to the introduction of macromolecules into cells. *Biotechniques* **6,** 742–751.
5. Willson, T. A. and Gough N. M. (1988) High voltage *E. coli* electrotransformation with DNA following ligation. *Nucleic Acids Res.* **16,** 11,820.

6. Hanahan, D. (1995) Techniques for transformation of *E. coli*, in *DNA Cloning. A Practical Approach*, vol. 1, 2nd ed. (Rickwood, D. and Hames, B. D., eds.) Oxford University Press, New York, pp. 1–36.
7. Powell, I. B., Achen, M. G., Hillier, A. J., and Davidson, B. E. (1988) A simple and rapid method for genetic transformation of *Lactic streptococci* by electroporation. *Appl. Environ. Microbiol.* **54,** 655–660.
8. Calvin, N. M. and Hanawalt P. C. (1988) High efficiency transformation of bacterial cells by electroporation. *J. Bacteriol.* **170,** 2796–2801.
9. Chassy, B. M. and Flickinger, J. L. (1987) Transformation of *Lactobacillus casei* by electroporation. *FEMS Microbiol. Lett.* **44,** 173–177.

CHAPTER 31

Preparation of Plasmid DNA Using Alkaline Lysis

Etienne Joly

1. Introduction

There are probably as many "miniprep" recipes as there are laboratories doing molecular biology. The following protocol is derived from the alkaline lysis recipe originally described by Birnboim and Doly (1), which was slightly modified in the 1982 edition of *Molecular Cloning: A Laboratory Manual (2)*. This technique relies of the lysing of bacteria by sodium hydroxide and SDS, followed by neutralization with a high concentration of potassium acetate, which results in the selective precipitation of the bacterial chromosomal DNA and other high-mol-wt cellular structures. The plasmid DNA, which stays in suspension, is then precipitated with isopropanol.

Short cuts have been added, and most liquid dispensing can be achieved using a repetitive pipeter. If handling only a few samples and using the shortest possible protocol, plasmid DNA can be obtained that is ready for restriction enzyme digestion in <10 min, with yields >20 µg plasmid from a 2-mL culture. After the initial characterization, it is possible to purify further some or all of the plasmid DNAs by RNase digestion and extraction with organic solvents. This further purified DNA is suitable for subcloning, sequencing, radiolabeling, and even transformation of eukaryotic cells. This protocol can be scaled up as the "maxiprep" to prepare large quantities of plasmid DNA. If an extremely clean preparation is required, the maxiprep can be separated from small traces of contaminating chromosomal DNA by banding on a cesium chloride/ethidium bromide gradient.

From: *Methods in Molecular Biology, Vol. 58: Basic DNA and RNA Protocols*
Edited by: A. Harwood Humana Press Inc., Totowa, NJ

2. Materials
2.1. Small-Scale Plasmid DNA Preparations
1. TB bacteria culture medium: For 1 L, mix 100 mL of KP solution to 900 mL of pre-TB. KP solution is made with 2.31 g of KH_2PO_4, 12.54 g of K_2HPO_4, and distilled water to 100 mL. Pre-TB is made with 12 g of bactotryptone, 24 g of bacto-yeast extract, 4 mL of glycerol, and water to 900 mL. These two solutions must be sterilized by autoclaving separately. Supplement with the appropriate antibiotic(s).
2. Sterile tubes: Must have a volume >10 mL to ensure good aeration.
3. 1.5-mL microfuge tubes, and a large metal rack. The wire-mesh kind that holds 13-mm diameter tubes works best for me.
4. Repetitive pipet with syringes: For example, Eppendorf combitips, which dispense multiples of 50, 100, and 250 µL. Each of these "disposable" syringes is dedicated to the handling of one solution, but can be reused hundreds of times.
5. Bacterial resuspension solution (BRS): 50 mM glucose, 50 mM Tris-HCl, pH 8.0, 10 mM EDTA. Keep at 4°C to prevent growth of contamination.
6. Lysis solution (LS): 200 mM NaOH, 1% SDS. Store at room temperature.
7. Neutralizing solution (KoAc): 3M potassium/5M acetate. For 100 mL, take 29.4 g of potassium acetate, add water to 88.5 mL, and 11.5 mL of glacial acetic acid. Store at room temperature.
8. Isopropanol.
9. 70% Ethanol.
10. TE: 10 mM Tris-HCl, pH 8.0, 1 mM EDTA.
11. RNase A: Purchased as a lyophilized powder (Sigma [St. Louis, MO] R9009). Make up as a solution in water at 10 mg/mL, and place in boiling water for 10 min to eliminate any residual DNase activity. Store aliquoted at –20°C.

2.2. Plasmid "Clean Up"
1. 0.4M ammonium acetate.
2. Phenol/chloroform: A 25:24:1 mix of TE-equilibrated phenol, chloroform, and isoamyl alcohol. Store at 4°C to slow down evaporation of chloroform.
3. Chloroform: A 24:1 mix of chloroform and isoamyl alcohol. Store at 4°C to slow down evaporation of chloroform.
4. 100% Ethanol.

2.3. Large-Scale Plasmid DNA Preparation
1. Large flasks: Flasks of >2 L in volume. Sterilize by autoclaving.
2. 500-mL centrifuge pots (and optionally 250 or 100 mL), and suitable rotors and centrifuges for these.

Plasmid DNA

3. $T_{50}E$: 50 mM Tris-HCl, pH 8.0, 10 mM EDTA.
4. 10-mL polypropylene disposable tubes.
5. Cesium chloride. Choose the molecular biology grade, but not necessarily the super-ultrapure reagent, which can cost more than twice as much.
6. Ethidium bromide: A 10 mg/mL solution in water. Ethidium bromide is highly mutagenic. Gloves should be worn when handling solutions that contain it, and it should be disposed of correctly (*see* ref. 2).
7. Ultracentrifuge tubes: Quick-seal tubes and heat sealer (Beckmann, Palo Alto, CA).
8. Syringes and needles: 2-mL syringes and 19-gage needles.
9. WSB: Water-saturated butanol.
10. 30-mL centrifuge tubes, such as Corex, and adequate rotor and centrifuge.

3. Methods

3.1. Small-Scale Plasmid DNA Preparations

For subcloning, it is often necessary to screen large numbers of bacterial colonies by restriction enzyme digestion of the plasmid DNA they contain. In these cases, it is necessary to be able to perform, rapidly and simultaneously, small-scale "minipreps" of their plasmid DNA.

1. Pick individual bacterial colonies and disperse into separate sterile tubes that contain 2 mL of TB (*see* Note 1). Close the tubes, and shake at 37°C overnight.
2. Transfer each culture to a labeled 1.5–mL tube, and spin for 30 s at 12,000g in the microfuge. The bacteria form a tight creamy pellet.
3. Decant the supernatant, and sit the tubes vertically in the tube rack for a few seconds (i.e., the time it takes to decant the others). Remove the last drop of liquid that collects at the bottom by aspiration with a fine, pulled, Pasteur pipet or a 200-µL Gilson tip at the end of the vacuum line.
4. Dispense 100 µL of BRS into each tube using the repetitive pipet. Close the tubes, and individually resuspend the bacteria in each tube by rubbing them across the metal rack (*see* Note 2).
5. Dispense 200 µL of LS into each tube using the repetitive pipet. Mix by inverting the tube and shaking by hand for a few seconds (*see* Notes 3 and 4). From turbid, the solution should rapidly become more transparent, indicating that the bacteria have been duly lysed. Allow at least 3 min for the lysis to occur, and make sure you leave the tubes standing for 30 s before opening them, so that the viscous liquid returns to the bottom.
6. Dispense 150 µL of KoAc into each tube using the repetitive pipet. Shake the tubes as in step 5. A white precipitate forms that contains the bacterial chromosomal DNA, most of the SDS, and a lot of other

undesirable bacterial components. Spin the tubes for 2–5 min at full speed in the microfuge.
7. While this is taking place, label a new set of tubes, and dispense 250 µL of isopropanol into each.
8. Remove the tubes from the microfuge, and carefully place them in a rack so that the white precipitate remains stuck to the tube wall. Remove the supernatant with a 1-mL adjustable pipet, avoiding the white precipitate as much as possible (*see* Note 5). This liquid phase is transferred to the new set of tubes, which contain isopropanol.
9. Close the tubes, and vortex them for a few seconds. An opalescent precipitate usually forms. Spin the tubes in the microfuge for 30 s at 12000*g*. The plasmid DNA precipitates as a white pellet.
10. Decant the supernatant, and wash the pellets by the addition of 750 mL of 70% ethanol, brief vortexing, and spinning for 30 s. Decant the ethanol, and respin for 10 s to collect the remaining ethanol at the bottom of the tubes. Carefully aspirate and leave the tubes to air-dry for 1–5 min.
11. Dispense 50 µL of TE into each tube, and resuspend the pellet (*see* Note 6).

If the plasmid grows at high copy number and your yields are correct, 2–3 µL of DNA solution should be sufficient for restriction enzyme digestion and agarose gel electrophoresis. The addition of RNase (1 µg/digest) to the digestion mix will remove the contaminating RNA prior to gel electrophoresis. Alternatively, it is possible to use a longer protocol that removes the RNA during the plasmid DNA preparation (*see* Notes 7 and 8).

3.2. Plasmid "Clean Up"

After checking plasmid DNAs by gel electrophoresis, they may require further work, such as DNA sequencing, subcloning, and DNA-mediated gene transfer. In these cases, it is advisable to purify the DNA further.

1. Add 50 µL of 4*M* ammonium acetate containing 200 µg/mL RNase A to each miniprep, and incubate at room temperature for 20 min.
2. Add 100 µL of phenol/chloroform to each DNA, vortex briefly, and spin at 12,000*g* for 2 min. Remove the top layer (which contains the DNA), and place in a fresh tube, which contains 100 µL of chloroform (*see* Note 9).
3. Vortex briefly, and spin at 12,000*g* for 2 min. Again remove the DNA in the top layer, and place in a second fresh tube, which contains 200 µL of 100% ethanol.
4. Vortex briefly to accelerate DNA precipitation, and spin for 5 min at room temperature.

Plasmid DNA

5. Rinse with 70% EtOH, respin briefly, aspirate the remaining EtOH, and air-dry for 5 min.
6. Resuspend in 50 µL of sterile TE.

3.3. Large-Scale Plasmid DNA Preparation

Large-scale plasmid DNA preparations, "maxipreps," can be carried out using a scale-up of the miniprep method.

1. Seed 250–500 mL of TB, and grow by shaking at 37°C overnight.
2. Transfer the culture to a suitable centrifuge pot (e.g., Beckman 355607 pots), and pellet at 5000g for 5 min.
3. Decant the supernatant, and drain the tube by inverting it for a few minutes or by aspiration with a clean Pasteur pipet at the end of a vacuum line.
4. Resuspend the bacteria in 20 mL of BRS.
5. Lyse the bacteria in 40 mL of LS. Shake by hand and leave for 10 min at room temperature.
6. Neutralize the solution with 30 mL of KoAC. Shake by hand, gently at the start and quite vigorously toward the end. A white precipitate should form.
7. Centrifuge at 6000g for 10 min at 4°C. Remove the liquid phase to a fresh centrifuge pot that contains 55 mL of isopropanol. Use a 25-mL pipet and avoid as much of the white precipitate as possible.
8. Vortex and centrifuge at 6000g for 5 min. at room temperature. Rinse the pellet with 10 mL of 70% EtOH, recentrifuge for 5 min, decant, and aspirate the last traces of alcohol with a Pasteur pipet on a vacuum line.
9. Resuspend the pellet in 4 mL of $T_{50}E$. Resuspension can be accelerated by shaking at 37°C (*see* Note 10).
10. Adjust the volume to exactly 4.5 mL (*see* Note 11), and transfer to a disposable 10-mL polypropylene tube. Add exactly 5 g of cesium chloride. Wait for it to dissolve, and then add 500 µL of ethidium bromide.
11. Centrifuge at 4000 or 5000 rpm for 10 min to remove the purple precipitate, and transfer to a 5-mL quick-seal ultracentrifuge tube using a Pasteur pipet.
12. Seal the tube (*see* Note 12) and spin at 200,000g (45,000 rpm) at 16°C for longer than 10 h. At the end of the centrifuge run, carefully remove the tube, and firmly clamp it. The plasmid comes to float as a red band around the middle of the tube (*see* Note 13). RNA sinks and therefore collects on the outside wall of the tube during centrifugation, whereas proteins float and collect on the inside wall.
13. Puncture the top of the tube with a syringe needle to create an "air hole." Using a needle placed on a syringe, puncture the wall of the tube, avoiding the RNA and protein deposits, and underneath the plasmid band. Bring the end of the needle just in contact with the bottom of the plasmid band, and slowly suck it into the syringe. Harvest as much of the band as possible, but keep the volume as low as possible, <1 mL.

14. Pull the needle out of the tube, but make sure that this is done above a suitable container, since the rest of the tube's contents will flow out. Transfer the DNA/ethidium bromide solution to a microfuge tube.
15. Fill the rest of the microfuge tube with WSB, vortex briefly, and spin for a few seconds. Ethidium bromide preferentially partitions into the butanol phase, which becomes a purple color. The butanol phase is removed and discarded. Repeat this extraction 4–7 times, depending on the quantity of plasmid present (*see* Note 14). As a rule of thumb, one should extract twice more after no pink color can be detected in the aqueous phase.
16. Transfer the aqueous layer to a 30-mL centrifuge tube, and make up to 5 mL with TE. Add 10 mL of 100% ethanol, vortex, and place at −20°C for 10 min or longer. Precipitate by spinning at 12,000*g* for 10 min at 4°C. Rinse the pellet with 70% ethanol, air-dry, and resuspend in 500 µL of TE (*see* Note 14).

4. Notes

1. Small culures, 2 mL, give optimal growth of the bacteria, and one can then pour the culture into a 1.5-mL Eppendorf tube the next morning, rather than having to pipet it. This saves time and pipet tips. Enough liquid will remain in the culture tubes to keep as a stock for freezing or seeding larger cultures.
2. Two or three firm rubs across the mesh at the base of a metal rack are very good for dissociating the bacterial pellet, and also give a musical dimension to the process. It is **crucial** that the bacterial pellet is fully resuspended before adding the lysis buffer.
3. Do not vortex at this stage, since this will shear the chromosomal DNA and result in a very messy preparation.
4. All tubes can be handled together by placing your free hand on top of all tubes to prevent them from flying across the room.
5. Some of the precipitate will float, so it is critical to use a pipet and disposable tips to recover the supernatant rather than pouring it.
6. I find that using a fresh 200-µL tip for each tube and leaving the tip in the tube is the most ergonomical (and economical) procedure. This tip can then be used to resuspend the DNA by pipetting up and down a few times, and finally to dispense a fraction of it for digestion with a restriction enzyme.
7. The following protocol is slightly longer, but results in DNA preparations that are cleaner and contain no RNA. This is achieved by replacing the glucose in the BRS with RNase A at 100 µg/mL. This solution is still kept at 4°C. The rest of the protocol is identical, apart from the following:
    a. The bacteria are resuspended in 300 µL of BRS-RNase, lysed with 300 µL of LS, and incubated for 15–20 min at room temperature in order for the RNase work.

Plasmid DNA

   b. The lysates are then neutralized with 300 µL of an alternative KoAc (2.55*M* potassium acetate, pH 4.8, with acetic acid).
   c. The plasmids are precipitated with 500 µL of isopropanol.
8. Another option for slightly cleaner DNA is to spin the KoAc precipitates at 4°C for 5–10 min, but the improvement is only marginal.
9. For phenol/chloroform extractions, I find that the most ergonomical way is to place the chloroform or the ethanol in the clean tubes before transferring the DNA. That way, the same tip can be used for dispensing into all tubes without worrying about crosscontamination.
10. A quicker, but dirtier alternative to banding on a cesium chloride gradient is to treat the DNA with RNase followed by phenol/chloroform extraction. Add RNase A to 100 µg/mL final and NaCl to 100 m*M* final. Incubate at 37°C for 30 min. Extract with 4 mL phenol-chloroform, then 4 mL chloroform, and finally precipitate with 8 mL of 100% EtOH. Vortex and spin at 8000 rpm for 10 min. Rinse once with 5 mL 70% EtOH, spin to collect residual liquid, air-dry, and resuspend in 1 mL TE. Analyze 1 µL on an agarose gel to estimate concentration and yield, and dilute preparation if necessary. A second alternative is to use a diatomateous earth column (*see* Chapter 33).
11. To band the plasmid DNA, it is important to establish the correct starting density of the cesium/ethidium bromide solution. Precision is required when weighing the cesium chloride and measuring the volumes of solutions. A 10-mL pipet is a convenient way to measure the total volume of the solution. If using different ultracentrifuge tubes, volumes can be easily adapted. To obtain the right concentration, the rule is 1 g of cesium chloride added to 1 mL of aqueous liquid.
12. If the density of cesium chloride solution is correct (i.e., 1.55 g/mL) a 5-mL quick-seal tube should weigh close to 9.5 g.
13. If plasmid quantities are low (<100 µg), the band is better visualized using a hand-held long-wave UV lamp. The exposure time should be kept as short as possible, since UV light damages DNA.
14. With a bit of experience, you will be able to estimate the quantity of DNA from the intensity of the band in the gradient, and adjust the volume of TE accordingly to your requirements (1 µg/mL is quite appropriate for most applications).

References

1. Birnboim, H. C. and Doly, J. (1979) A rapid alkaline extraction procedure for screening recombinant plasmid DNA. *Nucleic Acids Res.* **7,** 1513.
2. Maniatis, T., Fritsch, E. F., and Sambrook, J. (1982) *Molecular Cloning: A Laboratory Manual,* Cold Spring Harbor Laboratory, Cold Spring Harbor, NY, pp. 368,369.

CHAPTER 32

The Rapid Boiling Method for Small-Scale Preparation of Plasmid DNA

Adrian J. Harwood

1. Introduction

The classic alternative to the alkaline lysis method for plasmid DNA preparation (*1; see* Chapters 31 and 33) is that of Holmes and Quigley *(2),* and is commonly known as the rapid boiling method. This method is based on exactly the same principles as the alkaline lysis method. The cells are partially lysed, allowing plasmid molecules to escape, while most of the genomic DNA is trapped in the cell debris, which is then spun out. The remaining genomic DNA is removed by a denaturing step, in this case, high temperature rather than high pH, followed by rapid reannealing. As a consequence of their supercoiling, plasmid molecules are able to reassociate rapidly in these conditions, whereas genomic DNA remains denatured and is lost on the subsequent ethanol-precipitation step.

This is a rapid and convenient method for making large numbers of small-scale plasmid preparations (minipreps). The DNA is of good quality, and can be used directly for restriction enzyme digestion and subcloning. With only a small amount of further processing, it can be used as a template for DNA sequencing. The rapid boiling method, however, is less convenient for large-scale preparations, for which the alkaline lysis method is better suited.

From: *Methods in Molecular Biology, Vol. 58: Basic DNA and RNA Protocols*
Edited by: A. Harwood Humana Press Inc., Totowa, NJ

2. Materials

1. STET: 5% Triton X-100, 50 mM Tris-HCl, pH 8.0, 50 mM EDTA, pH 8.0, 8% sucrose. Can be stored at room temperature.
2. Lysozyme: dry powder. Store at –20°C.
3. Isopropanol:propan-2-ol.
4. 70% Ethanol.
5. TE: 10 mM Tris-HCl, pH 7.5, 1 mM EDTA.
6. A boiling water bath: A beaker of water heated by a bunsen burner will suffice. An opened bottom tube rack is required because the tubes must be directly placed in the water to achieve rapid heating.

3. Methods

1. Set up a culture for each miniprep by inoculating 2–3 mL of L-broth, containing an appropriate antibiotic (e.g., 100 µg/mL ampicillin) with a bacterial colony. Grow overnight at 37°C with vigorous shaking (*see* Note 1).
2. Before starting the miniprep, start boiling the water and make up a fresh 1 mg/mL lysozyme STET mix.
3. Fill a 1.5-mL labeled microfuge tube with an aliquot from each culture. Pellet the bacteria by centrifugation for 1 min at 12,000g (*see* Note 2). Carefully aspirate off the supernatant on a water pump using a drawn-out Pasteur pipet.
4. Vortex each pellet for a few seconds to break up the pellet. Add 200 µL of STET to each tube. The pellet should now easily resuspend by vortexing (*see* Note 3).
5. Immediately place the tubes in the open-bottom rack, and place in the boiling water for exactly 45 s. Ensure each tube is at least half submerged.
6. Centrifuge the tubes at 12,000g for 10 min. A large, sticky, and loose pellet should form.
7. Remove the pellet from each tube by "fishing" it out with a wooden toothpick. Since the pellet is quite slippery, it is useful to have a paper tissue at the top of the tube to catch the pellet and prevent it from slipping back down into the tube.
8. Add 200 µL of isopropanol to each tube, and centrifuge at 12,000g for 5 min.
9. Aspirate off the supernatant, and wash the pellet in 500 µL of 70% ethanol. Centrifuge the tube for 1 min to compact the pellet, and then aspirate off the 70% ethanol.
10. Air-dry the pellets for 10 min, and resuspend each in 100 µL of TE buffer. Vortex and shake for 10 min before use to ensure complete dissolution.
11. Use 2–10 µL (equivalent to 100 ng of plasmid for most vectors) per restriction enzyme digest and analyze by gel electrophoresis.

4. Notes

1. If the plasmid has a high copy number, growth time can be reduced to approx 6 h, enabling the whole process to be carried out in a day.
2. The short centrifugation time leaves a loose pellet that is easier to resuspend.
3. If the pellet does not readily resuspend, pipet the solution up and down to dislodge it. Do not suck the pellet directly into the pipet tip.

References

1. Holmes, D. S. and Quigley, M. (1981) A rapid boiling method for the preparation of bacterial plasmids. *Anal. Biochem.* **114,** 193–197.

CHAPTER 33

Plasmid Preparations with Diatomaceous Earth

Laura Machesky

1. Introduction

Plasmid purification by alkaline lysis is one of the most generally used molecular biology techniques. Traditionally, the cleanest plasmid DNA has been made by banding the DNA on a cesium chloride gradient. This, however, is time-consuming and expensive, so alternative plasmid purification methods have been sought. One that is becoming more widespread is the use of silica-based DNA-binding matrices. Such matrices are available commercially, but can also be easily made in the laboratory. In this chapter, a protocol for large-scale plasmid DNA preparation using diatomaceous earth is presented. This protocol is based on methods of Boom et al. *(1)* and of Carter and Milton *(2)*, as well as recommendations for homemade DNA purification kits in TIBS *(3)*. Although this protocol is a large-scale preparation, "maxiprep" and yields milligrams of DNA, it can also be adapted for the small-scale, "miniprep" *(2)* to yield microgram quantities.

2. Materials

All solutions are made with sterile, distilled water and should be stored at room temperature unless noted.

2.1. Preparation of Plasmid DNA

1. 2X TY: To make 1 L, dissolve 16 g of bacto-tryptone, 10 g of yeast extract, and 10 g of NaCl. Sterilize by autoclaving.
2. Centrifuge bottles: 500- and 250-mL centrifuge bottles that are autoclaved or washed in 1M NaOH.

3. Solution 1: 50 mM glucose, 25 mM Tris-HCl, pH 7.5, 10 mM EDTA.
4. Lysozyme solution: a 40 mg/mL solution of lysozyme made in solution 1. Make immediately before using.
5. Solution 2: 0.2M NaOH, 1% SDS. This solution should be made just before use.
6. Solution 3: 5M potassium acetate, pH 4.8. Glacial acetic acid is added to 3M potassium acetate until the desired pH is reached.
7. 70% Ethanol and 80% ethanol.
8. TE: 1 mM Tris-HCl, pH 7.5, 1 mM EDTA.

2.2. Column Purification of DNA

1. Resin buffer: 7M guanidine-HCl, pH 6.5. It is not necessary to adjust the pH, since it dissolves to this value.
2. Stock resin slurry: Mix 5 g of diatomaceous earth with 50 mL of resin buffer. This author uses acid-washed, calcined diatomaceous earth from Sigma (St. Louis, MO) (cat no. D-5384) (*see* Note 1).
3. Suitable columns: a column that fits into a 50-mL conical tube with a diameter of 1.5 cm (*see* Notes 2 and 3). Econo-columns from Bio-Rad (Hercules, CA) will work if shortened using a hot razor blade to cut off the tops.
4. Column wash solution: 50% ethanol, 100 mM NaCl, 10 mM Tris-HCl, pH 7.5, 5 mM EDTA.

3. Methods

3.1. Preparation of Plasmid DNA

1. Pick a plasmid-containing bacterial colony from a freshly streaked plate (*see* Note 4), and seed into a 4-L flask containing 500 mL of 2X TY medium/ antibiotics. Grow overnight in an orbital incubator shaking at 200 rpm and at 37°C.
2. Pellet the cells by centrifugation in a 500-mL centrifuge bottle at 5000g for 10 min. Drain the bottle well by tipping off fluid and then inverting on paper towels for a few minutes (*see* Note 5).
3. Resuspend the cell pellet in 18 mL of solution 1. Add 2 mL of lysozyme solution, and mix by stirring with a pipet. Leave for 10 min at room temperature.
4. Add 40 mL of solution 2. Swirl to mix the solution, and incubate for 5 min on ice.
5. Add 20 mL of solution 3. Swirl to mix, and incubate for at least 15 min on ice. A white precipitate will form.
6. Centrifuge mixture at 8000g for 10 min. Pour the supernatant through cheesecloth into a clean 250-mL centrifuge bottle.
7. Add 45 mL of isopropanol to the supernatant. Mix and centrifuge at 8000g for 10 min.
8. Carefully drain the tube, and wash the pellet with 70% ethanol. Aspirate the excess ethanol, and dry on the bench for 10–20 min. (*see* Note 6).

9. Dissolve the pellet in 2 mL of TE. This can take some time. This author adds the TE to the pellet and leaves it sitting on the bench for approx 10 min. This initiates the dissolution, but this author then helps the process by pipeting up and down with a 1-mL pipet tip. Mild vortexing also helps for small plasmids.

3.2. Column Purification of DNA

1. Add the DNA solution from Section 3.1., step 9 to a 15-mL conical tube. Add 10 mL of stock resin slurry, and shake the resin slurry before use to ensure that it is in solution. Mix the DNA and resin by inverting a few times.
2. Pour the resin–DNA slurry into the column, and suspend over a beaker or 50-mL conical tube.
3. Fit the column onto a vacuum line (one fitted to a water pump is adequate) to pull the buffer through the column (*see* Note 7). If necessary, use a smaller rubber tube as an adaptor to ensure a good fit to the vacuum line.
4. Rinse the tube that originally contained the DNA and slurry with 13 mL of column wash solution and add to the column. Use the vacuum to pull the wash solution through. Repeat.
5. Rinse the column with 5 mL of 80% ethanol, pulling through by the vaccuum. Leave the column on the vacuum for an additional 10 min for the column to dry (*see* Note 8).
6. Remove the column from the vacuum line, add 1.5 mL of water or TE preheated to 65–70°C, and wait 1 min.
7. Place the column inside a larger tube, e.g., a 50-mL conical tube, and elute the DNA by spinning at 1300g for 5 min in a clinical centrifuge or fixed-angle rotor. The eluate collects in the bottom of the larger tube and can then be transferred to a 1.5-mL tube for storage.

   A typical yield from a high copy number plasmid is between 2 and 5 mg of DNA (*see* Note 9). The DNA can be used for double-stranded sequencing, subcloning, and transfection of eukaryotic cells.

4. Notes

1. Other types of resins suggested by Hengen *(3)* are crushed flint glass from scintillation vials, Cellite™, or silical particles. He also suggests making columns from a syringe cartridge and filter paper or microcentrifuge tubes and siliconized glass filters.
2. For the maxipreps, it is important to have a wide column to prevent clogging of the column frit by the DNA solution.
3. It is possible to reuse some columns from the commonly used maxiprep kits by washing in NaOH or autoclaving *(3)*.
4. Although it is preferable to use a single colony from a freshly streaked plate, it is possible to cut a corner by taking a stab directly from a glycerol

stock of cells. If only DNA is available, it is possible to transform bacteria and then use the transformed cells to seed the culture directly. This last method runs the risk of obtaining a heterogeneous population of plasmid molecules.
5. Cell pellets may be frozen at –20°C and stored at this stage.
6. The pellets do not need to be absolutely dry, and carefully removing all of the ethanol with a Pasteur pipet a few times can substitute for the drying time.
7. It is not necessary to save the washes unless the DNA sample is very precious, since most of the DNA, up to 5 mg, will bind to the resin at this stage.
8. An alternative is to spin at 2000g in a clinical centrifuge for 10 min inside a larger tube, e.g., a 50-mL conical tube.
9. For a minipreparation protocol, *see* Carter and Milton *(2)*.

References

1. Boom, R., Sol, C. J. A., Salimans, M. M. M., Jansen, C. L., Wertheim van Dillen, P. M. E., and van der Noordaa, J. (1990) Rapid and simple method for purification of nucleic acids. *J. Clin. Microbiol.* **28,** 495,496.
2. Carter, M. J. and Milton, I. D. (1993) An inexpensive and simple method for DNA purifications on silica particles. *Nucleic Acids Res.* **21,** 1044.
3. Hengen, P. N. (1994) Methods and reagents: on the magic of mini-preps. *TIBS* **19,** 182,183.

PART V
PCR Techniques

CHAPTER 34

Polymerase Chain Reaction

Beverly C. Delidow, John P. Lynch, John J. Peluso, and Bruce A. White

1. Introduction

The melding of a technique for repeated rounds of DNA synthesis with the discovery of a thermostable DNA polymerase has given scientists the very powerful technique known as polymerase chain reaction (PCR). PCR is based on three simple steps required for any DNA synthesis reaction: (1) *denaturation* of the template into single strands; (2) *annealing* of primers to each original strand for new strand synthesis; and (3) *extension* of the new DNA strands from the primers. These reactions may be carried out with any DNA polymerase and result in the synthesis of defined portions of the original DNA sequence. However, in order to achieve more than one round of synthesis, the templates must again be denatured, which requires temperatures well above those that inactivate most enzymes. Therefore, initial attempts at cyclic DNA synthesis were carried out by adding fresh polymerase after each denaturation step *(1,2)*. The cost of such a protocol becomes rapidly prohibitive.

The discovery and isolation of a heat-stable DNA polymerase from a thermophilic bacterium, *Thermus aquaticus (Taq)*, enabled Saiki et al. *(3)* to synthesize new DNA strands repeatedly, exponentially amplifying a defined region of the starting material, and allowing the birth of a new technology that has virtually exploded into prominence. Not since the discovery of restriction enzymes has a new technique so revolutionized molecular biology. There are scores of journal articles published *per month* in which PCR is used, as well as an entire journal (at least one) devoted to it. To those who use and/or read about PCR every day, it is remarkable that this method is little more than 10 years old.

One of the great advantages of PCR is that, although some laboratory precaution is called for, the equipment required is relatively inexpensive and very little space is needed. The only specialized piece of equipment needed for PCR is a thermal cycler. Although it is possible to perform PCR without a thermal cycler—using three water baths at controlled temperatures—the manual labor involved is tedious and very time-consuming. A number of quality instruments are now commercially available. A dedicated set of pipets is useful, but not absolutely necessary. If one purchases oligonucleotide primers, all of the other equipment required for PCR is readily found in any laboratory involved in molecular biology. Thus, a very powerful method is economically feasible for most research scientists.

The reader is referred to vol. 15 of this series and to several reviews *(4,5)* for the wide variety of applications that use PCR. In this chapter, we outline the isolation of DNA and RNA as templates, the basic PCR protocol, and several common methods for analyzing PCR products.

2. Materials
2.1. Isolation of DNA and RNA Templates
2.1.1. Isolation of DNA

1. Source of tissue or cells from which DNA will be extracted.
2. Dounce homogenizer.
3. Digestion buffer: 100 mM NaCl, 10 mM Tris-HCl, pH 8.0, 25 mM EDTA, 0.5% SDS.
4. Proteinase K: 20 mg/mL. Store at –20°C.
5. a. Buffered phenol *(6,7)*: Phenol is highly corrosive; wear gloves and protective clothing when handling it. Use only glass pipets and glass or polypropylene tubes. Phenol will dissolve polystyrene plastics.
   b. Buffering solutions: 1M Tris base; 10X TE, pH 8.0 (100 mM Tris-HCl, pH 8.0, 10 mM EDTA); 1X TE, pH 8.0 (10 mM Tris, pH 8.0, 1 mM EDTA). To a bottle of molecular biology grade recrystallized phenol add an equal volume of 1M Tris base. Place the bottle in a 65°C water bath and allow the phenol to liquify (approx 1 h). Transfer the bottle to a fume hood and allow it to cool. Cap the bottle tightly and shake to mix the phases, **point the bottle away**, and vent. Transfer the mix to 50-mL screw-top tubes by carefully pouring or using a glass pipet. Centrifuge at 4000g for 5–10 min at room temperature to separate the phases.

PCR Basic Protocols

Remove the upper aqueous phase by aspiration. To the lower phase (phenol) add an equal volume of 10X TE, pH 8.0. Cap tubes tightly, shake well to mix, and centrifuge again. Aspirate the aqueous phase. Re-extract the phenol two or three more times with equal volumes of 1X TE, pH 8.0, until the pH of the upper phase is between 7 and 8 (measured using pH paper). Aliquot the buffered phenol, cover with a layer of 1X TE, pH 8.0, and store at –20°C.
6. $CHCl_3$: chloroform.
7. 70% Ethanol.
8. TE buffer, pH 8.0: 10 mM Tris-HCl, pH 8.0, 1 mM EDTA.
9. Phosphate-buffered saline (PBS): 20X stock (2.74M NaCl, 53.6 mM KCl, 166 mM Na_2HPO_4, 29.4 mM KH_2PO_4, pH 7.4). Make up in deionized distilled water, filter through a 0.2-μm filter, and store at room temperature. For use, dilute 25 mL of 20X stock up to 500 mL with deionized distilled water and add 250 μL of 1M $MgCl_2$. Sterile-filter and store at 4°C.
10. 7.5M Ammonium acetate.
11. RNase A. Prepare at 10 mg/mL in 10 mM Tris-HCl, pH 7.5, 15 mM NaCl. Incubate at 100°C for 15 min and allow to cool to room temperature. Store at –20°C.
12. 20% SDS: (w/v) in deionized water.

2.1.2. Isolation of RNA

2.1.2.1. ISOLATION OF RNA BY CsCl CENTRIFUGATION (SEE NOTE 1)

13. Source of tissue or cells from which RNA will be extracted.
14. 2-mL Wheaton glass homogenizer.
15. Guanidine isothiocyanate/β-mercaptoethanol solution (GITC/βME): 4.2M guanidine isothiocyanate, 0.025M sodium citrate, pH 7.0, 0.5% *N*-laurylsarcosine (Sarkosyl), 0.1M β-mercaptoethanol. Prepare a stock solution containing everything except β-mercaptoethanol in deionized distilled water. Filter-sterilize using a Nalgene 0.2-μm filter (Nalge Co., Rochester, NY) (*see* Note 2). Store in 50-mL aliquots at –20°C. To use, thaw a stock tube, transfer the required volume to a fresh tube, and add 7 μL of β-mercaptoethanol/mL of buffer. Guanidine isothiocyanate and β-mercaptoethanol are strong irritants handle them with care.
16. 1-mL tuberculin syringes with 21-g needles.
17. Ultraclear ultracentrifuge tubes; 11 × 34 mm (Beckman #347356).
18. Diethylpyrocarbonate (DEPC); 97% solution; store at 4°C.
19. DEPC-treated water *(6,7)*. Fill a baked glass autoclavable bottle to two-thirds capacity with deionized distilled water. Add diethyl pyrocarbonate to 0.1%, cap, and shake. Vent the bottle, cap loosely, and incubate at 37°C for at least 12 h (overnight is convenient). Autoclave on liquid cycle for 15 min to inactivate the DEPC. Store at room temperature.

20. 200 m*M* EDTA, pH 8.0. Use molecular biology grade disodium EDTA. Make up in deionized distilled water and filter through a 0.2-μm filter. Place in an autoclavable screw-top bottle. Treat with DEPC as described in the preceding step for DEPC water. Store at room temperature.
21. CsCl: molecular biology grade. For 20 mL, place 20 g of solid CsCl in a sterile 50-mL tube. Add 10 mL of 200 m*M* EDTA, pH 8.0 (DEPC-treated). Bring volume to 20 mL with DEPC water. Mix to dissolve. Filter through a 0.2-μm filter and store at 4°C.
22. TE buffer, pH 7.4: 10 m*M* Tris-HCl, pH 7.4, 1 m*M* EDTA. Make a solution of 10 m*M* Tris-HCl and 1 m*M* EDTA, pH 7.4, in DEPC water (*see* Note 3). Filter through a 0.2-μm filter, autoclave 15 min on liquid cycle, and store at room temperature.
23. TE-SDS: Make fresh for each use. From a stock solution of 10% SDS in DEPC water, add SDS to a concentration of 0.2% to an aliquot of TE, pH 7.4.
24. 4*M* NaCl. Make up in deionized distilled water and DEPC treat. Autoclave 15 min on liquid cycle and store at room temperature.
25. Polyallomer 1.5-mL microcentrifuge tubes, for use in an ultracentrifuge (Beckman #357448, Beckman Instrument Inc., Fullerton, CA).
26. RNasin RNase inhibitor, 40 U/μL (Promega, Madison, WI). Store at –20°C.
27. Beckman TL-100 table-top ultracentrifuge, TLS 55 rotor, and TLA-45 rotor.

2.1.2.2. Isolation of RNA by Guanidine/Phenol (RNAzol™) Extraction

28. RNAzol reagent (TEL-TEST, Inc., Friendswood, TX). This reagent contains guanidine isothiocyanate, β-mercaptoethanol, and phenol; handle with care.
29. Glass-Teflon homogenizer.
30. Disposable polypropylene pellet pestle and matching microfuge tubes (1.5 mL) (Kontes Life Science Products, Vineland, NJ).
31. CHCl$_3$ (ACS grade).
32. Isopropanol (ACS grade). Store at –20°C.
33. 80% Ethanol. Dilute 100% ethanol with DEPC-treated H$_2$O and store at –20°C.

2.1.3. Synthesis of Complementary DNAs (cDNAs) from RNA

34. Oligo dT$_{18-20}$ primer (Pharmacia, Piscataway, NJ). Dissolve 5 OD U in 180 μL of sterile water to give a concentration of 1.6 μg/μL.
35. Specific primer, optional. Choose sequence and obtain as for PCR primers.
36. MMLV reverse transcriptase (200 U/μL) with manufacturer-recommended buffer and 0.1*M* DTT.
37. Deoxynucleotides dATP, dCTP, dGTP, and dTTP. Supplied as 10 mg solids. To make 10 m*M* stocks: Resuspend 10 mg of dNTP in 10% less sterile

water than is required to give a 10 mM solution. Adjust the pH to approximate neutrality using sterile NaOH and pH paper. Determine the exact concentration by OD, using the wavelength and molar extinction coefficient provided by the manufacturer for each deoxynucleotide. For example, the A_m (259 nm) for dATP is 15.7×10^3; therefore a 1:100 dilution of a 10 mM solution of dATP will have an A_{259} of $(0.01M \times 15.7 \times 10^3$ OD U/M) $\times 1/100 = 1.57$. If the actual OD of a 1/100 dilution of the dATP is 1.3, the dATP concentration is $1.3/1.57 \times 10$ mM = 8.3 mM. Store deoxynucleotides at $-20°C$ in 50–100-μL aliquots. Make a working stock containing 125 μM of each dNTP in sterile water for cDNA synthesis or for PCR. Unused working stock may be stored at $-20°C$ for up to 2 wk.

2.2. Performing PCR (see Note 4)

38. Oligonucleotide primer: These are complementary to the 5' and 3' ends of the sequence to be amplified and can be obtained commercially. Store at $-20°C$.
39. Sterile UV-irradiated water (*see* Note 5). Sterile-filter deionized distilled water. UV irradiate for 2 min in a Stratagene (La Jolla, CA) Stratalinker UV crosslinker (200 mJ/cm^2) *(8)* or at 254 and 300 nm for 5 min *(9)*. Store at room temperature.
40. PCR stock solutions: Dedicate these solutions for PCR use only. Prepare the following three solutions, filter-sterilize, and autoclave 15 min on liquid cycle: 1M Tris-HCl, pH 8.3; 1M KCl; and 1M MgCl$_2$.
41. 10X PCR buffer: 100 mM Tris-HCl, pH 8.3; 500 mM KCl; 15 mM MgCl$_2$; 0.01% (w/v) gelatin. This buffer is available from Perkin-Elmer/Cetus. Per milliliter of 10X buffer combine 100 μL of 1M Tris-HCl, pH 8.3, 500 μL of 1M KCl, 15 μL of 1M MgCl$_2$, and 375 μL of UV-irradiated sterile water. Make up a 1% solution of gelatin in UV-irradiated sterile water. Heat at 60–70°C, mixing occasionally, to dissolve the gelatin. Filter the gelatin solution while it is still warm through a 0.2-μm filter, and add 10 μL of gelatin to each milliliter of 10X PCR buffer. Store PCR buffer in small aliquots (300–500 μL) at $-20°C$. As an extra precaution, the 10X buffer may be UV-irradiated before each use.
42. 10 mM Deoxynucleotide stocks (dATP, dCTP, dGTP, and dTTP), made up in UV-irradiated sterile water; *see* Section 2.1.4., item 5.
43. 1.25 mM Deoxynucleotide working stock. A 1.25 mM solution of each nucleotide made up in UV-irradiated sterile water.
44. Light mineral oil.
45. 7.5M Ammonium acetate, filter through a 0.2-μm filter and store at room temperature.
46. *Taq* DNA polymerase.

2.3. Analysis of PCR Products by Nested PCR (10)
47. Low-melting-point agarose.
48. Agarose gel electrophoresis reagents (Section 2.3.1.1., items 2–5).
49. Oligonucleotide primers complementary to internal portions of the DNA amplified (nested primers).

2.4. Analysis of PCR Products by Polyacrylamide Gel Electrophoresis of Directly Labeled PCR Products
50. α^{32}P-dCTP, 3000 Ci/mmol.
51. 3MM Filter paper.
52. X-ray film.

3. Methods
3.1. Isolation of DNA and RNA Templates
3.1.1. Isolation of DNA (7)

The following method works well for isolation of DNA for PCR from larger tissue samples or for bulk preparations of DNA from cultured cells.

1. Remove tissue into ice-cold PBS. Weigh tissue and mince with a razor blade. For cultured cells, collect by centrifugation, wash once in ice-cold PBS, and resuspend in 1 pellet vol of PBS.
2. Transfer tissue or cells to a Dounce homogenizer containing 12 mL of digestion buffer/g of tissue (per mL of packed cells).
3. Homogenize by 20 gentle strokes using a B pestle. Keep on ice.
4. Transfer the sample into a test tube, add proteinase K to a final concentration of 100 µg/mL, and incubate at 50°C overnight.
5. Extract the sample twice with an equal volume of phenol/CHCl$_3$ (1:1 by volume).
6. Extract twice with an equal volume of CHCl$_3$.
7. Add 0.5 vol of 7.5M ammonium acetate and 2 vol of 100% ethanol. Mix gently. The DNA should immediately form a stringy precipitate.
8. Recover the DNA by centrifugation at 12,000g for 15 min at 4°C.
9. Rinse pellet with 70% ethanol, decant, and air-dry.
10. Resuspend DNA in TE buffer, pH 8.0 (7–10 mL/g of tissue). Resuspension can be facilitated by incubation of sample at 65°C with gentle agitation.
11. Add SDS to final concentration of 0.1% and RNase A to 1 µg/mL. Incubate at 37°C for 1 h.
12. Re-extract with phenol/CHCl$_3$, precipitate, and resuspend DNA as described above in steps 5–10. Keep the DNA in ethanol at 4°C for long-term storage.

3.1.2. Isolation of RNA

There are a number of protocols now available for the isolation of RNA from cells or tissues. The following are two procedures we routinely use to isolate RNA from small tissue samples or from cultured cells. One procedure more rigorously removes DNA by centrifugation of the RNA through a CsCl cushion. The other relies on the extraction of RNA out of a guanidine solution and is less time-consuming.

3.1.2.1. Isolation of RNA by CsCl Centrifugation *(11)*

We have used this procedure for isolating RNA from whole rat ovaries (up to six ovaries, or about 150 mg of tissue, per sample), from ovarian granulosa cells and from nuclei of GH_3 pituitary tumor cells (nuclei from up to 5×10^7 cells). This procedure requires more time than the following guanidine/phenol extraction, but we found it gives cleaner RNA preparations from ovarian tissue, which contains not only DNA, but also substantial lipid deposits. The procedure is also recommended for preparing nuclear RNA because of the much higher DNA content of nuclei as opposed to whole cells or tissues.

1. Remove the tissue from the animal within several minutes of death. Place in ice-cold PBS and trim off fat and/or fascia if necessary. Cut large pieces of tissue into smaller pieces (2- to 3-mm cubes) (*see* Note 6).
2. Place the tissue in a 2-mL Wheaton glass homogenizer containing 1 mL of GITC/βME buffer. Homogenize by hand until no visible clumps remain (*see* Notes 7 and 8).
3. Transfer the sample to a 5-mL or 15-mL Falcon tube. To shear the DNA, draw the homogenate up into 1-mL tuberculin syringe with a 21-gage needle. Pass the homogenate up and down through the needle, avoiding foaming, until it becomes less viscous and can be released in individual drops (*see* Note 9).
4. Rinse Beckman Ultraclear centrifuge tubes with 0.3 mL of GITC/βME buffer and allow to dry inverted. Turn dried tubes up and place 875 μL of CsCl solution into the bottom of each tube.
5. Add 300 μL of GITC/βME to each tissue sample. Mix. Layer each entire sample (1.3 mL) on top of a CsCl cushion, taking care not to disrupt the boundary.
6. Fill each tube with sample and/or GITC/βME to within 2 mm of the top. Balance tubes to within 0.01 g with GITC/βME.
7. Load the tubes into the buckets of a Beckman TLS-55 rotor and centrifuge at 100,000*g* for 3 h at 16°C. This pellets the RNA, but not DNA (*see* Note 10).

8. Remove the tubes from the rotor buckets. Empty by rapid inversion and immediately place the inverted tubes in a rack or on a clean paper towel to drain-dry for about 15 min. Do not right the tubes until they dry.
9. Using a clean Kimwipe, remove the last traces of liquid from the sides of the tube, without touching the bottom. The RNA pellet will not be visible.
10. Add 400 µL of TE-SDS to the bottom of each tube, without allowing the solution to run down the sides. Cover the tubes and place in a rack on a rotary platform. Solubilize the RNA pellets by gently rocking for 20 min at room temperature.
11. Using a pipeter set at 200 µL, transfer each sample to a 1.5-mL microfuge tube (*see* Note 11). This requires two transfers. During each transfer, pipet the sample up and down in the Ultraclear tube and scrape the pipet tip across the bottom to ensure that the RNA is solubilized. Avoid foaming of the SDS during this procedure.
12. To the RNA sample in a 1.5-mL microfuge tube add 200 µL of buffered phenol and 200 µL of chloroform. Mix well.
13. Separate the phases by centrifuging at top speed in a table-top microfuge for 2 min.
14. Transfer the upper aqueous phase to a clean microfuge tube, add 400 µL of chloroform, mix, and spin as in step 13.
15. Again, transfer the aqueous phase to a clean tube and repeat the chloroform extraction.
16. Transfer the final clean aqueous phase to a Beckman ultramicrocentrifuge tube. Add 25 µL of 4*M* NaCl and mix. Add 1 mL of cold 95% ethanol and mix again. Precipitate at –20°C overnight (*see* Note 12).
17. Collect the RNA by centrifuging in a Beckman TLA-45 rotor at 12,000*g* for 30 min at 4°C.
18. Decant the supernatant and invert the tubes over a clean tissue (e.g., Kimwipe) to air-dry. The RNA should be visible as a translucent white pellet at the bottom of the tube.
19. Resuspend the pellet in 25–100 µL of TE, pH 7.4. The volume used will be determined by the size of the pellet. To prevent degradation, add 1 U/µL of RNasin ribonuclease inhibitor and mix gently.
20. Measure the OD_{260} and OD_{280} of 3–5 µL of RNA in a total of 0.4–1 mL of sterile water. The ratio of OD_{260}/OD_{280} should be close to 2.0. If this ratio is <1.7, the sample may contain residual phenol or proteins and should be re-extracted and precipitated. To obtain the concentration of RNA, use the following formula:

    [RNA] (µg/µL) = (OD_{260} × 40) × total vol OD'd (mL)/µL RNA OD'd
21. For short-term storage (several weeks), store RNA in aqueous solution at –20°C. For more stable long-term storage, store RNA in ethanol. Add NaCl

PCR Basic Protocols 283

to 0.25*M* to RNA in aqueous solution, add 2.5 vol of 95% ethanol, mix well, and store at –20°C. To recover the RNA, centrifuge it as in step 17.

3.1.2.2. ISOLATION OF RNA BY RNAZOL METHOD

RNA can be isolated quickly and with great purity using the RNAzol technique (TEL-TEST, Inc.), based on the method of Chomczynski and Sacchi *(12)*. This procedure is most useful for isolating RNA from many samples, especially small tissue specimens (<500 mg). The following protocol is from TEL-TEST *(13)*, with minor modifications we commonly employ.

1. Homogenize tissue samples in RNAzol (2 mL for each 100 mg of tissue) with several strokes of a glass-Teflon homogenizer. Samples of <50 mg should be homogenized directly in 1.5-mL Eppendorf tubes using Kontes polypropylene pestles. *Brief* sonication is helpful to break up any residual tissue clumps, but do not allow the homogenate to become heated (*see* Note 9).
2. Cells grown in suspension should be pelleted in culture media (5 min, 200g_{max}). After pouring off the supernatant, add 0.2 mL of RNAzol/10^6 cells and completely lyse the pellet by repeated pipetting and vortexing.
3. Cells grown on culture dishes can be lysed in the dish. After removing the medium, add RNAzol until the dish is well covered (e.g., 1.5 mL/3.5-cm culture dish). Scraping and/or repipetting will ensure complete lysis. Alternatively, attached cells can be collected by scraping them from the dish, then pelleted and lysed as in step 2.
4. Add 0.1 mL of $CHCl_3$ for each 1 mL of homogenate. Vortex rapidly for at least 15 s, until the homogenate is completely frothy white, and incubate on ice for 15 min. After the incubation, vortex again as before, then centrifuge for 15 min at 10,000g at 4°C.
5. There should now be two liquid phases visible in the tube. Carefully remove and save the upper aqueous phase that contains the RNA. The volume of this aqueous phase is approximately half of the volume of the homogenate. Do not transfer any of the interface. Pour the lower organic phase into a waste bottle and dispose of properly.
6. Precipitate the RNA by adding an equal volume of ice-cold isopropanol to the aqueous phase and incubate at –20°C for 45 min (*see* Note 13). Pellet the RNA by centrifuging at 12,000g at 4°C for 15 min (or 10,000g at 4°C for 30 min in a table-top microfuge). A white pellet of RNA is often (but not always) visible after this step.
7. Carefully decant the supernatant and wash the pellet with 80% ethanol (0.8 mL/100 µg of RNA). Vortex briefly to loosen pellet, then centrifuge for 10 min at 12,000g at 4°C. Remove supernatant and repeat the ethanol wash.

The RNA pellet is often not well attached to the wall of the tube, so the decanting should be performed gently.
8. Allow the pellet to air-dry until just damp (completely dried pellets are difficult to resuspend). Resuspend the pellet in approx 50 µL of TE buffer, pH 7.4, for each 100 µg of RNA by vortexing and by repipeting. A room temperature incubation (15–30 min) can help resuspend difficult pellets. Incubation at 60°C (10–15 min) may also be used for resuspension, but only if all else fails. We often obtain an OD 260/280 ratio of 2.0:2.1 by this method. Samples with a ratio of <1.7, should be re-extracted and precipitated, as described in Section 3.1.2., step 1. RNA isolated by this method should also be reprecipitated prior to enzymatic manipulation.

3.1.3. Synthesis of Complementary DNAs from RNA

In order to perform PCR on RNA sequences using *Taq* DNA polymerase, it is necessary to first convert the sequence to a complementary DNA (cDNA) because *Taq* has limited reverse transcriptase activity *(14)* (*see* Note 14). Several different kinds of primers can be used to make cDNAs. Oligo-dT will prime cDNA synthesis on all polyadenylated RNAs and is most often used for convenience, since these cDNAs can be used for amplification of more than one species of RNA. Random-primed cDNA synthesis similarly gives a broad range of cDNAs and is not limited to polyadenylated RNAs. Lastly, oligonucleotide primers complementary to the RNA(s) of interest may be used to synthesize highly specific cDNAs. We developed the following procedure for use with oligo-dT or RNA-specific primers. A procedure for using random primers to synthesize cDNAs may be found in ref. *15*.

1. Place up to 20 µg of RNA in a microfuge tube containing 4 µg of oligo dT or 200 pmol of specific primer and 5 µL of 10X RT buffer in a total volume of 36.5 µL (*see* Note 15). Mix gently.
2. Incubate at 65°C for 3 min. Cool on ice.
3. Add 5 µL of 100 mM DTT, 1 µL (40 U) of RNasin, and 5 µL of a deoxynucleotide mix containing 1.25 mM of each dNTP (final concentration 125 µM each). Add 2.5 µL (500 U) of MMLV reverse transcriptase and mix gently. The final volume is 50 µL.
4. Incubate at 37°C for 1 h.
5. This cDNA may be used directly in PCR reactions or may be modified further (*see* Note 16).

3.2. Performing PCR

The ideal way to perform PCR is in a dedicated room, using reagents and equipment also dedicated only to PCR. Such luxuries are often not

PCR Basic Protocols

available. Dedicated PCR reagents are essential. A set of dedicated pipets is very helpful, as are filter-containing pipet tips now available from several manufacturers. Gloves should always be worn when handling PCR reagents. An attempt should be made to keep concentrated stocks of target sequences (e.g., recombinant plasmids) away from PCR areas and equipment. Chance contamination can be very difficult to trace and to get rid of.

PCR cycles consist of three basic steps:

1. Denaturation, to melt the template into single strands and to eliminate secondary structure; this step is carried out at 94°C for 1–2 min during regular cycles. However, amplification of genomic DNA requires a longer initial denaturation of 5 min to melt the strands.
2. Annealing, to allow the primers to hybridize to the template. This step is carried out at a temperature determined by the strand-melting temperature of the primers and by the specificity desired. Typical reactions use an annealing temperature of 55°C for 1–2 min. Reactions requiring greater stringency may be annealed at 60–65°C. Reactions in which the primers have reduced specificity may be annealed at 37–45°C.
3. Extension, to synthesize the new DNA strands. This step is usually carried out at 72°C, which is optimal for *Taq* polymerase. The amplification time is determined by the length of the sequence to be amplified. At optimal conditions, *Taq* polymerase has an extension rate of 2–4 kb/min (manufacturer's information). As a rule of thumb, we allow 1 min/kb to be amplified, with extra time allowed for each kb >3 kb (*see* Note 17).

Between 20 and 30 cycles of PCR are sufficient for many applications. DNA synthesis will become less efficient as primers and deoxynucleotides are used up and as the number of template molecules surpasses the supply of polymerase. Therefore, following the last cycle, the enzyme is allowed to finish any incomplete synthesis by including a final extension of 5–15 min at 72°C. Following completion of the program, many cycling blocks have a convenient feature allowing an indefinite hold at 15°C, to allow preservation of the samples, particularly during overnight runs.

Ideally, PCR conditions should be optimized for each template and primer combination used. Practically, most researchers will use the manufacturer's recommended conditions unless the results obtained fall far short of expectations. Other than primer sequence, there are six variables that may be optimized for a given amplification reaction: annealing temperature, primer concentration, template concentration, $MgCl_2$ concentration, extension time, and cycle number (e.g., *see* ref. 15). Standard conditions are described in the following.

1. Prepare a master mix of PCR reagents containing (per 100 µL of PCR reaction): 10 µL of 10X PCR buffer, 100 pmol of upstream primer, 100 pmol of downstream primer, and 16 µL of 1.25 mM dNTP working stock (*see* Note 18). Bring to volume with sterile UV-irradiated water, such that, after addition of the desired amount of sample and *Taq* polymerase, the total reaction volume will be 100 µL. Make up a small excess (an extra 0.2–0.5 reaction's worth) of master mix to ensure that there is enough for all samples.
2. Aliquot the desired amount of sample to be amplified into labeled 0.5-mL microfuge tubes. We routinely amplify 5 µL (1/10) of a 50-µL cDNA made using up to 10 µg of RNA. Genomic DNA is usually amplified in amounts of 100 ng to 1 µg. Adjust the volumes with sterile UV-irradiated water so that all are equal.
3. To the master mix, add 0.5 µL of *Taq* polymerase (2.5 U) for each reaction. Mix well and spin briefly in a microfuge to collect all of the fluid (*see* Note 19).
4. Add the correct volume of master mix to each sample tube so that the total volume is now 100 µL. Cap and vortex the tubes to mix. Spin briefly in a microfuge.
5. Reopen the tubes and cover each reaction with a few drops of light mineral oil to prevent evaporation.
6. Put a drop of mineral oil into each well of the thermal cycler block that will hold a sample. Load the sample tubes (*see* Note 20).
7. Amplify the samples, according to the principles previously oulined.
   a. A typical cycling program for a cDNA with a 1-kb amplified region is 30 cycles of 94°C, 2 min (denaturation); 55°C, 2 min (hybridization of primers); and 72°C, 1 min (primer extension); followed by 72°C, 5 min (final extension); and 15°C, indefinite (holding temperature until the samples are removed).
   b. A typical cycling program for genomic DNA with a 2-kb amplified region is 94°C, 5 min (initial denaturation); followed by 30 cycles of 94°C, 2 min; 60°C, 2 min; and 72°C, 2 min; final extension 72°C, 10 min; hold 15°C.
8. Following PCR, remove the samples from the block and add 200 µL of chloroform to each tube. The mineral oil will sink to the bottom.
9. Without mixing, centrifuge the tubes for 30 s at top speed in a table-top microfuge.
10. Transfer the upper phase to a clean microfuge tube. Add 50 µL of 7.5M ammonium acetate and mix well (*see* Note 21).

PCR Basic Protocols

11. Add 375 µL of 95% ethanol and mix. Precipitate for 10 min at room temperature for concentrated samples, or for 30 min on ice for less concentrated products.
12. Centrifuge at top speed in a table-top microfuge for 15 min.
13. Decant the supernatant (*see* Note 22), air-dry the pellet, and resuspend in 20 µL of sterile water.

3.3. Analysis of PCR Products by Nested PCR

Both agarose and acrylamide gel electrophoresis may be used to analyze PCR products, depending on the resolution required and whether the sample is to be recovered from the gel. Agarose gel electrophoresis on minigels is fast and easy and allows quick estimates of the purity and concentration of a PCR product (*see* Chapter 3). DNA may be recovered much more quickly and efficiently out of agarose gels than out of polyacrylamide (Chapter 28). Agarose gel electrophoresis, followed by Southern blotting and hybridization of a specific probe, allows the detection of a given PCR product in a background of high nonspecific amplification *(3)*. It is also a means of proving that the amplified fragment is related to a known sequence *(3,16)*. Finally, Southern blotting can be used to detect PCR products that are still not abundant enough to be detected by ethidium bromide staining *(16)*.

Nested primers allow the definition of PCR products by reamplification of an internal portion of the DNA. The method takes advantage of the fact that DNA bands in a low-melting agarose gel may be excised and used directly in PCR reactions without further purification *(17,18)*.

1. Resolve amplified product(s) on a gel of 0.7–1% low-melting agarose, such as NuSieve (FMC BioProducts). Resolution may be improved by running the gel slowly (12–25 V) at 4°C.
2. Locate the band of interest using UV illumination of the ethidium bromide-stained gel. Excise the band and transfer it to a microfuge tube.
3. Melt the gel slice by incubation at 68°C for 5–10 min.
4. Transfer 10 µL directly into a tube containing a second PCR mix with 100 pmol each of the nested primers and reamplify.
5. Examine the product(s) of the second PCR by agarose gel electrophoresis as described in Chapter 3.

3.4. Analysis of PCR Products by Polyacrylamide Gel Electrophoresis of Directly Labeled PCR Products

Acrylamide gel electrophoresis provides better resolution and a much more precise estimate of product size (*see* Chapter 13). This is the method of choice for detecting directly labeled PCR products. Denaturing acrylamide gels containing urea may be used to analyze single-stranded products, as from asymetric PCR.

For very sensitive detection and relative quantitation of PCR products, the DNA fragments may be labeled by inclusion of radiolabeled nucleotide in the PCR mix, followed by acrylamide gel electrophoresis and autoradiography. To quantitate the bands, the autoradiograms may be scanned by densitometry, or the labeled bands themselves may be cut out of the gel and counted. We have used autoradiography of directly labeled PCR products to measure the relative levels of several mRNAs in rat ovarian granulosa cells *(19)* and in a pituitary tumor cell line *(11)*.

1. Prepare the gel and samples, run the gel, stain and photograph it (*see* Chapter 13). Remember that the gel is radioactive and handle it accordingly.
2. Cut a piece of 3MM filter paper larger than the gel. Lift the gel on the plastic wrap and place it on a flat surface. Smooth out any wrinkles in the gel.
3. Lay the filter paper on top of the gel. It should begin to wet immediately as the gel adheres to it. Turn over the gel, plastic wrap, and filter paper all at once. The gel now has a filter backing.
4. Dry the gel on a gel dryer for 30–45 min. To avoid contamination of the gel dryer, place a second layer of filter paper below the gel.
5. Wrap the dried gel in fresh plastic wrap. Place in a film cassette with X-ray film and expose for 4 h to 2 d.

4. Notes

1. In order to protect RNA from ubiquitous RNases, the following precautions should be followed during the preparation of reagents for RNA isolation and during the isolation procedure: Wear gloves at all times. Use the highest quality molecular biology grade reagents possible. Bake all glassware. Use sterile, disposable plasticware.
2. Guanidine solutions must be sterilized using Nalgene filters because they dissolve Corning filters *(6)*.
3. DEPC breaks down in the presence of Tris buffers and cannot be used to treat them *(6)*.
4. Wear gloves when preparing PCR reagents or performing PCR to prevent contamination.
5. UV irradiation of all solutions used for PCR that do not contain nucleotides or primers is recommended to reduce the chance that accidental contamination of stocks with PCR target sequences will interfere with sample amplifications *(8,9)*.

PCR Basic Protocols

6. To collect cultured cells for RNA isolation, pour the desired volume of cells in suspension into 50-mL tubes, or remove attached cells from plates by scraping. Avoid enzymatic detachment of plated cells because the enzyme preparations may contain contaminating nucleases. Spin the cells down at 250g for 4 min at 4°C. Resuspend in 1/10 the original volume of cold PBS and spin down again. For isolation of whole-cell RNA, proceed as in Note 10. For nuclear RNA, lyse the cells in 1 mL of PBS plus 0.5% NP-40, incubate for 3 min on ice, then collect the nuclei by centrifugation at 600g for 5 min at 4°C. Proceed to Note 10.
7. Avoid foaming of the sample during homogenization.
8. For cell or nuclear pellets, gently flick the side of the tube to loosen the pellet, then add 1 mL of GITC/βME, incubate on ice for several minutes, and allow the pellet to dissolve. Proceed as for tissue with shearing of the DNA (Section 3.1.2.1., step 3).
9. An alternative to shearing the DNA is to sonicate the sample in a 1.5-mL microfuge tube. We have used a Virsonic 50 cell disruptor (Virtis, Gardiner, NY) at a setting that delivers a pulse of 40–50% of maximal, for 10–30 s, depending on the viscosity of the sample. To use a sonicator, make sure the tip of the probe is placed all the way at the bottom of the sample tube to prevent foaming. Activate the sonicator only when the probe is immersed, and cool the sample between 10-s pulses if more than one is necessary. Rinse the probe in sterile water prior to use to protect the RNA sample, as well as afterward to remove the guanidine solution. Ear protection is recommended for the user.
10. The TLS-55 rotor holds four buckets. Although it can be used containing only two samples, all four buckets must be in place during centrifugation.
11. The samples cannot be extracted in the Ultraclear tubes because these tubes are not resistant to organic solvents.
12. Precipitation may also be carried out at –70°C for 1 h.
13. Prolonged isopropanol precipitation at –20°C can precipitate contaminants with the RNA. If the procedure must be halted here, store the samples at 4°C. Resume the isolation at step 4 by incubating the samples at –20°C for 45 min.
14. Several dual-function thermostable enzymes that have both reverse transcriptase and DNA polymerase activities are now commercially available (TetZ, Amersham, Arlington Heights, IL; TTh; *20*). The different activities rely on differing divalent cations and therefore can be regulated by buffer changes.
15. The lower limit for the amount of RNA required to synthesize a PCR-amplifiable cDNA is beyond the limit of normal detection. We have made cDNAs starting with as little as 0.4 µg of measurable RNA, and with even less of RNAs too dilute to obtain an OD measurement *(11,19)*. Others have

used PCR to detected specific mRNAs in RNA/cDNA samples prepared from single cells *(21)*.
16. For some applications, such as tailing, synthesis of a double-stranded cDNA may be required. Second-strand synthesis is achieved by addition of RNase H and DNA polymerase. Procedures for this may be found in manufacturers' information in kits sold for this purpose and in the following references *(6,7)*. Homopolymeric tailing is used to anchor a DNA sequence so that amplification may be performed from a known region out to the tailed end, which is unknown *(18)*. Procedures for tailing may also be found in refs. *6,7*. Finally, a procedure has been described for addition of a linker to the 3' end of a single-stranded cDNA (to the unknown 5' end of an RNA) using RNA ligase. This allows subsequent PCR to be carried out between a known region of the cDNA and the linker, and is used to help define the 5' ends of mRNAs. Other PCR applications require removal of the primers used to prepare the cDNA. This may be achieved by several rounds of ammonium acetate precipitation or by purification out of gels.
17. Although the efficiency may be reduced for very large target sequences, we have successfully amplified sequences >6 kb in length.
18. To label PCR products directly, alter the nucleotide mix to contain 625 μM dCTP, 1.25 mM dATP, 1.25 mM dGTP, and 1.25 mM dTTP. To this mix add 5–10 μCi (0.5–1 μL) of α^{32}P-dCTP (3000 Ci/mmol)/100 μL.
19. For PCR programs that are allowed an initial denaturation before addition of the polymerase ("hot-started," *22)*, the reactions are made up lacking a small volume and the *Taq*. The samples are loaded into the block and put through an initial denaturation step, then held at annealing temperature for 2 min. The temperature is then raised to 65°C and the *Taq* is added in a small volume of sterile UV water; the timing of this step is set to allow addition of enzyme to all of the samples. The samples are then allowed to extend at 72°C for the appropriate time and then are cycled normally.
20. More uniform heating and cooling may be achieved if the samples are clustered in the block and surrounded by blank tubes.
21. Ammonium acetate precipitation removes primers and unincorporated nucleotides *(23)*.
22. If the samples are radioactive, be sure to dispose of wastes properly.

Acknowledgments

We thank Puja Agarwal and Anna Pappalardo for expert technical assistance. Supported by Grant DK43064 from the NIH.

References

1. Saiki, R., Scharf, S., Faloona, F., Mullis, K. B., Horn, G. T., Erlich, H. A., and Arnheim, N. (1985) Enzymatic amplification of β-globin genomic sequences and restriction site analysis for diagnosis of sickle cell anemia. *Science* **230**, 1350–1354.

2. Mullis, K. B. and Faloona, F. A. (1987) Specific synthesis of DNA in vitro via a polymerase-catalyzed chain reaction. *Methods Enzymol.* **155**, 335–350.
3. Saiki, R. K., Gelfand, D. H., Stoffel, S., Scharf, S. J., Higuchi, R., Horn, G. T., Mullis, K. B., and Erlich, H. A. (1988) Primer-directed enzymatic amplification of DNA with a thermostable DNA polymerase. *Science* **239**, 487–491.
4. Erlich, H. A., Gelfand, D., and Sninsky, J. J. (1991) Recent advances in the polymerase chain reaction. *Science* **252**, 1643–1651.
5. Robert, S. S. (1991) Amplification of nucleic acid sequences: the choices multiply. *J. NIH Res.* **3(2)**, 81–94.
6. Sambrook, J., Fritsch, E. F., and Maniatis T. (1989) *Molecular Cloning: A Laboratory Manual.* Cold Spring Harbor Laboratory, Cold Spring Harbor, NY.
7. Ausubel, F. M., Brent, R., Kingston, R. E., Moore, D. D., Smith, J. A., Seidman, J. G., and Struhl, K. (eds.) (1987) *Current Protocols in Molecular Biology.* Wiley Interscience, New York.
8. Dycaico, M. and Mather, S. (1991) Reduce PCR false positives using the Stratalinker UV crosslinker. *Stratagene Strategies* **4(3)**, 39,40.
9. Sarkar, G. and Sommer, S. S. (1990) Shedding light on PCR contamination. *Nature* **343**, 27.
10. Zintz, C. B. and Beebe, D. C. (1991) Rapid re-amplification of PCR products purified from low melting point agarose gels. *Biotechniques* **11**, 158–162.
11. Delidow, B. C., Peluso, J. J. and White, B. A. (1989) Quantitative measurement of mRNAs by polymerase chain reaction. *Gene Anal. Tech.* **6**, 120–124.
12. Chomczynski, P. and Sacchi, N. (1987) Single-step method of RNA isolation by acid guanidinium thiocyanate-phenol-chloroform extraction. *Anal. Biochem.* **162**, 156–159.
13. Chomczynski, P. (1991) Isolation of RNA by the RNAzol method. *TEL-Test, Inc., Bull.* **2**, 1–5.
14. Myers, T. W. and Gelfand, D. H. (1991) Reverse transcription and DNA amplification by a *Thermus thermophilus* DNA polymerase. *Biochemistry* **30**, 7661–7666.
15. Zimmer, A. and Gruss, P. (1991) Use of polymerase chain reaction (PCR) to detect homologous recombination in transfected cell lines, in *Methods in Molecular Biology, vol. 7: Gene Transfer and Expression Protocols* (Murray, E. J., ed.) Humana, Clifton, NJ, pp. 411–418.
16. Ohara, O., Dorit, R. L., and Gilbert, W. (1989) One-sided polymerase chain reaction: the amplification of cDNA. *Proc. Natl. Acad. Sci. USA* **86**, 5673–5677.
17. Belyavsky, A. (1989) Polymerase chain reaction in the presence of NuSieve™ GTG agarose. *FMC Resolutions* **5**, 1,2.
18. Belyavsky, A., Vinogradova, T., and Rajewsky, K. (1989) PCR-based cDNA library construction: general cDNA libraries at the level of a few cells. *Nucleic Acids Res.* **17**, 2919–2932.
19. Delidow, B. C., White, B. A., and Peluso, J. J. (1990) Gonadotropin induction of c-fos and c-myc expression and deoxyribonucleic acid synthesis in rat granulosa cells. *Endocrinology* **126**, 2302–2306.
20. Tse, W. T. and Forget, B. G. (1990) Reverse transcription and direct amplification of cellular RNA transcripts by Taq polymerase. *Gene* **88**, 293–296.

21. Rappolee, D. F. A., Mark, D., Banda, M. J., and Werb, Z. (1988) Wound macrophages express TGF-α and other growth factors in vivo: analysis by mRNA phenotyping. *Science* **241**, 708–712.
22. Rauno, G., Brash, D. E., and Kidd, K. K. (1991) PCR: the first few cycles. *Perkin-Elmer Cetus Amplifications* **7**, 1–4.
23. Crouse, J. and Amorese, D. (1987) Ethanol precipitation: ammonium acetate as an alternative to sodium acetate. *BRL Focus* **9**, 3–5.

CHAPTER 35

Inverse Polymerase Chain Reaction

Daniel L. Hartl and Howard Ochman

1. Introduction

The inverse polymerase chain reaction (IPCR) was the first extension of the conventional polymerase chain reaction to allow the amplification of unknown nucleotide sequences without recourse to conventional cloning. In the conventional polymerase chain reaction (PCR), synthetic oligonucleotides complementary to the ends of a known sequence are used to amplify the sequence (1,2). The primers are oriented with their 3' ends facing each other, and the elongation of one primer creates a template for annealing the other primer. Repeated rounds of primer annealing, polymerization, and denaturation result in a geometric increase in the number of copies of the target sequence. However, regions outside the boundaries of the known sequence are inaccessible to direct amplification by PCR. Since DNA synthesis oriented toward a flanking region is not complemented by synthesis from the other direction, there is at most a linear increase in the number of copies of the flanking sequence.

Nevertheless, the applications of a procedure that could selectively amplify flanking sequences are numerous. They include the determination of genomic insertion sites of transposable elements, in which case the sequence of the transposable element can be used to choose oligonucleotide primer sequences; as well as the determination of the upstream and downstream genomic regions flanking cDNA coding sequences, in which case the 5' and 3' sequences of the cDNA can be used to choose the primers. Another class of applications includes the determination of the nucleotide sequences at the ends of cloned DNA fragments, particularly DNA fragments as large as those in yeast artificial chromosomes

From: *Methods in Molecular Biology, Vol. 58: Basic DNA and RNA Protocols*
Edited by: A. Harwood Humana Press Inc., Totowa, NJ

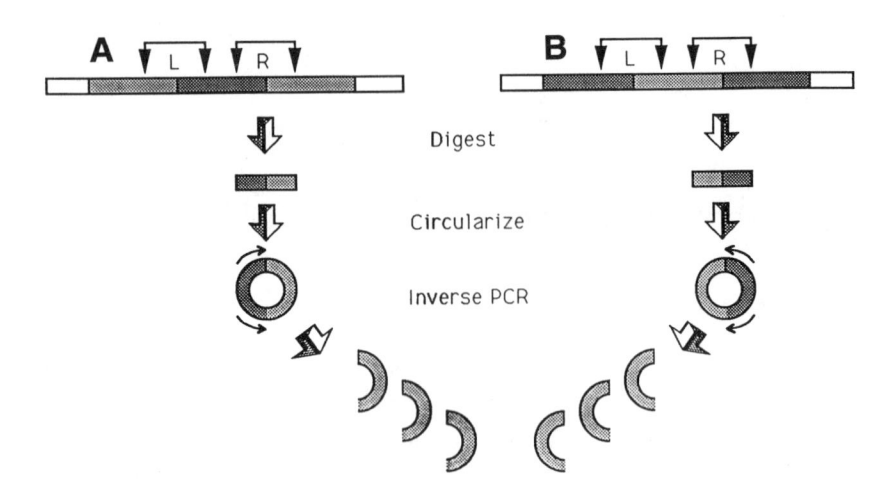

Fig. 1. Major applications of IPCR. The figure shows known DNA sequences by dark shading and unknown flanking sequences of interest by light shading. **(A)** The known sequence is contained within the unknown sequences; **(B)** the unknown sequence is contained within known sequences. The brackets indicate the possible positions of restriction sites that allow IPCR of either the left (L) or right (R) flanking sequence. After digestion with the appropriate restriction enzyme, molecules are circularized, bringing two positions in the known sequence into juxtaposition with the unknown sequence. The known sequences then serve as oligonucleotide priming sites (curved arrows) to amplify the unknown sequence.

(3,4) or bacteriophage P1 *(5,6)*; in these applications the primer sequences can be chosen based on the nucleotide sequence of the vector flanking the cloning site.

The possibility of amplifying flanking sequences was sufficiently interesting that several groups undertook the challenge, and successful protocols were developed independently and nearly simultaneously by three research groups *(7–9)*. All three methods are based on the same idea, which is outlined in Fig. 1. Panels A and B exemplify the two main types of application. The dark stippling denotes regions of known sequence, and the light stippling represents regions of unknown sequence. In panel A the known sequence (for example, a coding region) is continuous, and the problem is to amplify either the left (L) or right (R) flanking sequence in genomic DNA. In panel B the known sequence (for example, in a cloning vector) is interrupted, and the problem is to amplify the junction with the unknown sequence on either the left (L) or the right (R). For convenience of further discussion, we will call the known sequence the "core" sequence and the unknown sequence the "target" sequence.

The first step in the method is to identify restriction enzymes that will cleave within the core and target sequences in order to liberate DNA fragments containing core sequence juxtaposed with target sequence. The cutting sites of the restriction enzymes are denoted by the arrows in Fig. 1, and normally each junction requires a different enzyme. For each junction, a set of candidate restriction enzymes can be created by choosing enzymes that will cleave within the core sequence 50–1,000 bp from the junction. Final choice among the candidates is based either on trial and error or, more reliably, on a Southern blot *(10)* using the core sequence as a probe. Restriction enzymes with four-base cleavage sites are generally preferred, since they are more likely to cleave at an acceptable point within the target sequence.

After cleavage, the next step in the procedure is circularization, in which the ends of the fragment liberated by the restriction enzyme are ligated together. The circular molecule produced in this way has two core-target junctions, one of which was present in the original molecule, and a second one produced by the ligation. The ligation step is followed by a PCR using oligonucleotide primers that anneal to the core sequences near the core–target junctions. The "inverse" part of IPCR comes from the fact that the 3' ends of the primers are oriented outward, toward the target sequence (curved arrows in Fig. 1). Because of the circular configuration of the substrate, the product of each polymerization can serve as a template for annealing of the opposite primer. Hence, repeated rounds of annealing, polymerization, and denaturation result in a geometric increase in the number of copies of the target sequence.

IPCR has proven useful in recovering genomic sequences flanking viral and transposable element insertions *(7–9)*, in generating probes corresponding to the ends of cloned inserts for use in chromosome walking *(11–14)*, and in the direct cloning of unknown cDNA sequences from total RNA *(15)*. Since the time that IPCR was developed, a number of other methods have been proposed that can either serve as an alternative to IPCR or that are more suited to certain specialized applications. For example, one technique for human genomic DNA uses vector sequence for one primer and the Alu consensus sequence for the other primer in order to amplify the insert ends of YAC or cosmid clones *(16)*. The most relevant of these techniques are discussed briefly (*see* Notes).

2. Materials
2.1. Preparation of Restriction Fragments
1. Restriction enzymes and buffers: These are generally supplied together. Digests are carried out as indicated by the manufacturer and differ according to the particular enzyme.
2. $T_{0.1}E$: 10 mM Tris-HCl, 0.1 mM EDTA, pH 8.0.

2.2. Ligation to Form Circles
3. 10X Ligation buffer: 500 mM Tris-HCl, pH 7.4, 100 mM MgCl$_2$, 100 mM dithiothreitol, 10 mM ATP.
4. T4 DNA ligase: Supplied at a concentration 1,000 Weiss U/µL.
5. TE: 10 mM Tris-HCl, pH 7.5, 1 mM EDTA.

2.3. PCR Amplification
6. 10X PCR buffer: 100 mM Tris-HCl, pH 8.4, 500 mM KCl, 5 mM MgCl$_2$, and 0.1% gelatin.
7. dNTP mix: Combine equimolar amounts of each deoxynucleotide triphosphate (dNTP) to produce a 100 mM stock solution (25 mM of each dNTP).
8. *Taq* polymerase: supplied at a concentration of 5 U/µL (*see* Note 1).
9. Primers: The length of primers varies in each application but averages 20 nucleotides (*see* Note 2). Dilute stocks to 50 pmol/µL.

3. Methods
3.1. Preparation of Restriction Fragments

The single largest variable in this step is the quality and complexity of the DNA. Best results are usually obtained with DNA that has been extracted with phenol, and with DNA extracted from simple genomes, such as bacteria, yeast, nematodes, and *Drosophila* (*see* Notes 3–7).

1. Restriction enzyme digest for 2 h at 37°C according to the manufacturer's conditions. Usually 1–5 µg of genomic DNA is used for the initial digestions that are carried out in a volume of 20–50 µL. Since the DNA sample is precipitated prior to ligations, the reaction volume is not critical.
2. Phenol extract by the addition of an equal volume of equilibrated phenol. Invert the sample, spin briefly in a microfuge to separate phases, and remove the aqueous phase (*see* Note 8).
3. Chloroform:isoamyl alcohol (24:1) extract the aqueous phase from the previous extraction as described in step 2.
4. Ethanol precipitate by addition of 0.1 vol of 3M sodium acetate, pH 5.3, to the DNA sample, followed by 2 vol of absolute ethanol. Precipitate the

DNA at −20°C for 30 min and harvest by centrifugation for 20 min at 4°C. Wash the DNA pellet with 70% ethanol and dry under vacuum.
5. Resuspend the sample in $T_{0.1}E$ at 100 ng/µL.

3.2. Ligation to Form Circles

It is advisable to carry out three simultaneous reactions with different DNA concentrations (20, 100, and 500 ng/µL) in order to increase the probability that one of the dilutions will be favorable for the formation of monomeric circles.

1. Set up 3 tubes containing 0.2, 1.0, and 5.0 µL of digest.
2. Add 10 µL of 10X ligation buffer and water to final volume of 95 µL (i.e., 93.8, 93, and 89 µL of distilled H_2O, respectively).
3. Add 0.05 Weiss U of T4 DNA ligase/µL. Incubate at 14°C for least 2 h. (Ligations generally proceed overnight.)
4. Heat inactivate the ligation mixture at 65°C for 15 min and extract with phenol and chloroform as described in the previous section. Precipitate the DNA as described above and resuspend DNA in 10 µL of TE.

3.3. PCR Amplification

1. For each ligation product, add 4 µL of the DNA from the ligation (*see* Note 9) to a tube suitable for PCR. Add 5 µL of 10X PCR buffer, 0.4 µL of the dNTP mix, 1 µL of each primer, and water to a final volume of 50 µL. Layer 50 µL of mineral oil on top of the reaction mix.
2. Heat the reaction mixture at 95°C for 10 min. This facilitates amplification by introducing nicks into the circular templates. Add 0.2 µL of *Taq* polymerase.
3. Standard amplification conditions are 30 cycles of denaturation at 95°C for 40 s, primer annealing at 56°C for 30 s, and primer extension at 72°C for 100 s (*see* Note 10).
4. Run 1/10 of each sample on a 1% agarose gel to analyze amplification products. The IPCR products should be confirmed by Southern blotting and hybridization to a DNA probe derived from core (*see* Note 11). Once confirmed, the PCR product may then be either cloned (*see* Chapters 27 and 37) or sequenced directly (*see* Chapter 49).

3.4. Secondary Amplification of IPCR Products

If nonspecific products are amplified, it is sometimes helpful to carry out a second amplification. This can be done in two different ways.

3.4.1. Nested Primers

In this case, one of the original primers is replaced with another, which anneals within the core sequence closer to the core–target junction (i.e., still within the amplified fragment).

1. Take 1 μL of the first PCR reaction to be used as the amplification template.
2. Repeat original PCR with the alternate primers, but reduce the number of cycles to 15.

3.4.2. Gel Purification

Alternatively, the products of the initial PCR reaction can be separated in an agarose gel.

1. Run 5 μL of the first PCR reaction on a low melting point temperature agarose gel.
2. Extract a fragment of the correct size by poking with the narrow end of a Pasteur pipet. This should remove an agarose plug of 5–10 μL.
3. Dispense the agarose plug into 100 μL of distilled water and heat to 95°C for 5 min to melt the agarose.
4. Use 1 μL to repeat the PCR with either both of the original primers or a nested primer. Follow the protocol for the nested primer, as in Section 3.4.1.

4. Notes

1. When amplifying fragments from enteric bacteria, we use *Taq* polymerase purified from *Thermus aquaticus*; cloned thermostable polymerase (AmpliTaq) is used in all other applications.
2. Specificity of the PCR amplification depends heavily on the choice of oligonucleotide primers. We typically use oligonucleotide primers 20 nucleotides in length that are approx 50% in their G + C content, which yields a theoretical average melting temperature of 60°C using the empirical rule of thumb that the T_d approximately equals 2× (number of As plus Ts) + 4× (number of Gs plus Cs).
3. Choice of restriction enzyme is determined by the distribution of restriction sites in the target region, established, if necessary, by Southern blots *(10)*, and by the maximum size limit of PCR amplification. Fragments containing <3 kb of target sequence are preferable, since these amplify best. Fragments of satisfactory size containing either the left or right core–target junction (L and R in Fig. 1) can sometimes be obtained in a single restriction digestion by suitable choice restriction enzymes.
4. In some cases digestion with two different restriction enzymes is required in order to generate a core–target fragment of satisfactory size; if the restriction enzymes produce incompatible ends, then it is necessary to render the ends of the fragments blunt using the Klenow polymerase or T4 DNA polymerase prior to the ligation step.
5. In particular applications of IPCR, we have had good success using *Pst*I to recover sequences flanking insertion sequence IS*1* in *Escherichia coli (7)*, and using *Cla*I or *Taq*I to recover sequences flanking insertion sequence

IS*30* *(17)*. Most common YAC vectors allow recovery of the insert junction nearest the centromere using *Eco*RV and the recovery of the insert junction nearest the telomere using *Hinc*II *(12)*.

6. When using restriction enzymes that have convenient four-base cleavage sites within the core sequence, the usual experience in most laboratories is that IPCR is successful in about half the cases without performing preliminary Southern blots *(10)*. The success rate probably results from the fact that four-base restriction sites are sufficiently frequent (theoretically, one in every 256 base pairs) that chance favors the target sequence having a site neither too close to the core sequence nor too far from it. Failing in an initial attempt, subsequent trials should be based on preliminary data from Southern blots to identify a suitable restriction enzyme (or a pair of enzymes).
7. It should be emphasized again that the success of IPCR depends on the complexity of the starting material, with simpler genomes yielding more reliable results. The procedure has worked well with cloned material and with *E. coli*, *Caenorhabditis*, and *Drosophila*. Genome complexity increases the chance of spurious PCR products and decreases the efficiency of the circularization reaction. Recovery of flanking sequences from more complex genomes may be improved by enrichment of DNA fragments of a suitable size class. For example, restriction fragments of the desired size (determined from preliminary Southern blots), can be extracted from agarose gels with glass powder, electroelution, or DEAE membranes. There should be a total of about 1 µg of cleaved DNA in order to conduct circularization at several DNA concentrations and PCR.
8. In cases where the digested sample can be heat treated to inactivate the restriction endonuclease (68°C for 10 min), samples may be diluted and ligation can often proceed without intermediate purification procedures (steps 2–5).
9. Some reports indicate that the efficiency of IPCR is improved by the amplification of linear rather than circular molecules *(9)*. In this case, the circularized fragments are cleaved at a unique restriction site within the core sequence. However, finding the appropriate restriction enzyme introduces additional steps and potential complications.
10. When amplification products greater than about 3 kb are expected, the denaturation time should be increased to 60 s and the extension time to 150 s.
11. Do not use the PCR primers as a probe since they will hybridize to all products generated by these primers.
12. Alternatives to IPCR: Among the alternatives to IPCR that may be considered for recovering flanking sequences from genomic DNA are the following. Experimental details can be found in the original citations and additional general discussion in Ochman et al. *(17)*.
    a. Ligation-mediated PCR employs a double-stranded cassette consisting of a 24-mer and an 11-mer complementary to the 3' end of the 24-mer.

The blunt end of the cassette is ligated onto DNA fragments created by primer extension using an oligonucleotide that anneals to the core sequence. Experimental details can be found in refs. *18–20*.
  b. Vectoret (or "bubble") PCR has been used to recover the vector–insert junctions in YACs *(21)*. DNA from yeast cells containing a YAC is digested with restriction enzymes that cleave at a convenient site in the core sequence. A double-stranded oligonucleotide cassette ("vectoret") containing a "bubble" region of noncomplementary bases is ligated onto the end of the target sequence. Amplification proceeds from a primer specific to the YAC vector, which produces a newly synthesized strand complementary to one unpaired loop in the bubble, and polymerization in the reverse direction uses an oligonucleotide that anneals to this region in the newly synthesized strand.
  c. Oligo-cassette mediated PCR employs a double-stranded 28-mer that is ligated onto the target sequence after digestion with an appropriate restriction enzyme *(22)*. Approx 50 cycles of primer extension are carried out using a biotinylated primer complementary to the core region, and the biotinylated products are trapped with streptavadin-coated magnetic beads. The isolated product is then subjected to conventional PCR using a nested primer within the core sequence and a reverse primer for the oligo-cassette.

Acknowledgments

This work was supported by NIH grants GM40322 and HG00357 to D. L. Hartl and GM40995 to H. Ochman.

References

1. Saiki, R. K., Scharf, S. J., Faloona, F., Mullis, K. B., Horn, G. T., Erlich, H. A., et al. (1985) Enzymatic amplification of β-globin genomic sequences and restriction site analysis for diagnosis of sickle cell anemia. *Science* **230**, 1350–1354.
2. Saiki, R. K., Gelfand, D. H., Stoffel, S., Scharf, S. J., Higuchi, R. G., Horn, G. T., et al. (1988) Primer-directed enzymatic amplification of DNA with a thermostable DNA polymerase. *Science* **239**, 487–491.
3. Burke, D. T., Carle, G. F., and Olson, M. V. (1987) Cloning of large segments of exogenous DNA into yeast by means of artificial chromosome vectors. *Science* **236**, 806–812.
4. Hieter, P., Connelly, C., Shero, J., McCormick, M. K., Antonarakis, S., Pavav, W., et al. (1990) Yeast artificial chromosomes: promises kept and pending, in *Genetic and Physical Mapping*, vol. 1 (Davies, K. E. and Tilghman, S. M., eds.), Cold Spring Harbor Laboratory, Cold Spring Harbor, NY, pp. 83–120.
5. Sternberg, N. (1990) Bacteriophage P1 cloning system for the isolation, amplification, and recovery of DNA fragments as large as 100 kilobase pairs. *Proc. Natl. Acad. Sci. USA* **87**, 103–107.

6. Pierce, J. C. and Sternberg, N. L. (1993) Using the bacteriophage P1 system to clone high molecular weight (HMW) genomic DNA. *Meth. Enzymol.* (in press).
7. Ochman, H., Gerber, A. S., and Hartl, D. L. (1988) Genetic applications of an inverse polymerase chain reaction. *Genetics* **120**, 621–623.
8. Triglia, T., Peterson, M. G., and Kemp, D. J. (1988) A procedure for in vitro amplification of DNA segments that lie outside the boundaries of known sequences. *Nucleic Acids Res.* **16**, 8186.
9. Silver, J. and Keerikatte, V. (1989) Novel use of polymerase chain reaction to amplify cellular DNA adjacent to an integrated provirus. *J. Virol.* **63**, 1924–1928.
10. Southern, E. M. (1975) Detection of specific sequences among DNA fragments separated by gel electrophoresis. *J. Mol. Biol.* **98**, 503–517.
11. Garza, D., Ajioka, J. W., Carulli, J. P., Jones, R. W., Johnson, D. H., and Hartl, D. L. (1989) Physical mapping of complex genomes. *Nature* **340**, 577,578.
12. Ochman, H., Medhora, M. M., Garza, D., and Hartl, D. L. (1990) Amplification of flanking sequences by IPCR, in *PCR Protocols: A Guide to Methods and Applications* (Innis, M., Gelfand, D., Sninsky, J., and White, T., eds.), Academic, New York, pp. 219–227.
13. Silverman, G. A., Ye, R. D., Pollack, K. M., Sadler, J. E., and Korsmeyer, S. J. (1989) Use of yeast artificial chromosome clones for mapping and walking within human chromosome segment 18q21.3. *Proc. Natl. Acad. Sci. USA* **86**, 7485–7489.
14. Silverman, G. A., Jockel, J. I., Domer, P. H., Mohr, R. M., Taillon-Miller, P., and Korsmeyer, S. J. (1991) Yeast artificial chromosome cloning of a two-megabase-size contig within chromosomal band 18q21 establishes physical linkage between BCL2 and plasminogen activator inhibitor type-2. *Genomics* **9**, 219–228.
15. Huang, S., Hu, Y., Wu, C., and Holcenberg, J. (1990) A simple method for direct cloning cDNA sequence that flanks a region of known sequence from total RAN by applying the inverse polymerase chain reaction. *Nucleic Acids Res.* **18**, 1922.
16. Breukel, C., Wijnen, J., Tops, C., Klift, H. V., Dauwerse, H., and Meera Khan, P. (1990) Vector-Alu PCR: a rapid step in mapping cosmids and YACs. *Nucleic Acids Res.* **18**, 3097.
17. Ochman, H., Ayala, F. J., and Hartl, D. L. (1993) Use of the polymerase chain reaction to amplify segments outside the boundaries of known sequences. *Meth. Enzymol.* **218**, 309–321.
18. Pfeifer, G. P., Steigerwald, S. D., Mueller, P. R., Wold, B., and Riggs, A. D. (1989) Genomic sequencing and methylation analysis by ligation mediated PCR. *Science* **246**, 810–813.
19. Mueller, P. R. and Wold, B. (1989) In vivo footprinting of a muscle specific enhancer by ligation mediated PCR. *Science* **246**, 780–786.
20. Fors, L., Saavedra, R. A., and Hood, L. (1990) Cloning of the shark Po promoter using a genomic walking technique based on the polymerase chain reaction. *Nucleic Acids Res.* **18**, 2793–2799.
21. Riley, J., Butler, R., Ogilvie, D., Finniear, R., Jenner, D., Powell, S., et al. (1990) A novel, rapid method for the isolation of terminal sequences from yeast artificial chromosome (YAC) clones. *Nucleic Acids Res.* **18**, 2887–2890.
22. Rosenthal, A. and Jones, D. S. C. (1990) Genomic walking and sequencing by oligo-cassette mediated polymerase chain reaction. *Nucleic Acids Res.* **18**, 3095,3096.

CHAPTER 36

Polymerase Chain Reaction with Degenerate Oligonucleotide Primers to Clone Gene Family Members

Gregory M. Preston

1. Introduction

1.1. What Are Gene Families?

As more and more genes are cloned and sequenced, it is apparent that nearly all genes are related to other genes. Similar genes are grouped into families. Examples of gene families include the collagen, globin, and myosin gene families. There are also gene superfamilies. Gene superfamilies are composed of genes that have areas of high homology and areas of high divergence. Examples of gene superfamilies include the oncogenes, homeotic genes, and a newly recognized gene superfamily of transmembrane proteins related to the lens fiber cells *ma*jor *i*ntrinsic *p*rotein, or the MIP gene superfamily *(1)*. In most cases, the different members of a gene family carry out related functions.

1.2. Codon Usage

In particular organisms, certain amino acid codons are "preferred" and are utilized more often than others *(2)*. For instance, the four codons for alanine begin with GC. In the third position of this codon, G is rarely used in humans (~10.3% of the time) or rats (~8.0%), but is often used in *E. coli* (~35%). This characteristic of codon usage may be advantageously used when designing degenerate oligonucleotide primers.

From: *Methods in Molecular Biology, Vol. 58: Basic DNA and RNA Protocols*
Edited by: A. Harwood Humana Press Inc., Totowa, NJ

N-Terminal Sequence of CHIP28 Protein.
(M) ASEFKKKLFWRAVVAEFLATTLFVFISIGSALGFK

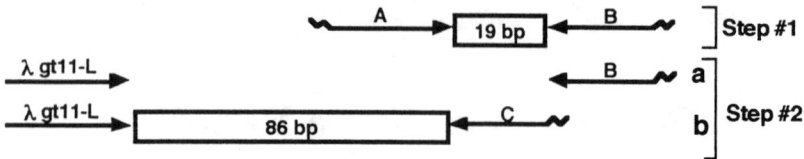

Fig. 1. Two-step strategy for cloning the 5'-end of the CHIP28 gene. Listed on top is the N-terminal amino acid sequence of purified CHIP28 protein *(8)*, with the addition of the initiating methionine (in brackets) deduced from the sequence of the 5'-clone, pPCR-2. Oligonucleotide primers are shown as lines, with arrowheads at their 3'-ends and ~ representing 5'-extensions containing restriction enzyme recognition sequences. In step 1, the sequence of the 19-bp separating degenerate primers A and B was determined. The sequence was used to make the nondegenerate primer C. In step 2, a nested anchor amplification was made with primers B and C, anchoring with primer λgt11-L. The 110-bp product obtained contained sequence corresponding to the N-terminal 22 amino acids, the initiating methionine, and 38 bp of 5'-untranslated sequence.

1.3. Advantages of PCR Cloning of Gene Family Members

There are several advantages of using PCR to identify members of a gene family over conventional cloning methods.

1. Either one or two degenerate primers can be used in PCR cloning. Since the sequences of conserved genes tend to diverge at their termini, the two degenerate primer approach usually can only provide the sequence of an internal fragment of a gene. Anchor PCR uses only a single degenerate primer and can be used to obtain sequence that stretches to the 5'- or 3'-terminus. For anchor PCR, the second primer site must be introduced to the sequence by ligation either into a cloning vector (Fig. 1, step 2) or to a synthetic linker sequence (such as a poly-G tail as in RACE-PCR). This approach, however, may lead to a higher background of unspecific PCR products. Since the second primer is present in every gene of the starting material, this may be a particular problem when working with genes with low abundance in the starting material. This disadvantage can often in part be ameliorated by using a nested amplification approach to amplify desired sequences preferentially. In nested PCR, the products of the primary PCR-amplification are reamplifed using a second primer, which anneals to a site that is internal to the original primer.

2. It is possible to carry out a PCR reaction on first-strand cDNAs made from a small amount of RNA and, in theory, from a single cell. Several single-stranded "minilibraries" can be rapidly prepared and analyzed by PCR from a number of tissues or cell cultures under different hormonal or differentiation stages. Therefore, PCR cloning can potentially provide information about the timing of expression of an extremely rare gene family member, or messenger RNA splicing variants, which may not be present in a recombinant library.
3. PCR cloning can be faster and cheaper than conventional methods and, in some cases, the only feasible cloning strategy. It takes at least 4 d to screen 300,000 plaques from a λgt10 library, whereas with PCR, an entire library containing 10^8 independent recombinants (~5.4 ng DNA) can be screened in one reaction.

It is recommended, however, that after a clone is obtained by PCR it should be used to isolate the corresponding clone from a library, since mutations can often be introduced during PCR cloning. Alternatively, sequencing two or more PCR clones from independent reactions will also meet this objective.

1.4. Degenerate Oligonucleotide Theory

As the genetic code is degenerate, primers targeted to particular amino acid sequences must also be degenerate in order to encode all possible permutations. Thus, a primer to a six amino acid sequence with 64 possible permutations can potentially recognize 64 different nucleotide sequences, only one of which will be in the target gene. If two such primers are used in a PCR reaction, then there are 64 × 64 or 4096 possible permutations. Since PCR-amplifies the product exponentially, it can overcome the fact that with highly degenerate primers, only a small fraction of the primers recognize the target gene. A problem, however, is that some of the other 4095 possible permutations are likely to recognize other gene products and cause a background of nonspecific product. This can be ameliorated by performing a second nested PCR amplification with a second set of primers and by using "guessmers" as primers. A guessmer primer is made by considering the preferential codon usage exhibited by many species and tissues, and therefore does not contain all the possible permutations in the amino acid sequence.

1.5. General Strategy

The strategy for gene cloning that this author adopts is first to use two degenerate primers on a cDNA library template to clone a fragment of the target gene. The sequence obtained from this fragment is then used to design specific primers that are oriented toward the 5'- and 3'-ends of the cDNA. The new primers in combination with those that anneal to sequences within the library vector are used to PCR-amplify the missing termini of the target gene. In both cases, a second round of amplification is used, either employing the original primers or nested primers to increase the amount of product. An example of how this strategy was used to clone CHIP28 is shown in Fig. 1.

2. Materials

1. Template: A heat-denatured cDNA library. Prior to PCR amplification, the DNA was heat-denatured at 99°C for 10 min. The author has found no significant difference in the PCR amplification of crude phage lysates and DNA isolated from the bacteriophage capsids.
2. Double distilled water filtered in 0.2-µ nitrocellulose filter (Millipore, Bedford, MA). Autoclave before storage.
3. Degenerate oligonucleotide primers: At 20 pmol/µL in double-distilled water. The primers you choose will correspond to your amino acid sequence (*see* Section 3.1.).
4. 10X PCR reaction buffer: 100 m*M* Tris-HCl (pH 8.3 at 25°C), 500 m*M* KCl, 15 m*M* MgCl$_2$, 0.1% w/v gelatin. This buffer should be incubated at 50°C to melt the gelatin fully and then filter-sterilized. Store at –20°C.
5. dNTP stock solution: A 1.25-m*M* solution of dATP, dCTP, dGTP, and dTTP.
6. Thermostable DNA polymerase: I use Amplitaq DNA polymerase (Perkin Elmer Cetus, Norwalk, CT) supplied at 5 U/mL.
7. Mineral oil.
8. Chloroform.
9. 7.5*M* Ammonium acetate (AmAc): AmAc is preferred over sodium acetate because nucleotides and primers tend not to precipitate with it. Dissolve in double-distilled water and filter through 0.2-µ membrane.
10. 100% Ethanol: Store at –20°C.
11. 70% Ethanol: Store at room temperature.
12. Spin dialysis columns: These are ideal for purifying PCR products. They are available from a number of manufacturers (e.g., Millipore and Centricon). The author prefers Millipore Ultrafree MC microfilter units (30,000 mol wt cutoff) owing to the convenience of being able to spin the samples in a tabletop microfuge.
13. TE: 10 m*M* Tris-HCl, pH 7.4, 1 m*M* EDTA. Autoclave.

Degenerate Oligonucleotide Primers

Table 1
The Degenerate Nucleotide Alphabet

Letter	Specifications
A	Adenosine
G	Guanosine
R	puRine (A or G)
K	Keto (G or T)
S	Strong (G or C)
B	Not A (G, C, or T)
H	Not G (A, C, or T)
N	aNy (A, G, C, or T)
C	Cytidine
T	Thymidine
Y	pYrimidine (C or T)
M	aMino (A or C)
W	Weak (A or T)
D	Not C (A, G, or T)
V	Not T (A, C, or G)
I	Inosine[a]

[a]Although inosine is not a true nucleotide, it is included in this degenerate nucleotide list, since many researchers have employed inosine-containing oligonucleotide primers in cloning gene family members.

3. Methods
3.1. Design of Degenerate Oligonucleotide Primers

The design of the degenerate oligonucleotides is fundamental to the sucess of the cloning. I use the following steps as a guide.

1. First write down the amino acid sequence, and then use it to predict the putative nucleotide sequence (or the complement of this sequence for a downstream primer), considering all of the possible permutations. For example, the degenerate nucleotide sequence of the amino acid sequence Leu-Ile-Gly-Glu is 5'-YTN-ATH-GGN-GAR-3' (*see* Table 1). If the amino acid sequence is relatively long, two or more degenerate primers may be designed. If only one is made, use the sequences with the highest GC content, since these primers can be annealed under more stringent conditions, such as higher temperature (*see* Note 1).
2. Determine the number of permutations in the deduced nucleotide sequence. There are 192 permutations ([2 × 4] × 3 × 4 × 2) in the sequence 5'-YTN-

ATH-GGN-GAR-3'. The degeneracy can be reduced by making a guessmer. If the primer is to a human gene, a potential guessmer for our example could be 5'-CTB-ATY-GGN-GAR-3', which only contains 64 permutations *(2)*. The 3'-end of a primer should contain all possible permutations in the amino acid sequence, since *Taq* DNA polymerase will not extend from a primer with a mismatch at the 3' end.

3. The degeneracy of a primer can be reduced further by replacing "unguessable" nucleotides with inosine, since this synthetic nucleotide will hydrogen bond with all four naturally occuring nucleotides. In our example, this would result in the sequence 5'-CTB-ATY-GGI-GAR-3', which contains only 12 permutations. Use of inosine, however, reduces the annealing temperature and may therefore result in a higher background on nonspecifc product.
4. If possible, restriction endonuclease sites should be incorporated at the 5'-ends of a primer to facilitate cloning into plasmid vectors. If different restriction sites are added to each primer, the PCR product can be cloned directionally. Note that not all restriction enzymes can recognize cognate sites at the ends of a double-stranded DNA molecule. This difficulty can be ameliorated by the addition a 2–4-nucleotide 5'-overhang before the beginning of the restriction enzyme site (*see* Note 2). The best restriction enzyme sites to use are *Eco*RI, *Bam*HI, and *Xba*I *(3)*. An unforeseeable pitfall that can occur is when the same restriction site chosen for the primer is also found within the amplified product, in which case, only part of the amplified product would be cloned.
5. The final consideration is the identity of the 3'-most nucleotide. The nucleotide on the 3'-end of a primer should preferably be G or C, and not N, I, or T, since thymidine (and inosine) cannot specifically prime on any sequence. Guanosines and cytidine are preferred, since their three H-bonds are stronger than an A:T base pair.

3.2. Degenerate Oligonucleotide PCR Cloning

The first step in cloning is to determine the nucleotide sequence that lies between the degenerate primers (Fig. 1, step 1). Newly synthesized cDNA can be used as template for these reactions, but the author prefers to use a cDNA library (*see* Note 3). Vector DNA should also be PCR-amplified with the same primers as a negative control.

1. Pipet into 0.5-mL microcentrifuge tubes in the following order: 58.5 µL of double-distilled water that has been autoclaved; 5.0 µL of heat-denatured library DNA; 5.0 µL of each degenerate primer; 10 µL of 10X PCR reaction buffer; 16 µL of dNTP stock solution; and 0.5 µL of thermostable DNA polymerase. If several reactions are being set up concurrently, a

Degenerate Oligonucleotide Primers

master reaction mix can be made up, consisting of all the reagents used in all of the reactions, such as the double-distilled water, primers, reaction buffer, dNTPs, and the polymerase. This reaction mix should be added last (*see* Notes 4 and 5).
2. Briefly vortex each sample, and spin for 10 s in a microfuge.
3. Overlay each sample with two to three drops of mineral oil.
4. Amplify using the following cycle parameters: step 1: 94°C, 60 s (denaturation); step 2: 50°C, 90 s (annealing; *see* Note 6); and step 3: 72°C, 60 s (extension). For 24 cycles, then: step 5: 72°C, 4 min and step 6: 10°C hold.
5. Remove the reaction tubes from the thermal cycler, and add 200 µL of chloroform. Spin for 10 s in a microfuge to separate oil-chloroform layer from the aqueous layer.
6. Carefully transfer the aqueous layer to a clean microfuge tube. If any of the oil-chloroform layer is also transferred, it must be removed by extracting the sample again with another 100 µL of chloroform.
7. Set up four secondary PCR-amplifications with 10 µL aliquots of the first reactions, using the same procedure and cycling parameters. Again extract all of the samples with chloroform to get rid of the oil, and pool all four reactions into a 1.5-mL microfuge tube.
8. Add 150 µL of 7.5*M* ammonium acetate to 300 µL of the PCR reaction. Vortex briefly to mix. Precipitate the DNA with 1 mL of 100% ethanol. Vortex the samples for 5–10 s and ice for 15 min.
9. Precipitate the DNA by spinning at 12,000*g* for 10 min at 4°C in a microfuge. Decant the supernatant, and wash in 500 µL of 70% ethanol. Vortex briefly, and spin another 5 min at 4°C. Decant the ethanol and allow the pellets to dry inverted at room temperature, or dry in a Speed-Vac for 2–10 min. Resuspend the DNA in 100 µL of double-distilled water.

The amplified DNA can be sequenced directly or cloned. If the degenerate primers incorporate novel restriction enzyme sites, the DNA should be digested, gel-purified, and cloned into an appropriate plasmid vector. Alternatively, the DNA could be blunt-end-cloned using the T-vector approach (*see* Chapter 37). If the amplified product is >400 base pairs, it can be purified on an agarose gel (*see* Chapter 28). If it is smaller than this size, a polyacrylamide gel should be used (*see* Chapter 13).

3.3. Anchor PCR to Clone the 5'- and 3'-cDNA Ends

The initial degenerate PCR approach will only isolate a sequence within the conserved gene. Anchor PCR can be used to obtain the remaining 5'- and 3'-regions of the gene. Again vector DNA should also be PCR-amplified with the same primers as a negative control.

1. Pipet into 0.5-mL microcentrifuge tubes in the following order: 53.5 µL of double-distilled water that has been autoclaved; 10 µL of heat-denatured library DNA; 5.0 µL of the internal primer; 5.0 µL of primer homologous to the vector; 10 µL of 10X PCR reaction buffer; 16 µL of dNTP stock solution; 0.5 µL of DNA polymerase.
2. Briefly vortex each sample, and then spin for 10 s in a microfuge.
3. Overlay each sample with two to three drops of mineral oil.
4. Amplify using the following cycle parameters: step 1: 94°C, 60 s (denaturation); step 2: 54°C, 60 s (annealing; *see* Note 6); and step 3: 72°C, 60 s (extension). For 31 cycles, then: step 4: 72°C, 4 min and step 5: 10°C hold.
5. Remove the reaction tubes from the thermal cycler and add 200 µL of chloroform. Spin for 10 s in a microfuge to separate oil-chloroform layer from the aqueous layer.
6. Carefully transfer the aqueous layer to a clean microfuge tube. If any of the oil-chloroform layer is also transferred, it must be removed by extracting the sample again with another 100 µL of chloroform.
7. Remove all unincorporated nucleotides and primers by spinning the sample through a spin dialysis column (*see* Note 6 and 7). The DNA will remain on top of the membrane, while nucleotides and primes will pass through. Wash the sample with an additional 400 µL of TE and respin to concentrate. Repeat wash one more time. Add TE to the remaining sample to bring the final volume of retained fluid to 50 µL. Transfer to a clean sterile tube.
8. Reamplify a 5-µL aliquot of the PCR reaction using the original primers, or if possible, replace the first primer with an second nested one. Purify the product as in Section 3.2., steps 8 and 9.

4. Notes

1. A critical parameter when attempting to clone by PCR is the selection of a primer annealing temperature. This is especially true when using degenerate primers. The primer melting temperature (T_m) is calculated by adding 2°C for A:T base pairs, 3°C for G:C base pairs, 2°C for N:N base pairs, and 1°C for I:N base pairs. Most references suggest you calculate the T_m and set the primer annealing temperature to 5–10°C below the lowest T_m. Distantly related gene superfamily members have been cloned using this rationale *(4)*. However, the author has found that higher annealing temperatures are helpful in reducing nonspecific priming, which can significantly affect reactions containing degenerate primers.
2. When designing primers with restriction enzyme sites and 5'-overhangs, note that this 5'-overhang should not contain sequence complementary to the sequence just 3' of the restriction site, since this would facilitate the production of primer dimers. Consider the primer 5'-ggg.aagctt. CCC AGCTAGCTAGCT-3', which has a *Hin*dIII site preceded by a 5'-ggg and

followed by a CCC-3'. These 12 nucleotides on the 5'-end are palindromic, and can therefore easily dimerize with another like primer. A better 5'-overhang would be 5'-cac.
3. A possible problem with using a library as template is that certain sequences do not clone as well as the majority and therefore may be underrepresented. If the initial cloning fails, either try a different library or amplify directly from the uncloned cDNA.
4. A nonionic detergent, such a Nonident P-40, can be incorporated in rapid, sample preparations for PCR analysis without significantly affecting *Taq* polymerase activity *(5)*. In some cases, such detergents are absolutely required in order to detect a specific product reproducibly *(6)* presumably owing to inter- and intrastrand secondary structure. Tetramethylammonium chloride has been shown to enhance the specificity of PCR reactions by reducing nonspecific priming events *(7)*.
5. All PCR reactions should be set up in sterile laminar flow hoods using either positive displacement pipeters or pipet tips containing filters to prevent the contamination of samples, primers, nucleotides, and reaction buffers by DNA. Similarly, all primers, nucleotides, and reaction buffers for PCR should be made up and aliquoted using similar precautions.
6. Removal of the primers from the previous reaction is essential before performing a nested PCR-amplification, but may not be necessary if the original primers are to be used.
7. Extraction on glass beads (*see* Chapter 28) can also be used if the PCR product is larger than 500 bp. Because the glass beads bind molecules smaller than this size only poorly, this effectively removes small contaminating PCR products.

Acknowledgments

I am grateful to my colleagues for their support and helpful discussions. I especially thank Eric Fearon for technical assistance, and Peter Agre for his consent to report these procedures in detailed form and for his generous support. This work was supported in part by NIH grant HL33991 to Peter Agre.

References

1. Pao, G. M., Wu, L.-F., Johnson, K. D., Höfte, H., Chrispeels, M. J., Sweet, G., Sandal, N. N., and Saier, M. H., Jr. (1991) Evolution of the MIP family of integral membrane transport proteins. *Mol. Microbiol.* **5,** 33–37.
2. Wada, K.-N., Aota, S.-I., Tsuchiya, R., Ishibashi, F., Gojobori, T., and Ikemura, T. (1990) Codon usage tabulated from the GenBank genetic sequence data. *Nucleic Acids Res.* **18,** 2367–2411.
3. Appendix of any New England Biolab enzyme catalog.

4. Zhao, Z.-Y. and Joho, R. H. (1990) Isolation of distantly related members in a multigene family using the polymerase chain reaction technique. *Biochem. Biophys. Res. Comm.* **167,** 174–182.
5. Weyant, R. S., Edmonds, P., and Swaminathan, B. (1990) Effects of ionic and nonionic detergents on the Taq polymerase. *BioTechnology* **9,** 308,309.
6. Bookstein, R., Lai, C.-C., To, H., and Lee, W.-H. (1990) PCR-based detection of a polymorphic Bam HI site in intron 1 of the human retinoblastoma (RB) gene. *Nucleic Acids Res.* **18,** 1666.
7. Hung, T., Mak, K., and Fong, K. (1990) A specificity enhancer for polymerase chain reaction. *Nucleic Acids Res.* **18,** 4953.
8. Preston, G. M. and Agre, P. (1991) Isolation of the cDNA for erythrocyte integral membrane protein of 28 kilodaltons: member of an ancient channel family. *Proc. Natl. Acad. Sci. USA* **88,** 11,110–11,114.

CHAPTER 37

Cloning PCR Products Using T-Vectors

Michael K. Trower and Greg S. Elgar

1. Introduction

The polymerase chain reaction (PCR) has revolutionized the way that molecular biologists approach the manipulation of nucleic acids through its ability to amplify specific DNA sequences *(1–3)*. It has numerous applications *(4)*, many of which require the cloning of amplified DNA products into vectors for further analysis. This is commonly achieved by so-called "sticky end" cloning, in which restriction endonuclease recognition sites are incorporated into the 5' ends of the PCR primers *(5)*. Following amplification, the DNA fragment is purified, digested with the appropriate enzyme(s), and then ligated into an identically restricted vector. Obtaining efficient cleavage at the extreme ends of linear PCR products can be difficult *(6)*, and moreover, their use can result in the restriction of sites that lie within the amplified DNA fragment. The one particular advantage to this method is that the PCR product can be force-cloned using designed restriction sites, so, for example, a DNA fragment, such as a leader signal sequence, can be fused in-frame to a structural gene for expression studies.

Blunt-end cloning is an alternative procedure for cloning PCR products when precise orientation is not required. The protocol is complicated by the inherent terminal transferase activity of *Taq* polymerase, which tends to add a template-independent single deoxyadenosine (A) residue to the 3' ends of the PCR product *(7)*. The PCR product, after purification, must therefore be "flush-ended" by treatment with a DNA polymerase having "proof-reading" 3' to 5' exonuclease activity, such as

From: *Methods in Molecular Biology, Vol. 58: Basic DNA and RNA Protocols*
Edited by: A. Harwood Humana Press Inc., Totowa, NJ

Klenow or T4 DNA polymerase, prior to ligation into a blunt-ended vector *(8)*. Unfortunately, it is common to find on screening that a large proportion of the plasmids isolated lack an insert even when blue/white β-galactosidase color selection is available.

Recently there have been reports of alternative procedures for cloning PCR products in which the terminal transferase activity of *Taq* polymerase is exploited *(9,10)*. These methods are up to 50 times more efficient than blunt-ended cloning. The protocols rely on the creation of cloning vectors (T-vectors), which once linearized have a single 3' deoxythymidine (T) at each end of their arms. This allows direct "sticky end" ligation of PCR products containing *Taq* polymerase-catalyzed A extensions without further enzymatic processing.

In the method of Smith et al. *(9)* the vector extensions are generated by endonuclease digestion of a specialized plasmid. By contrast, the procedure described by Marchuk et al. *(10)* utilizes the terminal transferase activity of *Taq* polymerase to add deoxythymidine (T) residues to blunt-ended restricted vectors (Fig. 1). The latter procedure has two general advantages: First, it can be used to generate T-vectors from many of the cloning vectors commonly found in molecular biology laboratories, and second, there is a very low background because of self ligation of the vector. It is this method that we describe in detail; covering primer design, setting up of a PCR reaction, product isolation, the steps involved in manufacturing T-vectors, and then their use for cloning the PCR products.

2. Materials

All reagents are prepared with sterile distilled water and stored at room temperature unless stated otherwise.

2.1. PCR Reaction and Product Isolation

1. PCR primers: synthetic oligonucleotides diluted to 10 m*M* (*see* Note 1). The design of these is critical to the success of the PCR reaction (*see* Notes 2 and 3). Store at –20°C.
2. 10X PCR buffer: 100 m*M* Tris-HCl, pH 8.3, 25°C, 500 m*M* KCl, 15 m*M* MgCl$_2$, 0.1% w/v gelatin. This buffer should be incubated at 50°C to fully melt the gelatin and then filter sterilized. Store at –20°C.
3. dNTP mix: a mix in distilled water containing 2 m*M* of each dNTP. Liquid stocks of dNTPs are available from companies such as Pharmacia (Piscataway, NJ) and Boehringer (Indianapolis, IN), and we find these convenient and reliable. Store at –20°C.
4. Mineral oil.

Cloning PCR Products

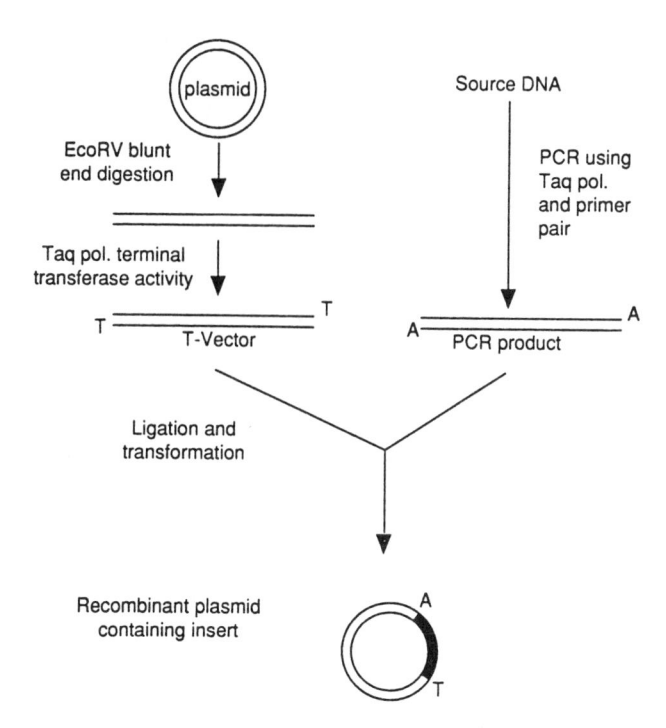

Fig. 1. Schematic diagram demonstrating the principle of cloning PCR products with plasmid T-vectors.

5. *Taq* polymerase (5 U/μL). Store at −20°C.
6. PCR stop mix: 25% Ficoll 400, 100 mM EDTA, 0.1% (w/v) bromophenol blue.
7. Agarose gels: Melt electrophoresis grade agarose in either TAE (50X TAE: 242 g Tris base, 57.1 mL glacial acetic acid, 100 mL 0.5M EDTA, pH 8.0 or TBE (10X TBE: 121 g Tris, 55 g orthoboric acid, 7.4 g/L EDTA) by gentle boiling (this can be carried out in a microwave). Cool until hand warm and pour into prepared gel former. Run small gels at around 100 V using the dye in the stop mix as an indicator of migration *(see* Note 4).
8. Phenol: Ultra-pure phenol is buffer-saturated in TE. Store at 4°C.
9. Chloroform: 29:1 mix of chloroform with isoamyl alcohol.

2.2. T-Vector Preparation, Ligation, and Transformation

10. Vector DNA: A suitable plasmid, such as pBluescript II (Stratagene, La Jolla, CA), prepared as a miniprep *(11)*. Store at −20°C.
11. Restriction endonucleases: *Eco*RV, *Sma*I, or other blunt-end cutter as required. Store at −20°C.

12. dTTP: 100 mM dTTP stock. Store as small aliquots at –20°C.
13. 10X Ligation buffer: 0.5M Tris-HCl, pH 7.6, 100 mM MgCl$_2$, 500 µg/mL bovine serum albumin (BSA). Store at –20°C.
14. DTT: 100 mM Dithiothreitol stock. Store as small aliquots at –20°C.
15. ATP: for ligation reactions. 10 mM stock. Store as small aliquots at –20°C.
16. T4 DNA ligase: as available from numerous commercial suppliers. The unit activity of T4 DNA ligase may be assessed in Weiss units (a pyrophosphate exchange assay), circle formation units, or by the supplier's own unit assay (such as λ/*Hin*dIII fragment ligation). It is therefore best to refer to product information, although generally 0.5 µL is sufficient because T4 DNA ligase is always added in excess. Store at –20°C.

3. Methods

3.1. PCR Reaction and Product Isolation

It is difficult to define a single set of conditions that will ensure optimal specific PCR amplification of the target DNA sequence. We describe a basic protocol that has been successful in our hands for most applications and that should be used first before attempting any variations. However, we advise readers who are not familiar with PCR to refer to Notes 5–8 and Chapter 35.

Although aliquots from the PCR reaction mix can be used directly for T-vector ligation, we have found an improvement in efficiency if the DNA products are first purified. There are a number of options available, as described in steps 6 and 7.

1. Prepare the PCR reaction mix as follows. To a 0.5-mL Eppendorf tube, add 5 µL of each PCR primer, 5 µL of 10X PCR buffer, 5 µL of dNTP mix, 1 µL of template DNA *(see* Note 9), and 29 µL of distilled water, giving a total volume of 50 µL. Overlay the mixed reaction mix with sufficient mineral oil to prevent evaporation.
2. Transfer the tube to a thermal cycler and heat at 95°C for 5 min. "Hot start" the reaction by the addition of 0.3 µL of *Taq* DNA polymerase.
3. Immediately initiate the following program for 30–35 cycles *(see* Note 10): denaturation, 95°C for 0.5 min; primer annealing, 50–60°C for 0.5 min; and primer extension, 72°C for 0.5–3 min.
4. After the final cycle, carry out an additional step of 72°C for 5 min. This will ensure that primer extension is completed to give full-length double-stranded product.
5. Add 1 µL of PCR stop mix to 5 µL of PCR product and run on a 1.5% agarose gel to determine the yield and specificity of the PCR reaction. The anticipated yield of PCR products is 10–50 ng of DNA/µL reaction.

6. Where a PCR produces a single band or a number of bands, all of which could be the correct product, the entire reaction mix may be phenol/chloroform extracted and precipitated. Resuspend in a volume of 10 µL of TE or distilled water *(see* Note 11).
7. If unwanted bands are also present, the whole reaction should be run on a low melting point agarose gel and the bands of interest excised with a clean scalpel blade. When excising gel slices it is preferable to use a UV box that emits longer wavelength UV light (365 nm), because this causes less damage to the DNA. The resulting gel slices may be purified in a variety of ways *(see* Note 11).

3.2. T-Vector Preparation, Ligation, and Transformation

Preparation of T-vectors involves first digestion of the cloning vector with a restriction enzyme that generates blunt ends and then addition of T-residues to the cut vector utilizing the inherent terminal transferase activity of *Taq* DNA polymerase. The simplicity of this system allows T-vectors to be prepared using any cloning vector with blunt-end restriction sites, such as *Eco*RV or *Sma*I. The pBluescript II series of vectors (Stratagene) are excellent for this purpose since both these restriction sites are located within the polylinker. The T-vector is prepared in batch so it will last for a number of cloning reactions. For example, 5 µg of T-vector DNA is sufficient for 100 cloning experiments.

The PCR product isolated following *in vitro* amplification is directly ligated into the prepared T-vector using the same conditions as for sticky-end cloning. The optimum amount of PCR product(s) to be added per ligation reaction is difficult to estimate because of variance in yield, complexity of the PCR product profile, and the efficiency of A addition to the DNA products by the *Taq*-polymerase-catalyzed terminal transferase activity. To ensure that the PCR products are within a range that should ensure successful cloning, we usually set up two ligations, one of which involves a 1:10 dilution of the purified DNA.

1. Digest 5 µg of vector DNA with a restriction enzyme that generates a unique blunt ended site, for example, *Eco*RV or *Sma*I, for 2 h at 37°C *(see* Note 12).
2. Run the digest on a 1% low-melting-point agarose gel. Excise the linear vector DNA under UV, phenol/chloroform extract, and ethanol precipitate *(see* Notes 11 and 13). Resuspend in a volume of 20 µL of water in a 0.5-mL Eppendorf tube.

3. Add 5 µL of 10X PCR buffer, 1 µL of 100 m*M* dTTP, 24 µL of distilled water, and 0.4 µL of *Taq* DNA polymerase. Overlay with 40 µL of mineral oil. Incubate at 72°C for 2 h. For convenience, a thermal cycler set at 72°C may be used.
4. Purify the T-vector by phenol/chloroform extraction and ethanol precipitation. Resuspend the prepared T-vector in 100 µL of water or TE, giving a concentration of 50 ng/µL.
5. Set up three tubes containing 1.0 µL of either undiluted PCR product; a 1:10 dilution of the PCR product or distilled water (a control that will indicate the background of vector self ligation). Add to each tube 1.0 µL of prepared T-vector, 1.0 µL of 10X ligation buffer, 1.0 µL of DTT, 1.0 µL of ATP, 4.5 µL of distilled water and 0.5 µL of T4 DNA ligase. Incubate overnight at 16°C.
6. Transform 5.0 µL of each ligation reaction into a suitable competent strain of *E. coli (see* Note 14).

Typically up to several hundred colonies may be isolated following transformation *(see* Notes 15–17). If a blue/white colony selection has been used, we generally find a ratio of about 50%. The white colonies or, if no method of nonrecombinant differentiation has been used, random colonies, are then screened for inserts. Typical results from a PCR cloning experiment are shown in Fig. 2.

4. Notes

1. We have found it unnecessary for primers of the size we have outlined to be gel or HPLC purified. We routinely precipitate the primers and resuspend them in 200–500 µL of sterile distilled water or TE buffer. Their concentration can be obtained using a spectrophotometer (1 OD_{260} = 20 µg of single-stranded oligonucleotide). The molarity is estimated, assuming an average molecular weight/base of 325 Daltons, and an aliquot of each stock is diluted to a concentration of 10 µ*M*.
2. The selection of a pair of oligonucleotide primers is the first step in preparing a PCR reaction. Although there are no hard and fast rules one can follow to absolutely ensure a given pair of PCR primers will result in the isolation of a desired DNA fragment, we have outlined some guidelines that should be taken into account when designing your oligonucleotides, and that will enhance your chances of achieving successful amplification of the target DNA sequence.
    a. Some sequence information (17–25 bp) is generally necessary at either end of the region to be amplified *(see* Note 3). The length of this segment should not exceed 3–4 kbp for practical purposes *(see* Note 10).

Cloning PCR Products

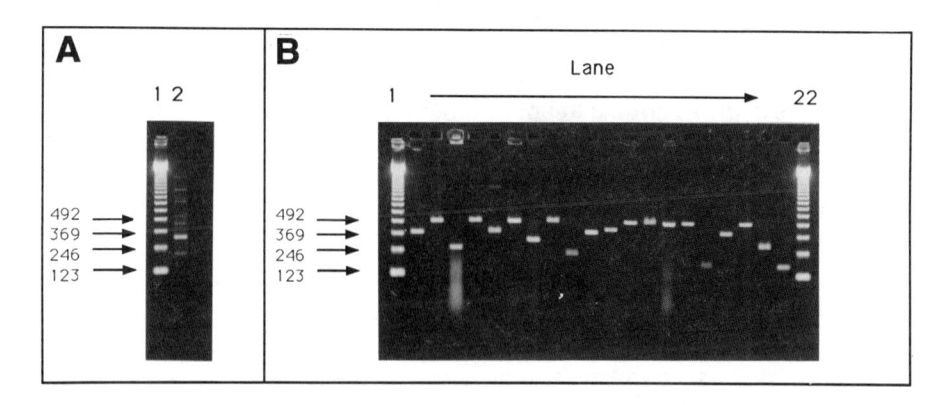

Fig. 2. **(A)** A pair of degenerate primers was used to PCR genomic DNA from a complex organism. Analysis by agarose gel electrophoresis of an aliquot of the PCR reaction revealed a complicated series of products with a dominant band of around 340 bp **(B)**. Since the anticipated size of the target sequence was 200–400 bp, the PCR reaction was purified as in Section 3.1., step 6, and then directly ligated into a prepared pBluescript II SK(+) T-vector. The ligation was transformed into *E. coli* strain XL1-Blue and selection was made on enriched media plates (with 100 µg/mL ampicillin) in the presence of X-gal and IPTG. Approximately 200 colonies were isolated with a blue/white ratio of around 50%. Nineteen white colonies and one blue were subjected to a PCR screen with flanking T3 and T7 Universal primers to determine both the size of the products cloned and efficiency of the method (*see* Chapter 39). Agarose gel electrophoresis (1.5% gel) (B) revealed that all nineteen white colonies carried cloned PCR products, with a predominant fragment of 500 bp. Because the polylinker of Bluescript II contributes an additional 166 bp, the actual insert size is approx 340 bp. This confirms cloning of the PCR products. (A) Product profile on a 1.5% agarose gel. Lane 1, 123-bp ladder (Life Technologies); lane 2, PCR products. (B) PCR screen of 19 white colonies (lanes 2–20) and one blue control colony (lane 21) with T3 and T7 Universal primers. lanes 1 and 22, 123-bp ladder.

   b. For PCR from a complex genomic DNA source we have found 20–24-mers to be long enough to give specific amplification of the target region; when the template is less complex, for example, from a plasmid template, the primer length may be reduced to a 17-mer. Longer primers can be prepared, but this is usually unnecessary and costly.
   c. The primers should match their target hybridization sites well, especially at their 3' ends. If possible, keep the G/C content of each primer to about 50% and try to avoid long stretches of the same base. Both primers should be approximately the same length.

d. Check that the two primers do not have significant complementarity to each other, particularly at their 3' ends, to avoid "primer dimers;" where two primers hybridize to one another forming a very effective substrate for PCR that subsequently may become the dominant product *(13)*.

e. Where the aim of the PCR is to clone a known gene, or its homolog, check the primer sequences against the EMBL/Genbank database to try to ensure they are unique to the template DNA.

f. It is worth bearing in mind that both length and G/C content of primers determine the optimum annealing temperature in the PCR reaction. There are now a number of computer programs available that will design primers for specific target sequences.

3. One of the difficulties in primer design when cloning from complex genomes is that the precise sequence of the target region is often unknown, and therefore must be predicted from reported sequences of the same gene in different organisms or from sequence information from similar genes. For structural genes, the known DNA sequence homologs should be first translated into protein and then aligned. There are a number of alignment programs available on mainframe computers, such as CLUSTAL *(14)*, that can assist in alignments and therefore allow identification of conserved regions. Within the conserved region, amino acids of low degeneracy are particularly sought after, such as those with one codon, methionine and tryptophan; and two codons, asparagine, aspartate, cysteine, glutamine, glutamate, histidine, lysine, phenylalanine, and tyrosine; which thereby lower the degeneracy of the primers. The amino acids arginine, leucine, and serine are encoded by six different codons and should be avoided if possible. When designing the primers one should also take into account the codon usage for the particular organism under study for which codon usage tables are available *(15)*. Finally, try to locate the most conserved sequence information available at the 3' ends of the primers.

4. When making agarose gels it is worth remembering that TAE gels are easier to use with kits, such as GeneClean. For products of 100 bp or less, it may be necessary to pour gels of >2%. Products above 300–400 bp can be run on 1.5% gels. It is useful to run PCR products against a low-mol-wt ladder, such as a 123-bp ladder (Life Technologies Inc., Gaithersburg, MD) or pBR322/*Msp*I digest.

5. There are many modifications that can be made to the basic PCR reaction mix in an attempt to improve the specificity of the amplification reaction. However, even after taking precautions, multiple PCR products may be generated because of nonspecific priming. The parameter that will give the greatest variation in product yield and complexity is the annealing temperature. By increasing the annealing temperature, the specificity of the reaction is increased because of more stringent binding of the primers.

However, if the temperature of annealing is raised too high, the yield of product will decrease and eventually disappear. By experimenting with a number of different annealing temperatures it is usually possible to select the optimum PCR conditions for a particular pair of primers. Improvements in specificity may also be achieved by altering the Mg^{2+} concentration in the PCR buffer over the range 0.5–5 mM final concentration. Another specificity enhancer is DMSO, which may be included in the range 5–10% (v/v). If all else fails, a second set of primers, which are located internally to the first pair (nested), may be used to reamplify the products obtained from the initial amplification *(2)*.

6. Even after optimizing the PCR reaction there may be a number of seemingly specific bands that may not correspond to the desired product. We have found that this is most often caused by specific double-priming of one of the primer pair so as to produce a PCR product. We therefore routinely run parallel reactions that contain only one primer. By running these reactions side by side on a gel it is possible to determine which products are formed by the interaction of two primers and which are formed from one primer only.
7. *Taq* polymerase lacks a 3'–5' proofreading exonuclease activity *(16)*. Therefore, during enzymatic DNA amplification, errors can accumulate in the PCR product at rates as high as 2×10^{-4} *(3)*. Errors occurring early in the PCR will be present in a substantial number of the amplified molecules and therefore in many of the clones. In situations where unknown sequences are being isolated, we suggest determining the sequence of clones from at least three independent PCR reactions.
8. The exquisite sensitivity of the amplification reaction makes it liable to contamination by minute amounts of exogenous DNA. This is a particular problem for PCR reactions involving complex genomes when small amounts of template are present and which therefore need extra rounds of amplification. Measures can be taken to avoid this contamination, including the use of gloves, pipet tips with filters, UV treating solutions, and tubes, but obviously not DNA, and if necessary carrying out the preparation work in a lamina flow hood.
9. The source of template DNA can vary from genomic DNA, cDNA, and DNA prepared from YACs, cosmids, λ, plasmids, and M13 phages, and may be single- or double-stranded. As a guide, use more DNA as the complexity of the source increases. For complex genomes use a DNA concentration of 10–100 µg/mL, whereas for plasmid DNA preparations use 1–10 µg/mL. The amplification efficiency is reduced in the presence of too much template. We have also successfully amplified by directly picking or aliquoting from single clone λ phage stocks or plaques, plasmid colonies and glycerol stocks, and M13 plaques.

10. As a very rough guide to annealing temperature, allow 4°C for G or C and 2°C for every A or T. Thus, for an 18-mer with 10(G + C), annealing temperature is [10 × 4] + [8 × 2] = 56°C. The time allowed for the primer extension step is based on a rate of approx 1 kb/min; for practical purposes using *Taq* polymerase the maximum product size that can be visualized on a gel, purified, and cloned is around 3–4 kb.
11. Two commonly used methods for purifying PCR products either directly or following agarose gel electrophoresis are:
    a. Phenol/chloroform extraction. Add an equal volume of phenol, mix and centrifuge 13,000g for 10 min, transfer the aqueous phase (upper phase) to a fresh tube, and add an equal volume of chloroform. Mix, respin, and then to the transferred aqueous phase (top) add 1/10 vol 3*M* potassium acetate and 2.5 vol of ethanol. Leave at either –20°C or on dry ice for 10 min, and then spin 13,000g for 10 min. Remove the supernatant and wash pellet with 200 µL of 70% ethanol. Dry the pellet either in the air or under vacuum and resuspend in the desired volume of TE or water. If extracting from low-melting-point agarose, weigh the agarose slice, add an equal volume of water, and heat to 65°C to melt the agarose completely before adding the phenol.
    b. DNA binding matrices in kits such as GlassMax (Gibco-BRL) may also be used to purify PCR products, again either directly or from agarose. Follow the protocols specified by the manufacturer.
12. *Eco*RV is preferred to *Sma*I if available because it is more stable at 37°C and tends to give more efficient cleavage of vector DNA. If *Sma*I is the only blunt-end cleavage site available, then digestion should be carried out by incubation at 25°C.
13. By gel purifying the T-vector any uncut vector is removed, reducing the background of colonies that lack inserts.
14. Competent *E. coli* can be either prepared by the method of Hanahan *(17; see* Chapter 29) or purchased directly from commercial suppliers.
15. If the cloning vector of choice has β-galactosidase blue/white selection, inclusion of IPTG and X-Gal in the plating mix on transformation will result in recombinant clones giving rise to white colonies. This is caused by insertional inactivation of the β-galactosidase gene, thereby preventing it from cleaving the chromogenic lactose analog X-Gal, which in turn prevents blue colony formation. White colonies do not definitively indicate recombinants because some may be caused by frameshift events or by misligation of noncompatible ends *(18)*. Conversely, some blue colonies may be recombinant clones that have left the β-galactosidase gene in frame in such a way that the enzyme still retains activity.

16. The lack of any colonies on a plate following transformation may be attributable to either low competency of the *E. coli* or failure of the PCR product to ligate to the T-vector. For each set of transformations a control plasmid should also be included to monitor the transformation efficiency of the *E. coli*. In addition, confirm the concentration of both T-vector and PCR product, and whether successful ligation has taken place, by running samples on an agarose gel.
17. An extremely high background of nonrecombinant colonies, either as blue colonies or as determined by the screening protocol, is most likely because of failure of the T addition to the vector arms. Repeat this step ensuring all the components are included and that the *Taq* polymerase is active. An indicator of successful T-vector preparation can be ascertained by its ligation in the absence of PCR product, which following transformation should result in the presence of only a small number of nonrecombinant (blue) colonies.

Acknowledgments

We would like to thank Sydney Brenner of the Department of Medicine, University of Cambridge for the opportunity to devise this protocol. At the time M. K. T. was supported by a fellowship from E. I. du Pont de Nemours and Co. Inc., Wilmington, DE, USA, and G. S. E. was supported by The Jeantet Foundation.

References

1. Saiki, R. K., Scharf, S. J., Faloona, F. A., Mullis, K. B., Horn, G. T., Erlich, H. A., and Arnheim, N. (1985) Enzymatic amplification of β-globin genomic sequences and restriction site analysis for diagnosis of sickle cell anemia. *Science* **230**, 1350–1354.
2. Mullis, K. B. and Faloona, F. A. (1987) Specific synthesis of DNA *in vitro* via a polymerase-catalyzed chain reaction. *Methods Enzymol.* **155**, 335–350.
3. Saiki, R. K., Gelfand, D. H., Stoffel, S., Scharf, S. J., Higuchi, R., Horn, G. T., Mullis, K. B., and Erlich, H. A. (1988) Primer-directed enzymatic amplification of DNA with a thermostable DNA polymerase. *Science* **239**, 487–491.
4. Erlich, H. A., Gelfand, D., and Sininsky, J. J. (1991) Recent advances in the polymerase chain reaction. *Science* **252**, 1643–1651.
5. Scharf, S. J., Horn, G. T., and Erlich, H. A. (1986) Direct cloning and sequence analysis of enzymatically amplified genomic sequences. *Science* **233**, 1076–1078.
6. Kaufman, D. L. and Evans, G. A. (1990) Restriction endonuclease cleavage at the termini of PCR products. *Biotechniques* **9**, 304,305.
7. Clark, J. M. (1988) Novel non-templated nucleotide reactions catalyzed by procaryotic and eucaryotic DNA polymerases. *Nucleic Acids Res.* **18**, 9677–9686.
8. Hemsley, A., Arnheim, N., Toney, M. D., Cortopassi, G., and Galas, D. J. (1989) A simple method for site-directed mutagenesis using the polymerase chain reaction. *Nucleic Acids Res.* **17**, 6545–6551.

9. Mead, D. A., Pey, N. K., Herrnstadt, C., Marcil, R. A., and Smith, L. (1991) A universal method for the direct cloning of PCR amplified nucleic acid. *Biotechnology* **9,** 657–663.
10. Marchuk, D., Drumm, M., Saulino, A., and Collins, F. S. (1991) Construction of T-vectors, a rapid and general system for direct cloning of unmodified PCR products. *Nucleic Acids Res.* **19,** 1154.
11. Sambrook, J., Fritsch, E. F., and Maniatis, T., eds. (1989) *Molecular Cloning. A Laboratory Manual,* 2nd ed. Cold Spring Harbor Laboratory, Cold Spring Harbor, NY, pp. 1.25–1.28.
12. Güssow, D. and Clackson, T. (1989) Direct clone characterisation from plaques and colonies by the polymerase chain reaction. *Nucleic Acids Res.* **17,** 4000.
13. Mullis, K. B. (1991) The polymerase chain reaction in an anemic mode: how to avoid cold oligodeoxyribonuclear fusion. *PCR Meth. Appl.* **1,** 1–4.
14. Higgins, D. G. and Sharp, P. M. (1988) CLUSTAL: a package for performing multiple sequence alignment on a microcomputer. *Gene* **73,** 237–244.
15. Wada, K., Wada, Y., Ishibashi, F., Gojobori, T., and Ikemura, T. (1992) Codon usage tabulated from the GenBank genetic sequence data. *Nucleic Acids Res.* **20(Suppl.),** 2111–2118.
16. Tindall, K. R. and Kunkel, T. A. (1988) Fidelity of DNA synthesis by the *Thermus aquaticus* DNA polymerase. *Biochemistry* **27,** 6008–6013.
17. Hanahan, D. (1985) Techniques for transformation of *E. coli,* in *DNA Cloning: A Practical Approach* (Glover, G. M., ed.), vol. 1, pp. 109–135.
18. Wiaderkiewicz, R. and Ruiz-Carillo, A. (1987) Mismatch and blunt to protruding-end joining by DNA ligases. *Nucleic Acids Res.* **15,** 7831–7848.

CHAPTER 38

Direct Radioactive Labeling of Polymerase Chain Reaction Products

Tim McDaniel and Stephen J. Meltzer

1. Introduction

Radioactively labeled polymerase chain reaction (PCR) products are being used in an increasing number of molecular biology research techniques. Among these are PCR-based polymorphism assays, such as linkage analysis *(1)* and detecting allelic loss in cancer cells *(2)*. Other uses of radioactive PCR include generating probes for Southern and Northern blotting *(3)* and screening for polymorphisms and point mutations by the single-strand conformation polymorphism (SSCP) technique *(4)*.

The two most common methods for radioactively labeling PCR products, adding radioactive deoxynucleotide triphosphates (dNTPs) to the PCR mixture *(1)* and end-labeling primers prior to the reaction *(5)*, are problematic in that they risk radioactive contamination of the thermal cycler and extend the radioactive work area. The technique described here *(6)*, based on the methods of O'Farrell et al. *(7)* and Nelkin *(8)*, overcomes these problems because labeling is carried out after the PCR is completed, totally separate from the PCR machine. The technique is rapid, requires no purification of PCR products (e.g., phenol-chloroform extraction), and allows visualization of a tiny fraction of the product in under 2 h after labeling.

The method employs the Klenow fragment of *E. coli* DNA polymerase I, exploiting both the 5'–3' synthetic and 3'–5' excision (proofreading) functions of the enzyme. Briefly, unlabeled PCR product is mixed with Klenow fragment in the absence of dNTPs. Under these conditions, the enzyme lacks necessary substrates for its 5'–3' synthetic function, and thus engages solely in its 3'–5' excision activity, nibbling away bases at the 3' end of the PCR fragments. After allowing this reaction to proceed a short time, dNTPs, including [^{32}P]-dCTP, are added, allowing Klenow fragment to fill in and thereby label the recessed 3' ends.

2. Materials

1. PCR product.
2. Klenow fragment (1 U/μL).
3. 10X Klenow buffer: 500 mM Tris-HCl, pH 7.8, 100 mM MgCl$_2$, 10 mM β-mercaptoethanol.
4. [α-^{32}P]dCTP (deoxycytidine 5'-[α-^{32}P]triphosphate, triethyl ammonium salt, aqueous solution, 10 mCi/mL, 3000 Ci/mmol).
5. dNTP solution (400 μM each: dGTP, dATP, and dTTP).
6. Sequenase stop solution: 95% formamide, 20 mM EDTA, .05% bromophenol blue, 0.05% xylene cyanol FF.
7. DNA sequencing apparatus (including glass plates, spacers, and combs).

3. Methods

1. Digest the 5' ends of the PCR product by mixing 2 μL of PCR product (from a 100 μL reaction), 2 μL of 10X Klenow buffer, 1 μL of Klenow fragment, and 15 μL of H$_2$O in a microfuge tube.
2. Let stand for 10 min at room temperature.
3. Initiate fill-in of the digested ends by adding 0.5 μL of [α-^{32}P]dCTP and 1 μL of dNTP solution.
4. Let stand 10 min at room temperature *(see* Note 1).
5. Destroy the enzyme by incubating at 70°C for 5 min.
6. At this point, labeled DNA can be incorporated into whatever assay requires it, or can be visualized by running on a denaturing polyacrylamide gel as follows.
7. Mix 8 μL of the final labeling reaction mixture with 2 μL of sequenase stop solution.
8. Denature double-stranded DNA by heating for 5 min at 90°C.
9. Load samples into prerun 8% acrylamide, 50% urea sequencing gel and run 1^1/$_2$ h at a voltage that maintains a surface temperature of 50°C (this voltage varies among sequencing apparatuses).
10. Autoradiograph the gel 1^1/$_2$ h at –70°C with an intensifying screen. (*See* Note 2).

4. Notes

1. To ensure complete fill-in of the 3' ends, it may be desirable to chase the reaction with 1 µL of 400 µ*M* dCTP for 5 min just before destroying with the 70°C incubation. However, we have found this step to be unnecessary.
2. To verify the size of the labeled PCR products, run them along with DNA size markers that have been labeled by the same method. To label marker DNA, run the aforementioned protocol substituting 2 µL (2 µg) of marker DNA for the PCR product.

Acknowledgments

Supported by ACS grant #PDT-419, the Crohn's and Colitis Foundation of America, and the Department of Veterans Affairs.

References

1. Peterson, M. B., Economou, E. P., Slaugenhaupt, S. A., Chakravarti, A., and Antonarakis, S. E. (1990) Linkage analysis of the human HMG 14 gene on chromosome 21 using a GT dinucleotide repeat as polymorphic marker. *Genomics* **7**, 136–138.
2. Bookstein, R., Rio, P., Madreperla, S. A., Hong, F., Allred, C., Grizzle, W. E., and Lee, W.-H. (1990) Promoter deletion and loss of retinoblastoma gene expression in human prostate carcinoma. *Proc. Natl. Acad. Sci. USA* **87**, 7762–7766.
3. Schowalter, D. B. and Sommer, S. S. (1989) The generation of radiolabeled DNA and RNA probes with polymerase chain reaction. *Anal. Biochem.* **177**, 90–94.
4. Orita, M., Suzuki, Y., Sekiya, T., and Hayashi, K. (1989) Rapid and sensitive detection of point mutations and DNA polymorphisms using the polymerase chain reaction. *Genomics* **5**, 874–879.
5. Hayashi, S., Orita, M., Suzuki, Y., and Sekiya, T. (1989) Use of labeled primers in polymerase chain reaction (LP-PCR) for a rapid detection of the product. *Nucleic Acids Res.* **18**, 3605.
6. McDaniel, T. K., Huang, Y., Yin, J., Needleman, S. W., and Meltzer, S. J. (1991) Direct radiolabeling of unpurified PCR product using Klenow fragment. *BioTechniques* **11**, 7,8.
7. O'Farrell, P. H., Kutter, E., and Nakanishi, M. (1980) A restriction map of the bacteriophage T4 genome. *Mol. Gen. Genet.* **179**, 421–435.
8. Nelkin, B. D. (1990) Labeling of double stranded oligonucleotides to high specific activity. *BioTechniques* **8**, 616–618.

CHAPTER 39

A Rapid PCR-Based Colony Screening Protocol for Cloned Inserts

Michael K. Trower

1. Introduction

Following transformation of a ligation reaction into competent *E. coli* cells, successful subclones are conventionally identified by two methods. The first involves preparing "mini-prep" plasmid DNA from a number of colonies and then identifying the desired recombinant plasmid on the basis of either its unique restriction enzyme digest pattern or by direct DNA sequencing. The second, often used when large numbers of putative recombinants are involved, is by colony hybridization with a labeled probe. In this chapter an alternative PCR-based method for direct screening of transformants is described that is both facile and rapid, completely circumventing time-consuming DNA plasmid preparations. In its simplest form described below, transformed colonies are directly toothpicked into a small volume of PCR reaction mix that includes primers that flank the cloning site. This is usually achieved with Universal primers from the vector polylinker *(1)*. Following in vitro amplification, aliquots of each reaction are analyzed by agarose gel electrophoresis, which reveals both the presence and size of cloned inserts.

2. Materials

The protocol described is based on the use of thermostable microtiter plates, although it can be adapted to 200–500 µL volume thin-walled tubes if required. All reagents are prepared with sterile distilled water and unless stated otherwise are stored at room temperature.

From: *Methods in Molecular Biology, Vol. 58: Basic DNA and RNA Protocols*
Edited by: A. Harwood Humana Press Inc., Totowa, NJ

1. PCR primers: Synthetic oligonucleotide primers (17–21-mers) prepared as 10 µM stocks. Store at –20°C.
2. 10X PCR buffer: 100 mM Tris-HCl, pH 9.0, 500 mM KCl, 15 mM MgCl$_2$, 1% Triton X-100. Store at –20°C.
3. dNTP mix: stock containing 2 mM dATP, dCTP, dGTP, and dTTP. Store at –20°C.
4. *Taq* DNA polymerase (5 U/µL) (Amplitaq from Applied Biosystems, Foster City, CA). Store at –20°C.
5. Theromostable 96-well microtiter plate *(see* Note 1).
6. Enriched culture medium.
7. Sterile cocktail sticks.
8. Sterile culture microtiter plates.
9. Mineral oil.
10. 10X Gel loading dye: 25% Ficoll 400, 100 mM EDTA, 0.1% bromophenol blue.

3. Methods

1. Determine the number of screening reactions to be performed *(see* Note 2). Prepare a stock PCR reaction mix containing, for each colony to be amplified: 0.5 µL of forward and reverse Universal primers *(see* Note 3), 1.0 µL of dNTP stock, 1.0 µL of 10X PCR buffer, 0.04 µL *Taq* DNA polymerase, and 6.96 µL of distilled water.
2. Aliquot 10 µL of the stock PCR reaction mix into each designated well of the microtiter plate.
3. For each colony to be screened, pipet 100 µL of an enriched culture medium (supplemented with antibiotic or other selective agent), into the wells of a second sterile microtiter plate. Carefully number the wells so that after analysis of the amplified products, a particular clone can be readily identified.
4. Pick each colony with a sterile cocktail stick and swirl it first in the PCR reaction mix and then its corresponding aliquot of fresh culture medium *(see* Notes 4 and 5). Place the culture microtiter plate in a 37°C incubator for at least 6 h.
5. Seal the thermostable plate according to the manufacturer's instructions. If a thermal cycler is being used without a heated lid, overlay each reaction with sufficient mineral oil to prevent evaporation (typically 40–50 µL).
6. Transfer the plate to a thermal cycler with a microtiter plate block, and heat at 95°C for 1 min. Then initiate the following program for 30–35 cycles: Denaturation, 95°C for 30 s; primer annealing, 50–55°C for 30 s (for 17-mer oligonucleotides use 50°C); primer extension, 72°C for 30 s– 3 min (allow for a rate of 1 kbp DNA polymerized/min) *(see* Note 6). After

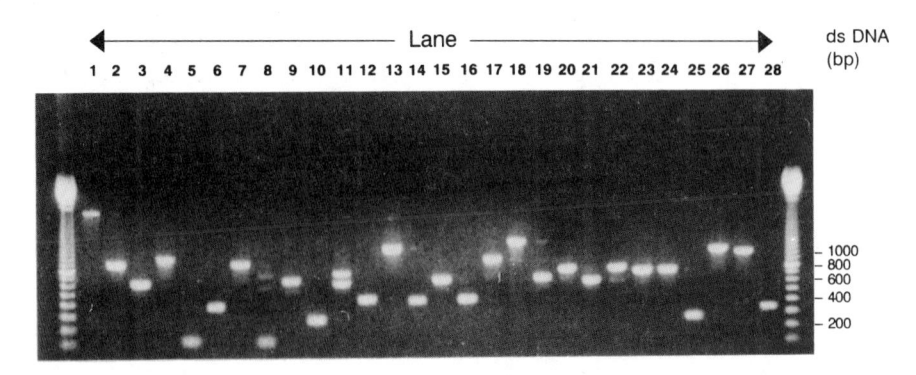

Fig. 1. A mixture of DNA fragments of varying length were ligated into the *Srf*I site of pCRscript SK(+) and the products transformed into *E. coli* strain XL1-Blue. Selection was made on ampicillin plates (100 µg/mL) containing X-gal and IPTG, and 28 colonies were analyzed by PCR (lanes 1–28) with T3 and KS Universal polylinker primers, using the protocol described. The results showed the presence of an insert in each clone and the differences in sizes can be clearly resolved. Two nonrecombinants are present (lanes 5 and 8), generating a control band of 123 bp derived from the vector polylinker. A double pick resulting in the generation of two PCR products is revealed in lane 11.

the final cycle, carry out an additional step of 72°C for 5 min to ensure full-length double-stranded products.
6. Add 1 µL of 10X gel-loading dye to each reaction and analyze by agarose gel electrophoresis (1% gel) (*see* Notes 7 and 8).
7. Positive clones can then be taken from the culture microtiter plate and used to inoculate cultures in order to prepare plasmid DNA for further analyses.

The same Universal primers used for PCR can be employed depending on their distance from the cloning site (between 10 and 50 bases is suitable) as initial primers for sequencing the cloned DNA fragments (*see* Note 9). Typical results from a PCR cloning experiment are shown in Fig. 1.

Modifications to the above protocol allow DNA sequences altered by site-directed mutagenesis to be screened *(2)*; also the direct sequencing of the *unpurified* PCR products generated may be undertaken *(3)*.

4. Notes

1. Such plates are available for a number of thermal cyclers, including those from Hybaid (Middlesex, UK), MJ Research (Watertown, MA) (PTC100 and 200), and Perkin-Elmer (Norwalk, CT) (9600), which can operate without the need for an oil overlay.

2. This may vary between 10 and 96 (the capacity of a microtiter plate), and will depend on the complexity of the DNA used for subcloning and whether a single or multiple species of DNA fragment is sought after.
3. A particular advantage of using flanking Universal primers is that even in the absence of an insert, a dsDNA product is generated, demonstrating that the PCR reaction was successful. Other primer combinations may be used, including one Universal primer and an insert-specific primer that will define orientation of the cloned DNA fragment. Alternatively, in the absence of any Universal primers, or when the desired cloned fragment is >4 kbp in size, two insert-specific primers may be employed.
4. The screening protocol can be adapted for recombinant M13 phage clones. The only modification is that the culture medium is supplemented with 1/100 vol of an overnight culture of *E. coli*. The host strain must carry the F' episome so that the phage can infect the cells; common laboratory strains that are suitable include JM109, TG1, NM522, and XL1-Blue.
5. If large numbers of colonies are to be screened, they may be first picked into the culture microtiter plates and grown overnight. A 96-pronged hedgehog may then be used to inoculate from the culture wells to microtiter plates containing PCR reaction mix.
6. For practical purposes, the upper limit of the screening protocol is 3–4 kb. For inserts smaller than 500 bp a rapid two-step cycling program may be invoked involving 30–35 cycles of denaturation at 95°C for 30 s or primer annealing at 50–55°C for 30 s. The *Taq* DNA polymerase has sufficient time to generate a product during the ramping phase between the primer annealing and denaturation temperatures.
7. Nonrecombinant PCR products on screening will be the size of the distance between the two primers in the cloning vector. Recombinant PCR products, however, will be increased in size by this value if a single restriction site was used for subcloning. This contribution should be taken into account when estimating insert size.
8. Alternatively, remove an aliquot of each reaction, add gel-loading dye, and submit to agarose gel electrophoresis. If a single band of known sequence is the product of the PCR reaction, then further confirmation can be provided by restriction mapping of a small aliquot (2–5 µL) in 20–50 µL of digestion mix.
9. Even if a single band is visible after PCR, it is possible that there is more than one species of product present. It is always wise, therefore, to sequence a number of recombinant clones to ensure that the desired fragment is present.

References

1. Güssow, D. and Clackson, T. (1989) Direct clone characterization from plaques and colonies by the polymerase chain reaction. *Nucleic Acids Res.* **17,** 4000.
2. Trower, M. K. (1992) A rapid PCR dependent microtitre plate screening method for DNA sequences altered by site-directed mutagenesis. *DNA Sequence–J. DNA Sequenc. Map.* **3,** 233–235.
3. Trower, M. K., Burt, D., Purvis, I. J., Dykes, C. W., and Christodoulou, C. (1995) Fluorescent dye-primer cycle sequencing using *unpurified* PCR products; development of a protocol amenable to high-throughput DNA sequencing. *Nucl. Acids Res.* **23,** 2348,2349.

CHAPTER 40

Use of Polymerase Chain Reaction to Screen Phage Libraries

Lei Yu and Laura J. Bloem

1. Introduction

Isolating a clone from a cDNA or genomic library often involves screening the library by several rounds of plating and filter hybridization. This is not only laborious and time-consuming, but also is prone to artifacts, such as false positives commonly encountered in filter hybridization. These problems can be alleviated by using polymerase chain reaction (PCR) in the early rounds of screening prior to conventional filter hybridization. The advantages of PCR screening are threefold: (1) Positive clones are identified by DNA bands of correct sizes in gel, thus avoiding the confusion from false positive spots in filter hybridization; (2) it saves time, especially in initial rounds of screening; and (3) screening of multiple genes can be performed in the same PCR by using appropriate primers for these genes. After the complexity of the phage pool is reduced and the existence of true positives in the pool confirmed, individual clones can be isolated by conventional methods.

The basic method consists of three steps, as shown in Fig. 1:

1. Lifting membrane filters from library plates;
2. Rinsing off phage particles from the filters; and
3. Using the phage eluate for PCR.

After a positive signal is identified from a particular plate, another round of PCR screening may be carried out by repeating these steps with smaller sectors of that plate to reduce further the size of the phage pool.

From: *Methods in Molecular Biology, Vol. 58: Basic DNA and RNA Protocols*
Edited by: A. Harwood Humana Press Inc., Totowa, NJ

```
                    PLATE PHAGE LIBRARY
                            │
                            ▼
            ┌─────────► LIFT FILTERS
            │               │
            │               ▼
            │       RINSE OFF PHAGE PARTICLES
   CUT FILTER              │
   INTO SECTIONS           ▼
       ▲           PCR WITH PHAGE ELUATE
       │                    │
       └──────────────► GEL ELECTROPHORESIS
                            │
                            ▼
                    FILTER HYBRIDIZATION
```

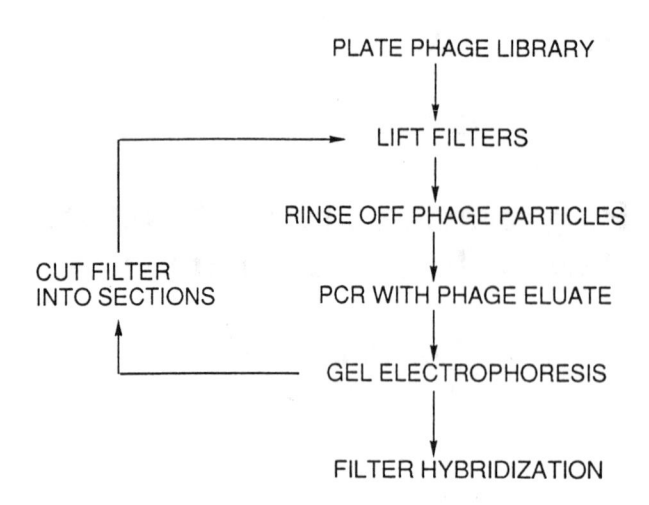

Fig. 1. Basic method of using PCR to screen phage libraries.

Alternatively, another filter can be lifted from the positive plate and used in conventional filter hybridization.

2. Materials

1. Phage dilution buffer: 100 mM NaCl, 8 mM MgSO$_4$, 50 mM Tris-HCl, pH 7.5, and 0.1% gelatin. Sterilize by autoclaving and store at room temperature.
2. Filter membranes (*see* Note 1).

3. Method

See Notes 2–4.

1. Begin with plated phage cDNA or genomic library with appropriate host bacterial strain on 150-mm agar plates with top agarose using standard protocols *(1)*. After the phage plaques have grown to the desired size at 37°C, chill the plates at 4°C for at least 1 h.
2. Place a nitrocellulose filter on the plate, making sure there are no air bubbles trapped between the filter and the agar. The filter will be wet in a few seconds if the plate is fresh. For plates stored for a period of time and for later rounds in multiple lifting, longer time may be required to wet the filter completely.
3. Carefully lift the filter with a pair of flat-ended forceps, making sure not to rip the top agar layer. Place the filter, with the phage side down, onto a sterile 150-mm Petri dish (either the bottom part or the lid) containing 3 mL of phage dilution buffer. Rinse off the phage particles by lifting the

Screening Phage Libraries

filter up and down a couple of times. Finally, lift the filter and let the solution drip for a while (usually 10–20 s is sufficient). Then discard the filter.

4. Transfer 20 µL of phage eluate to a PCR tube, close the cap tightly, place in a boiling water bath for 5 min, chill on ice, and use as template in a 100 µL PCR (*see* Chapter 34 for conditions of the standard PCR and Notes 5 and 6). Save some phage eluate in an Eppendorf tube if you consider doing more PCR later on (*see* Note 7).

5. Analyze the PCR products by agarose gel electrophoresis (*see* Note 8). If a plate has one or more positives, its corresponding lane on the gel will have a band of expected size. This plate is then considered a positive plate.

6. At this point, one can take the positive plates and proceed to screen them for positive phage plaques by conventional filter hybridization. Alternatively, one can carry out another round of PCR screening as described herein to reduce further the complexity of the phage pool on a positive plate.

7. Put a new nitrocellulose filter on a positive plate (*see* Note 9). With a sterile scalpel, cut the filter and the agar underneath it into several sections. Sections can be any number or shape, although we found that the radial-shaped (pie-shaped) sections between four and eight per plate can be handled easily.

8. Lift each filter sector and rinse in phage dilution buffer as in step 2. Owing to the smaller size of the filter sector, rinsing can be done either on smaller Petri dishes, such as the 100-mm ones, or on a piece of plastic film such as Saran Wrap™. An aliquot of 0.2–0.5 mL of phage dilution buffer is used for rinsing depending on the size of the filter sector.

9. Do PCR with the section eluate as in step 4.

10. After a sector containing the positive phage signal is identified by PCR, another membrane filter can be lifted from this sector and used in conventional filter hybridization. Alternatively, the top agar of this sector can be scraped off for replating as described in the following.

11. To replate the phage from a positive section, use a rubber policeman to scrape off the top agar into a 50-mL conical tube containing 20 mL of phage dilution buffer. The rubber policeman should be rinsed extensively with sterile water between each scraping to prevent cross-contamination among sections.

12. The tube is vortexed vigorously and the top agar is soaked for 2 h to overnight. The tube is then centrifuged at 8000*g* for 15 min to pack the agar. Save a portion of the supernatant in an Eppendorf tube and discard the conical tube containing the agar. Determine the phage titer and plate at desired density for another round of PCR screening or conventional filter hybridization.

4. Notes

1. Different brands of noncharged membrane all work well for this procedure, be they nitrocellulose or nylon. However, we have experienced difficulty with charged membranes, presumably because the charges on the membrane interfere with rinsing off of the phage particles.
2. The main advantage of using PCR to screen a phage library is to reduce complexity before starting conventional filter hybridization. Multiple plates of a library can be processed and screened by the method described here. To screen one million phage plaques, for example, plating at 50,000 pfu/plate will give 20 plates. Although considerable time and effort would be needed to screen these plates by conventional filter hybridization, they can be easily screened in a few hours by PCR screening. Furthermore, by cutting a positive plate into eight sections and performing another round of PCR screening, positive phage plaques can be located to a pool of approx 6250 pfu. This pool can be readily handled by conventional filter hybridization.
3. Before screening a library, it is highly recommended to do a PCR with an aliquot of the whole library. This serves two purposes: to ensure that the desired clone is present in the library and to ascertain that the conditions used for PCR can amplify the expected signal. It can be done by heat denaturing 5 µL of the library and using it in a 100-µL PCR. It is also prudent to sequence the amplified fragment to confirm its identity and the specificity of PCR conditions. This can be done either before or in parallel with the PCR screening of the library.
4. Another useful thing to do before screening the library is to estimate the abundance of the desired clone in the library. This can be combined with the whole library amplification by PCR (*see* Note 3) by amplifying various library dilutions in addition to the undiluted library. If the desired clone is present in a pool of 20,000 clones, for example, it would not be necessary to screen one million clones.
5. From time to time, PCR may fail with no apparent reason. To prevent such an occasional "system failure" from being interpreted as negative, it is desirable to include controls in PCR. Two types of controls may be used: internal and parallel. For internal control, primers can be included in the same PCR tube that will amplify vector sequence, a stretch of the λ arm if the library is in a λ vector. Alternatively, the library can be spiked with a known clone and the primers for this clone can be used in the same PCR. If an internal control is not desired, a PCR to amplify a known clone in an adjacent tube can be used as a parallel control.
6. In simultaneous screening of multiple genes, several pairs of primers can be used in the same PCR, provided that the expected PCR products are of

different lengths and can be distinguished from one another on a gel. The major limitation of this technique is the need for sequence information for making primers before library screening can be undertaken. Nondegenerate primers based on definitive sequence information are best for PCR screening. It may be possible to use degenerate primers to search for related genes, although we have not tried this approach.

7. Because the phage eluate can be stored, the same set of plates can be used for several screenings, either for the same gene or for different genes.
8. If screening for a small fragment, it may be necessary to amplify more than the standard 30 rounds. For a 200-bp fragment, we amplified 35 rounds to achieve a high enough DNA concentration for band visualization on a 3% agarose gel.
9. Because of the sensitivity of PCR, the plated library can be screened several times without losing the positive signal. Also, plates can be stored at 4°C for several weeks before subsequent lifts. We have made five lifts from the same plates, both fresh and after 3-wk storage, and the phage eluate amplified well. The same goes for the phage eluate, which can be stored for several weeks or longer before amplification.

Reference

1. Sambrook, J., Fritsch, E. F., and Maniatis,T. (1989) *Molecular Cloning: A Laboratory Manual,* (2nd ed.), Cold Spring Harbor Laboratory, Cold Spring Harbor, NY.

PART VI
DNA Sequencing

CHAPTER 41

Cloning into M13 Bacteriophage Vectors

Qingzhong Yu

1. Introduction

The bacteriophage M13 has been developed as a cloning vector system to obtain single-stranded DNA templates that are used for the dideoxy chain termination method of sequencing DNA (1,2). General aspects of bacteriophage M13 as a cloning vector system are reviewed in ref. 3, and the preparation of foreign DNA fragments (insert DNA) for M13 cloning is described in Chapters 27 and 28. This chapter describes the preparation of M13 vectors and the ligation of foreign DNA fragments (inserts) into them.

2. Materials

2.1. Preparation of Replicative Form (RF) M13 DNA

1. L-broth: 1% bacto-tryptone, 0.5% bacto-yeast extract, 1% NaCl. Sterilize by autoclaving.
2. M9 minimal agar: M9 salts: 12.8 g of $Na_2HPO_4 \cdot 7H_2O$, 3 g of KH_2PO_4, 0.5 g of NaCl, 1.0 g of NH_4Cl, 20 mL of 20% glucose in H_2O to 1 L. Sterilize by autoclaving.
3. Bacteriophage M13: A single blue plaque from a plate of freshly transformed bacteria.
4. *E. coli* JM103 or JM109: a colony grown on an M9 minimal agar plate.
5. Solution 1: 50 m*M* glucose, 25 m*M* Tris-HCl, pH 8.0, 10 m*M* EDTA, pH 8.0, autoclaved and stored at 4°C.
6. Solution 2: 0.2*M* NaOH, 1% SDS, Mix together equal volumes of 0.4*M* NaOH and 2% SDS stocks before use.
7. Solution 3: 60 mL of 5*M* potassium acetate, 11.5 mL of glacial acetic acid, 28.5 mL of H_2O. Store at 4°C, but place on ice just before use.

8. TE: 10 mM Tris-HCl, pH 8.0, 1 mM EDTA, pH 8.0.
9. Phenol-chloroform: Mix equal volumes of TE-saturated phenol (nucleic acid grade) and chloroform (AR grade), stored at 4°C in a dark glass bottle.
10. Chloroform: AR grade.
11. Ethanol: 100% ethanol, stored at –20°C.
12. RNase: 1 mg/mL DNase-free pancreatic RNase A.

2.2. Preparation of M13 Vectors

1. Appropriate restriction enzymes: Store at –20°C.
2. 10X restriction enzyme (RE) buffers: These are usually supplied with restriction enzymes.
3. 3M Sodium acetate, pH 5.2. Store at 4°C.
4. Calf intestinal alkaline phosphatase (CIP). Store at 4°C.
5. 10X CIP dephosphorylation buffer: 500 mM NaCl, 100 mM Tris-HCl, 100 mM MgCl$_2$, 10 mM dithiothrietol.
6. Proteinase K: Store at –20°C as a 10 mg/mL stock.
7. 10% SDS.
8. 0.5M EDTA, pH 8.0.
9. 10X TBE: 108 g of Tris-base, 55 g of boric acid, 9.5 g of EDTA · 2H$_2$O, water to 1 L.
10. 0.8% Agarose gel in 1X TBE.

2.3. Ligation of Inserts into M13 Vectors

1. T4 DNA ligase: Store at –20°C.
2. 10X ligation buffer: 500 mM Tris-HCl, 100 mM MgCl$_2$, 10 mM dithiothrietol.
3. 10 mM ATP: Store at –20°C.

3. Methods

A number of M13 vectors have been constructed *(1,4)* and are commercially available. It therefore may be more convenient to purchase them than to prepare them oneself. Sometimes, however, you may need an M13 vector with an unusual cloning site to fit your cloning strategy.

3.1. Preparation of RF M13 DNA

1. Inoculate 5 mL of L-broth in a 20-mL sterile culture tube (e.g., universal) with a single bacterial colony of an appropriate M13 phage host (e.g., JM103 or JM109) picked from a freshly streaked M9 minimal agar plate. Incubate at 37°C in an orbital shaking incubator overnight (*see* Note 1).
2. Add 50 µL of the bacterial culture to 2 mL of L-broth in a 5-mL culture tube (e.g., bijou). Inoculate this culture with M13 bacteriophage by touching a single blue plaque from a transformation plate with a sterile tooth-

pick and washing its end in the culture (*see* Note 2). Incubate the infected culture at 37°C for 4–5 h in an orbital incubator.

3. Transfer 1.0–1.5 mL of the culture to a microfuge tube, and centrifuge at 12,000g for 2 min at room temperature. Remove the supernatant to a fresh tube, being careful not to disturb the pellet. Single-stranded M13 (ss-M13) DNA may also be prepared from the supernatant (*see* Chapter 43).
4. Remove any remaining supernatant by aspiration (*see* Note 3). Resuspend the pellet in 100 µL of solution 1 by pipeting up and down or by vigorous vortexing. Leave at room temperature for 5–10 min.
5. Add 200 µL of solution 2. Close the tube, and mix the contents by inverting the tube rapidly five times. Do not vortex. Place the tube on ice for 5 min.
6. Add 150 µL of ice-cold solution 3. Vortex the tube gently in an inverted position for 10 s. Place on ice for 5 min.
7. Centrifuge at 12,000g for 5 min, and transfer the supernatant to a fresh tube.
8. Add an equal volume of phenol:chloroform, and mix by vortexing for 20–30 s. Spin as in step 7, and transfer the aqueous phase (top layer) to a fresh tube.
9. Add an equal volume of chloroform, and mix by vortexing for 20–30 s. Spin as in step 7, and transfer the aqueous phase (top layer) to a fresh tube.
10. Add 2 vol of ethanol, mix by vortexing, and stand for 5 min at room temperature. Spin as in step 7, and remove the supernatant by gentle aspiration (*see* Note 4).
11. Wash the pellet with 1 mL of 70% ethanol. Spin for 2 min with tubes in the same orientation in centrifuge as before so as not to disturb the pellet. Remove the supernatant as in step 10. Air-dry for 10 min.
12. Dissolve the pellet in 20 µL of TE, pH 8.0, containing RNase (20 µg/mL) to remove RNA, and vortex briefly.

The double-stranded RF M13 DNA is now ready for analysis by digestion with restriction enzymes.

3.2. Preparation of M13 Vectors

The RF M13 DNA must be digested by a restriction enzyme and then dephosphorylated by treatment with CIP to reduce the background of nonrecombinant molecules formed by recircularization of the vector alone during ligation. If the insert DNA can be generated with incompatible cohesive ends by using two different restriction enzymes, then a double-digested vector can be prepared that does not require dephosphorylating (*see* Note 5).

1. Digest RF M13 DNA with a single restriction enzyme in the following reaction: 10 µL of RF M13 DNA (approx 400 ng), three- to fivefold excess restriction enzyme, 2 µL 10X appropriate buffer, and water made up to 20 µL total volume. Incubate for 2 h at 37°C.
2. Remove 2 µL of the reaction and analyze the extent of digestion by electrophoresis through 0.8% agarose gel, using undigested M13 DNA as a marker (*see* Note 3). If digestion is incomplete, add more restriction enzyme, and continue the incubation.
3. When digestion is complete, extract the M13 DNA with phenol:chloroform and precipitate DNA with 0.1 vol of $3M$ NaAc and 2.5 vol of ethanol at –20°C for 1 h or longer.
4. Recover the DNA by centrifugation at 12,000g for 10 min in a microfuge. Wash the pellet with 70% ethanol and vacuum dry. Dissolve the pellet in 20 µL of TE.
5. To 20 µL of linerized M13 vector add: 5 µL of 10X CIP buffer, 1 unit of CIP, and water made up to 50 µL total volume. Incubate for 30 min at 37°C (*see* Note 6).
6. At the end of the incubation period, add the following to the reaction: 2.5 µL of 10% SDS (final concentration 0.5%), 0.5 µL of $0.5M$ EDTA, pH 8.0 (final concentration 5 mM), and 0.5 µL of 10 mg/mL proteinase K (final concentration 100 µg/mL). Incubate for 30 min at 56°C to inactivate the CIP (*see* Note 7).
7. Cool the reaction to room temperature, and extract with phenol:chloroform twice and once with chloroform (*see* Section 3.1., steps 8 and 9). Add 0.1 vol of $3M$ sodium acetate, pH 7.0 (*see* Note 8), mix well, and add 2.5 vol of ethanol. Mix and place at –20°C for 1 h or longer.
8. Recover the DNA as in step 4. Dissolve the pellet in 20 µL of TE. It is now ready to use for ligation with inserts.

3.3. Ligation of Inserts into M13 Vectors

1. Set up ligation reaction in the following order: 1 µL of M13 vector (approx 20 ng), 1 µL of 10 mM ATP, 1 µL of 10X ligation buffer, 1–4 µL of insert DNA (three- to fivefold molar excess), 1 U of T4 DNA ligase for cohesive termini (*see* Note 9), and water made up to 10 µL total volume. Include two control reactions, one with water replacing the insert DNA and the other containing an appropriate amount of a test DNA that has been previously successfully cloned into the M13 vector. Incubate at 14°C overnight. Then store at –20°C, or use directly for transformation.
2. After ligation, analyze 1 µL of each ligation reaction by electrophoresis through 0.8% agarose gel to check that the ligation has been successful (*see* Note 10). Use the same amounts of the M13 vector, and insert DNA without ligase as controls.

3. Transform 5 µL of the remaining ligation sample into competent bacteria of the appropriate strain of *E. coli*, e.g., JM103 or JM109 (*see* Chapters 29 and 30).

4. Notes

1. M9 minimal medium selects for the presence of the F' episome in bacterial strains, such as JM103 or JM109. They only grow slowly in M9 minimal medium and do not survive prolonged storage at 4°C. It is therefore better to streak a master culture of the bacteria on the M9 minimal agar plate every month, from which to seed L-broth cultures. The master stock of bacteria should be stored at −70°C in L-broth containing 15% glycerol.
2. Bacteriophage M13 can diffuse a considerable distance through the top agar on the transformation plate. Therefore, it is important to pick up a single plaque that is well separated from its neighbors to avoid cross-contamination with others. For good yields of RF M13 DNA, it is better to pick a plaque from a freshly transformed plate.
3. Since RF M13 DNA remains inside the infected bacteria and M13 phage progeny containing ss-M13 DNA is released to the medium, removing the remaining supernatant minimizes ss-M13 DNA contamination. Even so, ss-M13 DNA contamination may sometimes occur and can confuse analysis by restriction enzyme digestion. It is important therefore always to run an undigested DNA control when analyzing RF M13 DNA by restriction enzyme digest and gel electrophoresis. Since ss-M13 DNA cannot be digested by restriction enzymes, its mobility will not change between digested and undigested lanes.
4. When recovering M13 DNA, sometimes the M13 DNA pellet is not visible. Therefore, it is necessary to remove the supernatant carefully and leave a small amount (20–30 µL) behind in the tube to minimize the loss of the DNA. Do likewise when washing with 70% ethanol, and then air-dry.
5. A double-digested vector can be generated that has incompatible cohesive ends and therefore gives little background owing to vector recircularization. The RF M13 DNA can be digested simultaneously with two different restriction enzymes if they work in the same enzyme buffer. To monitor the extent of the digestion, set up two reaction mixtures with each restriction enzyme alone as controls. After the RF M13 DNA has been completely digested, follow Section 3.2., steps 3 and 4. The final pellet is dissolved in 20 µL of TE and is ready to use for ligation. If the two restriction enzymes cannot be used simultaneously, then after digestion with one of the enzymes, the DNA should be recovered as in Section 3.2., steps 3 and 4 and resuspended in 10 µL of TE. The DNA should then be digested with the second enzyme and again recovered in the same way as after the

first. To follow the digestion of the second restriction enzyme, set up a parallel digest of the enzyme and uncut RF M13 DNA.

The small fragment removed by the double digest may cause background problems if it religates into the vector. If such problems are encountered, the vector DNA should be gel-purified (see Chapter 28).

6. For blunt or recessed termini, use 1 U/2 pmol of CIP and incubate for 15 min at 37°C. Then add the same amount of CIP, and continue the incubation for a further 45 min at 55°C.
7. The presence of active CIP will inhibit the subsequent ligation of M13 vectors with insert DNA. An alternative method to the one described is to inactivate the CIP by heating at 70°C for 10 min in the presence of 5 mM of EDTA and then extracting with phenol:chloroform.
8. Since EDTA precipitates from solution at acid pH when its concentration exceeds 5–10 mM, the commonly used 3M NaAc at pH 5.2 must be replaced with 3M NaAc at pH 7.0 during ethanol precipitation.
9. The efficiency of ligation of blunt termini is lower than for ligation of cohesive termini. To improve the efficiency of the ligation of blunt termini, a higher concentration of insert DNA and T4 DNA ligase is required. Use 5 U in reaction for blunt termini.
10. The extent of ligation of M13 vectors with inserts can be checked by electrophoresis through 0.8% agarose gel, using unligated M13 vectors and inserts as control. Normally, the ligated M13 vectors with inserts will migrate slowly through the gel owing to the increased molecular weight, sometimes giving fuzzy or smeared bands, which indicates the extent of the ligation.

Acknowledgment

The author wishes to thank D. Cavanagh for advice in the preparation of the manuscript.

References

1. Messing, J. (1983) New M13 vectors for cloning. *Meth. Enzymol.* **101,** 20–79.
2. Sanger, F., Nicklen, S., and Coulson, A. R. (1977) DNA sequencing with chain-terminating inhibitors. *Proc. Natl. Acad. Sci. USA* **74,** 5463–5467.
3. Messing, J. (1993) M13 cloning vehicles, in *DNA Sequencing Protocols, Methods in Molecular Biology,* vol. 23 (Griffin, H. G. and Griffin, A. M., eds.), Humana, Totowa, NJ, pp. 9–22.
4. Yanisch-Perron, C., Vieira, J., and Messing, J. (1985) Improved M13 phage cloning vectors and host strains: nucleotide sequences of the M13mp18 and pUC19 vectors. *Gene* **33,** 103–119.

CHAPTER 42

Ordered Deletions Using Exonuclease III

Denise Clark and Steven Henikoff

1. Introduction

An important manipulation in molecular genetics is to make ordered deletions into a cloned piece of DNA. The most widely used application of this method is in DNA sequencing. Ordered deletions can also be used in delineating sequences that are important for the function of a gene, such as those required for transcription. The principle behind using deletions for sequencing is that consecutive parts of a fragment cloned into a plasmid vector are brought adjacent to a sequencing primer site in the vector. Deletions are generated by digesting DNA unidirectionally with *Escherichia coli* exonuclease III (ExoIII) *(1)*. ExoIII digests one strand of double-stranded DNA by removing nucleotides from 3' ends if the end is blunt or has a 5' protrusion. A 3' protrusion of 4 bases or more is resistant to ExoIII digestion.

We present two procedures, outlined in Fig. 1, for preparation of the template for ExoIII digestion. Procedure I begins with double-stranded plasmid DNA. The DNA is digested with restriction endonucleases A and B, where A generates a 5' protrusion or blunt end next to the target sequence and B generates a 3' protrusion next to the sequencing primer site. The linearized plasmid DNA is digested with ExoIII, with aliquots taken at time-points that will yield a set of deletions of the desired lengths. Procedure II *(2)* begins with single-stranded phagemid DNA. A nicked double-stranded circle is generated by annealing a primer to the phagemid so that its 5' end is adjacent to the insert and synthesizing the second strand with T4 DNA polymerase.

From: *Methods in Molecular Biology, Vol. 58: Basic DNA and RNA Protocols*
Edited by: A. Harwood Humana Press Inc., Totowa, NJ

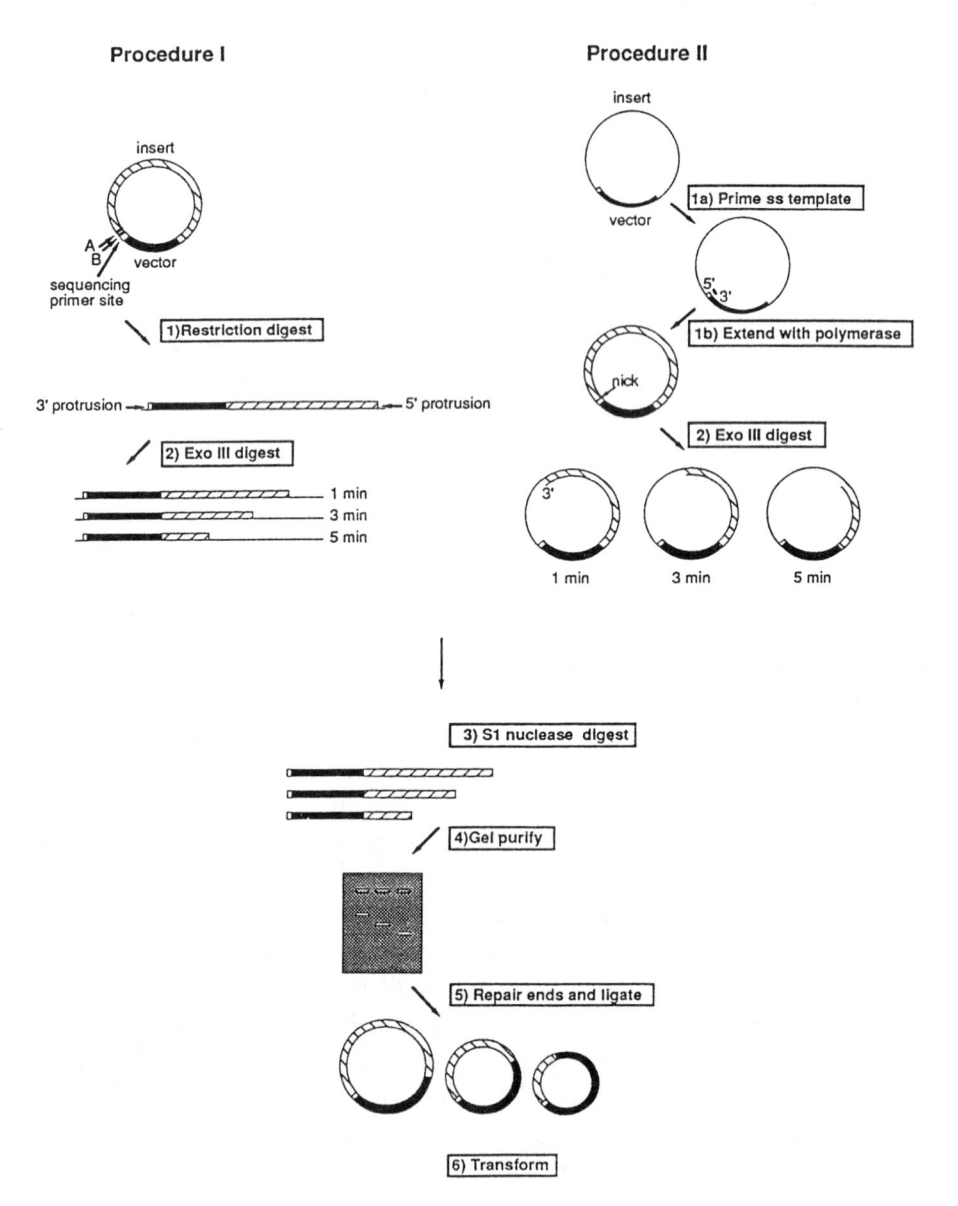

Fig. 1. Outline of the method.

The 3' end of the nicked strand is then digested with ExoIII. The remaining undigested single strands are digested with S1 nuclease. The deletion reactions are examined by electrophoresis in a low melting point agarose gel and the desired deletion products are isolated in gel slices

(3,4). This step serves to eliminate DNA products that do not contain the desired deletion. Repair of ends with Klenow DNA polymerase and ligation to circularize are performed on the DNA in diluted agarose. Following transformation into *E. coli*, clones are selected for sequencing.

An advantage of Procedure I is that deletions can be made into the insert from either end, so that both strands of the insert can be sequenced starting with a single plasmid clone. However, the requirement for multiple unique restriction sites in the vector polylinker cannot always be met, which is where Procedure II is advantageous, since no restriction sites are necessary. This also means that, with Procedure II, a set of nested deletions can be made from any fixed point using a custom primer. A disadvantage of Procedure II is that to sequence both strands of an insert, two clones with the insert in either orientation are required. Nevertheless, the two procedures can be used together to obtain the sequence of both strands from a single parent clone, reducing the need for multiple unique restriction sites. When planning a sequencing strategy, it is important to realize that although either single- or double-stranded templates for sequencing are possible for Procedure I, only double-stranded templates are possible for Procedure II. The reason is that for most phagemid vectors the primer for synthesis of the second strand of the nicked circle anneals to a site on the other side of the insert from the primer site that would be used for sequencing a single-stranded template.

2. Materials
2.1. General

Solutions are made with distilled water that when used for enzyme buffers is sterilized by autoclaving. Store solutions at room temperature, unless otherwise specified. Store enzymes as directed by the manufacturer.

2.1.1. Restriction Enzyme Digestion

1. Ten micrograms of double-stranded DNA: The fragment should be cloned into the multiple cloning site (polylinker) of a plasmid or phagemid vector, such as Bluescript (Stratagene, La Jolla, CA). DNA can be prepared using an alkaline lysis procedure (*5*; *see* Note 1). A 10-mL liquid culture incubated overnight in LB media with the appropriate selection (e.g., ampicillin) should yield a sufficient amount of DNA.
2. Restriction enzymes and buffers as supplied by manufacturer. Select a restriction enzyme (A in Fig. 1) that cuts the plasmid only in the polylinker adjacent to the insert and yields a 5' protrusion or blunt end. Restriction

enzyme B must also cut at a unique site in the polylinker and yield a 4-base 3' protrusion to make the end of the linearized fragment nearest the sequencing primer site resistant to ExoIII. If such a site is not available, a 5' protrusion can be generated that is filled in with α-phosphorothioate nucleotides using Klenow fragments *(6)*.
3. Phenol:chloroform solution: 1:1 mix; store at 4°C.
4. 10*M* Ammonium actetate.
5. Ethanol: both 95% and 70% (v/v).
6. 10X ExoIII buffer: 660 m*M* Tris-HCl, pH 8.0, 6.6 m*M* MgCl$_2$.

2.1.2. Procedure II: Synthesis of a Nicked Circular Plasmid

7. Template: 2.5 µg single-stranded phagemid DNA, 0.2–2.0 µg/µL. A small-scale preparation (2–3 mL culture) should yield a sufficient amount of DNA *(5)*.
8. Primer: α-T3 20-mer (5' CCCTTTAGTGAGGGTTAATT 3'), or the equivalent, e.g., reverse hybridization 17-mer (5' GAAACAGCTATGACCAT 3') at 4 pmol/µL; store at –20°C.
9. T4 DNA polymerase: 1–10 U/µL, cloned or from T4-infected cells.
10. 10X TM: 660 m*M* Tris-HCl, pH 8.0, 30 m*M* MgCl$_2$.
11. 2.5 m*M* dNTPs: 2.5 m*M* each of the 4 deoxynucleoside triphosphates; store at –20°C.
12. BSA: 1 mg/mL bovine serum albumin (nuclease-free); store at –20°C.

2.2. Exonuclease III and S1 Nuclease Digestion

13. *E. coli* exonuclease III: 150–200 U/µL (1 U = amount of enzyme required to produce 1 nmol of acid-soluble nucleotides in 30 min at 37°C).
14. S1 buffer concentrate: 2.5*M* NaCl, 0.3*M* potassium acetate (titrate to pH 4.6 with HCl, not acetic acid), 10 m*M* ZnSO$_4$, 50% (v/v) glycerol.
15. S1 nuclease: e.g., 60 U/µL from Promega (1 U = amount of enzyme required to release 1 µg of perchloric acid-soluble nucleotides per minute at 37°C). Mung bean nuclease may also be used.
16. S1 mix: 27 µL S1 of buffer concentrate 173 mL of H$_2$O, 1 µL (60 U) of S1 nuclease. Prepare just before use and store on ice.
17. S1 stop: 0.3*M* Tris base (no HCl), 50 m*M* EDTA.

2.3. Gel Purification, End Repair, Ligation, and Transformation

18. Agarose: Low melting point agarose. Use a grade appropriate for cloning, e.g., Seaplaque GTG agarose (FMC).
19. Ethidium bromide: 10 mg/mL; store in a dark container.
20. 50X TAE gel running buffer: 1 L = 242 g of Tris base, 57.1 mL of glacial acetic acid, 100 mL of 0.5*M* EDTA, pH 8.0. When diluted this gives a 1X buffer of 40 m*M* Tris-acetate, 1 m*M* EDTA.

21. 10X Gel-loading buffer: 0.25% bromophenol blue, 50% (v/v) glycerol in distilled water.
22. 10X Klenow buffer: 20 mM Tris-HCl, pH 8.0, 100 mM MgCl$_2$.
23. Klenow enzyme: the large fragment of the *E. coli* DNA polymerase (2 U/µL). Store at –20°C.
24. Klenow mix: For every 10 µL of 10X Klenow buffer, add 1 U of Klenow enzyme. Prepare just before use and store on ice.
25. dNTPs: 0.125 mM of each deoxyribonucleoside; store at –20°C.
26. 10X Ligase buffer: 0.5M Tris-HCl, pH 7.6, 100 mM MgCl$_2$, 10 mM ATP; store at –20°C.
27. PEG: 50% (w/v) polyethylene glycol 6000–8000 fraction, store at 4°C.
28. DTT: 0.1M dithiothreitol; store at –20°C.
29. T4 DNA ligase: (1 U/µL). Store at –20°C.
30. Ligation cocktail (1 mL): 570 µL of H$_2$O, 200 µL of ligase buffer, 200 µL of PEG, 20 µL of 0.1M DTT, 10 U of T4 ligase. Prepare just before use and store on ice.
31. Host *E. coli* cells: Competent recA⁻ cells, e.g., DH5α.
32. Growth media: LB medium *(5)*.

3. Methods
3.1. Preparation of ExoIII Substrate
3.1.1. Procedure I: Restriction Enzyme Digestion

1. Digest 10 µg of plasmid to completion in a 100 µL volume. At least 200 ng is required for each ExoIII digestion time point (*see* Note 1).
2. To extract the DNA, vortex in 100 µL of phenol:chloroform and remove the aqueous layer to a fresh tube.
3. Add 25 µL of 10M ammonium acetate and 200 µL of 95% ethanol. Chill on ice for 15 min and pellet the DNA in a microfuge (12,000g) for 5 min.
4. Wash pellet with 70% ethanol and air dry. Resuspend the DNA pellet in 90 µL of distilled H$_2$O and 10 µL of 10X ExoIII buffer (i.e., 100 ng/µL).

3.1.2. Procedure II: Synthesis of a Nicked Circular Phagemid

1. Mix in a volume of 22 µL: approx 2.5 µg single-stranded phagemid DNA, 4 µL of 10X TM, 1 µL (4 pmol) of α-T3 primer (or equivalent).
2. Heat to 75°C for 5 min, then cool slowly over 30–60 min. Evaporation and condensation can be minimized by placing a piece of insulating foam over the tube in an aluminum tube-heating block. Remove a 2 µL aliquot for subsequent gel analysis.
3. Mix in a volume of 20 µL: 2 µL of DTT, 4 µL of 2.5 mM dNTPs, 4 µL of BSA, and 5 U of T4 DNA polymerase.

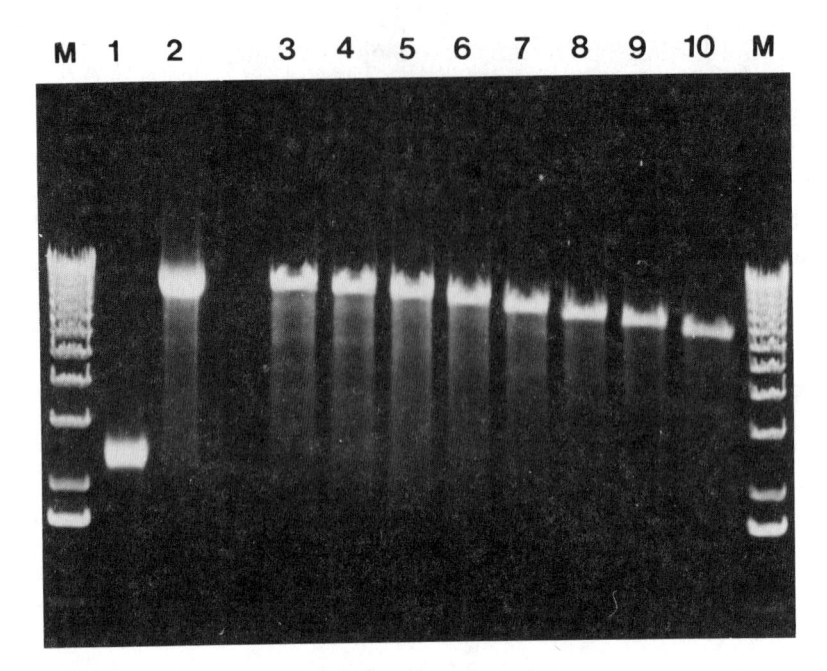

Fig. 2. Low-melting-point agarose gel electrophoresis of a set of deletions prepared following Procedure II. The size of the phagemid is 9.5 kb. Lane 1: Single-stranded phagemid. Lane 2: Double-stranded nicked circle after overnight extension with T4 DNA polymerase. Depending on the gel, one occasionally sees a minor band below the nicked circle that may be the result of extension of single-stranded linear molecules that are primed by their own 3' end (2). Lanes 3–10: ExoIII digestion of the material in Lane 2 for 1, 2, ..., 8 min. Lane M: 1-kb ladder (BRL), with the 7-kb band comigrating with the 8 min time-point.

4. Prewarm to 37°C and add to the primed DNA at 37°C. Incubate 2–8 h. An example of a nicked circle is shown in Fig. 2. When the plasmid is as large as this one, we recommend supplementing the extension reaction with an additional 0.4 mM dNTPs and 0.1 U/µL polymerase after 4–6 h, and then incubation overnight. Extension can be monitored by removing a 0.5-µL aliquot and electrophoresing it on an agarose gel.
5. Inactivate polymerase by heating for 10 min at 70°C and store on ice.

3.2. ExoIII Digestion

Deletions separated by 200–250 bases are required for obtaining contiguous sequence with some overlap, therefore to sequence a 4-kb insert, 16–20 time-points are desirable (see Note 2). The following protocol is written to give 16 time-points.

Exonuclease III 355

1. Prepare 16 7.5-µL aliquots (*see* Note 3) of S1 nuclease mix on ice, one for each time-point.
2. Place 40 µL of DNA (from Section 3.1.4. or 3.2.5.) into a single tube. Warm the DNA to 37°C in a heating block (*see* Note 3). Add 1–2 µL of ExoIII (5–10 U/µL), and mix rapidly by pipeting up and down.
3. Remove aliquots of 2.5 µL at 30-s intervals and add into the S1 mix tubes on ice. Mix by pipeting up and down and hold on ice until all aliquots are taken.
4. Remove samples from ice and incubate at room temperature for 30 min.
5. Add 1 µL of S1 stop and heat to 70°C for 10 min to inactivate the enzymes.

3.3. Gel Purification, End Repair, Ligation, and Transformation

1. Add 1 µL of 10X gel loading buffer to each sample and load onto a 0.7% low melting point agarose gel in 1X TAE gel running buffer containing 0.5 µg/mL ethidium bromide. Electrophorese for about 2 h at 3–4 V/cm until the deletion time-points can be resolved (*see* Note 4). Using a 300-nm wavelength ultraviolet light source, remove the desired bands from the gel with a small spatula or scalpel.
2. Dilute gel slices in approx 2 vol of distilled water and melt at 68°C for 5 min. Transfer 1/3 of the diluted DNA to a new tube containing 0.1 vol Klenow mix at room temperature and mix by pipeting up and down. Incubate at 37°C for 3 min. Add 0.1 vol of dNTPs and mix. Incubate for 5 min at 37°C (*see* Note 5).
3. Add an equal volume of ligation cocktail, mix, and incubate at room temperature for at least 1 h. The yield of transformants is greatest if ligations are incubated overnight.
4. Transform *E. coli* by combining half of each ligation with 2–3 vol of competent cells that have been thawed on ice. Incubate for 30 min on ice, heat shock at 42°C for 90 s, and further incubate the cells in 0.2 mL of LB medium for 60 min at 37°C. Plate using the appropriate selection and incubate at 37°C overnight. The yield from each transformation can be up to 500 colonies, depending on ligation time and competence of the cells. Store the remaining ligation mixtures at –20°C.
5. Set up cultures to prepare plasmid DNA from one colony from each time-point and verify that each one contains a deletion of expected size before sequencing.

4. Notes

1. ExoIII digestion occurs at nicks in the DNA as efficiently as at ends. This gives rise to a background of undesired clones. Nicks can be introduced by impurities in the DNA and by restriction enzymes. Gel purification eliminates this background. As a result of the gel purification step, plasmid DNA

can be prepared with a standard alkaline lysis procedure, including a phenol:chloroform extraction *(4)*. If nicking proves to be a problem, supercoiled DNA may be column purified using the premade columns from Qiagen, Inc. (Chatsworth, CA). Restriction enzymes can also occasionally have excessive nicking or "star" activity. Often this problem can be overcome by trying different suppliers.
2. At 37°C, ExoIII digests 400–500 bases/min, with the rate changing directly with temperature about 10%/1°C in the 30–40°C range *(7)*. It is advisable to perform a test run of the ExoIII digestion on your DNA, picking a few time-points. The three reasons for this are:
   a. You can confirm that both enzymes cut to completion if using Procedure I (*see also* Note 4);
   b. You can get an accurate rate of digestion for your particular batch of ExoIII; and
   c. You can confirm that there is not excessive nicking of your DNA (*see* Note 3).
3. When processing a large number of time-points, it is much more efficient to use a conical bottom microtiter plate with a lid rather than microfuge tubes. These plates can be used for the S1 nuclease step, collecting and diluting gel slices, end repair, ligation, and transformation.
4. Occasionally, an ExoIII resistant fragment is seen. This results from incomplete digestion by enzyme A (Fig. 1). Provided this is not a dominant band, the gel isolation of the digested fragment should eliminate this as a problem. Alternatively, a second fragment that digests at a higher rate may be present. This results from incomplete digestion by enzyme B and subsequent ExoIII digestion of the fragment from both ends. Again, if this is not a dominant product, gel isolation should alleviate this problem.
5. An alternative method for in-gel ligation incorporates end repair and ligation into one step. For 1 mL of cocktail, mix 560 µL of water, 200 µL of ligation buffer, 200 µL of PEG, 20 µL of DTT, 20 µL each of dNTP (2.5 m*M*), 1 U of T4 DNA polymerase, and 10 U of T4 DNA ligase. Mix one-third of the DNA in the diluted gel slice with an equal volume of the cocktail. Incubate at room temperature 1 h to overnight. Transformation is done following step 6.

References

1. Henikoff, S. (1984) Unidirectional digestion with exonuclease III creates targeted breakpoints for DNA sequencing. *Gene* **28**, 351–359.
2. Henikoff, S. (1990) Ordered deletions for DNA sequencing and in vitro mutagenesis by polymerase extension and exonuclease III gapping of circular templates. *Nucleic Acids Res.* **18**, 2961–2966.
3. Nakayama, K. and Nakauchi, H. (1989) An improved method to make sequential deletion mutants for DNA sequencing. *Trends Genet.* **5**, 325.

4. Steggles, A. W. (1989) A rapid procedure for creating nested sets of deletions using mini-prep plasmid DNA samples. *Biotechniques* **7,** 241,242.
5. Sambrook, J., Fritsch, E. F., and Maniatis, T. (1989) *Molecular Cloning: A Laboratory Manual.* Cold Spring Harbor Laboratory, Cold Spring Harbor, NY.
6. Putney, S. D., Benkovic, S. J., and Schimmel, P. R. (1981) A DNA fragment with an α-phosphorothioate nucleotide at one end is asymmetrically blocked from digestion by exonuclease III and can be replicated in vivo. *Proc. Natl. Acad. Sci. USA* **78,** 7350–7354.
7. Henikoff, S. (1987) Unidirectional digestion with exonuclease III in DNA sequence analysis. *Meth. Enzymol.* **155,** 156–165.

CHAPTER 43

M13 Phage Growth and Single-Stranded DNA Preparation

Fiona M. Tomley

1. Introduction

M13 bacteriophage has a single-stranded DNA (ssDNA) genome, and has proven an extremely useful vector from which to derive single-stranded templates for sequencing and site-directed mutagenesis. During infection of its host cell, the phage DNA replicates as a double-stranded intermediate from which the ssDNA containing phage particles are produced. Infected cells do not lyse, but instead phage particles are continuously released. Cells infected with M13 phage, however, have a longer replication cycle, which means that as the infection proceeds, the areas of slower-growing cells can be visualized as turbid plaques on the lawn of unaffected *E. coli (1)*. Recombinant M13 phage can be cloned from well-separated plaques and used as a source of ssDNA.

This chapter describes a miniprep method for the production of M13 ssDNA that is suitable for sequencing. There are two elements to the method: Packaged particles are recovered from the medium of cells growing in liquid culture and then phenol is used to remove their coat proteins *(2,3)*. This method can be easily used to purify multiple samples at the same time and takes a single day, provided that an overnight culture of host bacteria is available.

2. Materials

1. An *E. coli* strain suitable for propagating bacteriophage M13 vectors. Recommended strains include JM101, JM103, JM107, JM109, TG1, and TG2.

2. 2X TY broth: 1% tryptone, 1% yeast, 100 m*M* NaCl. Sterilize by autoclaving in 50-mL aliquots.
3. Sterile culture tubes for growing up 1.5-mL cultures, e.g., disposable 10-mL Falcon tubes or glass/plastic universal bottles and an orbital incubator for shaking the tubes vigorously at 37°C.
4. A microfuge and 1.5-mL microfuge tubes.
5. PEG/NaCl: 20% PEG, 2.5*M* NaCl. Make up 100 mL at a time and sterilize by filtration. Store at 4°C.
6. TE: 10 m*M* Tris-HCl, 1 m*M* EDTA, pH 8.0. For convenience, store 1*M* Tris-HCl, 100 m*M* EDTA, pH 8.0, in 1-mL aliquots at –20°C. When needed, thaw, dilute to 100 mL, check the pH, filter, and store at room temperature.
7. Phenol: Use high-grade redistilled phenol stored at –20°C in aliquots of 100–500 mL. To equilibrate, warm to room temperature, then melt at 65°C, and add hydroxyquinoline to 0.1%. To the melted phenol, add an equal volume of 0.5*M* Tris-HCl, pH 8.0, mix for a few minutes to allow phases to separate, and take off the upper layer (buffer). Repeat extractions with 0.1*M* Tris-HCl, pH 8.0, until the pH of the phenolic phase gets up to around 7.8. Extract once with TE, remove the aqueous phase, and then store the phenol at 4°C in a dark bottle under a layer of fresh TE (around 20 mL for 100 mL phenol). Fresh phenol should be prepared once a month.
8. Chloroform: Use high-grade chloroform that has not been exposed to the air for long periods of time. Mix 24:1 with isoamyl alcohol, and store in a tightly capped bottle at room temperature.
9. 3*M* sodium acetate, pH 5.2: Dissolve solid sodium acetate in water, adjust the pH to 5.2 with glacial acetic acid, dispense into aliquots, autoclave, and store at room temperature.
10. Ethanol: For convenience, store in tightly capped bottles at –20°C.

3. Method

1. Pick a single colony from a freshly streaked plate of a suitable *E. coli* host strain, dispense into 10 mL of 2X TY in a 250-mL conical flask, and grow with shaking at 37°C overnight.
2. Dilute the overnight culture by 1 in 100 in 2X TY to give sufficient fresh culture for 1.5 mL/sample. For each sample, prepare a 1.5-mL aliquot of culture in a sterile tube.
3. Carefully touch a sterile toothpick into a well-separated single plaque, and wash the end of the toothpick in a 1.5-mL aliquot of fresh cells (*see* Notes 1 and 2). This is sufficient to transfer infected cells from the plaque to the new culture. Incubate the 1.5-mL cultures with vigorous shaking at 37°C for 4.5–5.5 h (*see* Notes 3 and 4).
4. Transfer cultures to 1.5-mL microfuge tubes, and spin for 5 min in a microfuge.

M13 Phage Growth

5. Transfer supernatants to clean tubes, making sure the pellet is undisturbed. Add 200 µL of PEG/NaCl to each, vortex well, and leave at room temperature for at least 15 min.
6. Spin for 5 min in microfuge to pellet the phage particles. Remove the PEG-containing supernatant with a pipet, respin the tubes for a few seconds, and carefully remove all remaining traces of supernatant from around the phage pellet using a drawn-out Pasteur pipet (*see* Note 5). The phage pellet should be visible at the bottom of the tube.
7. Add 100 µL of TE to each phage pellet, and vortex vigorously to ensure that they are properly resuspended. Then add 50 µL of phenol, vortex well, and leave for a few minutes. Vortex again and spin for 2 min to separate the phases.
8. Carefully remove the upper aqueous layer into a fresh 1.5-mL tube, being careful to leave behind all the precipitated material at the interface.
9. Add 50 µL of chloroform, vortex, and spin for 1 min to separate the phases (*see* Note 6).
10. Carefully remove the upper aqueous layer into a fresh 1.5-mL tube, add 0.1 vol of 3M sodium acetate and 2.5 vol of absolute ethanol, and precipitate the DNA.
11. Spin in microfuge for 5 min, and remove the ethanol by aspiration, being careful not to disturb the DNA pellet, which is often barely visible at this stage. Add 200 µL of 70% ethanol, spin for 2 min, remove the ethanol very carefully, and dry the pellet either under vacuum for a few minutes or by leaving the open tubes on the bench until dried by evaporation.
12. Dissolve the pellet in 30 µL of TE. At this stage, a few microliters (2–5 µL) can be removed and run on a minigel to check the quality and yield of DNA (*see* Note 7). The remainder of the DNA should be stored at –20°C until required.

4. Notes

1. If plaques have been picked and regrown on plates as colonies (e.g., because of the need to screen inserts by hybridization), cultures for ssDNA preps are prepared by touching the toothpicks onto the colonies and washing in the 1.5-mL fresh cells.
2. Once plaques have been obtained, it is best to grow phages and purify the DNA as quickly as possible, since there is an increase in plaque contamination and a deterioration in quality of DNA obtained if plaques are stored at 4°C.
3. Do not grow up 1.5-mL cultures for extended periods of time, since this results in increased cell lysis causing higher levels of contaminating host chromosomal DNA and increases the probability of mutant phage arising.

4. Do not grow at temperatures above 37°C, because although the host cells will grow, the yields of M13 phages drop rapidly with increasing temperature.
5. Make sure all the PEG is removed, since this causes high backgrounds in sequencing.
6. The chloroform extraction can be omitted, but the highest-quality templates are obtained if it is included.
7. A yield of ssDNA of between 5 and 10 µg/mL is normal. If the final pellet is suspended in 30 µL of TE, approx 4–8 µL are ample for obtaining high-quality sequence using standard dideoxy methods.

References

1. Marvin, D. A. and Hohn, B. (1969) Filamentous bacterial viruses. *Bacteriol. Rev.* **33,** 172–209.
2. Bankier, A. and Barrell, B. G. (1983) Shotgun DNA sequencing, in *Techniques in the Life Sciences (Biochemistry), vol 85: Techniques in Nucleic Acid Biochemistry* (Flavell, R. A., ed.), Elsevier, Amsterdam, pp. 1–34.
3. Sambrook, J., Fritsch, E. F., and Maniatis, T. (1989) Small-scale preparation of single-stranded bacteriophage M13 DNA, in *Molecular Closing, A Laboratory Manual.* 2nd ed., Cold Spring Harbor Laboratory, Cold Spring Harbor, NY, pp. 4.29–4.30.

CHAPTER 44

Preparation of ssDNA from Phagemid Vectors

Michael K. Trower

1. Introduction

Our ability to generate single-stranded DNA (ssDNA) templates has been invaluable for procedures involving DNA sequencing and site-directed mutagenesis. Conventionally, this is carried out by subcloning the DNA region under analysis into vectors that are based on the single-stranded DNA bacteriophages, the most popular being M13. However, although these vectors are suitable for the production of ssDNA, they are less convenient to use than plasmids, giving lower yields of double-stranded dsDNA and lacking a positive selectable marker. The development of the so-called phagemid vectors combined the advantages of both plasmids and bacteriophages and stemmed from the work of Zinder *(1),* who demonstrated that a plasmid carrying that intergenic region of f1 filamentous phage is packaged as ssDNA when the host cells are super-infected with helper phage. The first generation of phagemids, the pEMBL vectors, however, gave poor and irreproducible ssDNA yields. This problem was resolved in two main ways: First, deletions within the intergenic region enhanced the replication of phagemids *(2),* and second, the packaging ratio of phagemid to helper phage was greatly improved by the use of helper phage with weak replication origins *(3).* As a result of these investigations, a new generation of phagemid vectors are available that impart both the stability and convenience of plasmid dsDNA cloning vectors and that may be readily mobilized to generate high yields of ssDNA.

From: *Methods in Molecular Biology, Vol. 58: Basic DNA and RNA Protocols*
Edited by: A. Harwood Humana Press Inc., Totowa, NJ

2. Materials

Solutions are stored at room temperature unless stated otherwise.

1. Phagemid transformed into an appropriate *E. coli* host containing an F' episome, such as the common laboratory strains JM109, NM522, TG1, or XL1-Blue (*see* Notes 1–3).
2. Helper phage: M13K07 or VCSM13. These typically have a titer of 10^{11}–10^{12} pfu/mL.
3. TBG medium: To make 1 L, add 12 g of tryptone, 24 g of yeast extract, and 4 mL of glycerol in a total volume of 882 mL of distilled water and autoclave. When cool, add 100 mL of sterile $0.17M$ KH_2PO_4, $0.72M$ K_2HPO_4 solution, and 18 mL of 20% (w/v) glucose solution. The 20% glucose solution should be prepared and autoclaved separately.
4. Kanamycin: a 75 mg/mL stock in water. Store at –20°C.
5. PEG/NaCl solution: 20% (w/v) polyethylene glycol 6000 or 8000, $2.5M$ NaCl.
6. TE buffer: 10 mM Tris-HCl, pH 7.4; 1 mM EDTA. Autoclave.
7. Phenol: double-distilled and saturated with TE buffer. Store at 4°C.
8. $3M$ Sodium acetate, pH 5.2.

3. Methods

1. From a freshly streaked plate pick an isolated colony into 2 mL of TBG medium. Add approx 10^7–10^8 pfu/mL of either VCSM13 or M13KO7 helper phage. This gives a multiplicity of infection (moi) of 10–20 (*see* Notes 4 and 5). If the plasmid can be selected with an antibiotic, such as ampicillin (100 µg/mL), this should also be included. Grow with vigorous aeration at 37°C for 2 h.
3. Add 2 µL of kanamycin (75 mg/mL), since both VCSM13 and M13KO7 phage confer resistance against this antibiotic and then continue growth overnight. Since the helper phage does not lyse the infected cells, the overnight culture should be saturated.
4. Transfer 1.4 mL of the overnight culture to a 1.5-mL Eppendorf tube and centrifuge at 13,000g for 5 min. Pour the supernatant into a fresh tube taking care not to transfer any cells. Respin for 5 min once more and again transfer the supernatant to a fresh tube. If the ssDNA cannot be processed at this time, the supernatant may be stored for 24 h at 4°C, although it must be then recentrifuged before continuing with the protocol. The pellet can be discarded.
5. Add 200 µL of PEG/NaCl solution, mix and stand for 20 min on ice, and then centrifuge for 5 min. At this stage a small pellet should be visible. Discard the supernatant and then recentrifuge at 13,000g for 2 min. Carefully remove any residual PEG/NaCl solution with a micropipet tip.

6. Resuspend the pellet in 150 µL of TE buffer and add an equal volume of TE-saturated phenol. Vortex for 10–15 s and then stand the tubes for 5 min. Revortex and centrifuge at 13,000g for 5 min. Remove the upper phase and transfer to a fresh tube. Do not be concerned if the upper phase is very cloudy. This is quite common with these preparations.
7. Add 150 µL of chloroform, vortex for 10–15 s, and stand for 5 min. Revortex and then spin for 3 min. Remove the upper phase once more and again transfer to a fresh tube.
8. Add 15 µL of 3M sodium acetate, pH 4.8–5.2, and 300 µL of ethanol. Place the tube in either dry ice for 10 min or at –20°C overnight. Centrifuge at 13,000g for 10 min and pour off the liquid. Recentrifuge for 2 min, remove any remaining liquid with a micropipet tip, and then either vacuum or air dry. A small pellet should be visible. Redissolve the pellet in 50 µL of water or TE. If not used immediately store at –20°C.

The yield of ssDNA is usually around 10 µg of DNA/mL of broth. Confirmation of template isolation and yield can be achieved by running a small aliquot (2–5 µL) on a 1% agarose gel. Two bands, the helper phage ssDNA and the phagemid ssDNA, should be visible following ethidium bromide staining. The helper phage ssDNA should not interfere with sequencing or mutagenesis reactions.

4. Notes

1. The *E. coli* strain used must contain an F' episome to allow the helper phage to infect the host cells.
2. Phagemids include the pUC118/9 *(3)*, pBluescript II (Stratagene), and pGEMZf (Promega) series of cloning vectors.
3. The orientation of the f1 origin of replication determines which DNA strand of the phagemid is generated following helper phage infection. This is usually denoted for a given phagemid by the designation (+) or (–) and in the pBluescript II phagemids refers to the recovery of the sense and antisense strands of the *lacZ* gene, respectively. Such information is clearly critical when designing both sequencing and mutagenic primers to ensure that they will hybridize to the rescued ssDNA template.
4. Moi refers to the ratio of helper phage to cells present in the culture. For example, a moi of 10 means there are 10 helper phages for each bacterial cell. Generally a moi of 10–20 generates good yields of the phagemid ssDNA.
5. R408 helper phage may also be used, but the antibiotic addition in step 3 should not be carried out since R408 lacks resistance against kanamycin.

Acknowledgments

I would like to thank Sydney Brenner of the Department of Medicine, University of Cambridge, for providing me with the opportunity to devise this protocol, and I was supported by a fellowship from E. I. du Pont de Nemours and Co. Inc., Wilmington, DE.

References

1. Dotto, G. P., Enea, V., and Zinder, N. D. (1981) Functional analysis of bacteriophage f1 intergenic region. *Virology* **114,** 463–473.
2. Zagursky, R. J. and Berman, M. L. (1984) Cloning vectors that yield high levels of single-stranded DNA for rapid DNA sequencing. *Gene* **27,** 183–191.
3. Vieira, J. and Messing, J. (1983) Production of single-stranded plasmid DNA. *Meth. Enzymol.* **153,** 3–11.

CHAPTER 45

A Rapid Plasmid Purification Method for Dideoxy Sequencing

Annette M. Griffin and Hugh G. Griffin

1. Introduction

The dideoxy chain-termination method of DNA sequence analysis involves the synthesis of a DNA strand from a single-stranded template *(1)*. The enzymatic synthesis is initiated at the site where an oligonucleotide primer anneals to the template. Traditionally, sequencing was performed on templates produced by single-stranded phage M13 *(2,3)*, or phagmid *(4)*. Sequencing plasmid DNA has the advantage of eliminating the need to subclone fragments into M13, and also enables both strands to be sequenced from the same plasmid by the use of reverse primers. The most consistently satisfactory sequencing results for most purposes are probably obtained by using linear amplification techniques (cycle sequencing) either in manual procedures *(5)* or in automated sequencing machines *(6)*. However, plasmid DNA that has been denatured can also serve as satisfactory template DNA for manual, nonamplified procedures *(7)*. It is particularly important when preparing plasmid DNA for sequencing to give the utmost care and attention to the DNA isolation and purification techniques. Most problems that occur with plasmid sequencing are related to poor quality template.

CsCl-ethidium bromide gradient preparations of plasmid DNA *(8)* can be used as template for sequencing but do not necessarily provide better results than miniprep DNA. Many plasmid miniprep methods are available for preparing the template for sequencing *(9)*. Denaturation is usu-

From: *Methods in Molecular Biology, Vol. 58: Basic DNA and RNA Protocols*
Edited by: A. Harwood Humana Press Inc., Totowa, NJ

ally achieved with the use of alkali or by boiling (in the presence of primer). Prior denaturation is not necessary for linear amplification sequencing because the plasmid is heat denatured during the thermal cycling process.

This chapter presents a miniprep method that involves an alkaline lysis procedure *(10)* followed by DNA purification by GeneClean®—a commercial kit for the purification of DNA. The GeneClean kit is based on the principle that DNA will bind to a silica matrix, whereas DNA contaminants will not bind and can be removed by repeated washing of the silica-DNA matrix *(11)*. The DNA can be used directly as a template for cycling sequencing or can be denatured by alkali and sequenced using Sequenase® DNA polymerase Version 2.0.

2. Materials
2.1. DNA Preparation

1. GET buffer: 50 mM glucose, 10 mM EDTA, 25 mM Tris-HCl, pH 8.0. This solution should be autoclaved for 15 min at 10 lb/in^2 (68.9 kPa) and stored at 4°C.
2. Lysozyme buffer: 20 mg/mL lysozyme in GET buffer. Make up fresh.
3. Alkaline SDS: 200 mM NaOH, 1% SDS. Make up fresh as follows: 200 µL of 5M NaOH, 500 µL of 10% SDS in 4.3 mL water.
4. 3M sodium acetate, pH 4.8. Adjust pH to 4.8 with glacial acetic acid.
5. Phenol-chloroform: This is prepared by mixing equal amounts of phenol and chloroform. If desired, the mixture can be equilibrated by extracting several times with 0.1M Tris-HCl, pH 7.6, although this is not essential.

2.2. DNA Purification (Using GeneClean-II Kit)

6. GeneClean-II kit. This is available from BIO 101 Inc. (La Jolla, CA), or from Stratech Scientific (Luton, Bedfordshire, England). Alternatively, a homemade version can be used (*see* Chapter 28).

2.3. Alkali Denaturation

7. 100% Ethanol.
8. 70% Ethanol.

2.4. Primer-Annealing

9. Plasmid reaction buffer: (5X concentrate) 200 mM Tris-HCl, pH 7.5, 100 mM MgCl$_2$, 250 mM NaCl.
10. Primer, 0.5 pmol/µL.

3. Methods
3.1. DNA Preparation

1. Harvest the cells from 5 mL of overnight broth culture by centrifugation at 12,000g for 5 min (see Note 1).
2. Resuspend the pellet in 100 µL of lysozyme buffer and transfer to a 1.5-mL microcentrifuge tube, then incubate on ice for 30 min (see Notes 2–5).
3. Add 200 µL of alkaline SDS. Vortex, then incubate on ice for 5 min.
4. Add 150 µL of 3M sodium acetate, pH 4.8. Vortex, then incubate on ice for 20 min.
5. Centrifuge at 12,000g in a microcentrifuge for 10 min and transfer 0.4 mL of supernatant to a new tube.
6. Extract twice with an equal volume of phenol-chloroform.

3.2. DNA Purification (Using GeneClean-II Kit)

1. Add 3 vol of NaI stock solution from the GeneClean-II kit (see Notes 6 and 7) to the phenol-chloroform extracted miniprep. Use a 2-mL microcentrifuge tube if necessary.
2. Add 5 µL of GLASSMILK from the GeneClean kit suspension. Mix, and incubate for 5 min at room temperature.
3. Centrifuge for 5 s in a microcentrifuge.
4. Remove the supernatant and wash the pellet three times with NEW WASH from the GeneClean kit.
5. Carefully remove the last traces of NEW WASH from the pellet.
6. Elute the DNA in 100 µL of water.

3.3. Alkali Denaturation

Alkali denaturation is not necessary for cycle sequencing.

1. To 86 µL of the purified DNA, add 4 µL of 5M NaOH, and 10 µL of 2 mM EDTA, then incubate at 37°C for 30 min (see Note 8).
2. Neutralize by adding 10 µL of 3M sodium acetate, pH 4.8.
3. Precipitate with 2 vol of 100% ethanol at –70°C for 15 min.
4. Centrifuge for 2 min in a microcentrifuge to pellet precipitated DNA.
5. Wash pellet with 70% ethanol.
6. Dry under vacuum and redissolve the pellet in 7 µL of water.

3.4. Primer-Template Annealing Reaction

This procedure is not necessary for cycle sequencing.

1. In a small microcentrifuge tube, set up the following reaction: 1 µL of primer (0.5 pmol/µL), 2 µL of 5X plasmid reaction buffer, and 7 µL of denatured plasmid DNA (see Note 9).

2. Place in a 65°C waterbath for 2 min. Allow to cool slowly to 30°C over a period of about 30 min. Place on ice (*see* Note 10).
3. Sequence with Sequenase Version 2 T7 DNA polymerase as described in Chapter 46. Cycle sequencing is performed as described (*5, 6, 9* and Chapter 50) (*see* Notes 11 and 12).

4. Notes

1. It is important to remove all the broth supernatant from the cell pellet following centrifugation of the bacterial culture.
2. It is important to ensure that the bacterial pellet is fully resuspended in lysozyme buffer. This can be achieved by vigorous vortexing.
3. Some authors point out that lysozyme is unnecessary for alkaline lysis-based minipreps *(8)*. We have found that although lysozyme is not essential, better lysis is achieved by its use.
4. An RNase step is not necessary in this protocol.
5. This protocol is designed for use with the high copy-number pUC series of plasmids *(12)* and their derivatives. If lower copy-number plasmids are to be used (such as pBR322 and derivatives) it may be necessary to scale up the procedure to achieve the same yield of plasmid DNA.
6. Miniprep plasmid DNA is often contaminated by small oligodeoxyribonucleotides and ribonucleotides that can act as primers and produce faint background bands, stops, and other artifacts in the gel. Inhibitors of DNA polymerase may also be present. In our experience, purification of the DNA prep by the GeneClean procedure, combined with the use of Sequenase® DNA polymerase Version 2.0 for sequencing, alleviates these problems.
7. Other commercial kits, such as US Bioclean (available from United States Biochemical Corp. [Cleveland, OH], or from Amersham Life Science [Little Chalfont, England]) or noncommercial procedures and protocols based on similar principles as GeneClean *(9, 11)* may also work satisfactorily.
8. Double-stranded supercoiled DNA is best denatured by the alkaline denaturation method. Linear DNA can be successfully denatured by boiling (in the presence of primer).
9. A small excess of template DNA may be helpful when sequencing plasmid DNA with Sequenase DNA polymerase. Aim to use about 5 µg.
10. Annealing can also be achieved by warming the annealing reaction mix to 37°C for 15–30 min.
11. The use of longer primers (25–29 nucleotides) may help reduce artifactual bands when sequencing denatured double-stranded DNA.
12. Sequenase DNA polymerase Version 2.0 works well for double-stranded DNA sequencing.

References

1. Sanger, F., Nicklen, S., and Coulson, A. R. (1977) DNA sequencing with chain-terminating inhibitors. *Proc. Natl. Acad. Sci. USA* **74,** 5463–5467.
2. Messing, J., Gronenborn, B., Müller-Hill, B., and Hofschneider, P. H. (1977) Filamentous coliphage M13 as a cloning vehicle: insertion of a *Hin*dIII fragment of the *lac* regulatory region in M13 replicative form *in vitro*. *Proc. Natl. Acad. Sci. USA* **74,** 3642–3646.
3. Messing, J. (1983) New M13 vectors for cloning. *Meth. Enzymol.* **101,** 20–78.
4. Zagursky, R. J. and Berman, M. L. (1984) Cloning vectors that yield high levels of single-stranded DNA for rapid DNA sequencing. *Gene* **27,** 183–191.
5. Blakesley, R. W. (1993) Cycle sequencing, in *DNA Sequencing Protocols. Methods in Molecular Biology Series,* vol. 23 (Griffin, H. G. and Griffin, A. M., eds.), Humana, Totowa, NJ, pp. 209–217.
6. Rosenthal, A. and Jones, D. S. C. (1993) Linear amplification sequencing with dye terminators, in *DNA Sequencing Protocols. Methods in Molecular Biology Series,* vol. 23 (Griffin, H. G. and Griffin, A. M., eds.), Humana, Totowa, NJ, pp. 281–296.
7. Griffin, H. G. and Griffin, A. M. (1993) Plasmid sequencing, in *DNA Sequencing Protocols. Methods in Molecular Biology Series,* vol. 23 (Griffin, H. G. and Griffin, A. M., eds.), Humana, Totowa, NJ, pp. 131–135.
8. Sambrook, J., Fritsch, E. F., and Maniatis, T. (1989) *Molecular Cloning—A Laboratory Manual.* Cold Spring Harbor Laboratory, Cold Spring Harbor, NY.
9. Griffin, H. G. and Griffin, A. M., eds. (1993) *DNA Sequencing Protocols. Methods in Molecular Biology,* vol. 23, Humana, Totowa, NJ.
10. Birnboim, H. C. and Doly, J. (1979) A rapid alkaline extraction procedure for screening recombinant plasmid DNA. *Nucleic Acids Res.* **7,** 1513–1523.
11. Vogelstein, B. and Gillespie, D. (1979) Preparation and analytical purification of DNA from agarose. *Proc. Natl. Acad. Sci. USA* **76,** 615–619.
12. Vieira, J. and Messing, J. (1982) The pUC plasmids, an M13 mp7-derived system for insertion mutagenesis and sequencing with synthetic universal primers. *Gene* **19,** 259–268.

CHAPTER 46

DNA Sequencing Using Sequenase Version 2.0 T7 DNA Polymerase

Carl W. Fuller, Bernard F. McArdle, Annette M. Griffin, and Hugh G. Griffin

1. Introduction

The dideoxy chain-termination method of DNA sequence analysis involves the synthesis of a DNA strand by enzymatic extension from a specific primer using a DNA polymerase *(1)*. Several different enzymes are available for this purpose, each having different qualities and properties.

Bacteriophage T7 DNA polymerase is a two-subunit protein *(2–4)*. The smaller subunit, thioredoxin (mol wt 12,000) is the product of the *E. coli trxA* gene. The larger subunit (mol wt 80,000) is the product of the T7 gene 5. The polymerase has two known catalytic activities, DNA-dependent DNA polymerase and 3'–5' exonuclease activity, that are active on both single-stranded and double-stranded DNA. The level of exonuclease is so high that this polymerase is not suitable for DNA sequencing. Tabor and Richardson *(5,6)* reported an oxidation procedure that inactivates the exonuclease activity with little effect on the polymerase activity *(7,8)*. This chemically modified form of T7 DNA polymerase is known commercially as Sequenase™ Version 1.0 T7 DNA polymerase. A genetically engineered form of T7 DNA polymerase lacking 28 amino acids in the exonuclease domain of the catalytic subunit has also been produced *(9,10)*. This enzyme also has excellent proper-

From: *Methods in Molecular Biology, Vol. 58: Basic DNA and RNA Protocols*
Edited by: A. Harwood Humana Press Inc., Totowa, NJ

ties for DNA sequencing and has the commercial name Sequenase Version 2.0 T7 DNA polymerase.

The chemically modified and genetically engineered enzymes have similar properties and are used in similar ways *(5–11)*. Thus, only protocols and results for the Sequenase Version 2.0 T7 DNA polymerase will be discussed here.

1.1. Properties of Modified T7 DNA Polymerase

The methods used and the quality of the results of a DNA sequencing experiment depend on the properties and capabilities of the DNA polymerase. These properties include the speed and processivity of the polymerase, its ability to displace strands annealed to the template to read through template secondary structures, and the ability to use nucleotide analogs as substrates. Low-exonuclease forms of T7 DNA polymerase, unlike the native, high-exonuclease activity form, have potent strand-displacement activity *(9, 12)*, giving modified T7 DNA polymerase the ability to sequence templates with secondary structures.

Modified T7 DNA polymerase catalysis is rapid (300 nucleotides/s) and highly processive, remaining bound to the primer-template for the polymerization of hundreds or thousands of nucleotides without dissociating *(5)*. This property helps eliminate background terminations that might interfere with reading sequence.

DNA sequencing, as currently practiced, makes extensive use of nucleotide analogs for labeling, resolution of compressions, and for chain termination. The rate of incorporation of α-thio dNTPs for labeling sequences with ^{35}S is essentially equal to that of normal dNTPs *(7,12)*. Similarly, the dGTP analogs used to eliminate compression artifacts in sequencing gels (including 7-deaza-dGTP and dITP) are readily incorporated by modified T7 DNA polymerase *(7,13)*. Among the DNA polymerases used for sequencing, it requires the lowest concentration ratio of dideoxy- to deoxy-nucleotide, indicating the high reactivity with these chain-terminating nucleotides. It has also been found to work well with dye-labeled nucleotide terminators for fluorescent sequencing *(14)*.

One important aspect of the reactivity of terminators is the variability of termination rate that shows up in sequence results as variation in band intensities. Band intensities obtained with modified T7 DNA polymerase vary over a relatively small range of about fourfold. For comparison, band intensities in sequencing gels generated using Klenow enzyme have been observed to vary over a 14-fold range (7,15–18). Tabor and Richardson reported observations made when sequencing reactions were run in the presence of Mn^{2+} instead of (or in addition to) Mg^{2+} (15,16). When modified T7 DNA polymerase is used with Mn^{2+}, a lower concentration ratio of dideoxy- to deoxy-nucleotide is required and the bands generated are considerably more uniform than those generated with Mg^{2+} (16,17). In the presence of Mn^{2+}, over 95% of the bands fall within 10% of the mean corrected band intensity. This high degree of uniformity in band intensity improves the interpretation of sequences read by machine (16,19,20). Modified T7 DNA polymerase in the presence of Mn^{2+} can be used in conduction with either fluorescent labeled primers or specifically designed fluorescent labeled dideoxy terminators to generate sequencing data (14,20).

1.2. Use of Pyrophosphatase in Sequencing Reactions

All DNA polymerases, including T7 DNA polymerase, catalyze pyrophosphorolysis (21,22). Pyrophosphorolysis is the reversal of the polymerization wherein the 3' terminal base of the DNA and pyrophosphate react to form a deoxynucleoside-triphosphate, leaving the DNA one base shorter. Under the conditions normally used for DNA synthesis (i.e., high dNTP concentrations and low pyrophosphate concentration), the forward reaction (polymerization) is greatly favored over the reverse reaction (pyrophosphorolysis). However, pyrophosphorolysis does occur under some conditions (particularly when dITP is used in place of dGTP) given enough time (7). The rate of pyrophosphorolysis varies with the neighboring sequence, reducing the intensity of some bands but leaving others unchanged (12,21,22). The addition of inorganic pyrophosphatase (from yeast) to DNA sequencing reactions completely stabilizes all band intensities, even when incubating the reactions for 60 min and when dITP is used in place of dGTP. The addition of pyrophosphatase is beneficial or at worst harmless for all T7 sequencing protocols.

1.3. Glycerol in DNA Sequencing Reactions

A glycerol content of 5% or more in sequencing reactions has been shown to be advantageous in several circumstances. The simplest case is the use of DNA polymerase, which is stored at a concentration that does not require dilution prior to use. In addition, reactions can be run successfully at increased temperatures in the presence of 5–20% glycerol, conditions that may improve sequencing results. For example, the temperature of the labeling reaction can be increased to 37°C without significant loss of enzyme activity further increasing the stringency of primer annealing *(23)*. Termination reactions can also be run for longer periods of time (10–30 min) and at warmer temperatures (37–50°C) in the presence of glycerol. Polymerization will be more rapid at the elevated temperatures, and some template secondary structures may be eliminated at temperatures above 45°C (*see* Note 1).

1.4. Glycerol-Tolerant Sequencing Gels

The presence of glycerol (typically 25 µg or more per lane) in the samples applied to DNA sequencing gels causes conspicuous distortions in the shape and spacing of the bands *(7,24)*. This distortion occurs about 300–500 bases from the primer, rendering this region of the sequence unreadable. This phenomenon is important because glycerol is widely used for the storage of enzymes, such as the DNA polymerases used for DNA sequencing. It is known that glycerol and boric acid react to form a charged complex that migrates through the sequencing gel *(25)*. When boric acid in TBE buffer is replaced by a different weak acid that does not react with glycerol, no distortion is seen. One good substitute for boric acid is taurine (aminoethanesulfonic acid). A 20-fold concentrate of Tris/taurine/EDTA buffer can be made easily. This is simply used in place of the usual TBE (5- or 10-fold concentrate) in conventional sequencing gels *(23)*.

1.5. Annealing Template and Primer

Chain-termination sequencing methods originally used single-stranded template DNAs. This requirement was met through the use of bacteriophage M13 as a cloning vector. Single-stranded phage DNA is simply mixed with an oligonucleotide primer and annealed briefly at 37–70°C. If double-stranded plasmid DNA is to be used for sequencing template, it must first be denatured. Several early methods for sequencing plasmid DNAs were developed, but none proved satisfactory until the introduction of alkaline denaturation methods *(26–31)*. The original alkaline denaturation method typically involves four steps: mixing the DNA with alkali, neutralizing with concentrated acetate buffer, precipitation with ethanol, and redissolving the DNA in reaction buffer. The precipitation step serves both to concentrate the DNA and to separate it from the salts added in prior steps. These steps work quite well to generate single-strand template suitable for sequencing, but the precipitation and redissolving steps are time-consuming and laborious.

Two simple and efficient denaturation methods that do not require precipitation of plasmid DNA are described here. In the first method, sodium hydroxide is added directly to a mixture of concentrated, purified plasmid DNA and primer, denaturing the double-stranded DNA *(32)*. After a brief incubation at 37°C, a carefully measured equimolar amount of hydrochloric acid is added to the mixture, thereby neutralizing the alkali. Then concentrated reaction buffer is added (fixing the pH at an appropriate value) and the mixture incubated briefly at 37°C to allow the primer to anneal to the appropriate sequence within the template. The sodium hydroxide and hydrochloric acid combine to form sodium chloride, which is normally a component of the DNA sequencing reaction mixtures. As long as the concentration of sodium chloride is kept below approx $0.2M$, the polymerase works well and high-quality sequence is obtained.

In the second method, double-stranded plasmid DNA and primer are mixed with a glycol reagent (50% glycerol, 50% ethylene glycol) so that the final concentration of glycols is approx 40%. This concentration is sufficient to decrease the melting temperature of most plasmids to <90°C. The mixture then is incubated at 90–100°C for 5 min, denaturing the plasmid. Following denaturation, buffer is added and the mixture is incubated briefly at 37°C to allow the primer to anneal to the appropriate sequence within the template. A glycerol-tolerant sequencing gel is required when using this method *(33; see* Note 1).

1.6. The Two-Step Reaction Protocol

After annealing template and primer, the polymerase is typically used both for labeling sequencing products as well as sequence-specific terminations. The most commonly used method for sequencing with modified T7 DNA polymerase consists of two sequential steps; the labeling step and the termination step *(7,18)*. This method differs from the one originally used by Sanger, wherein labeling and termination were achieved in a single reaction that necessitated a "chase" step to reduce background *(1)*.

In the labeling step, the primer is extended using limiting amounts of the deoxynucleoside triphosphates, including one radioactively labeled dNTP. Most of the nucleotide present, including the labeled nucleotide, is incorporated into relatively short (<50 nucleotides) DNA chains. Thus, the label is efficiently and quantitatively incorporated into the DNA. (The labeling step can be omitted entirely when labeled primers are used). The termination step is initiated by transferring four equal aliquots of the labeling step reaction mixture to termination reaction vials. These vials have been prefilled with a supply of all the deoxynucleoside triphosphates and one of the four dideoxynucleoside triphosphates. The temperature and concentrations of nucleotides are chosen so that polymerization will be rapid and processive, so that the resulting sequence will have low background and the reaction time is minimized. The reactions are stopped by the addition of EDTA and formamide, denatured by heating, and loaded onto electrophoresis gels.

2. Materials

2.1. Annealing Template and Primer

1. Reaction buffer (5X concentrate): 200 mM Tris-HCl, pH 7.5, 100 mM MgCl, 250 mM NaCl.
2. Primer: 0.5–2.5 pmol/µL.

2.2. Labeling Step

3. Modified T7 DNA polymerase, either chemically or genetically modified *(12,18)*. These should be adjusted to a concentration of 13 U/µL or approx 1.0 mg/mL.
4. Yeast inorganic pyrophosphatase, 5 U/mL in 10 mM Tris-HCl, pH 7.5, 0.1 mM EDTA, 50% glycerol. Note: pyrophosphatase can be premixed with modified T7 DNA polymerase at a ratio of about 0.004 U pyrophosphatase for each 3 U of polymerase. (One unit of pyrophosphatase hydrolyses 1 µM of pyrophosphate/min at 25°C.)
5. Enzyme dilution buffer: 10 mM Tris-HCl, pH 7.5, 5 mM DTT, 0.5 mg/mL BSA.
6. Glycerol enzyme dilution buffer: 20 mM Tris-HCl, pH 7.5, 2 mM DTT, 0.1 mM EDTA, 50% glycerol. Can be used to keep glycerol concentration high (*see* Section 1.3.).
7. Dithiothreitol solution: 0.1M.
8. Labeled dATP—either [α-^{32}P]-dATP, [α-^{33}P]-dATP, or [α-^{35}S]-dATP is required for autoradiographic detection of the sequence. The specific activity for ^{35}S, ^{33}P, or ^{32}P should be 1000–1500 Ci/mmol.
9. Labeling mix for labeled dATP (dGTP reactions): 1.5 µM dGTP, 1.5 µM dCTP, and 1.5 µM dTTP (*see* Note 2).

2.3. Termination Step

10. ddG termination mix: 80 µM dGTP (or 7-deaza-dGTP), 80 µM dATP, 80 µM dCTP, 80 µM dTTP, 8 µM ddGTP, 50 mM NaCl.

11. ddA termination mix: 80 μM dGTP (or 7-deaza-dGTP), 80 μM dATP, 80 μM dCTP, 80 μM dTTP, 8 μM ddATP, 50 mM NaCl.
12. ddT termination mix: 80 μM dGTP (or 7-deaza-dGTP), 80 μM dATP, 80 μM dCTP, 80 μM dTTP, 8 μM ddTTP, 50 mM NaCl.
13. ddC termination mix: 80 μL M dGTP (or 7-deaza-dGTP), 80 μM dATP, 80 μM dCTP, 80 μM dTTP, 8 μM ddCTP, 50 mM NaCl.
14. Stop solution: 95% formamide, 20 mM EDTA, 0.05% bromophenol blue, 0.05% xylene cyanol FF.

2.4. Sequencing Plasmid DNA

15. Plasmid reaction buffer: (5X concentrate) 200 mM Tris-HCl, pH 7.5, 100 mM MgCl$_2$, 250 mM NaCl.
16. NaOH solution: 1.00 ± 0.03M (**must** be accurate).
17. HCl solution: 1.00 ± 0.03M (**must** be accurate).
18. Plasmid denaturing reagent: 10 mM Tris-HCl, pH 7.5, 1 mM EDTA, 50% glycerol, 50% ethylene glycol
19. 20X glycerol-tolerant gel buffer: 1.78M Tris, 0.58M taurine, 0.01M Na$_2$EDTA · 2H$_2$O.

2.5. Improving Band Intensities Close to the Primer

20. Mn buffer: 0.15M sodium isocitrate, 0.1M MnCl$_2$.

2.6. Extending Sequences Beyond 400 Bases from the Primer

21. Sequence extending mix (for dGTP reactions): 180 μM dGTP, 180 μM dATP, 180 μM dCTP, 180 μM dTTP, 50 mM NaCl.

2.7. Compressions

22. Labeling mix for labeled dATP (7-deaza-dGTP reactions): 1.5 μM 7-deaza-dGTP, 1.5 μM dTTP, and 1.5 μM dCTP; (dITP reactions): 3.0 μM dITP, 1.5 μM dCTP, and 1.5 μM dTTP.

Termination mixes for dITP reactions:

23. ddG termination mix: 160 μM dITP, 80 μM dATP, 80 μM dCTP, 80 μM dTTP, 1.6 μM ddGTP, 50 mM NaCl.
24. ddA termination mix: 160 μM dITP, 80 μM dATP, 80 μM dCTP, 80 μM dTTP, 8 μM ddATP, 50 mM NaCl.
25. ddT termination mix: 160 μM dITP, 80 μM dATP, 80 μM dCTP, 80 μM dTTP, 8 μM ddTTP, 50 mM NaCl.
26. ddC termination mix: 160 μM dITP, 80 μM dATP, 80 μM dCTP, 80 μM dTTP, 8 μM ddCTP, 50 mM NaCl.

27. Sequence extending mix (for dITP reactions): 360 µM dITP, 180 µM dATP, 180 µM dCTP, 180 µM dTTP, 50 mM NaCl.

3. Methods
3.1. Annealing Template and Primer

For each set of four sequencing lanes, a single annealing (and subsequent labeling) reaction is used. *See* Section 3.4. for denaturing and sequencing plasmid DNA templates.

1. In a centrifuge tube, combine the following: 1 µL of primer (0.5 pmol), 2 µL of reaction buffer, 1 µg of DNA (denatured if dsDNA), and 7 µL of ddH$_2$O to a total volume of 10 µL.
2. Warm the capped tube to 65°C for 2 min, then allow the temperature of the tube to cool slowly to room temperature (about 30 min).
3. After annealing, place the tube on ice and use within 4 h.

3.2. Labeling Step

1. Dilute modified T7 DNA polymerase to working concentration (1.6 U/µL) using the enzyme dilution buffer.
2. To the annealed template-primer add the following (on ice): 10 µL of template-primer, 1 µL of DTT (0.1M), 2 µL of labeling mix, 0.5 µL of [α-^{35}S]dATP (*see* Note 3), 2 µL of diluted enzyme (always add enzyme last). Total volume is 15.5 µL.
3. Mix thoroughly and incubate for 2–5 min at room temperature (*see* Note 4).

3.3. Termination Step

Steps 1 and 2 are best done before beginning the labeling reaction.

1. Have on hand 4 tubes labeled G, A, T, and C.
2. Place 2.5 µL of the ddGTP termination mix in the tube labeled G. Similarly fill the A, T, and C tubes with 2.5 µL of the ddATP, ddTTP, and ddCTP termination mixes, respectively. Cap the tubes to prevent evaporation.
3. Prewarm the tubes at 37°C at least 1 min before adding labeling reaction aliquot.
4. When the labeling incubation is complete, remove 3.5 µL and transfer it to the tube labeled G. Continue incubation of the G tube at 37°C. Similarly transfer 3.5 µL of the labeling reaction to the A, T, and C tubes, returning them to the 37°C bath. Continue the incubations for a total of 3–5 min.
5. Add 4 µL of stop solution to each of the termination reactions, mix thoroughly, and store on ice until ready to load the sequencing gel. Samples labeled with ^{35}S can be stored at –20°C for 1 wk with little degradation. Samples labeled with ^{32}P should be run the same day.

Sequencing with Sequenase DNA Polymerase

6. When the gel is ready for loading, heat the samples to 75–80°C for 2 min and load immediately on the gel. Use 2–4 µL in each lane.

3.4. Sequencing Plasmid DNA

It is important to use high-quality plasmid DNA for best sequencing results. Preparations that contain large fractions of nicked or cut plasmid DNAs may produce sequences with high backgrounds and false bands resulting from the strand breaks. Double-stranded plasmid DNA must be denatured prior to sequencing. Two convenient and rapid methods of denaturing plasmids are presented here.

1. Denature double-stranded templates with either of the fast denaturing protocols below:
   a. Alkaline denaturation (without precipitation): Combine the following in a microcentrifuge tube: 1–8 µL of DNA (1–3 µg), 2 µL of 1.0M NaOH, 1 µL of primer (0.5–5 pmol), H_2O to adjust to a total volume of 11 µL. Mix thoroughly and incubate for 10 min at 37°C. Place the mixture on ice. Then add 2 µL of 1.0M HCl and 2 µL of plasmid reaction buffer for a total volume of 15 µL. Continue with annealing in step 2 below.
   b. Heat/glycol denaturation: Combine the following in a microcentrifuge tube: 1–7 µL of DNA (1–3 µg), 5 µL of plasmid denaturing reagent, 1 µL of primer (0.5–5 pmol), H_2O to adjust to a total volume of 13 µL. Mix thoroughly (the mixture is viscous). Incubate at 90–100°C for 5 min. Chill the mixture in an ice water bath. Then add 2 µL of plasmid reaction buffer to bring to a total volume of 15 µL. Continue with annealing in step 2 below (*see* Note 1).
2. Anneal by incubating the template/primer/buffer mixture at 37°C for 10 min. Chill on ice.
3. Proceed to the labeling step (Section 3.2.). Using these protocols, the volume of the labeling reaction will be larger than outlined above. Therefore, transfer 4.5 µL of labeling reaction mix to each termination tube (G, A, T, and C), and continue with the protocol (*see* Note 1).
4. Electrophorese through a gel made with glycerol-tolerant gel buffer.

3.5. Running Sequencing Reactions in 96-Well Microassay Plates

Heat-resistant plates with 96 U-bottom wells are ideal for running several sets of termination reactions at the same time. Care should be taken not to allow reactions to evaporate to dryness. Unless reagents are specifically designed for sequencing reactions using a plate format, annealing and labeling reactions are best performed in small centrifuge vials while termination reactions are run in the plate as follows:

1. Prepare the plate by adding 2.5 µL of termination mixes to appropriate wells in the plate.
2. When the labeling reactions are nearly finished, place the 96-well plate on a 37°C heating block or in a 37°C water bath to prewarm the termination reactions.
3. When the labeling reactions are done, transfer 3.5 µL of the labeling reaction product into each of the four termination wells, mixing as usual.
4. Allow the termination reactions to incubate at 37°C until all of the reactions have run at least 2 min and no more than 10 min.
5. Remove the plate from the 37°C bath and add 4 µL of stop solution to each reaction.
6. Immediately prior to loading the gel, denature the reaction products by heating the plate on a 75°C heating block for 2 min.

3.6. Improving Band Intensities Close to the Primer

The addition of Mn^{2+} to the reaction buffer changes the ratio of reactivities of dNTPs to ddNTPs by approximately fivefold, even when both Mg^{2+} and Mn^{2+} are present (15,16). Thus, adding Mn^{2+} to the sequencing reaction buffer has the same effect as increasing the relative concentrations of all four ddNTPs. This makes terminations occur closer to the priming site, emphasizing sequences close to the primer. This is useful since reactions that contain less than the normal amount of template DNA result in weak bands near the primer. Running the sequencing reactions in the presence of Mn^{2+} will usually restore the sequences close to the primer even when less template is available. To use Mn^{2+}, simply add 1 µL of Mn buffer (Section 2.5.) to the labeling reaction prior to the addition of diluted enzyme.

Sequence close to the primer also can be obtained by using less deoxynucleotide in the labeling step. Simply diluting the labeling mix described in Section 2.2. to a fourfold lower concentration of deoxynucleotide to 0.38 µM results in average extensions from the primer of only a few nucleotides.

3.7. Extending Sequences Beyond 400 Bases from the Primer

Sequence 300–400 bases from the primer should be readily obtained following the above procedure. To extend the sequence further, use the sequence-extending mix to increase the ratio of deoxy- to dideoxy-nucleotides in the termination step, thereby increasing the average length of extensions. This method can extend the length of extensions thousands of bases without necessarily sacrificing information closer to the primer. It is important to remember that although it is possible to extend the primers in sequencing reactions to lengths of thousands of bases, the current gel technology is incapable of resolving DNA molecules greater than about 700 bases in length. To use extending mix, reduce the amount of each termination mix from 2.5 to 1.5 µL and add 1 µL of extending mix. This increases the ratio of deoxy- to dideoxy-nucleotide concentration about 2.5-fold. Experiment with other ratios as needed.

3.8. Compressions

The final analysis step in DNA sequencing involves the use of a denaturing polyacrylamide electrophoresis gel to separate DNA molecules by size. Separation based solely on size requires complete elimination of secondary structure from the DNA, typically by using high concentrations of urea in the gel and running at elevated temperatures. Compression artifacts are caused whenever stable DNA secondary structures exist in the gel during electrophoresis. Compression artifacts are recognized by anomalous band-to-band spacing on the gel. Several sequence bands run closer together than normal, unresolved, and the spacing of several bands above is larger than normal. It is important to look for both the unresolved (compressed) bands and the wider-spaced ("expanded") bands to confidently recognize a compression artifact.

Compression artifacts are overcome in one of two ways; either by the use of an analog of dGTP during the sequencing reactions, or by changing to a stronger denaturing condition in the gel. The use of nucleotide analogs for dideoxy-sequencing is simple and usually quite effective. Recommended nucleotide mixtures for sequencing with dITP *(7,34)* or 7-deaza-dGTP *(13)* are described in Section 2. When these analogs replace dGTP, the product DNA forms only weaker base pairs with dC, which are more readily denatured during gel electrophoresis. Even the most difficult compression artifacts are resolved by dITP, whereas 7-deaza-dG may not resolve all compressions *(7,34)*. It is essential, however, to run dGTP sequences alongside dITP sequences since the analog sequences are more prone to other artifacts.

A good alternative is to use gels with stronger denaturing composition. We routinely use gels containing both $7M$ urea (not $8M$) and 20–40% formamide. Polymerization of these gels requires the use of 3–5 times more *N,N,N',N'*-tetramethylethylenediamine (TEMED) than gels without formamide.

3.9. Resolving Bands in All Four Lanes

Another frequently observed artifact in sequencing gels is "bands in all four lanes," also known as "stops" or "pauses" in the sequence. This can be distinguished from compression artifacts by observing that the band-to-band spacing on the gel does not change. Stops can be the result of one or a combination of several different factors. Polymerases can pause at sites of exceptional secondary structure in the DNA template, particularly when nucleotide analogs such as dITP are used. Impurities in the DNA template preparation, the temperature at which the various steps of a sequencing reaction are performed, or even incomplete mixing of reactants can lead to ineffective polymerization. Generally, both the labeling and termination reaction mixtures must be carefully and completely mixed by repeated pumping of the pipet. The labeling step should be run at low temperature (4–20°C) and for <5 min. The termination reactions should be run by prewarming the termination mixes at 37–45°C prior to adding labeling reaction products and continued for 5–20 min. A few suggestions for reducing stops are presented (*see* Note 4).

4. Notes

1. The use of the increased glycerol in sequencing reactions in the form of plasmid denaturing reagent, glycerol enzyme dilution buffer, or otherwise added glycerol will necessitate the use of a glycerol-tolerant gel buffer in your sequencing gel system since glycerol severely distorts ordinary sequencing gels. See 20X glycerol-tolerant gel buffer in Sections 1.4. and 2.4.
2. The concentration of the labeling mixes for labeled dATP given here are equivalent to 1:5 dilution of those in the commercial kit.
3. The amount of labeled nucleotide can be adjusted according to the needs of the experiment. Nominally, 0.5 µL of 10 µCi/µL and 10 µM (1000 Ci/mmol) should be used.
4. Suggestions for reducing stops (bands in all four lanes).
   a. Reduce the temperature and/or duration of the labeling step (e.g., 4°C, 5 min).
   b. Reduce the concentration of dNTPs (to one-half or one-fourth of the normal amount) to keep extensions during this step from reaching the pause site.
   c. Using more polymerase (i.e., 3–8 times more than usual) on difficult templates can be effective at resolving stops. Be certain that at least the

recommended amount of polymerase is carefully and completely mixed (by repeated pumping of the pipet) into the labeling reaction.

d. The addition of 0.5 µg of *E. coli* ssb protein during the labeling reaction can sometimes be effective in eliminating the stops. When using ssb, it is necessary to inactivate it prior to running the gel. Add 0.1 µg of proteinase K and incubate at 65°C for 20 min after adding the stop solution.

e. A chase step using high concentrations of dNTPs and terminal deoxynucleotidyl transferase can reduce stops in sequencing reactions, especially those involving dITP *(34)*. Prepare a 1 m*M* stock solution of all four dNTPs in TE buffer. Dilute terminal deoxynucleotidyl transferase (TdT) in this stock solution to a concentration of 2 U/L and keep on ice (dilute only enough for immediate use). Sequence according to the two-step protocol using the either the dGTP or the dITP labeling and termination mixes, respectively. **Do not** add stop solution. After the termination reaction is complete, add 1 µL of the TdT dilution to each termination reaction tube and incubate 30 min at 37°C. Add 4 µL of stop solution to each reaction tube. Heat denature and load on a sequencing gel as usual.

f. For difficult plasmid templates, denaturing by boiling in the presence of glycerol has been observed to be very effective in resolving stops. *See* Section 3.4., step 1b for plasmid sequencing.

g. Make a new primer either for the other strand or to prime 50 bases farther from the site of the stop.

References

1. Sanger, F., Nicklen, S., and Coulson, A. R. (1977) DNA sequencing with chain-terminating inhibitors. *Proc. Natl. Acad. Sci. USA* **74,** 5463–5467.
2. Modrich, P. and Richardson, C. C. (1975) Bacteriophage T7 deoxyribonucleic acid replication in vitro. A protein of *Escherichia coli* required for bacteriophage T7 DNA polymerase activity. *J. Biol. Chem.* **250,** 5508–5514.
3. Modrich, P. and Richardson, C. C. (1975) Bacteriophage T7 deoxyribonucleic acid replication in vitro. Bacteriophage T7 DNA polymerase: an enzyme composed of phage and host-specific subunits. *J. Biol. Chem.* **250,** 5515–5522.
4. Hori, K., Mark, D. F., and Richardson, C. C. (1979) Deoxyribonucleic acid polymerase of bacteriophage T7. Characterization of the exonuclease activities of the gene 5 protein and the reconstituted polymerase. *J. Biol. Chem.* **254,** 11,598–11,604.
5. Tabor, S. and Richardson, C. C. (1987) Selective oxidation of the exonuclease domain of bacteriophage T7 DNA polymerase. *J. Biol. Chem.* **262,** 15,330–15,333.
6. Tabor, S. and Richardson, C. C. (1990) U.S. Patent 4,946,786.
7. Tabor, S. and Richardson, C. C. (1987) DNA sequence analysis with a modified bacteriophage T7 DNA polymerase. *Proc. Natl. Acad. Sci. USA* **84,** 4767–4771.
8. Tabor, S. and Richardson, C. C. (1989) U.S. Patent 4,795,699.

9. Tabor, S. and Richardson, C. C. (1989) Selective inactivation of the exonuclease activity of bacteriophage T7 DNA polymerase by in vitro mutagenesis. *J. Biol. Chem.* **264,** 6447–6458.
10. Tabor, S. and Richardson, C. C. (1990) U.S. Patent 4,942,130.
11. Patel, S. S., Wong, I., and Johnson, K. A. (1991) Pre-steady-state kinetic analysis of processive DNA replication including complete characterization of an exonuclease deficient mutant. *Biochemistry* **30,** 511–525.
12. Fuller, C. W. (1988) Sequenase® Version 2.0; a new genetically engineered enzyme for DNA sequencing. *Editorial Comments* **15(2),** United States Biochemical Corp., Cleveland, OH, pp. 1–4.
13. Mizusawa, S., Nishimura, S., and Seela, F. (1986) Improvement of the dideoxy chain termination method of DNA sequencing by use of deoxy-7-deazaguanosine triphosphate in place of dGTP. *Nucleic Acids Res.* **14,** 1319–1324.
14. Lee, L. G., Connell, C. R., Woo, S. L., Cheng, R. D., McArdle, B. F., Fuller, C. W., Halloran, N. D., and Wilson, R. K. (1992) DNA sequencing with dye-labeled terminators and T7 DNA polymerase: effect of dyes and dNTPs on incorporation of dye terminators and probability analysis of termination fragments. *Nucleic Acids Res.* **20(10),** 2471–2483.
15. Tabor, S. and Richardson, C. C. (1989) Effect of manganese ions on the incorporation of dideoxynucleotides by bacteriophage T7 DNA polymerase and *Escherichia coli* DNA polymerase I. *Proc. Natl. Acad. Sci. USA* **86,** 4076–4080.
16. Tabor, S. and Richardson, C. C. (1990) U.S. Patent 4,962,020.
17. Fuller, C. W. (1989) Variation in band intensities in DNA sequencing. *Editorial Comments* **16(3),** United States Biochemical Corp., Cleveland, OH, pp. 1–8.
18. Fuller, C. W. (1988) DNA Sequencing; Sequenase® users observations. *Editorial Comments* **14(4),** United States Biochemical Corp., Cleveland, OH, pp. 1–7.
19. Kristensen, T., Voss, H., Schwager, C., Stegemann, J., Sproat, B., and Ansorge, W. (1988) T7 DNA polymerase in automated dideoxy sequencing. *Nucleic Acids Res.* 16, 3487–3496.
20. McArdle, B. F. and Fuller, C. W. (1991) Using Sequenase in automated DNA sequencing. *Editorial Comments* **17(4),** United States Biochemical Corp., Cleveland, OH, pp. 1–8, 17.
21. Tabor, S. and Richardson, C. C. (1990) DNA sequence analysis with a modified bacteriophage T7 DNA polymerase. Effect of pyrophosphorolysis and metal ions. *J. Biol. Chem.* **265,** 8322–8328.
22. Ruan, C. C., Samols, S. B., and Fuller, C. W. (1990) Role of pyrophosphorolysis in DNA sequencing. *Editorial Comments* **17(1),** United States Biochemical Corp., Cleveland, OH, pp. 1–27.
23. Pisa-Williamson, D. and Fuller, C. W. (1992) Glycerol tolerant DNA sequencing gels. *Editorial Comments* **19(2),** United States Biochemical Corp., Cleveland, OH, pp. 1–7.
24. Fuller, C. W. (1989) The effect of excess glycerol on DNA sequencing gels. *Editorial Comments* **16(2),** United States Biochemical Corp., Cleveland, OH, p. 19.
25. Cotton, F. A. and Wilkinson, G. (1980) *Advanced Inorganic Chemistry,* John Wiley, New York, p. 298.

26. Chen, E. J. and Seeburg, P. H. (1985) Supercoil sequencing: a fast and simple method for sequencing plasmid DNA. *DNA* **4,** 165–170.
27. Hattori, M. and Sakaki, Y. (1986) Dideoxy sequencing method using denatured plasmid templates. *Anal. Biochem.* **152,** 232–238.
28. Haltiner, M., Kempe, T., and Tjian, R. (1985) A novel strategy for constructing clustered point mutations. *Nucleic Acids Res.* **13,** 1015–1026.
29. Lim, H. M. and Pene, J. J. (1988) Optimal conditions for supercoil DNA sequencing with the *Escherichia coli* DNA polymerase I large fragment. *Gene Anal. Tech.* **5,** 32–39.
30. Toneguzzo, F. Glynn, S., Levi, E., Mjolsness, S., and Hayday, A. (1988) Use of a chemically modified T7 DNA polymerase for manual and automated sequencing of supercoiled DNA. *Biotechniques* **6,** 460–469.
31. Hsiao, K. (1991) A fast and simple procedure for sequencing double-stranded DNA with Sequenase. *Nucleic Acids Res.* **19,** 2787.
32. Sequenase™ Quick Denature™ Plasmid Sequencing Kit, first edition (1993) United States Biochemical Corp., Cleveland, OH.
33. Gough, J. A. and Murray, N. E. (1983) Sequence diversity among related genes for recognition of specific targets in DNA molecules. *J. Mol. Biol.* **166,** 1–19.
34. Fawcett, T. W. and Bartlett, G., (1990) An effective method for eliminating "artifact banding" when sequencing double-stranded DNA templates. *BioTechniques* **9,** 46–49.

CHAPTER 47

Pouring Linear and Buffer-Gradient Sequencing Gels

Paul Littlebury

1. Introduction

The products of sequencing reactions are separated on thin, low percentage (usually 6%) polyacrylamide gels. Normally, these gels are 40–50 cm in length, 20 cm wide, and between 0.3 and 0.4 mm thick. Longer gels (to enable more sequence to be read from a single run) up to 100 cm in length can be poured, as can wider gels (to enable more clones to be loaded onto a single gel). However, these larger gels are more difficult to handle after electrophoresis, i.e., during fixation, drying, and autoradiography. The methods given in this chapter deal only with the "standard" size of sequencing gels, but the principles are the same for larger gels, except that more gel mix will be needed, and different sizes of combs, spacers, and glass plates may also be needed.

Sequencing gels are poured between two plates of glass, separated by two thin spacers, ensuring that a constant thickness of gel is maintained (changes in the thickness of the gel will alter the migration rate of the samples). The gel is also loaded and run while still in this mold, since the glass plates provide support for the fragile gel. As a consequence, a notch that is several centimeters deep and almost the width of the plate must be cut in one of the plates to provide an electrical contact with the running buffer at the anode. Sample wells are also formed at this end by inserting a comb, made from material of the same thickness as the spacers, into the gel immediately after pouring. Both the combs and the spacers can be bought commercially, or they can be made using a modeling card (Plastikard™)

From: *Methods in Molecular Biology, Vol. 58: Basic DNA and RNA Protocols*
Edited by: A. Harwood Humana Press Inc., Totowa, NJ

that is cut according to requirements. A typical comb for a 20-cm wide gel would have 50 teeth (2 mm wide; 3 mm deep) to enable 12 clones to be loaded/gel. This chapter describes how to pour both linear and buffer gradient gels.

Linear gels are the simplest to pour, but usually only 200–250 bases of sequence can be read/clone, since the resolution of the gel varies along its length. At the bottom of the gel, bands are widely spaced and well resolved, but toward the top they coalesce and cannot be read. Buffer gradient gels *(1)* increase the amount of readable sequence in comparison to a linear gel by slowing the migration rate of DNA molecules within the bottom 30% of the gel. This retains the small molecules on the gel and enables the molecules at the top of the gel to be run for longer periods of time, and hence they are better separated. The gradient is achieved by having a higher ionic concentration in the bottom of the gel. Buffer gradient gels are, therefore, poured using two gel mixes of differing ionic concentration, the mix with the higher ionic concentration being poured first, followed by the mix with the lower ionic concentration. As a consequence, pouring buffer gradient gels takes a little practice, since care must be taken when pouring the gel to ensure that the gradient forms properly in order to avoid distortion of the resultant sequence.

2. Materials

1. 10X TBE: 108 g of Trizma base (Tris), 55 g of boric acid, and 9.3 g of EDTA (di-sodium salt). Make up to 1 L with deionized water. Discard when a precipitate forms.
2. 40% Acrylamide (19:1): 380 g of acrylamide (Electran-BDH), 20 g of *N,N'*-methylene bis-acrylamide (Electran-BDH). Make up to 1 L with deionized water. Add 20 g of Amberlite MB1 resin, and stir gently for 10 min. Filter and store at +4°C (*see* Note 1).
3. 0.5X TBE gel mix: 75 mL of 40% acrylamide (19:1), 25 mL of 10X TBE, and 230 g of urea (ultrapure grade). Make up to 500 mL with deionized water. Filter and store at +4°C for up to 1 mo.
4. 5X TBE 6% gel mix: 30 mL of 40% acrylamide (19:1), 100 mL of 10X TBE, and 92 g of urea (ultrapure grade). Make up to 200 mL with deionized water. Filter and store at +4°C for up to 1 mo.
5. 25% Ammonium persulfate: 25% (w/v) in deionized water. Can be stored at +4°C for several months.
6. TEMED: *N,N,N',N'*-tetramethyl-1,2-diaminoethane. Store at +4°C.
7. Dimethyl dichlorosilane solution: for silanizing glass plates.
8. Glass plates: 50 × 20 cm.

9. Spacers: 50 × 0.5 cm × 0.35 mm thick Plastikard.
10. Combs: 5 × 15 cm × 0.35 mm Plastikard.
11. Waterproof tape and "Bulldog" or "Foldback" clips.

3. Methods
3.1. Preparation of Sequencing Plates

1. Wash one pair of plates in warm water, and then rinse them with distilled water. If the plates are cleaned after each use, there will be no need to use detergents on them. Allow the plates to air-dry.
2. Working in a fume hood, silanize one surface of each plate (*see* Note 2). To do this, spread approx 2 mL of silanizing solution over the selected surface using paper tissue, taking care to ensure that the entire surface of the glass plate is covered. The plates are then left to dry in the fume hood for 5–10 min.
3. Wipe each plate with a small quantity of ethanol. Ensure that the plates are well polished and free of dust, which may encourage the formation of air bubbles when the gel is poured (*see* Note 3).
4. Lay one spacer along each of the long edges of one plate (usually the unnotched one). Lay the second plate directly onto the first plate so that the two silanized sides form the inner surfaces of the mold assembly.
5. Clamp the plates together on one side using Foldback clips, and seal along the other side and the bottom of the plates with gel-sealing tape. Transfer the clips to the sealed side, and then seal the remaining side. If desired, a second layer of tape could be used along the bottom and the corners of the plates to reduce the risk of leaks further.

3.2. Pouring a Linear Sequencing Gel

1. Allow 50 mL of 0.5X TBE 6% gel mix to warm to room temperature.
2. Add to this gel mix 100 µL of 25% ammonium persulfate and 100 µL of TEMED. Mix by swirling.
3. Take the gel mix up into a 50-mL syringe (without a needle), and inject the mix between the plates, maintaining a steady flow. The rate of flow of gel mix into the mold can be controlled by altering the angle at which the plates are held.
4. Once the gel mold has been filled to the top, insert the comb just past the teeth. Clamp the edges of the plates at the top and bottom to ensure that a uniform thickness of gel is maintained. Leave the gel to polymerize for 30 min (*see* Note 4).

3.3. Pouring a Buffer Gradient Gel

1. Warm the gel mix to room temperature. For a buffer gradient gel, 50 mL of 0.5X TBE gel mix and 10 mL 5X TBE gel mix are required.

2. Add 100 µL each of TEMED and 25% ammonium persulfate to the 0.5X TBE gel mix, and 20 µL of each to the 5X TBE gel mix.
3. Using a 25-mL pipet, take up 7 mL of 0.5X TBE gel mix, followed by 7 mL of 5X TBE gel mix (*see* Note 5). The two gel mixes should be mixed together at the interface by introducing a few air bubbles into the pipet and letting them rise through the gel solutions.
4. Pour this mix into the mold either down one side or down the center.
5. As soon as the pipet is empty, lower the mold to the horizontal to stop the flow of gel mix. Take the remaining 0.5X TBE gel mix into a 50-mL syringe, and fill the gel mold to the top, washing the gradient mix to the bottom. Insert the comb, clamp the plates, and leave the gel to polymerize for 30 min.

4. Notes

1. Acrylamide is a potent neurotoxin, the effects of which are cumulative. Gloves and a mask should always be worn when weighing out solids and when handling solutions containing acrylamide. Gels should be poured in plastic trays to contain spillages and leaks. Polymerized acrylamide is considered to be nontoxic, although gloves should still be worn, since there may be residual unpolymerized acrylamide present.
2. Many workers silanize only one of the glass plates, the theory being that the gel will stick to the nonsilanized plate. In practice, the gel seems to stick to either plate with equal frequency, and silanizing both plates seems to aid the pouring of the gels.
3. Bubbles, should they occur, can be removed in one of two ways. Raise the gel assembly to a vertical position and lightly tap the top of the glass to dislodge the bubble, which should then rise to the top of the gel. Alternatively, an old piece of used X-ray film can be used to push the bubble to the side, where it will not affect the running of the samples.
4. The rate at which the gel polymerizes can be altered by changing the amount of TEMED added to the gel mix. It is often helpful, when first pouring buffer gradient gels, if the amount of TEMED is reduced by 10–20 µL. This will give the inexperienced worker a little more time to pour the gel, and remove any bubbles that may have occurred during pouring.
5. When using buffer gradient gels, it is possible to change the gradient simply by altering the relative ionic concentration of the two mixes. For example, if more 5X TBE mix is added, the gradient becomes steeper.

Reference

1. Biggin, M. D., Gibson, T. J., and Hong, G. F. (1983) Buffer gradient gels and ^{35}S label as an aid to rapid DNA sequence determination. *Proc. Natl. Acad. Sci. USA* **80,** 3963–3965.

CHAPTER 48

Electrophoresis of Sequence Reaction Samples

Alan T. Bankier

1. Introduction

The underlying principle of DNA sequencing by either the Sanger *(1)* or Maxam and Gilbert method *(2)*, is the ability to fractionate and resolve long, single-stranded DNA molecules that differ in length by only one nucleotide. Denaturing polyacrylamide gels have been reported to give interpretable separation of molecules up to 0.6 kb in length, on 1-m long gels *(3)*.

More practical systems for extended readings employ shorter buffer gradient gels (*see* Chapter 47), multiple loadings (giving run lengths of around 3 and 6 h), and reduced acrylamide concentration down to as low as 3.5%. Even using these techniques, a more reasonable estimate of the limit to accurate sequence that can be obtained on a routine basis is 450–500 nucleotides/template. The redundant nature of shotgun procedures, where gel running tends to be the rate-limiting step, make it more efficient to accept a reading of up to 350 nucleotides from each single run on a buffer gradient gel or a wedge gel *(4–6)*. Following electrophoresis, the gel is subjected to autoradiography, where the position of each band can be visualized by virtue of the radioactive nucleotide incorporated during polymerization. Samples visualized by chemiluminescence currently use similar exposure to film for detection *(7)*. Fluorescence-based automated procedures use direct in-gel detection without the need for gel handling and autoradiography (*see* Chapter 51).

From: *Methods in Molecular Biology, Vol. 58: Basic DNA and RNA Protocols*
Edited by: A. Harwood Humana Press Inc., Totowa, NJ

Direct autoradiography of wet gels, covered with Saran Wrap®, is possible with ^{32}P-labeled sequence reactions, but a considerable improvement in resolution is observed when the gel is dried. Particles from radioactive emissions in the lower part of the gel radiate outward in all directions and the resulting exposure on the film, positioned on the upper surface of the gel, appears broad and diffuse. If the gel is dried before autoradiography, this radiant "spread" is dramatically reduced. When ^{35}S label is used, drying is essential because of the lower energy of emission. Drying is most conveniently carried out on a heated, vacuum gel drier, and takes <1 h.

Having negotiated all of the pitfalls of subcloning, risked one's health handling hazardous materials, fiddled with almost nonexistent volumes of reagents, and successfully manipulated a very fragile gel that often seems bent on self-destruction to produce an autoradiograph, perhaps the hardest part remains. Reading sequencing films accurately is a skilled process. Although some broad guidelines can be learned from a chapter like this, there is no substitute for practical experience gained from comparing deduced sequences with known sequences and reconciling any differences. Several film scanners are available, with software intended to interpret sequencing data, notably the Amersham Autoreader (Amersham, Arlington Heights, IL). The accuracy of these devices is continually improving, but they are totally dependent on being "fed" good quality input. At the end of it all, it can be very difficult to accept that the film may be truly unreadable.

2. Materials

1. A gel electrophoresis apparatus. This need not be too complicated, since satisfactory results can be obtained using very cheap and simple systems.
2. A high voltage power supply capable of at least 2000 V (DC output). For a 20 × 50 cm × 0.3-mm thick gel, the running conditions are around 1500 V at 30 mA.
3. 10X TBE: 108 g Tris base, 55 g boric acid, 9.3 g Na$_2$EDTA dissolved and made up to 1 L in deionized water.
4. Formamide dye mix: 100 mL deionized formamide, 0.1 g xylene cyanol FF, 0.1 g bromophenol blue, 2 mL 0.5M Na$_2$EDTA. There seems to be little or no deterioration of results using dye mix that has been stored at ambient temperatures for several months, but it is recommended that aliquots be stored frozen. Formamide is hazardous and may cause irritation to skin and eyes.

Polyacrylamide Gel Electrophoresis

5. Narrow or flattened point, disposable pipet tips for gel loading. A small volume Hamilton syringe with a fine gage needle can also be used. Experienced users frequently prefer to use drawn-out capillaries fitted to a mouth pipet, but this practice should be avoided on safety grounds.
6. Whatman (Maidstone, UK) 3MM paper, cut just larger than the size of the gel.
7. Saran Wrap or equivalent, nonporous food wrap.
8. Slab gel drier, Savant (Hicksville, NY) SGD4050 or BioRad (Richmond, CA) SE1125B.
9. A high-vacuum pump. Oil pumps require considerable maintenance and are quickly damaged by acetic acid if the gels are fixed before drying. Teflon diaphragm pumps are cheaper and are low maintenance (e.g., Vacuubrand GMBH).
10. Fuji RX or equivalent, X-ray film. Some films are considerably faster (and more expensive) but may not be suitable for automatic processors.
11. Light-tight film cassettes for autoradiography.
12. Chemicals for film processing.

3. Methods

The process described in this chapter can be broadly categorized under three headings: gel running, autoradiography, and film reading.

3.1. Gel Running

1. Remove any sealing tape from the bottom of the polymerized gel, remove the well former from the top of the gel, and wash the well(s) under running deionized water (*see* Note 1).
2. If using a "shark's tooth" comb, carefully insert it until the teeth just enter the gel surface evenly across the full width of the gel.
3. Assemble the gel in the electrophoresis apparatus with the well recess of the glass plate facing the upper buffer chamber and fill the upper and lower chambers with 1X TBE.
4. Thoroughly flush the top of the well with TBE, using a Pasteur pipet or syringe, to remove any unpolymerized acrylamide. It is not absolutely necessary to prerun the gel, but if desired, put the cover on the apparatus, connect it to a high voltage power supply, and run it at around 30 V/cm for 30 min (*see* Note 2).
5. If the samples have been stored at –20°C, allow them to thaw and add 2 µL of formamide dye mix to each. Immediately before loading, denature the samples and place them on ice. Samples prepared in tubes should be denatured for 3 min at 70°C.
6. Isolate the gel running apparatus from the power supply and flush the wells at the top of the gel to remove the urea that will have diffused out of the gel.

7. Using a narrow polypropylene tip fitted to a small volume pipet, load around 2 µL of sample into each well. The same tip can be used repeatedly by rinsing the tip once in the bottom buffer chamber. If difficulty is encountered with air pockets trapping some residual sample in the tip, it is easier to discard it into the radioactive waste and to use a clean tip (*see* Notes 3 and 4).
8. When all the samples are loaded, put the cover on the apparatus and connect it to a high voltage power supply. Electrophorese the gel at around 30 V/cm, until the bromophenol blue reaches or has migrated off the bottom of the gel (*see* Notes 5 and 6).

3.2. Gel Drying and Autoradiography

1. When the run is complete, disconnect the electrophoresis unit from the power supply, discard the top buffer, and dispose of the radioactive bottom buffer in the recommended manner. Remove the gel/glass assembly from the unit and remove any remaining sealing tape.
2. With the notched plate uppermost, pry the two glass plates apart by inserting a thin spatula between them and carefully twisting or levering it. Should the gel prefer to stick to the notched plate, simply remove the spatula, turn over the whole gel/glass assembly, and pry off the backplate. If the gel persists in sticking equally to both plates, the whole assembly is best submerged in water or 10% acetic acid where the gel is free to float off the upper glass plate when pried open. The gel can be held in position during subsequent handling, using a piece of firm plastic netting (*see* Note 7).
3. Submerge the gel in around 1 L of 10% acetic acid in a seed tray for 10–15 min to dialyze out the urea. If the tray does not have a ridged base, gently agitate the solution periodically. Remove the gel, still on the glass plate, and drain off as much liquid as possible (*see* Note 8).
4. Transfer the gel to 3MM paper by placing a sheet smoothly and evenly on top of the gel by "rolling" the curved paper across from one end, or outward from the center. Gently ensure complete direct paper/gel contact over the whole gel and then peel the paper back and away from the glass.
5. Cover the gel with a layer of Saran Wrap and cut off any excess 3MM paper or plastic wrap. Trim the top and sides of the gel to fit into a slab gel drier and dry the gel for 15 min under vacuum at 80°C.
6. Remove the plastic wrap from the dry gel and place it in a light-tight X-ray film cassette in direct contact with X-ray film. Exposure times will vary depending on which label has been used. If the entire sample has been loaded and the label was ^{35}S, exposure times between 18–36 h should be adequate. When using ^{32}P, the ideal exposure can be as short as 1 h (*see* Note 9).
7. Process the X-ray film according to the supplier's recommendations.

Polyacrylamide Gel Electrophoresis

3.3. Film Reading

In principle, the interpretation of a single sequence from four tracks on an autoradiograph should be simple. Because the individual molecules are separated according to size, all that is required is that the positions of successive bands from bottom to top are noted. If the next band is in the A track, then the next base in the sequence must be an A.

Several different problems may disrupt this idealized situation. It may not be clear which band comes next, particularly if the bands in question are in the outer tracks. Two or more bands may appear to occupy the same vertical position, or spurious bands may even prevent an absolute assignment at a particular position. Some positions expected within an even "ladder" of bands may appear to be vacant. Superimposed on this will be the effects of poor template quality, less than perfect sequence reactions, and gel running and autoradiography problems.

Probably the most important lesson to learn when reading sequence films is when *not* to read. Most mistakes are made by trying to read sequences of inferior quality or by reading too far up the gel into regions of poor resolution. Eventually sequence-reading mistakes will have to be reconciled with other, more accurate readings, which can be very time consuming. Very often, some weeks or months later, when the film is read again, the first impression is one of disbelief that it could have been read at all.

Guidelines on reading sequence films range from the general to a description of the very specific artifacts expected from particular cloning or sequencing strategies. Only the more relevant points will be listed here and then in a general way. Many of the artifacts can be attributed to specific reagents and can be verified using a control system.

3.3.1. General Tips on Reading Sequences

1. Since many of the anomalous mobility problems associated with gels are predominantly related to high G/C sequences, it makes sense to have these two tracks adjacent to each other. The most commonly used track orders are ACGT and TCGA.
2. Before starting, make a general assessment of the quality of the data and decide whether or not it is worth reading. This overall inspection should consider the sharpness of the bands, the presence of background smear, and whether extra bands are present.

3. Ensure an optimum exposure. Too short an exposure can conceal the presence of weak bands. Too long an exposure will increase the background smear and accentuate artifact bands. If ^{32}P is used, over-exposure will dramatically reduce resolution.
4. Mark out or mask the four lanes to avoid confusion with other tracks, and mark the end-point. This may be the point where the quality or resolution falls below the required standard, or where a short insert meets the vector.
5. Reading the spaces between bands can be as important as reading the bands themselves. It should always be considered whether or not there is "space" for the band in question. Where there is ambiguity regarding the exact order of bands, it can often be deduced from the comparative spacing of consecutive bands within a track in the immediate vicinity. One should have some idea of how many bands to expect in a particular gap.
6. If consistent difficulty is encountered in interpreting the correct register of bands, a sharkstooth comb may prove to be beneficial. When these are used, there is often a small overlap between adjacent tracks, making order interpretation simpler.
7. Do not stop reading in "mid-sequence." Once started, read the entire sequence from bottom to top. It is quite common, after a break, to recommence reading from the wrong point, particularly when the sequence is of a repetitive nature.
8. Speckles and streaks on autoradiographs usually arise from particles or small bubbles in the gel and can generally be avoided. Ensure that the glass plates are scrupulously clean before pouring the gel, and avoid using tissues that drop particles to wipe down their surfaces.
9. Patches of fuzziness can be attributed to regions of the gel not being in direct, intimate contact with the film, or with differential polymerization of the gel matrix. These patches can also be caused by regions of overheating in the gel by running it too fast.

3.3.2. Common Problems and Artifacts in Sequences

1. A high background smear is generally specific to the template DNA. Further purification of the template, using phenol or chloroform extractions and ethanol precipitation, is not always successful.
2. Variations in band intensity usually are an intrinsic property of the enzymes. These are generally base sequence-specific and occur in a predictable manner. Sequenase used in the presence of manganese produces a very even band intensity.
3. "Pile-ups," where bands appear in the same position in all four tracks, result from nonspecific terminations. Typical causes are secondary structure within the template that the polymerase has difficulty passing through (different polymerases and different batches of the same polymerase can cope

Polyacrylamide Gel Electrophoresis 399

with this to varying degrees), and renaturation of a double-stranded template causing a block to polymerization. This is quite common in sequencing PCR products. Both of these problems can often be reduced by destabilizing the structure, for example, by increasing the temperature of the extension reaction or reducing the salt concentration in the reaction. Failing this the polymerase has to be changed, since the most likely cause will then be poor quality enzyme.

4. Faint extra bands, or "shadow" bands, in other tracks can have one of several causes. If the shadow bands appear only in adjacent tracks, a sample may have spilled over between wells. A random distribution of faint bands can indicate a second background sequence. This may be caused by a contaminating plaque inadvertently picked off the agar plate, a lower abundance template resulting from a deletion during phage growth, or nonspecific priming. Finally, some faint artifact bands are polymerase-specific and appear at base sequence-dependent positions.
5. A generally clean, but very weak, sequence is caused by low template or primer concentration, or by inefficient priming resulting from too high an annealing temperature.
6. Compressions, where band spacings are reduced or nonexistent at a specific point in a sequence, followed by a region of abnormally large spacing, result from secondary structure during gel electrophoresis.
7. An uneven distribution of band intensity is indicative of an incorrect dideoxy:deoxy nucleotide ratio. A peak of intensity at a shorter length implies too high a dideoxy concentration, and a high intensity at longer lengths too low a dideoxy nucleotide concentration. When using a two-step labeling procedure, as recommended for Sequenase, a similar intensity problem is caused by variations in template concentration, such as low template concentrations producing an increase in intensity of the longer molecules.
8. No signal at all in a single track is usually caused by accidentally missing the addition of a single ingredient. No signal in a set of four tracks, on the other hand, would indicate that the primer or buffer was omitted from the annealing, no template was recovered from the DNA preparation, the primer is inappropriate, or a deletion has occurred during growth of the recombinant that encompassed the priming site.

4. Notes

1. It is important to rinse the wells immediately after removing the slot former, because any unpolymerized acrylamide may subsequently polymerize and leave an uneven gel surface. Flushing the wells just before loading removes any urea that may have dialyzed out of the gel, making it easier to load the samples as a tight band.

2. The presence of "smiling," where samples toward the edge of the gel migrate slower than those in the center, is a consequence of variations of heat loss (and hence temperature) across the width of the gel. This can be minimized in two simple ways: first, by not loading samples in the outermost 2 cm on each side of the gel, because this is where the temperature difference is greatest, and second, by placing an aluminum sheet on the outer surface of the glass plate, because this evens out the temperature across the gel. More sophisticated and expensive methods of temperature control are available, but except for particular situations, such as in automated systems, it is unlikely they will be needed.
3. If possible, it is better to load the sample around 1–2 mm from the bottom of the well, giving a sharp sample band, rather than to let it trickle down from the top.
4. An extended denaturation period (20 min at 80°C) can be used with microtiter trays to reduce the volume so that the entire sample can be loaded. The duration can be altered to adjust this volume. For example, fan-assisted ovens will require a much shorter time. Only a fraction of samples denatured in tubes can be loaded since their volume is not reduced considerably. Do not attempt to overload the wells. As a guide, the sample height should be less than the well width, but ideally much less.
5. The conditions described in this method refer to running the gel at a constant voltage. In practice, though, it is best to run denaturing gels at constant power because the gel temperature is the most important factor to control. An internal gel temperature of 60–70°C should be attained. For 20 × 50 cm × 0.3 mm thick gels a constant power of 35–40 watts is sufficient. In cases where compressions still exist this may need to be even higher to increase the gel temperature.
6. The precise time at which to stop electrophoresis will, of course, depend on the length of the primer and the distance from the primer that sequence is to be read, as well as the concentration of acrylamide in the gel. On 6% gels the M13 cloning sites correspond approximately to the position of the bromophenol blue marker dye when the universal primer is used.
7. When opening the glass plates, do not lever against the "ears" of the notched plate. These tend to be fairly fragile and easily broken.
8. There is no need to fix gels before they are dried. Without fixing they will take 45–60 min to dry on a vacuum gel drier at 80°C. If no drying facilities are available, it is possible to air dry them overnight.
9. The introduction of thin gels and ^{35}S label was designed to improve resolution. Using single-sided film can improve the resolution of autoradiographs, although care must be taken to ensure that the emulsion side faces the dried gel and the Saran Wrap is removed.

References

1. Sanger, F., Nicklen, S., and Coulson, A. R. (1977) DNA sequencing with chain-terminating inhibitors. *Proc. Natl. Acad. Sci. USA* **74,** 5463–5467.
2. Maxam, A. M. and Gilbert, W. (1977) A new method for sequencing DNA. *Proc. Natl. Acad. Sci. USA* **74,** 560–564.
3. Ansorge, W. and Barker, R. (1984) System for DNA sequencing with resolution of up to 600 base pairs. *J. Biochem. Biophys. Meth.* **9,** 33–47.
4. Bankier, A. T. and Barrell, B. G. (1983) Shotgun DNA sequencing, in *Techniques in Life Sciences, Nucleic Acid Biochemistry* vol. B5 (Flavell, R. A., ed.), Elsevier, Ireland, pp. 1–34.
5. Biggin, M. D., Gibson, T. J., and Hong, G. F. (1983) Buffer gradient gels and ^{35}S label as an aid to rapid DNA sequence determination. *Proc. Natl. Acad. Sci. USA* **80,** 3963–3965.
6. Bankier, A. T. and Barrell, B. G. (1989) Sequencing single-stranded DNA using the chain termination method, in *Nucleic Acids Sequencing: A Practical Approach* (Howe, C. J. and Ward, E. S., eds.), IRL, Oxford, pp. 37,38.
7. Beck, S. (1993) DNA sequencing by chemiluminescent detection, in *Methods in Molecular Biology, vol. 23: DNA Sequencing Protocols* (Griffin, H. G. and Griffin, A. M., eds.), Humana, Totowa, N. J., pp. 235–242.

CHAPTER 49

Direct Sequencing of PCR Products

Janet C. Harwood and Geraldine A. Phear

1. Introduction

The development of the polymerase chain reaction (PCR) has allowed the rapid isolation of DNA sequences utilizing the hybridization of two oligonucleotide primers and subsequent amplification of the intervening sequences by *Taq* polymerase. There are many applications of this technique. One of the most useful is the screening of large numbers of samples in the search for mutations at a defined locus, for example in clinical studies or in the analysis of cultured cell lines *(1–3)*. In the absence of PCR, this can only be achieved by isolating DNA from each individual, and making and screening a library. The use of PCR means that whereas previously it may have taken a month to examine each individual sample, it is now possible to examine many samples in a few days. In addition, by using redundant primers, it is even possible to isolate related novel genes.

To make full use of this technique, it is necessary to be able to sequence the amplified fragments rapidly. There is, however, a major technical problem encountered when sequencing PCR products. The oligonucleotide primers used for the PCR reaction also act as primers in the sequencing reaction. This makes the sequencing gels unreadable owing to the overlaying of multiple sequences. A number of methods have been used to address this problem. Initially, PCR products were cloned into a vector suitable for sequencing *(4)*. This, however, has two disadvantages: (1) It is time-consuming, and (2) each clone represents only a single molecule from the population of PCR products. Therefore, several clones must be sequenced to avoid artifacts resulting from the infidelity of *Taq* polymerase. By sequencing the PCR product directly, these errors are minimized, since the template comprises a sample of the whole population of molecules generated by the PCR reaction.

Two general approaches to direct sequencing have been taken. First, the double-stranded template may be purified from the oligonucleotide primers. A number of methods have been devised to do this. They include gel purification and isolation by the use of size exclusion of DNA molecules using such materials as GeneClean™ or Centricon™ columns. Second, the double-stranded PCR product can be used as a template to make single-stranded DNA. This can be achieved by asymmetric PCR *(5,6)*. Alternatively, the PCR reaction may be modified so that one of the primers is biotinylated and then the single-stranded template can be purified from it by binding the biotinylated strand to streptavidin coated magnetic beads *(7)*. Although many of these methods have been used successfully, they all involve additional manipulations to the PCR template prior to sequencing.

The protocol presented here is a simple and rapid way of sequencing PCR products directly. It is based on two previously described methods *(4,8)*. It is ideal for sequencing PCR products that are <2 kb in length, where an internal sequencing primer is available. The double-stranded template is denatured in the presence of an end-labeled primer by boiling, and then snap-chilling in ice/water. This prevents reassociation of the template, favoring primer–template annealing. Subsequent sequencing is a modification of the dideoxy-nucleotide sequencing technique *(9)*.

The end-labeled oligonucleotide used as a sequencing primer is situated internally to the PCR primers, i.e., is a "nested" primer. This has several advantages.

1. There is no need to remove the PCR primers from the template. The products of the sequencing reaction primed by the excess of PCR primers will not be detected, since the PCR primers are not radioactively labeled. This minimal purification of the template material reduces loss of material and is sufficient for sequencing purposes.
2. The sequencing primer is a third oligonucleotide that is "nested" to the PCR primer and specific to the desired PCR product. It will not anneal to nonspecific PCR products, and therefore, sequence specificity is maximized.
3. The sensitivity of the technique is such that only 150 ng of template are required/sequencing reaction.

Direct Sequencing of PCR Products

The one drawback of this protocol is the requirement of a "nested" sequencing primer. The method, however, can be adapted to sequence PCR products using one of the PCR primers, provided the oligonucleotides used in the PCR reaction are removed from the template first.

This protocol is optimized to sequence a PCR product of approx 250 bp using a 20-mer as a primer. It is applicable for different lengths of PCR products and primers. However, the quantity of material used must be adjusted to maintain the molar ratio of 1 pmol of template to 3 pmol of primer (*see* Note 1).

2. Materials

All solutions should be made to the standard required for molecular biology. Use molecular-biology-grade reagents and sterile distilled water. The reagents for sequencing are available commercially as kits.

2.1. End-Labeling Oligonucleotide Primers

1. γ-[^{32}P] ATP: 5000 Ci/mmol, Amersham (Arlington Heights, IL). This is a high-energy β-emitter. Therefore, adequate safety precautions must be taken (*see* Note 2).
2. 10X polynucleotide kinase buffer: 0.5M Tris-HCl, pH 7.6, 0.1M MgCl$_2$, 50 mM DTT, 1 mM spermidine, 1 mM EDTA, pH 8.0. Store in aliquots at –20°C.
3. Nested oligonucleotide primer specific to PCR sequence: 0.1 µg/µL in water (*see* Note 3).
4. Polynucleotide kinase (10,000 U/mL, New England Biolabs, Beverly, MA).
5. TE: 10 mM Tris-HCl, 1 mM EDTA, pH 8.0.
6. Sephadex G50 (Pharmacia, Piscataway, NJ) column in TE (*see* Note 4).

2.2. Purification and Sequencing of the PCR Product

1. 3M Sodium acetate, pH 5.2.
2. 100% Ethanol.
3. 5X Sequenase buffer: 200 mM Tris-HCl, pH 7.5, 100 mM MgCl$_2$, 250 mM NaCl.
4. 0.1M DTT. Store at –20°C.
5. Labeling mix: A 7.5-µM solution of each deoxynucleotide triphosphate (dNTP). To make up the stock labeling mix: 7.5 µL 1 mM dATP, 7.5 µL 1 mM dCTP, 7.5 µL 1 mM dGTP, 7.5 µL 1 mM dTTP, and 970 µL water. Store at –20°C. Dilute eightfold in water for the working labeling mix.
6. Sequenase (Version 2.0 USB, 14 U/µL): dilute eightfold to 1.75 U/µL in cold TE just before use.

7. Termination mixes: A termination mix is required for each nucleotide (i.e., A, C, G, and T). These comprise solutions of 80 μM of each dNTP and 8 μM of the respective dideoxynucleotide (ddNTP) in 50 mM NaCl.
   The termination mixes can be made as follows (all amounts in μL):

	A mix	C mix	G mix	T mix
10 mM dATP	8	8	8	8
10 mM dCTP	8	8	8	8
10 mM dGTP	8	8	8	8
10 mM dTTP	8	8	8	8
1 mM ddATP	8	—	—	—
1 mM ddCTP	—	8	—	—
1 mM ddGTP	—	—	8	—
1 mM ddTTP	—	—	—	8
1M NaCl	50	50	50	50
water	910	910	910	910

8. Formamide dye mix: 95% formamide (v/v), 20 mM EDTA, 0.05% bromophenol blue (w/v), 0.05% xylene cyanol (w/v).
9. Sequencing gel: A 0.4-mm thick, 6% denaturing polyacrylamide gel.
10. Fix solution: 10% acetic acid/10% methanol (v/v). The fix can be stored at room temperature and reused 4–5 times.
11. Autoradiographic film: e.g., Kodak XAR5.

3. Methods

3.1. End-Labeling the Sequencing Primer

We describe labeling of 200 ng of oligonucleotide. This is sufficient for ten sequencing reactions. The labeling reaction can be adapted for 100 ng (*see* Note 5).

1. Dry down 100 μCi γ-[^{32}P] ATP in a 1.5-mL Eppendorf tube.
2. Place on ice, and add sequentially 5 μL of water, 1 μL of 10X polynucleotide kinase buffer, 200 ng (2 μL) of oligonucleotide, and 2 μL of polynucleotide kinase (20 U).
3. Incubate at 37°C for 1 h. Add 90 μL of TE to the reaction. Spin the labeled oligonucleotide through a Sephadex G50 spun column, and collect it in a 1.5-mL Eppendorf tube.
4. Dry the sample down and resuspend in water to 10 ng/μL. Freeze the sample at –20°C until required.

The labeled oligonucleotide is stable for several weeks at –20°C, but its use is limited by its radioactive half-life.

3.2. Preparation and Sequencing of the PCR Product

The yield of the PCR product should be estimated on an agarose gel before sequencing the template. At least 150 ng of PCR product is required for each sequencing reaction.

1. Spin the whole PCR reaction through a Sephadex G50 spun column in TE. Take care not to load any paraffin oil. If necessary, this can be removed by chloroform extraction. Collect the eluate in a 1.5-mL Eppendorf tube. This procedure removes the excess of dNTPs from the PCR reaction.
2. Measure the volume of the reaction, and add 0.1 vol of $3M$ sodium acetate, pH 5.2, and 2.5 vol of ethanol to precipitate the PCR product. Cool on dry ice for 15 min or leave at $-20°C$ overnight.
3. Spin in a microfuge at $12,000g$ for 15 min at 4°C. Remove the supernatant, and briefly dry the pellet. Resuspend the pellet in sterile distilled water to give a final concentration of 150 ng/µL.
4. Mix 1 µL (150 ng) of prepared PCR product, 2 µL (20 ng) of labeled primer, 2 µL of 5X sequenase buffer, and 5 µL of water in a 0.5-mL Eppendorf tube.
5. Anneal the template to the oligonucleotide by boiling the tube for 5 min and then immediately plunging it into ice/water (*see* Notes 6 and 7). Leave in ice/water for 5 min.
6. Spin for a few seconds in a microfuge to collect any liquid on the tube walls. Add 1 µL of $0.1M$ DTT, 2 µL of working labeling mix, and 2 µL of diluted sequenase. Incubate at room temperature for 5 min (*see* Note 8).
7. While the labeling reaction is proceeding, label four 1.5-mL Eppendorf tubes, one for each termination mix, i.e., A, C, G, and T. Add 2.5 µL of the relevant termination mix to each tube.
8. At the end of the first 5-min incubation, add 3.5 µL of the labeling reaction to each termination mix. Incubate at 37°C for 5 min.
9. Add 4 µL of formamide dye mix. Store at $-20°C$ until required. These reactions are fairly stable for a week at $-20°C$.
10. Heat the sequencing reactions to 95°C for 3 min before loading onto a sequencing gel. Run the gel until the bromophenol blue has just run off the bottom. This will allow the excess of labeled primer to run off the gel, but it will be possible to read the sequence within 20 bp of the primer.
11. Transfer the gel to a piece of 3MM paper (Whatman, Clifton, NJ). Alternatively it may be fixed on the glass sequencing gel plate. Place the gel in fix solution for at least 10 min. Remove the gel from the fix solution. Cover the surface of the gel in cling film and dry.

12. Autoradiograph at −70°C overnight. As a rough guide, if the reaction has worked well, it should be possible to obtain a reading of 50 cps on a dry gel using a series-900 mini-monitor (Mini Instruments, Ltd.) at the position where the xylene cyanol runs.

A typical result is shown in Fig. 1A. A number of common problems that may arise are discussed in Notes 9–14 (*see* Fig. 2).

4. Notes

1. The limit in template size for sequencing using this method is about 2 kb. It is important to alter the amount of template when using larger templates to keep the molar ratio of template to primer constant. For example, 1 pmol of a 250-bp template corresponds to 150 ng but for a 1.6-kb template this would be 1 µg.
2. It is possible to end-label the sequencing primer using [^{35}S]-γ-ATP or [^{33}P]-γ-ATP. This gives improved resolution toward the top of the gel and reduces the radiation exposure to the experimenter. The end-labeling reaction with [^{35}S]-γ-ATP takes 4 h and the reduced signal intensity means that autoradiography may take several days.
3. It is important that the sequencing primer does not contain sequences that are likely to form any secondary structure, that it does not contain a long run of G-residues, and that there are no mismatches at the 3' end.
4. Sephadex G50 spun columns can be made quickly and simply. Add Sephadex G50 to TE, pH 8.0, and allow it to stand at room temperature overnight. Plug a 1-mL disposable syringe with polymer wool, fill it with the Sephadex G50, and allow the TE to drain out (the syringe should be full to the top with Sephadex G50 before it is spun). Put the syringe into a centrifuge tube (a 15-mL Falcon tube 2095 will do). Spin for 3 min at 200*g* (1000 rpm in a bench centrifuge). The column should pack to 1 mL volume. Load the sample on to the top of the column and spin it again (at the same speed for 3 min) collecting the sample in an uncapped Eppendorf tube.
5. To label 100 ng (15 pmol) of oligonucleotide primer, use 50 µCi of γ-ATP. It is not necessary to dry the label down in this case. The final reaction volume remains at 10 µL, but the amount of polynucleotide kinase should be reduced to 1 µL (10 U).
6. For most PCR products 5 min of boiling are sufficient, but in some cases, for example, human DNA and DNA rich in guanine and cytosine residues, better denaturation may be achieved by boiling for as long as 10 min.
7. It is essential to plunge the annealing reaction into ice water without delay, since slow cooling of the primer–template mix will allow the template strands to reanneal.

Direct Sequencing of PCR Products

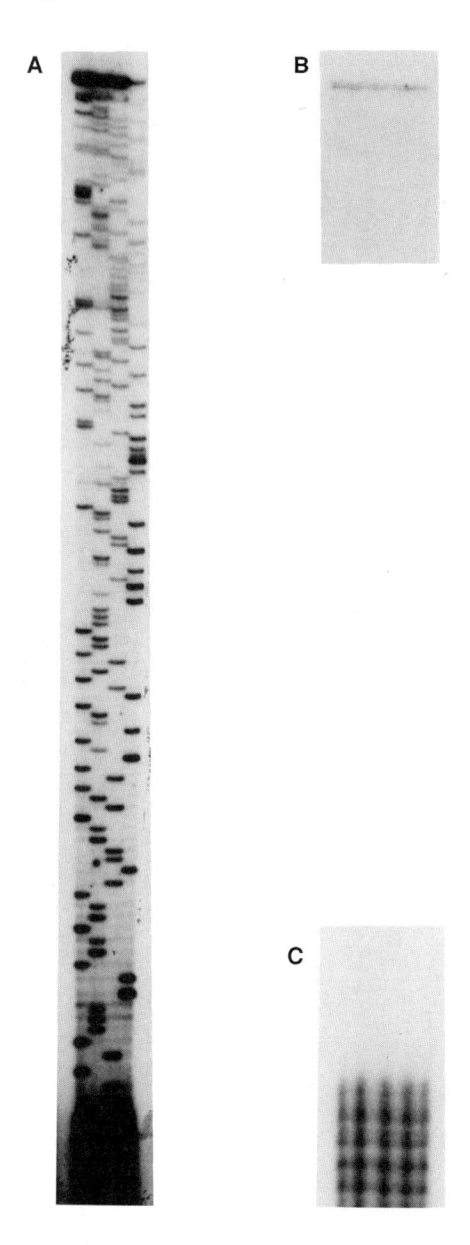

Fig. 1. **(A)** Autoradiograph of a PCR product sequenced using a ^{32}P-labeled nested primer. **(B)** The upper region of the gel showing the fully extended product produced by primer annealing and extension, but not termination. **(C)** The bottom of the sequencing gel showing the free end-labeled primer, which has not extended.

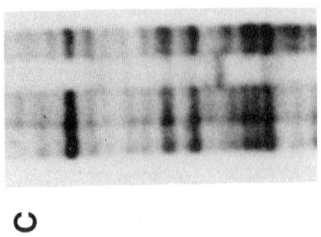

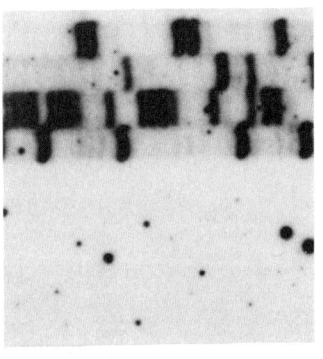

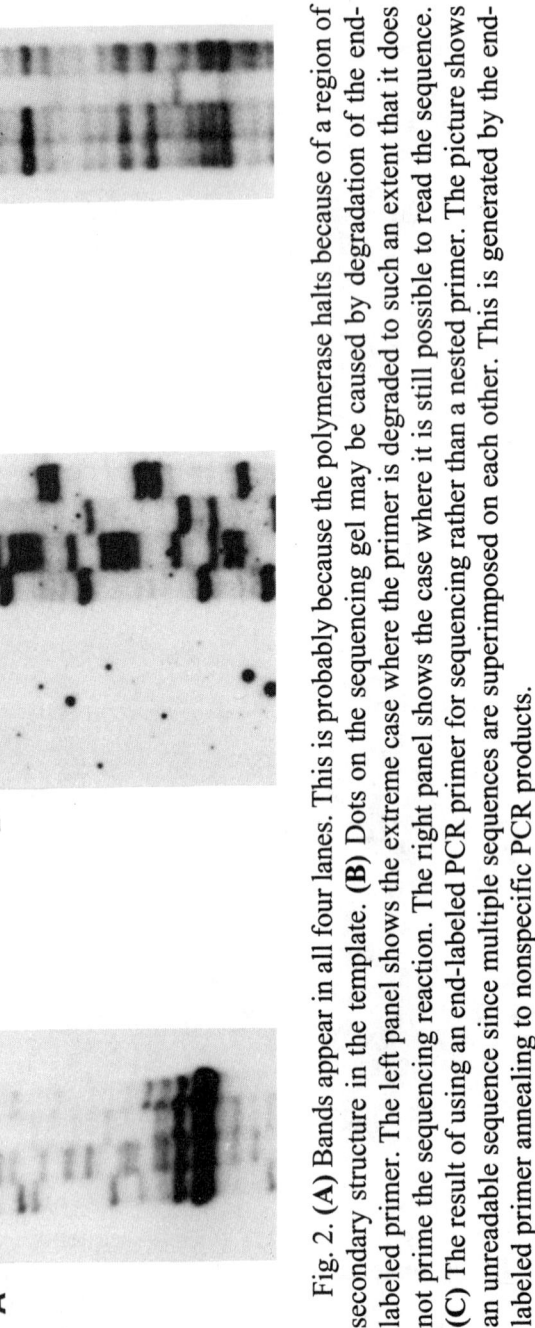

Fig. 2. (**A**) Bands appear in all four lanes. This is probably because the polymerase halts because of a region of secondary structure in the template. (**B**) Dots on the sequencing gel may be caused by degradation of the end-labeled primer. The left panel shows the extreme case where the primer is degraded to such an extent that it does not prime the sequencing reaction. The right panel shows the case where it is still possible to read the sequence. (**C**) The result of using an end-labeled PCR primer for sequencing rather than a nested primer. The picture shows an unreadable sequence since multiple sequences are superimposed on each other. This is generated by the end-labeled primer annealing to nonspecific PCR products.

8. An eightfold dilution of the labeling mix will allow the sequence to be read as close as 20 bp to the primer. Addition of manganese buffer (0.15M sodium isocitrate, 0.1M MnCl$_2$) to the annealing reaction allows the sequence to be read very close to the primer. One microliter of this buffer is added to the annealed primer–template mix at step 4 of Section 3.2. To read further from the primer, a more concentrated labeling mix may be used.
9. Loss of sequence, but a strong band at the top of the gel results from extension, but no termination of the template. This is likely to be the result of degradation of old termination mixes or their improper dilution. Therefore, fresh solutions should be made (Fig. 1B).
10. Loss of sequence, but a strong band at the bottom of the gel is the result of failure of the primer to anneal to the template (Fig. 1C). The strong band is the free end-labeled primer. This may result from mutations in the template where the primer anneals. Try another primer. An alternative explanation is that the template is not fully denatured (*see* Notes 6 and 7).
11. Faint sequences may be the result of (a) insufficient template or (b) poor labeling of the primer. In the latter case the γ-[^{32}P] ATP may be too old or the efficiency of the kinase reaction may not be high enough to achieve the required incorporation (10^8 cpm/µg).
12. Secondary structure may cause premature termination of the sequence in all lanes (Fig. 2A). This may be solved by incorporating nucleotide analogs, such as dITP or 7-deaza-2'dGTP *(10)*, into the sequencing mixes, or by sequencing using *Taq* polymerase *(11,12)*. More recently a method has been reported that couples PCR and sequencing using end-labeled primers *(13)*, which may prove to be valuable in this respect.
13. The sequence may become too faint to read after a short distance. This is caused by either the labeling mix being too dilute or the concentration of dideoxynucleotide in the termination mixes being too high.
14. Black spots on the autoradiograph (Fig. 2B) are usually the result of degradation of the primer, probably occurring during the labeling reaction. This is not a serious problem, unless the primer is degraded to such an extent that it is very short and does not prime the sequencing reaction.

References

1. Harwood, J., Tachibana, A., and Meuth, M. (1991) Multiple dispersed spontaneous mutations: a novel pathway of mutation in a malignant human cell line. *Mol. Cell Biol.* **11,** 3163–3170.
2. Phear, G. A., Armstrong, W., and Meuth, M. (1989) Molecular basis of spontaneous mutation at the *aprt* locus of hamster cells. *J. Mol. Biol.* **209,** 577–582.
3. Gibbs, R. A., Nguyen, P.-N., McBride, L. J., Koepf, S. M., and Caskey, C. T. (1989) Identification of mutations leading to the Lesch-Nyhan syndrome by automated DNA sequencing of *in vitro* amplified cDNA. *Proc. Natl. Acad. Sci. USA* **86,** 1919–1923.

4. Scharf, S. J., Horn, G. T., and Ehrlich, H. A. (1986) Direct cloning and sequence analysis of enzymatically amplified genomic sequences. *Science* **233,** 1076–1078.
5. Ward, M. A., Skandalis, A., Glickman, B. W., and Grosovsky, A. J. (1989) Rapid generation of a specific single-stranded template DNA from PCR-amplified material. *Nucleic Acids Res.* **20,** 8394.
6. Loh, E. Y., Elliott, J. F., Cwirla, S., Lanier, L. L., and Davis, M. M. (1989) Polymerase chain reaction with single-sided specificity: analysis of T cell receptor δ chain. *Science* **243,** 217.
7. Hultman, T., Stahl, S., Hornes, E., and Uhlen, M. (1989) Direct solid phase sequencing of genomic and plasmid DNA using magnetic beads as solid support. *Nucleic Acids Res.* **17 (13)** 4936,4937.
8. Wong, C., Dowling, C. E., Saiki, R. K., Higuchi, R. G., Ehrlich, H. A., and Kazazian, H. H. (1987) Characterisation of β-thalassemia mutations using direct genomic sequencing of amplified genomic DNA. *Nature* **330,** 384.
9. Sanger, F., Nicklen, S., and Coulson, A. R. (1977) DNA sequencing with chain-terminating inhibitors. *Proc. Natl. Acad. Sci. USA* **74,** 5463–5467.
10. Barnes, W. M., Bevan, M., and Son, P. H. (1983) Kilo sequencing: creation of an ordered nest of asymmetric deletions across a large target sequence carried on phage M13. *Meth. Enzymol.* **101,** 98–122.
11. Innis, M. A., Myambo, K. B., Gelfand, D. H., and Brow, M. A. D. (1988) DNA sequencing with *Thermus aquaticus* polymerase and direct sequencing of polymerase chain reaction-amplified DNA. *Proc. Natl. Acad. Sci. USA* **85,** 9436–9440.
12. Peterson, G. (1988) DNA sequencing using Taq polymerase. *Nucleic Acids Res.* **22,** 10,915.
13. Ruano, G. and Kidd, K. K. (1991) Coupled amplification and sequencing of genomic DNA. *Proc. Natl. Acad. Sci. USA* **88,** 2815–2819.

CHAPTER 50

Thermal Cycle Dideoxy DNA Sequencing

Barton E. Slatko

1. Introduction

Since DNA sequencing has rapidly become standard practice in many laboratories, a large variety of new cloning vectors, sequencing strategies, and techniques have been developed to allow more efficient sequencing of a large variety of DNA templates. Despite a current focus on the automation of DNA sequencing procedures, other methods are still required until automated DNA sequencing is in more general use. One recent addition to this repertoire of sequencing methods is termed thermal cycle sequencing.

Thermal cycle DNA sequencing protocols are based on the dideoxynucleotide chain termination method of Sanger et al. *(1)*. In the reaction, an appropriate primer DNA molecule is annealed to a complementary single-stranded stretch of DNA. This primer template complex is incubated with a highly thermostable DNA polymerase, such as the exonuclease-deficient DNA polymerase from *Thermococcus litoralis*, Vent$_R$™ (exo⁻) DNA polymerase *(2,3)*, or from *Thermus aquaticus (Taq)* polymerase *(4–7)* in the presence of deoxynucleotide triphosphates (dNTPs) and dideoxynucleotide triphosphates (ddNTPs). Four separate reaction mixes, each with all four dNTPs and one of the four ddNTPs, generate four different sets of fragments, each set corresponding to terminations at specific nucleotide residues. Repetitive cycles of denaturation, annealing, and chain extensions from small amounts of template molecules in the presence of primer excess, Vent$_R$™ (exo⁻) DNA polymerase, dNTPs,

From: *Methods in Molecular Biology, Vol. 58: Basic DNA and RNA Protocols*
Edited by: A. Harwood Humana Press Inc., Totowa, NJ

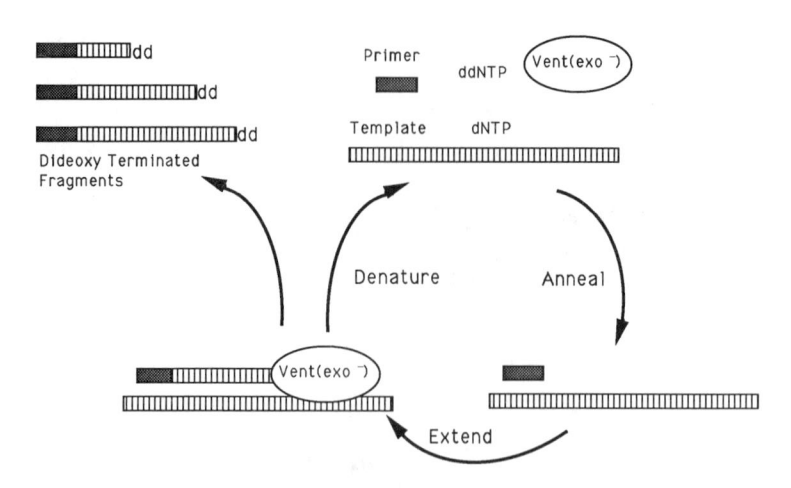

Fig. 1. Diagram of the thermal cycle sequencing reaction, demonstrating the annealing, extension, and denaturation steps that generate the dideoxy-terminated DNA fragments. Reprinted with permission from New England Biolabs, Inc.

and ddNTPs achieve a linear amplification of reaction products and a strong sequencing signal (Fig. 1). In each cycle, the reaction is raised to 95°C to denature double-stranded templates or secondary structure regions of single-stranded DNA templates, lowered to 55°C for the annealing of the primer to the template, and subsequently raised to 72°C for the elongation step of enzymatic synthesis. Subsequent cycles of denaturation, annealing, and extension occur in which the excess primer anneals to the identical denatured template molecules as in the first cycle.

Thermal cycle sequencing offers several important advantages over previously developed techniques *(4–8)*.

1. The reactions are rapid, easy to perform, efficient, and useful for both manual and automated DNA sequencing.
2. The method requires much less template than does a standard reaction, owing to the linear amplification of labeled product.
3. There is no need to denature ("collapse") double-stranded DNA templates before initiating the sequencing reactions.
4. There is no separate annealing step preceding the reactions.
5. The use of a highly thermostable DNA sequencing enzyme allows sequencing at high temperatures, an advantage for obtaining DNA sequence information from DNA templates that have high degrees of secondary structure and that may be recalcitrant with lower temperature sequencing methods.

DNA Sequencing

The method can be used to sequence single-stranded DNA templates, such as those derived from M13, fl, fd phage, or from phagemid vectors, and double-stranded templates, such as plasmid DNA, PCR products, or large linear double-stranded DNA, such as bacteriophage λ. In addition, sequencing directly from phage plaques and bacterial colonics is also feasible *(8–10)*.

In order to visualize the dideoxy-terminated chains on the gel, various means of labeling have been developed. Conventionally, ^{32}P, ^{33}P, and ^{35}S have been used for incorporation into the nascent chain by using α-labeled deoxynucleotide triphosphates in the reaction mixture. Alternatively, a second approach utilizes end-labeled primers, wherein T4 polynucleotide kinase can be used to transfer a γ-[^{32}P] (or γ-[^{33}P]) from rATP to the 5' end of a primer. Similarly, primers can be modified for chemiluminescent DNA sequencing *(11)*, or end-labeled with fluorescent dyes for automated DNA sequencing *(8,12)*. 5' end-labeling with ^{32}P or ^{33}P is used in thermal cycle sequencing when using lesser amounts of template DNA (*see* Section 3.2.). 5' end-labeled primers are also recommended for sequencing large templates (such as λgt11) or for templates that yield less than optimal results with incorporated label techniques.

The sequencing reaction products, DNA strands of varying lengths, are separated on a denaturing polyacrylamide gel. The gel is subsequently processed for exposure to X-ray film, when utilizing radioactively labeled DNA molecules or when using chemiluminescent (nonradioactive) detection. The resultant autoradiogram is analyzed for sequence information. The data is collected "in real time" when using automated DNA sequencers and no "postprocessing" of the gel is required.

This chapter provides a set of methods for sequencing nanogram amounts of single-stranded and double-stranded DNA templates by thermal cycle sequencing, using 5' end-labeled primers (^{32}P, ^{33}P, or modified for chemiluminescent detection) or by ^{35}S, ^{32}P, or ^{33}P [dATP] radiolabel incorporation.

2. Materials

1. 10X Vent$_R$™ (exo$^-$) DNA polymerase sequencing buffer: 100 mM (NH$_4$)$_2$SO$_4$, 100 mM KCl, 200 mM Tris-HCl, 50 mM MgSO$_4$, pH 8.8 (room temperature). Store at –20°C.
2. Vent$_R$™ (exo$^-$) DNA polymerase deoxy/dideoxy sequencing mixes: Make up in 1X Vent$_R$™ (exo$^-$) DNA polymerase sequencing buffer (*see* Note 1).

µM Concentrations

	A Mix	C Mix	G Mix	T Mix
ddATP	900	—	—	—
ddCTP	—	400	—	—
ddGTP	—	—	360	—
ddTTP	—	—	—	720
dATP	30	30	30	30
dCTP	100	37	100	100
dGTP	100	100	37	100
dTTP	100	100	100	33

Store at –20°C.
3. 30X Triton X-100: 3% Triton X-100. Store at –20°C.
4. Radiolabel: (see Note 2). Use α-[^{35}S]-dATP, 500 Ci/mmol; α-[^{32}P]-dATP, 400 Ci/mmol, or α-[^{33}P]-dATP, 3000 Ci/mmol for labeled dATP incorporation as described in Section 3.1. Use 3000 Ci/mmol γ-[^{32}P]-rATP or 3000 Ci/mmol γ-[^{33}P]-rATP for end-labeled primers as described in Section 3.2.). Store at –20°C.
5. Vent$_R$™ (exo⁻) DNA polymerase: 2 U/µL (New England Biolabs, Inc., Beverly, MA). Store at –20°C.
6. Stop/loading dye solution: Deionized formamide containing 0.3% xylene cyanol FF, 0.3% bromophenol blue, and 0.37% EDTA, pH 7.0. Store at –20°C.

3. Methods

3.1. Thermal Cycle Sequencing with Labeled dATP Incorporation

1. Label four microcentrifuge tubes A, C, G, T. Using the Vent$_R$™ (exo⁻) DNA polymerase deoxy/dideoxy sequencing mixes, add 3 µL of A mix to the bottom of the tube A, and 3 µL of the C, G, and T mixes to the bottoms of the tubes C, G, and T, respectively.
2. Mix together the following in a 0.5-mL microcentrifuge tube: 0.04 pmol of single-stranded template DNA or 0.1 pmol of double-stranded template DNA (see Notes 3 and 4), 0.6 pmol primer for a single-stranded template DNA or 1.2 pmol primer for a double-stranded template DNA, 1.5 µL of 10X Vent$_R$™ (exo⁻) DNA sequencing buffer, 1 µL of 30X Triton X-100 solution and distilled water to a total vol of 12.0 µL. Mix the solution by gentle pipeting.
3. Individually process each template/primer tube through this step. When all sets of reactions are complete, proceed to step 4. To the tube containing the template, primer, buffer, Triton X-100, and water, add 2 µL of radiolabel. Add 2–4 U of Vent$_R$™ (exo⁻) DNA polymerase and mix the solution by

gentle pipeting. Immediately distribute 3.2 µL of this reaction to the deoxy/dideoxy tube labeled A, and mix the solution by gentle pipeting. Changing pipet tips each time, repeat this addition to the C, G, and T tubes.
4. Overlay each reaction with one drop of sterile mineral oil. Place the tubes in the thermal cycler, which has been preset for reaction times and temperatures (see Note 5). Start the thermal cycler.
5. After completion of the thermal cycling, add 4 µL of stop/loading dye solution to each tube, beneath the mineral oil. The reactions are now complete and ready to be electrophoresed in appropriate denaturing sequencing gels (see Note 6). The reactions may be stored at −20°C.

3.2. Thermal Cycle Sequencing with 5' End-Labeled Primers

1. Label four microcentrifuge tubes A, C, G, and T. Using the Vent$_R$™ (exo$^-$) DNA polymerase deoxy/dideoxy sequencing mixes, add 3 µL of A mix to the bottom of the tube A, and 3 µL of the C, G, and T mixes to the bottoms of the tubes C, G, and T, respectively.
2. Mix together the following in a 0.5-mL microcentrifuge tube: 0.004 pmol of single-stranded template DNA or 0.01 pmol of double-stranded template DNA (see Notes 3, 4, and 7), 0.6 pmol of end-labeled primer for single-stranded template DNA or 1.2 pmol of end-labeled primer for double-stranded template DNA (see Note 8), 1.5 µL of 10X Vent$_R$™ (exo$^-$) DNA polymerase sequencing buffer, 1 µL of 30X Triton X-100 solution, and distilled water to a total volume of 14.0 µL. Mix the solution by gentle pipeting.
3. Individually process each template/primer tube through this step. When all sets of reactions are complete, proceed to step 4. Add 2–4 U of Vent$_R$™ (exo$^-$) DNA polymerase to the tube containing the template, primer, buffer, Triton X-100, and water and mix the solution by gentle pipeting. Immediately distribute 3.2 µL of this reaction to the deoxy/dideoxy tube labeled A, and mix the solutions by gentle pipeting. Changing pipet tips each time, repeat this addition to the C, G, and T tubes.
4. Overlay each reaction with a drop of sterile mineral oil. Place the tubes in the thermal cycler that has been preset for reaction times and temperatures (see Note 5). Start the thermal cycler.
5. After completion of the thermal cycling, add 4 µL of stop/loading dye solution to each tube beneath the mineral oil. The reactions are now complete and ready to be electrophoresed in appropriate denaturing sequencing gels (see Note 6). The reactions may be stored at −20°C.

Typical results are shown in Fig. 2. Potential problems are described in Notes 9 and 10.

418   Slatko

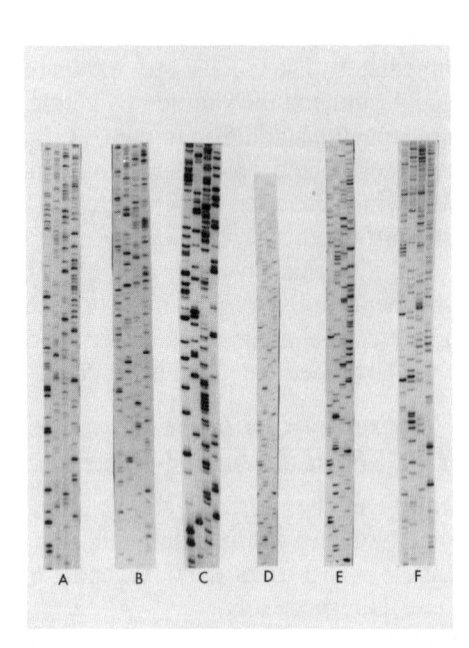

Fig. 2. Autoradiograms of Vent$_R$™ (exo⁻) DNA polymerase thermal cycle sequencing reactions, as described in the text. **(A)** DNA sequence obtained directly from a bacterial colony containing pUC19 double-stranded plasmid DNA with a ^{32}P 5' end-labeled (kinased) M13 forward sequencing primer (NEB, New England Biolabs, Inc. #1224); **(B)** DNA sequence obtained directly from an M13mp18 phage plaque with a ^{32}P 5' end-labeled (kinased) M13 forward sequencing primer (NEB #1224); **(C)** DNA sequence obtained from 0.01 pmol double-stranded PCR template with a ^{32}P 5' end-labeled (kinased) 20-mer primer; **(D)** DNA sequence obtained from 0.004 pmol (10 ng) M13mp18 single-stranded DNA template (NEB #404C) sequenced with the M13 forward universal primer (NEB #1224) and α-[^{35}S] dATP incorporation; **(E)** DNA sequence obtained from 0.01 pmol (20 ng) double-stranded pUC19 double-stranded DNA template (NEB#304-1) sequenced with the M13 forward universal primer (NEB #1224) and α-[^{35}S] dATP incorporation; **(F)** DNA sequence obtained from a minipreparation of double-stranded cosmid DNA template with a ^{32}P 5' end-labeled primer. All lanes loaded left to right: G, C, A, T. All exposures overnight without an intensifying screen, except for (D) and (E), which were 36 h exposures.

4. Notes

1. Improved sequencing of high secondary structure regions may be accomplished by substituting an equal molar amount of 7-deaza dGTP (7-deaza-2'-deoxyguanosine-5'-triphosphate) for the deoxyguanosine triphos-

phate or by substituting an equal molar amount of 7-deaza dATP (7-deaza-2'-deoxyadenosine-5'-triphosphate) for the deoxyadenosine triphosphate. dITP (2'-deoxyinosine-5'-triphosphate) is not recommended as a substitute for deoxyguanosine triphosphate because it is not incorporated as efficiently and because it tends to show shadow banding (premature terminations) in all four sequencing lanes.
2. For sequencing small amounts of DNA template (0.004–0.01 pmol) or for sequencing directly from bacterial colonies or phage plaques, use ^{33}P or ^{32}P end-labeled primers. For all other applications, including sequencing PCR products, ^{35}S, ^{33}P, or ^{32}P can be effectively utilized. Because ^{35}S is not efficiently incorporated in the kinase reaction, it is not recommended for sequencing smaller amounts of template DNA. When preparing end-labeled primers by the kinase reaction, it is recommended to use the higher specific activity (3000 Ci/mmol) rATP, since it provides a stronger signal. In all protocols it is recommended that all radioactive material be used within two radioactive decay half-lives.
3. The following template and preparations have given good results:
    a. Plasmid templates should be purified by CsCl gradient, standard minipreparation methods *(13)*, mini-column chromatography procedures, or by more recent methods, such as Insta-Prep™ (5 prime→3 prime, Inc., Boulder, CO).
    b. Single-stranded M13 templates should be purified by PEG precipitation methods *(13)* or solid support purification procedures *(14)*.
    c. λ (λgt10, 11, and so forth) and cosmid templates should be purified by CsCl gradient protocols or by minipreparation methods *(13)*.
    d. Double-stranded and single-stranded (asymmetric) PCR products must have the excess primers and nucleotides removed after completion of the PCR reaction. Phenol/chloroform extract and alcohol precipitate the PCR product from the reaction. Resuspend the dried pellet in 50 µL water. Drop dialyze the reaction *(15)* using a PVDF membrane (0.025 µ) (Millipore Corp., Bedford, MA) (1 h against 200 mL of water), or purify the product by gel purification, glass bead procedures, or by exclusion methods (i.e., "spin" columns).
    e. For direct sequencing from bacterial colonies *(8,10)* resuspend the colony in 12 µL of freshly prepared colony/plaque lysis solution (10 m*M* Tris-HCl, pH 7.5, 1 m*M* EDTA, 50 mg/mL proteinase K; prepare fresh each time. A proteinase K stock solution may be prepared and aliquots stored at –20°C.). Briefly vortex the solution, incubate 15 min at 55°C, and then 15 min at 80°C. Place on ice 1 min and microcentrifuge 3 min. Use 9 µL of the supernatant as the template in the end-labeled primer sequencing reaction.

f. For direct sequencing of M13 or λ plaques *(8–10)*, transfer with a toothpick the top agar containing a plaque, to a 1.5-mL centrifuge tube containing 12 µL of freshly prepared colony/plaque lysis solution (10 m*M* Tris-HCl, pH 7.5, 1 m*M* EDTA, 50 mg/mL proteinase K; prepare fresh each time. A proteinase K stock solution may be prepared and aliquots stored at –20°C). Briefly vortex the solution, incubate 15 min at 55°C, and then 15 min at 80°C. Place on ice 1 min and microcentrifuge 3 min. Use 9 µL of the supernatant as the template in the end-labeled primer sequencing reaction.
4. Molar ratio calculations: The following formulae can be used to calculate pmol amounts of primer and template:

Single-stranded DNA (M13 phage, ss PCR products, oligonucleotide primers, and so on.)*

pmol = (# µg DNA × 10^6 pg/µg)/(# bases × 330 pg/pmol)

Double-stranded DNA (ds PCR products, plasmids, λ DNA, restriction fragments, and so on.)**

pmol = (# µg DNA × 10^6 pg/µg)/(# base pairs × 660 pg/pmol base pair)

* 40 µg = 1 OD_{260}; # µg oligonucleotide = # OD_{260} × 40 µg/OD_{260}
** 50 µg = 1 OD_{260}; # µg oligonucleotide = # OD_{260} × 50 µg/OD_{260}

Useful numbers:

0.04 pmol of a 7.25-kb single-stranded DNA template = ~100 ng
0.004 pmol of a 7.25-kb single-stranded DNA template = ~10 ng

0.1 pmol of a 3-kb double-stranded DNA template = ~200 ng
0.01 pmol of a 3-kb double-stranded DNA template = ~20 ng

5. Thermal cycler parameters: For most thermal cyclers, including the Techne (Duxford, UK) PHC-2 and the Perkin-Elmer-Cetus (Norwalk, CT) 480 thermal cycler, use 20 cycles:
   95°C   20 s
   55°C   20 s
   72°C   20 s
It is possible to alter the cycling conditions of the reaction and still achieve excellent results. One may reduce the number of cycles in the reaction, especially if the amount of template DNA being used exceeds the recommendations of the protocol guideline. However, increasing the number of cycles to more than 30 has not been successful in sequencing smaller amounts of DNA; it often increases the shadow banding on the resultant autoradiograms. It is also possible to alter the cycle temperatures and/or times for the reaction. Templates may be sequenced using only a two-step

DNA Sequencing

cycle, a 95°C denaturation step plus a 72°C annealing and extension step, using primers with apparent melting temperatures (T_m values) of as low as 55°C. The signal intensities with this approach may be weaker than with the more standard three-step recommendation. It is also possible to lower the annealing/extension temperature, corresponding to the primer T_m, to perform sequencing cycles consisting of two steps, for example, a 95°C denaturation step and a 55°C annealing and extension step. If weak sequence signal is observed when using the standard three-step method, the recommended annealing temperature may be too high for the primer being used. The annealing step can be lowered to correspond to the individual primer T_m. Successful use of some thermal cyclers requires longer reaction times (increasing each step from 20 s to 1 min in each cycle).

6. A brief (5 s) microcentrifugation of the completed reaction helps ensure the aqueous layer fully separates from the oil layer. Immediately before loading the sequencing gel, heat the completed samples at 80°C for 2 min. For each loading, insert the pipet tip underneath the oil covering the reaction and remove a 2.5 µL sample. It may be necessary to briefly touch the pipet tip to a tissue to remove residual oil before loading the sample on the gel. Alternatively, one can remove the oil before loading the reaction on a gel by using silicone oil to cover the reactions followed by precipitation of the reaction products from the oil *(16)* or by using Parafilm™ to remove the oil *(17)*. Another approach is to use a "hot-top" apparatus, set 15°C hotter than the warmest step in the cycle reaction (~115°C), which eliminates evaporation and precludes the requirement for oil on the reactions. It should be emphasized that we have never observed any sequence aberration caused by overlay oil contaminating the reaction in the sequencing gel lane.

7. As much as 10-fold greater quantity of template DNA can be successfully used in this protocol.

8. For end-labeling with γ-[^{32}P] or γ-[^{33}P] rATP, the following protocol works well:
   a. Mix the following in a 0.5-mL microcentrifuge tube: 13.5 µL water, 2.5 µL 10X T4 polynucleotide kinase buffer (10X buffer = 500 mM Tris-HCl, pH 7.6, 100 mM MgCl$_2$, 50 mM DTT), 10.5 pmol primer (1 µL of stock primer solution; stock = 2.5 µg 24-mer primer resuspended in 30 µL of water), 7 µL γ-[^{32}P] rATP (11.5 pmol), and 1 µL (10 U) T4 polynucleotide kinase (10,000 U/mL).
   b. Allow the reaction to proceed at 37°C for 30 min.
   c. Terminate the reaction at 95°C for 5 min. Briefly microcentrifuge at room temperature. Store at –20°C.

   This protocol provides 10.5 pmol of 5' end-labeled primer in a 25 µL reaction. The final concentration of primer is thus 0.4 pmol/µL.

9. The following sequencing patterns are characteristic of Vent$_R$™ (exo$^-$) DNA polymerase:
   a. The first C of a run of Cs is darker than the following Cs.
   b. The second A of a run of As is darker than the preceding (and/or following) As.
   c. G following an A tends to be darker than other Gs.
   d. T following an A tends to be darker than other Ts.
10. When DNA sequencing results are less than optimal, several factors might be considered. Often the problem lies with template preparation. A phenol/chloroform extraction step or alcohol precipitation followed by a 70% alcohol rinse will often eliminate the problem. A second common problem is the use of too much or too little primer and/or template in the reaction; the pmol amounts in the protocols have been empirically determined to provide optimal results.

Another potential problem involves suboptimal performance of thermal cyclers. Some may require longer reaction times, may also cause problems, and are often indicated by one or two lanes of a reaction that are lighter than the others or by failure of one lane. If one or two lanes of a reaction are lighter than the others, or if failure of one lane occurs, a problem with the cycler is indicated.

References

1. Sanger, F., Nicklen, S., and Coulson, A. R. (1977) DNA sequencing with chain inhibitors. *Proc. Natl. Acad. Sci. USA* **74,** 5463–5467.
2. Kong, H., Kucera, R., and Jack, W. (1992) Characterization of a DNA polymerase from the hyperthermophile Archaea *Thermococcus litoralis. J. Biol. Chem.* **268,** 1965–1975.
3. Perler, F., Comb, D., Jack, W., Moran, L., Qiang, B.-Q., Kucera, R., Benner, J., Slatko, B., Nwankwo, D., Hempstead, K., Carlow, C., and Jannasch, H. (1991) Novel intervening sequences in an Archaea DNA polymerase gene. *Proc. Natl. Acad. Sci. USA* **89,** 5577–5581.
4. Adams, S. and Blakesley, R. (1991) Linear amplification sequencing. *Focus (BRL)* **13(2),** 56–57.
5. Carothers, A. M., Urlaub, G., Mucha, J., Grunberger, D., and Chasin, L. A. (1989) Point mutation analysis in a mammalian gene: rapid preparation of total RNA, PCR amplification of cDNA and *Taq* sequencing by a novel method. *BioTechniques* **7,** 494–499.
6. Craxton, M. (1991) Linear amplification sequencing: a powerful method for sequencing DNA. *Methods*, a companion to *Meth. Enzymol.* **3,** 20–26.
7. Murray, V. (1989) Improved double-stranded DNA sequencing using the linear polymerase chain reaction. *Nucleic Acids Res.* **17,** 8889.
8. Sears, L. E., Moran, L. S., Kissinger, C., Creasey, T., Sutherland, E., Perry-O'Keefe, H., Roskey, M., and Slatko, B. (1992) Thermal cycle sequencing and

alternative manual and automated DNA sequencing protocols using the highly thermostable Vent$_R$™ (exo⁻) DNA polymerase. *BioTechniques* **13**, 626–633.
9. Krishnan, B. R., Blakesly, R. W., and Berg, D. E. (1991) Linear amplification DNA sequencing directly from single phage plaques and bacterial colonies. *Nucleic Acids Res.* **19**, 1153.
10. Young, A. and Blakesley, R. (1991) Sequencing plasmids from single colonies with the dsDNA cycle sequencing system. *Focus (BRL)* **13(4)**, 137.
11. Creasey, A., D'Angio, L., Dunne, T. S., Kissinger, C., O'Keeffe, T., Perry-O'Keefe, H., Moran, L. S., Roskey, M., Schildkraut, I., Sears, L. E., and Slatko, B. (1991) Application of a novel chemiluminescence-based DNA detection method to single-vector and multiplex DNA sequencing. *BioTechniques* **11**, 102–109.
12. *Biosystems Reporter* (Applied Biosystems, Inc., Foster City, CA) (1991) **13**, 1–2.
13. Slatko, B., Heinrich, P., Nixon, B. T., and Eckert, R. (1991) Preparation of templates for DNA sequencing unit 7.3, in *Current Protocols in Molecular Biology* (Ausubel, F., Brent, R., Kingston, R., Moore, D., Seidman, J., Smith, J., and Struhl, K., eds.), Wiley, New York.
14. Hultman, T., Bergh, S., Moks, T., and Uhlen, M. (1991) Bidirectional solid phase sequencing of in vitro amplified plasmid DNA. *BioTechniques* **10**, 84–93.
15. Silhavy, T., Berman, M., and Enquist, L. (1984) *Experiments With Gene Fusions*. Cold Spring Harbor Laboratory, Cold Spring Harbor, NY.
16. Ross, J. and Leavitt, S. (1991) Improved sample recovery in thermocycle sequencing protocols. *BioTechniques* **11**, 618–619.
17. Whitehouse, E. and Spears, T. (1991) A simple method for removing oil from cycle sequencing reactions. *BioTechniques* **11**, 616–618.

CHAPTER 51

Using the Automated DNA Sequencer

Zijin Du and Richard K. Wilson

1. Introduction

Efficient completion of large DNA sequencing projects has been greatly facilitated by the development of fluorescence-based dideoxynucleotide sequencing chemistries and instruments for real-time detection of fluorescence-labeled DNA fragments during gel electrophoresis (*1–6*). Besides eliminating the use of radioisotopes, these systems automate the task of reading sequences and provide computer-readable data that may be directly analyzed or entered into a sequence assembly engine. In this chapter, we describe DNA sequencing using the first commercially available fluorescent instrument, the Applied Biosystems, Inc. Model 373A Automated DNA Sequencer. This sequencer, originally introduced in 1987 as the Model 370A, utilizes a multispectral approach in which four distinct fluorescent tags are detected and differentiated in a single lane on the sequencing gel *(6)*. The four tags are incorporated during the DNA sequencing reactions and may be present on either the 5' end of the sequencing primer ("dye primers") or on the dideoxynucleotide triphosphate ("dye terminators"). Since the 373A requires only one lane per sample, up to 36 samples may be analyzed per run. By comparison, the Pharmacia ALF sequencer supports only a one-dye, four-lane approach with dye-labeled primer, and can analyze up to ten samples per run. The ability of the 373A to use both dye-primer and dye-terminator chemistries greatly facilitates the completion of projects that require both shotgun and primer-directed sequencing phases.

From: *Methods in Molecular Biology, Vol. 58: Basic DNA and RNA Protocols*
Edited by: A. Harwood Humana Press Inc., Totowa, NJ

We have used the 373A for DNA sequence analysis of M13, plasmid, and phagemid subclones, as well as templates produced by PCR. A sequencing protocol using modified T7 DNA polymerase and fluorescent dye-primers is described for use with single-stranded M13 and phagemid DNA. A second sequencing protocol described here utilizes *Taq* DNA polymerase and thermal cycling, and may be used with dye-primers and both single- and double-stranded templates. A third sequencing protocol utilizes T7 DNA polymerase and fluorescent dye-terminators and may be used with single-stranded M13 and phagemid DNA. This third method requires that special hardware be installed in the 373A. An alternative dye-terminator chemistry using *Taq* DNA polymerase is available from Applied Biosystems, although it has not been reliable in our hands and hence is not included in this chapter.

2. Materials

1. A thermal cycler is required for the linear amplification sequencing method and for the PCR template preparation methods (*see* Note 1). In our hands, the GeneAmp 9600 Thermal Cycler from Perkin-Elmer (Norwalk, CT) and the PTC-200 thermal cycler from MJ Research (Boston, MA) have worked very well. Both offer the advantages of a 96-well microtiter plate format, significantly faster ramping times, and a heated plate cover that eliminates the need to overlay reactions with mineral oil. The Techne MW-1 thermal cycler also works well for linear amplification sequencing reactions in 96-well microtiter plates (M. Craxton, personal communication).
2. The automated fluorescent DNA sequencing system described here is the Applied Biosystems, Inc. (Foster City, CA) Model 373A. The current configuration includes a Macintosh IIci computer with 8 MB RAM and a 100 MB fixed disk drive.
3. *Thermus aquaticus* (*Taq*) DNA polymerase may be purchased from one of several enzyme suppliers. It is our experience that the Sequi-therm enzyme from Epicentre (Madison, WI) gives the best results.
4. Modified T7 DNA polymerase (Sequenase) should be purchased from United States Biochemical (Cleveland, OH); version 1.0 enzyme is best for dye-primer sequencing, version 2.0 enzyme containing pyrophosphatase is best for dye-terminator sequencing.
5. Deoxynucleotides (dNTPs) and dideoxynucleotides (ddNTPs):
   a. Stocks: 100 mM dNTP solutions and 5 mM ddNTP solutions may be purchased from Pharmacia (Piscataway, NJ). Prepare 20 mM stocks of each dNTP in TE buffer. Store at $-20°C$.
   b. For PCR: Prepare a mix containing 1.25 mM of each dNTP in TE buffer. Store at $-20°C$.

DNA Sequencing Reactions

c. For dye-primer sequencing with modified T7 DNA polymerase:
  i. Prepare 8 mM dNTP mixes: 100 µL each of 100 mM dATP, dCTP, dGTP, and dTTP in TE buffer to a final vol of 1.25 mL (store at –20°C).
  ii. Prepare 50 µM ddNTP solutions:
  ddA: 2 µL of 5 mM ddATP + 198 µL of TE buffer.
  ddC: 2 µL of 5 mM ddCTP + 198 µL of TE buffer.
  ddG: 4 µL of 5 mM ddGTP + 396 µL of TE buffer.
  ddT: 4 µL of 5 mM ddTTP + 396 µL of TE buffer.
  iii. Sequencing mixes: Combine equal volumes of the 8 mM dNTP mix and one of the four 50 µM ddNTP solutions. For 100 reactions, prepare 100 µL of A and C mixes and 200 µL of G and T mixes.
d. For dye-primers sequencing using linear amplification:
  i. Prepare stock dNTP mixes (sufficient for 200 reactions) (*see* Note 2):
  dATP mix: 1.25 µL of 20 mM dATP, 5.0 µL of each 20 mM dCTP, dGTP, dTTP, 184.75 µL of TE.
  dCTP mix: 1.25 µL of 20 mM dCTP, 5.0 µL of each 20 mM dATP, dGTP, dTTP, 184.75 µL of TE.
  dGTP mix: 2.50 µL of 20 mM dGTP, 10.0 µL of each 20 mM dATP, dCTP, dTTP, 367.50 µL of TE.
  dTTP mix: 2.50 µL 20 mM dTTP, 10.0 µL each 20 mM dATP, dCTP, dGTP, 367.50 µL of TE.
  ii. Prepare ddNTP solutions (sufficient for 167 reactions):
  ddATP (3.0 mM): 100 µL of 5 mM ddATP + 67 µL of TE.
  ddCTP (1.5 mM): 50 µL of 5 mM ddCTP + 117 µL of TE.
  ddGTP (0.25 mM): 16.7 µL of 5 mM ddGTP + 317.3 µL of TE.
  ddTTP (2.5 mM): 167 µL of 5 mM ddTTP + 167 µL of TE.
  iii. Prepare dNTP/ddNTP working mixes (sufficient for 100 reactions):
  50 µL of dATP mix + 50 µL of 3.0 mM ddATP.
  50 µL of dCTP mix + 50 µL of 1.5 mM ddCTP.
  100 µL of dGTP mix + 100 µL of 0.25 mM ddGTP.
  100 µL of dTTP mix + 100 µL of 2.5 mM ddTTP.
e. For dye-terminator sequencing with modified T7 DNA polymerase:
  i. 2 mM [αS]dNTPs, 10 mM Tris-HCl, pH 7.2, 0.1 mM EDTA.
  ii. ZL 1.1.1.2 dye-terminator mix: 2.2 µM ddT-6fam, 9.0 µM ddC-5zoe, 6.0 µM ddA-lou, 16.0 µM ddG-nan (purchase from ABI; store at –20°C).
6. Oligonucleotide dye primers: For M13 and phagemid sequencing, universal, reverse, T7, and SP6 primers are available from Applied Biosystems. The sequences of these primers are: (–21 M13) 5' TGT-AAA-ACG-ACG-GCC-AGT 3', (M13RP1) 5' CAG-GAA-ACA-GCT-ATG-ACC 3'. (*See* Note 3.)

7. 40% A&B: 380 g of acrylamide, 20 g of bisacrylamide, distilled water to 1 L. Deionize by stirring for 1 h with Amberlite MB-1 (50 g/L), filter, store at 4°C in an opaque bottle.
8. 20X TBE buffer: 324 g of Tris base, 55 g of boric acid, 18.6 g of EDTA, distilled water to 1 L.
9. 15% Ammonium peroxysulfate.
10. 5X *Hin*d/DTT (+Mn^{2+}; make fresh):
    a. 50 m*M* Tris-HCl, pH 7.5, 300 m*M* NaCl, 5 m*M* DTT.
    b. Immediately before sequencing, add 3 µL of 1*M* MnCl$_2$ to 197 µL of 5X *Hin*d/DTT buffer.
11. 5*M* ammonium acetate, pH 7.4.
12. 5X LASR buffer: 400 m*M* Tris-HCl, pH 8.9, 100 m*M* (NH$_4$)$_2$ SO$_4$, 25 m*M* MgCl$_2$.
13. 10X Mn^{2+} buffer: 50 m*M* MnCl$_2$, 150 m*M* sodium isocitrate.
14. 10X MOPS buffer: 400 m*M* MOPS, pH 7.5, 500 m*M* NaCl, 100 m*M* MgCl$_2$.
15. 9.5*M* ammonium acetate, pH 7.4.

3. Methods
3.1. DNA Sequencing Reactions
3.1.1. T7/mT7 DNA Polymerase Sequencing Method with Fluorescent Dye Primers

In this procedure, reactions are performed using the modified T7 DNA polymerase chemistry *(7,8)* as adapted for sequencing with fluorescent dye primers *(9)*. The reactions are conveniently performed in 96-well U-bottom plates (Falcon, Becton Dickinson Co., Rutherford, NJ, No. 3911), either by hand or using an automated pipeting station *(10)*. Since the fluorescent dyes used in the G and T reactions produce a weaker signal, the reaction volumes are doubled.

1. Annealing reactions should be set up as follows:

	A	C	G	T
5X Hind/DTT (+Mn^{2+}; make fresh)	1 µL	1 µL	2 µL	2 µL
Template DNA	3 µL	3 µL	6 µL	6 µL
Individual dye primer (0.4 pmol/µL)	1 µL	1 µL	2 µL	2 µL

The reactions are heated to 55°C for 3–5 min, then cooled slowly to room temperature for 10–15 min. If the reactions are performed in a 96-well plate, a dry block heater may be modified to effectively heat all 96 wells *(11,12)*.

2. To each annealing reaction is added:

	A	C	G	T
8 mM dNTPs + 50 µM ddXTP mix	2 µL	2 µL	4 µL	4 µL
Modified T7 DNA polymerase (1.5 U/µL)	1.5 µL	1.5 µL	3 µL	3 µL

The reactions are incubated at 37°C for 5–10 min.

3. After extension and termination, the four reactions must be stopped before they may be combined and concentrated for electrophoresis. A simple method for stopping the reactions is to place 6 µL of 5M ammonium acetate, pH 7.4, and 120 µL of 95% ethanol in 1.5-mL microcentrifuge tubes (one tube for each template). For each template set, 8 µL of the A reaction is transferred to the tube and two quick cycles of up-and-down pipeting are used to rinse the tip and mix the sample with the ethanol solution. Without changing the pipet tip, 8 µL of the C reaction and 16 µL of the G and T reactions are transferred to the tube, with up-and-down pipeting following each addition. The ethanol solution effectively stops further enzymatic activity. This procedure is repeated for each template set. The combined reactions are placed at –20°C for 30 min to precipitate the DNA products.

4. The DNA is pelleted by centrifugation at 13,000g for 15 min at room temperature, washed once with 300 µL of 70% ethanol (room temperature), and dried briefly under vacuum. The dried sample may be stored at –20°C for several days.

3.1.2. Linear Amplification Sequencing Method (Taq DNA Polymerase) with Fluorescent Dye Primers

1. Sequencing reactions should be set up in either 0.5 mL microcentrifuge tubes or in 0.2-mL Micro-Amp™ tubes and should contain the following components:

	A	C	G	T
5X LASR buffer	1 µL	1 µL	2 µL	2 µL
dNTP/ddXTP mix	1 µL	1 µL	2 µL	2 µL
Individual dye primer (0.4 pmol/µL)	1 µL	1 µL	2 µL	2 µL
Template DNA (approx 250 ng/µL)	1 µL	1 µL	2 µL	2 µL
Taq DNA polymerase (0.7 U/µL)	1 µL	1 µL	2 µL	2 µL
Total volume	5 µL	5 µL	10 µL	10 µL

2. If necessary, overlay each reaction with 50 µL of light mineral oil. The thermal cycler is preheated to 95°C, and the samples are placed in the pre-

heated block. With the Perkin-Elmer instruments, thermal cycling is performed using the following parameters (*see* Note 4):

DNA thermal cycler	9600 thermal cycler
95°C for 30 s	95°C for 4 s
55°C for 30 s	55°C for 10 s
70°C for 60 s	70°C for 60 s
for 15 cycles, then	for 15 cycles, then
95°C for 30 s	95°C for 4 s
70°C for 60 s	70°C for 60 s
for an additional 15 cycles	for an additional 15 cycles

At the end of thermal cycling, the samples are maintained at 4°C.
3. As described for the modified T7 DNA polymerase chemistry, the four sequencing reactions must be stopped before they may be combined and concentrated for electrophoresis. Again, a simple method for stopping the reactions is to transfer each sample to a 1.5-mL microcentrifuge tube containing the ammonium acetate/ethanol solution. After ethanol precipitation, the dried samples may be stored at –20°C for several days.

3.1.3. mT7 DNA Polymerase Sequencing Method with Fluorescent Dye Terminators

This procedure requires that the 373A DNA sequencer be equipped with a special five-color (531/545/560/580/610) filter wheel. The sequencing reactions require single-stranded DNA, either M13 or phagemid.

1. For each sample, set up an annealing reaction in 1.5 mL microcentrifuge tubes as follows: 2 µL of 10X Mn^{2+} buffer, 2 µL of 10X MOPS buffer, x µL of ssDNA (2.5 µg), 0.5 µL of primer (3.2 pmol/µL), and y µL of ddH_2O, for a total volume of 11 µL.
2. Incubate at 55°C for 5 min, then cool to room temperature and let stand for 7 min. Briefly centrifuge to collect the condensation.
3. Add the following reagents to the annealing reaction: 4 µL of 2 mM [αS]dNTPs, 4 µL of ZL 1.1.1.2 dye-terminator mix, and 1 µL of sequenase/ pyrophosphatase (13 U/µL).
4. Mix gently and incubate at 37°C for 10 min.
5. To each reaction, add 20 µL of 9.5M ammonium acetate and 90 µL of ethanol. Mix gently and place on ice for 10 min.
6. Pellet the reaction products by centrifugation at 13,000g for 15 min at room temperature. Wash twice with 300 µL of 70% ethanol, and dry briefly under vacuum. Typically, a DNA pellet will not be observed.

DNA Sequencing Reactions

3.2. Preparation of Sequencing Gels

1. Careful cleaning of the glass plates (supplied by Applied Biosystems, Inc.) is crucial with the fluorescent sequencing since dust and scratches will cause light scatter. The plates should be carefully washed with a dilute solution of Alconox using a soft paper wiper (a large Kimwipe™ works well), rinsed with warm water, and then thoroughly rinsed with distilled water. The plates are then rinsed with absolute ethanol, and the inner surfaces are wiped with isopropanol and allowed to air dry. Spacers and combs should also be washed and dried thoroughly. It is important to maintain each plate with a designated "inner" side to minimize gel bubbles caused by small scratches—to do this, the "outer" side of the plate may be marked on one corner by a glass etcher, with tape, or permanent ink.
2. Lay the unnotched plate, outer side down, on a flat stable platform (a styrofoam tube rack works well). Lay gel spacers along each long edge of the plate. Gently place the notched plate over the unnotched plate and spacers. Make sure the plate edges are all flush and that the spacers are properly positioned between the plates. Using two large binder clamps on each long side, clamp the assembly together.
3. Prepare the gel solution by combining the following in a 250-mL beaker: 22 g of urea, 8 mL of 40% A&B, 2.5 mL of 20X TBE, and distilled water to 50 mL. Warm the solution at 55°C for a few minutes to dissolve the urea, and stir until the solution becomes clear. Filter the gel solution through a 0.45-µm membrane (*see* Note 5).
4. To the gel solution, add 245 µL of 15% ammonium peroxysulfate and 30 µL TEMED. Swirl the beaker to mix and fill a 60-mL syringe with the solution. With the gel plate assembly flat on the platform, inject the gel solution between the plates along the notched edge of the top plate. Start at one side of the assembly and slowly move the syringe across the top, allowing capillary action to pull the solution into the form. If bubbles begin to form, tap on the top of the plates with a finger to force gel solution across the trouble spot. When the front edge of the gel solution flows to the open end of the gel form, cease injecting solution and insert the straight-edged casting comb into the notch at the top of the gel form. Place three binder clamps across the top of the notched plate. Allow the gel at least 2 h to polymerize.
5. Remove the clamps and the casting comb. There will be a thin strip of polyacrylamide along the inner bevel of the notched plate; remove this by running the edge of the casting comb along the bevel. Rinse the outside surface of the plates with water to remove any gel material, rinse again with absolute ethanol, and allow to air dry. Wipe the outside of the plates with Texwipe™ glass cleaner or isopropanol. Place the gel assembly in the electrophoresis chamber of the 373A DNA sequencer.

6. Restart the Macintosh computer. Make sure that the ABI Data Collection and Data Analysis programs are launched automatically on startup (automatic startup of Multifinder, Data Collection, and Data Analysis should be preconfigured under "Set startup"). On the 373A sequencer, perform the plate check function to ensure that the glass plates have been properly cleaned and are free of dirt or gel material. The scan of the plates should be observed in the "Scan" window of the Data Collection program. Four flat lines, one for each "color" of the four fluors, should be observed. If blue peaks are observed, these indicate dust or smudges on the glass plates; carefully clean the dirty area of the plates with isopropanol or Texwipe™ glass cleaner and repeat the plate check as needed. If four-color peaks are observed, these indicate light scattering caused by contaminants present in the gel itself. In this latter case, the peaks may disappear as the gel is prerun. However, if contaminant peaks persist after cleaning the plates and prerunning the gel, it is recommended that the gel be replaced or that samples not be loaded into the wells that correspond to the problem area. If wells are skipped, be sure to check sample tracking when the run is completed to ensure that the contaminant peak has not been tracked and analyzed.
7. Set the upper buffer chamber and the lucite alignment brace in place. Carefully insert the sharkstooth comb between the glass plates, centering the comb with the numbers printed on the alignment brace. The teeth of the comb should just barely pierce the top of the gel. Tighten the two screw clamps on the upper chamber to secure the alignment brace. Fill the upper and lower buffer chambers with 1X TBE. Using a Pasteur pipet, flush urea out of the sample wells. Connect the buffer chamber electrodes to the 373A and close the electrophoresis chamber door. Prerun the gel at 30 W constant power and 40°C for 30 min. At some point during the prerun, a new Sample Sheet (under the "File" menu) should be opened, and the appropriate data entered.

3.3. Electrophoresis and Data Collection

1. Dissolve the dried sequencing reaction products in 5 µL of loading solution. Heat at 98°C for 3–5 min.
2. Carefully flush urea out of all sample wells using a Pasteur pipet. Load all odd-numbered samples into the odd-numbered wells (e.g., 1, 3, 5). Close the electrophoresis chamber door and electrophorese samples into the gel at 30 W constant power and 40°C for 2–5 min. Open the electrophoresis chamber door, and load the even-numbered samples into the even-numbered wells (e.g., 2, 4, 6). Close the electrophoresis chamber door and begin electrophoresis at 30 W constant power and 40°C for 10–14 h. Because of salt effects as samples enter the gel, this odd/even loading scheme will leave a small space between lanes, thereby facilitating proper lane tracking by the analysis software.

DNA Sequencing Reactions

3. Using the mouse, click on the "Collect" button to begin data collection. The computer will collect data for 10–14 h. During this time, it is recommended that the host Macintosh not be used for any other function. See Notes 6–14 for problems that may occur during the gel run.
4. At the completion of the run, open the electrophoresis chamber door and siphon some of the buffer from the upper buffer tank. Disconnect the electrode cables and unlatch and lower the laser stop. Carefully lift the upper buffer tank and gel assembly out of the instrument. Buffer remaining in the upper tank can be discarded. The glass plates should be carefully separated, cleaned, and stored in a safe location. Avoid using a razor blade to remove the gel; instead, press a paper towel against the gel, then peel the towel and gel off of the plate and discard. Rinse the upper and lower buffer tanks and invert on paper towels to dry.

3.4. Data Analysis and Assessment

At the conclusion of data collection, the Macintosh computer will automatically launch the Data Analysis program. At this time, a gel file and a Results folder containing individual sample and sequence files will be created. Occasionally, problems will arise during the initial analysis phase that result in the absence of analyzed data. If this occurs, data can be salvaged using the recovery procedures described in Note 15. Data can be viewed by opening any of the sample files, selecting "Analyzed data" from the "Window" menu, and using the custom view magnifier located on the floating "Analysis" controller. If the host Macintosh has been connected to a color printer or plotter, the analyzed data may be printed. Alternatively, data may be archived by transferring sample files over a network connection to another Macintosh, or a Sun computer using a communications program such as NCSA Telnet. If data must be retrieved later for further work on a Macintosh, sample files should be transferred with the MacBinary protocol selected.

Typically, data analysis begins immediately following the large four-color mass that represents unincorporated dye-labeled sequencing primer. However, the occasional sample will be analyzed from an earlier point if an extraneous fluorescent signal is detected. These samples may be reanalyzed manually; a new starting coordinate may be determined by viewing the raw data for that sample. "Call bases" is selected from the "Analysis" menu, and the new starting coordinate is entered in the appropriate box. The "Analysis queue," located under the "Window" menu, may be opened to monitor the analysis process. Note that several samples may be reanalyzed simultaneously.

Clues as to data quality may be found in several places. The most obvious is the appearance and range of the analyzed data. Fairly uniform signal intensity, decreasing slightly over the length of the read, should be observed. Although base calls will be made for the entire read, experience has shown that the accuracy of machine base assignment falls off significantly after 400 bases. In a sample of fairly high quality, "no calls" ("N") should be infrequent in the first 400 bases. Background noise, visible as small peaks along the baseline, should not be so high as to interfere with base calling. Excessive noise often is caused by impure sequencing template (see Notes 16–18). Additional information as to data quality can be obtained from the "File info" window. Here, the signal and base spacing data is displayed. Typically, signal strength for a sample of high quality should be greater than 40–50 for the lowest of the four dyes. Base spacing should be 9–11; if base spacing falls outside of this range, accurate base calling can be affected. Base spacing can be optimized by adjusting gel and running buffer composition and electrophoresis conditions. The gel file, which contains a computer-generated color "autoradiograph" of the entire run, is also useful for assessment of data quality. Here, straight and separate lanes should be observed, with easily visible bands extending late into the run. Gel-related problems are often most apparent on observation of the data contained in the gel file. Tracking may be assessed by observing the white tracker lines that should bisect all visible samples. If a tracker line does not seem to properly follow sample bands, the sample should be tracked manually. This can be done by choosing "Track a lane" from the "Analysis" menu and clicking on several points along the sample lane. At the conclusion of retracking, the sample will be analyzed automatically.

4. Notes

1. Template production using PCR:
   (i) PCR is performed using bacteriophage lysate, DNA (0.01–1.0 µg), or bacterial colonies or bacteriophage plaques that have been picked and transferred to a small amount of water or TE buffer. The reactions should contain the following components: x µL of distilled water, 5 µL of 10X PCR buffer, 8 µL of 1.25 mM dNTPs, 2.5 µL of primer 1 (20 µM), 2.5 µL of primer 2 (20 µM), y µL of template (typically 1–5 µL), and 0.2 µL of *Taq* DNA polymerase (5 U/µL) for a total volume of 50 µL.
   (ii) If necessary, overlay each reaction with 80 µL of light mineral oil. The thermal cycler is preheated to 95°C, and the samples are placed in the pre-

heated block and incubated at 95°C for 2 min. With the Perkin-Elmer instruments, thermal cycling is performed using the following parameters:

DNA thermal cycler	9600 thermal cycler
94°C for 30 s	92°C for 10 s
55°C for 30 s	55°C for 60 s
72°C for 60 s	72°C for 60 s
for 35 cycles	for 35 cycles

At the conclusion of thermal cycling, the samples are maintained at 4°C.

(iii) The samples are removed from the reaction tubes and transferred to 1.5-mL microcentrifuge tubes containing 8 µL of 3M sodium acetate, pH 4.8, and 20 µL of 40% (w/v) PEG-8000, 10 mM MgCl$_2$. If the samples were covered with mineral oil, it is important to avoid transferring any of the oil. The samples are mixed by vortexing and allowed to stand at room temperature for 10 min.

(iv) The DNA is pelleted by centrifugation at 13,000g for 15 min at room temperature. All of the supernatant is carefully and completely removed by aspiration. The DNA pellets are washed twice with 250 µL of 100% ethanol (room temperature) and dried briefly under vacuum.

(v) Each sample is dissolved in 20 µL of 10 mM Tris-HCl, pH 8.0, 0.1 mM EDTA. One to two microliters of each sample may be analyzed on an agarose gel. For DNA sequence analysis using the linear amplification sequencing method described below, 1 µL of the PCR-prepared DNA is used in the A and C reactions and 2 µL in the G and T reactions.

This method can be used to directly prepare template from cosmid and λ phage clones.

2. Nucleotide analogs, such as 7-deaza-dATP and 7-deaza-dGTP, are useful for resolving some sequence compressions and may be purchased as 10 mM solutions from Boehringer-Mannheim Biochemicals (Indianapolis, IN). 7-deaza-dATP and 7-deaza-dGTP may be substituted for dATP and dGTP in all mixes. For dATP mix, use 2.5 µL of 7dc-dATP and 10 µL of 10 mM 7dc-dGTP; for dCTP mix, use 10 µL each of 10 mM 7dc-dATP and 10 mM 7dc-dGTP; for dGTP mix, use 5 µL of 5 mM 7dc-dGTP and 10 µL of 10 mM 7dc-dATP; for dTTP mix, use 20 µL each of 10 mM 7dc-dATP and 10 mM 7dc-dGTP.

3. Some preparations of pUC18 and pUC118 have deleted one nucleotide at the 3' end of the priming site for the M13RP1 primer; for sequencing subclones based in these vectors, a reverse primer that lacks the 3' C should be used, (T7) 5' TAA-TAC-GAC-TCA-CTA-TAG-GG 3', (SP6) 5' ATT-TAG-GTG-ACA-CTA-TAG 3'.

4. If the fluorescent dye-labeled SP6 sequencing primer is used, the annealing temperature should be reduced from 55 to 50°C. Similarly, for other sequencing primers, the melting temperature (T_m) should be calculated, and the thermal cycling conditions adjusted accordingly.
5. If sequencing gels are to be run daily, a working gel solution may be prepared and stored at 4°C. Combine 110.3 g of urea, 40 mL of 40% A&B, 13 mL of 20X TBE, and 125 mL of distilled water. Dissolve the urea and filter as above. If stored in an opaque bottle, this solution can be used for 2–3 wk. Use 50 mL/gel.
6. No scan lines appear in the Scan window. Make sure "Start scan" button on the 373A has been pressed. Main menu window should read "Stop scan." If scan has been started and no scan lines appear, abort the run and restart. If scan lines still do not appear, reset the 373A by pressing the delete key on main keypad and selecting "total reset" on the LCD menu (*Note*: This reset will erase two settings—the current date and time, and the PMT voltage; re-enter these values after the reset has been completed). If resetting the 373A does not solve the problem, check the cable connecting the 373A to the host computer. If none of the above solves the problem, ABI field service should be contacted.
7. Scan lines are approx 25% full scale. When the scan window is viewed at the beginning of a run, the flat scan lines should 25–33% of full scale. If the scan lines are lower than this, PMT voltage should be increased. The PMT voltage setting is accessed by selecting "calibration" and then "configure" on the LCD menu, and moving through the next few selections. The proper PMT voltage setting will differ by machine and may change as the laser ages. Typically a setting of 540–560 works well.
8. Electrophoresis power failure. When electrophoresis power fails, the 373A will alert the user with a loud beep. Stop the run or prerun and restart electrophoresis. If no current value is observed, open the electrophoresis chamber door and check electrode cable connections. Check the platinum wires and their connections. If the electrophoresis power still fails, press the delete key on the 373A main keypad and select "elect only" on the LCD. If none of the above solves the problem, ABI field service should be contacted.
9. Stage motor failure. The electric motor that drives the scanning stage has a finite lifespan. Failure of this unit is often preceded by screeching sounds. This problem is not easily remedied by the user; ABI field service should be contacted.
10. Laser failure. The argon laser also has a finite lifespan, although there appears to be a wide range to this lifespan. To ensure against sudden laser failure, a weekly diagnostic test should be performed. Under the "Self test"

menu on the 373A LCD, press "more" until the "laser" test option is displayed. Press "laser" and enter "40" as the target power (e.g., 40 mW); record the actual laser output and the current drawn by the laser. As the laser ages, it will draw more current to produce the necessary power. When laser failure is imminent, the amount of current the laser draws will increase noticeably. At this point, ABI field service should be contacted.
11. Incomplete polymerization. Characterized by distorted patterns in the gel file. The sequencing gel should be allowed to polymerize for no less than 2 h. If sufficient polymerization time had been allowed, preparation of fresh 15% APS and TEMED may remedy the problem.
12. Sample crosstalk. This can be seen by observing the gel file, and can be caused by failure to properly insert the sharkstooth comb. The comb should be inserted so that the points of the teeth pierce about 1–2 mm into the gel. Additionally, leakage can occur between wells if the comb was inserted, partially removed, and reinserted. If the comb damages part of the gel, avoid loading in that region or replace the gel. Sample leakage also may occur with incompletely polymerized gels (*see* Note 11).
13. Sample lanes overlap. This can occur when samples are not loaded in alternating lanes as described above. If samples overlap, lane tracking can be impeded. Additional problems can occur if too much time elapses between the second loading. Here, the fluorescence signal that accompanies the primer/salt front preceding the reaction products can interfere with proper detection and tracking. No more than 5 min should elapse before the even-numbered samples are loaded.
14. Gel runs too fast or too slow. Unincorporated dye primers or dye terminators should become visible in the Gel view after 60–80 min. If these peaks are seen much earlier or later, data analysis can be severely affected. Check the electrophoresis conditions on the 373A; we routinely run gels at 30 W constant power. The gel and buffer compositions also should be checked. Polyacrylamide gel compositions other than 6% are incompatible with the 373A.
15. Computer problems. Problems with data analysis can occur when the fixed disk drive of the host Macintosh computer becomes over 75% full. To avoid these problems, sample files should be backed-up, either on a floppy disk or by transfer to a remote fixed disk, and promptly removed. Additionally, the host Macintosh should be viewed as a dedicated computer and should not be used for word processing, graphics, or any other purpose. Placing various INITs and cDevs (control panel devices) on the computer can cause severe problems. To avoid problems caused by disk fragmentation, a disk repair program, such as Apple's Disk First Aid, should be run at least once a month. If problems arise with data analysis, check the "log

file" located under the "Windows" menu in either the Data Collection or Data Analysis programs. This file contains an event list of data collection and analysis and can provide clues as to intermittent computer problems. If data analysis has failed and an error message is present, complete recovery of data is possible using one of several recovery procedures. A complete list of error messages and their causes is available from Applied Biosystems, along with detailed instructions for data recovery; it is highly recommended that all 373A users obtain this list and post it near the sequencer.

16. Sequence data terminates early. Several factors can be responsible for truncated sequence data. If an RNase step was used in the template preparation, the presence of contaminating DNase will result in truncated sequence data. This problem often is accompanied by an increase in background peaks. Reactions performed with insufficient template DNA result in truncated data; a significantly decreased signal strength also will be observed in the File info window. Too much template in the sequencing reaction can also produce truncated data. If certain resins are used to remove RNA from the template preparation, contaminants may copurify with DNA that intensify this problem. Here, the proper amount of DNA necessary for obtaining high quality sequence data should be titrated. Another common cause of truncated data is incorrectly formulated dNTP/ddNTP mixes.

17. Low signal strength. This problem may be caused by insufficient template DNA or insufficient primer in the sequencing reaction. Check that the concentration of primer is correct. For some sequencing primers, the standard 55°C annealing temperature may be too high, and should be reduced. This is especially true for the linear amplification sequencing method. For example, the SP6 primer gives very good results with an annealing temperature of 50°C, but almost no signal is observed when the primer is used with a 55°C annealing temperature. Low signal also can be caused when sequencing reagents are incorrectly formulated or stored.

18. High background. DNA that contains a large amount of RNA will produce sequence data with high background. This is especially true for the linear amplification sequencing method. Another common cause of high background is the presence of more than one template in the DNA preparation. This often happens because more than one colony or plaque has been inadvertently picked, and can be avoided by carefully picking only single colonies or plaques. Other background problems can originate when contaminated reagents are used for either template preparation or sequencing reactions.

Acknowledgments

The authors wish to thank R. Waterston and J. Sulston and all of the members of the *C. elegans* genome sequencing project for advice, support, and technical assistance.

References

1. Smith, L. M., Sanders, J. Z., Kaiser, R. J., Hughes, P., Dodd, C., Connell, C. R., Heiner, C., Kent, S. B. H., and Hood, L. E. (1986) Fluorescence detection in automated DNA sequencing. *Nature* **321,** 674–679.
2. Ansorge, W., Sproat, B. S., Stegemann, J., Schwager, C., and Zenke, M. (1987) Automated DNA sequencing: ultrasensitive detection of fluorescent bands during electrophoresis. *Nucl. Acids Res.* **15,** 4593–4602.
3. Prober, J. M., Trainor, G. L., Dam, R. J., Hobbs, F. W., Robertson, C. W., Zagursky, R. J., Cocuzza, A. J., Jensen, M. A., and Baumeister, K. (1987) A system for rapid DNA sequencing with fluorescent chain-terminating dideoxynucleotides. *Science* **238,** 336–341.
4. Brumbaugh, J. A., Middendorf, L. R., Grone, D. L., and Ruth, J. L. (1988) Continuous, on-line DNA sequencing using oligodeoxynucleotide primer with multiple fluorophores. *Proc. Natl. Acad. Sci. USA* **85,** 5610–5614.
5. Kambara, H., Nishikawa, T., Katayama, Y., and Yamaguchi, T. (1988) Optimization of parameters in a DNA sequenator using fluorescence detection. *Biotechnology* **6,** 816–821.
6. Connell, C. R., Fung, S., Heiner, C., Bridgham, J., Chakerian, V., Heron, E., Jones, B., Menchen, S., Mordan, W., Raff, M., Recknor, M., Smith, L., Springer, J., Woo, S., and Hunkapiller, M. (1987) Automated DNA sequence analysis. *BioTechniques* **5,** 342–348.
7. Tabor, S. and Richardson, C. C. (1987) DNA sequence analysis with a modified bacteriophage T7 DNA polymerase. *Proc. Natl. Acad. Sci. USA* **84,** 4767–4771.
8. Tabor, S. and Richardson, C. C. (1989) Effect of manganese ions on the incorporation of dideoxynucleotides by bacteriophage T7 DNA polymerase. *Proc. Natl. Acad. Sci. USA* **86,** 4076–4080.
9. Wilson, R. K., Chen, C., and Hood, L. (1990) Optimization of asymmetric polymerase chain reaction for fluorescent DNA sequencing. *BioTechniques* **8,** 184–189.
10. Wilson, R. K., Chen, C., Avdalovic, N., Burns, J., and Hood, L. (1990) Development of an automated procedure for fluorescent DNA sequencing. *Genomics* **6,** 626–634.
11. Koop, B. F., Wilson, R. K., Chen, C., Halloran, N., Sciammis, R., and Hood, L. (1990) Sequencing reactions in microtiter plates. *BioTechniques* **9,** 32–37.
12. Seto, D. (1991) A temperature regulator for microtiter plates. *Nucleic Acids Res.* **19,** 2506.

CHAPTER 52

Terminal Labeling of DNA for Maxam and Gilbert Sequencing

Eran Pichersky

1. Introduction

There are two basic enzymatic activities that are used to end-label DNA with radioactive phosphate (^{32}P). The enzyme T4-polynucleotide kinase will use the substrate ATP to add a phosphate group (the gamma-phosphate of the ATP molecule) preferentially to the 5' ends of the molecule. The enzyme DNA polymerase will "fill-in" recessed 3' ends with the complimentary nucleotides, and can also create recessed 3' ends from "blunt" ends or even 3' overhanging ends by its 3'–5' exonuclease activity, then fill in such ends as well. However, the polymerase enzyme usually used for the fill-in reaction, the Klenow fragment of *E. coli* DNA Pol I, has very low 3'–5' exonuclease activity, and therefore its fill-in activity on blunt and 3' overhanging ends could in principle be ignored for the purpose of end-labeling DNA for sequencing (*see* Note 1).

A linear double-stranded DNA molecule produced by digestion with a restriction enzyme, almost always the starting material for end-labeling DNA (since virtually all restriction enzymes work on double-stranded DNA), has in effect two 5' ends and two 3' ends. Thus, labeling with the T4 DNA polynucleotide kinase will label the two 5' ends (at opposite ends of the double-stranded DNA molecule), and labeling with the Klenow enzyme will produce two labeled 3' ends, again at opposite ends of the double-stranded molecule. Note, however, that each *strand* is labeled on one end only.

There are two alternatives to obtaining labeled DNA suitable for chemical sequencing. One method is to denature the double-stranded molecule and isolate the two strands (each of which is labeled at one end only) separately. Although it is in principle sometimes possible (for example, by a long electrophoretic run on a denaturing gel), the alternative method is much easier and therefore isolating end-labeled single-stranded DNA for sequencing is almost never done. The alternative is to digest the double-stranded DNA molecule, labeled at both ends, with another restriction enzyme that cuts in between the two ends, producing two fragments (hopefully of unequal length) each labeled at only one end of the four ends of the double-stranded molecule. The two fragments can then be separated on a nondenaturing acrylamide gel, reisolated, and used directly for sequencing. Although such a fragment is a double-stranded molecule, only one strand in this molecule is labeled, and that strand is labeled on one end only. All the chemical reactions are carried out as usual. The sample is then denatured and electrophoresed (*see* Chapter 53); the degradation products of the unlabeled strand do not show up on the autoradiograph and otherwise have no effect on the sequencing process *(1)*.

An important issue for consideration is the starting material for the labeling reaction. The DNA of interest may be cloned in a plasmid, or it may be a recombinant phage, or a PCR product. We find that CsCl gradient-purified DNA is the best source, although plasmids obtained by various "mini-prep" procedures are also suitable. The DNA has to be cut with a restriction enzyme, producing a minimum of one linear fragment or more. If the source of the DNA produces more than one fragment on restriction digest and one wishes to sequence only one particular fragment, it would appear to be advantageous to isolate this fragment prior to the end-labeling reaction, for example by electroelution from an agarose gel. However, each purification step results in some loss of DNA (and it is always advantageous to reisolate the fragments again after the labeling reaction, to remove the unincorporated label and, in cases where more than one fragment is labeled, to separate the fragments from each other). In addition, fragments that have been eluted from agarose gels do not label as well as fragments still in the original restriction enzyme reaction tube. Thus, for the highest yield and specific activity, it is best to do all the steps in one reaction tube without any successive purification steps or buffer changes. The fragments are then reisolated by separating them

DNA End-Labeling

on acrylamide gels, meaning that they should be in the range of 100–1000 nucleotides (*see* Note 2).

How can a DNA fragment labeled on one end only be produced in a single reaction tube without any buffer changes? To label DNA on one end only we take advantage of the fact that the Klenow fragment of *E. coli*'s DNA polymerase I will label a recessed 3' end ("fill-in" reaction) but will not label a recessed 5' end (=overhanging 3' end) or a blunt end (in principle, the enzyme will label the latter two ends also, but to a much lower extent than the recessed 3' end, so the fragment is practically labeled at one end only; *see* Notes 1 and 3). Thus, if a restriction site giving a recessed 3' end is present and we wish to label it for sequencing, all that is required is to find a second site nearby that is cut by an enzyme producing ends that will not label. For example, the restriction enzymes *Pst*I, *Sac*I, *Sph*I (3' overhang), and *Dra*I, *Hae*III, and *Rsa*I (blunt end) produce sites that under the conditions we use are practically unlabeled in the reaction described below. It should be noted that because the labeled fragments need to be separated on acrylamide gels, there is an upper size limit to the fragment that should be produced (fragments > 1 kb will not be resolved well on the gel, and will also not diffuse out efficiently). When there is no known useful site nearby the site that is being labeled, we simply use a frequent 4-bp cutter, such as *Hae*III, on the assumption that a site does occur at some not-too-great distance from the site being labeled. If that is not the case, additional 4-bp cutters can be used until the appropriate one is found. Also, when a small fragment is cloned into the polylinker site of a plasmid, one can take advantage of the restriction sites of the polylinker, so that two enzymes are chosen, one at each end of the cloned fragment, with one enzyme producing a 3' recessed end and the other a 3' overhanging end. In this way the fragment can be labeled on one end only, and by choosing a different pair of enzymes in a separate reaction, the other end of this fragment can be labeled as well.

2. Materials

1. DNA: CsCl-purified plasmid, 1 µg/µL.
2. Restriction enzymes.
3. Restriction enzyme buffer (supplied by the manufacturer).
4. Klenow fragment of DNA Pol I.
5. [α-^{32}P]dATP; Specific activity = 3000 Ci/mmol.
6. 10 m*M* dCTP.

7. 10 mM dTTP.
8. 10 mM dGTP.
9. Loading dye: 50 mM EDTA, 10 mM HCl, pH 7.5, 0.1% (w/v) bromophenol blue, 0.1% (w/v) xylene cyanol FF, 50% glycerol.
10. 10X TBE buffer: 0.5M Tris-HCl, 0.5M boric acid, 10 mM EDTA, pH 7.5 *(2)*.
11. Acrylamide.
12. Bisacrylamide.
13. 100% ethanol.
14. 0.5M NH$_4$Acetate, 1 mM EDTA, pH 7.5.
15. Glass pipets.
16. Glasswool.
17. ddH$_2$O (dd = double-distilled).

3. Method

1. Digest 1 µg of DNA in 30 µL total volume that includes 1–3 U of each enzyme (e.g., 2 U *Eco*RI and 2 U *Hae*III). Incubate at 37°C for 30 min.
2. Mix the following: 1 µL of dTTP, 1 µL of dCTP, 1 µL of dGTP, 1 µL of ^{32}P(α)-dATP, and 1 U of Klenow enzyme, then add to the reaction tube and incubate for 20–30 min at 37°C.
3. Stop the reaction by adding 20 µL of loading dye (**Note:** This is the same dye used for agarose gels, but is not the dye used in sequencing gels).
4. Load sample on a 15-cm long, 0.3-mm thick nondenaturing 5% acrylamide gel (20:1 acrylamide:bisacrylamide, 1X TBE buffer) with well width of 3–4 cm. Run the gel until the bromephenol blue dye is at the bottom. Stop the gel, take one glass plate off, then cover the gel with Saran Wrap™. In the darkroom, put an X-ray film over the covered gel, mark the film's position on the gel, and expose for 5–10 min. Then develop the film.
5. By putting the developed X-ray film against the gel, locate the position of the labeled DNA fragments, cut and remove the appropriate gel region, and place it in a 1.5-mL Eppendorf tube. Do not cut the gel piece into smaller pieces. Overlay with 0.6 mL 0.5M NH$_4$Acetate, 1 mM EDTA, pH 7.5 solution, shake well, and make sure the entire gel piece is submerged, then incubate at 37°C for a minimum of 8 h (or overnight).
6. Pipet out the contents of the tube into a second tube, passing the solution through a pipet fitted with glasswool at the bottom. Add 1 mL 100% ethanol to the tube, shake well, and immediately centrifuge for 10 min. Decant the supernatant, fill the tube again with ethanol, spin for 2 min, then decant again and aspirate the rest of the solution with a drawn Pasteur pipet. Dry in vacuum for 10 min. Add ddH$_2$O (50–100 µL, depending on the amount of radioactivity) and resuspend the sample. The DNA is ready for sequencing (*see* Chapter 53).

DNA End-Labeling

4. Notes

1. The labeling reaction with the Klenow enzyme is done directly in the restriction digest tube, without any change of buffer. The restriction digests are done with either low-, medium-, or high-salt buffers, according to the requirement of the enzymes, and all these buffers also contain Mg^{2+}. The Klenow enzyme works well in all these buffers, so for labeling all that is required is the addition of the cold and radioactive nucleotides, and the Klenow enzyme. It is also possible to do the entire process at once—add the restriction enzymes *and* the nucleotides and the Klenow enzyme together. However, I prefer to do the labeling reaction for only 20–30 min, and the restriction of 1 µg of DNA should usually be allowed to proceed for about 1 h. Hence, the labeling mix should be added about 30 min after the start of the restriction reaction.

   I prefer to use ^{32}P-dATP when possible, since this is the standard radioactive nucleotide used for all other purposes in my lab. Almost all 3' recessed sites can be labeled with dATP. It is essential to add the other nucleotides that occur in the site as cold nucleotides. If they occur upstream to the position of the A nucleotide, the latter nucleotide cannot be added until the others are. If they occur downstream to the position of the A nucleotide, it is nevertheless advantageous to include them in the reaction because they are present at much higher concentrations than the ^{32}P-dATP, so once the A nucleotide is added, the positions downstream are added with almost 100% efficiency, and the Klenow enzyme cannot go back and remove the A nucleotide with its 3'–5' exonuclease activity. As a matter of routine I add all three cold nucleotides (dTTP, dCTP, dGTP) even when one or another of these nucleotides do not occur at the site. The unneeded nucleotide has no effect on the labeling reaction. Another beneficial effect of the cold nucleotides is that at the other end (usually a 3' overhanging end or a blunt end), even if the 3'–5' exonuclease activity of the Klenow enzyme produced a 3' recessed end, it would instantaneously be filled in with a cold nucleotide, except, of course, when that position requires an A nucleotide. It would seem that in this latter case one would get a higher background in the sequencing autoradiograph, but I have not found this to be a problem, possibly because of the slowness of the 3'–5' exonuclease activity of the Klenow enzyme and the relative shortness (20 min) of the fill-in reaction.

2. I use 5% nondenaturing acrylamide gels to purify labeled fragments. The fragments to be isolated are in the 100–1000 bp range. It is not worth the effort to sequence shorter fragments, although it is possible to isolate such fragments on acrylamide gels, and fragments >1 kb do not resolve well on such gels. Although large fragments can be isolated from agarose gels, in

my hands I have never been able to sequence fragments that have been directly isolated from agarose gels. (The DNA could go through agarose gel prior to labeling if it is subsequently reisolated from acrylamide gels prior to sequencing, but fragments sequenced directly after isolation from agarose gels do not yield readable sequence.)

To elute the fragments, I simply incubate the gel slice, without any further maceration, in the elution buffer overnight at 37°C. The DNA fragment diffuses out quite well, although larger fragments diffuse more slowly. After overnight incubation, tubes containing fragments of approx 1 kb will have at least 50% of the radioactivity in solution, and that fraction is higher for smaller fragments. The solution is then passed through glasswool to remove large pieces of acrylamide gel. The remaining acrylamide (monomers or small polymers) does not interfere with any of the sequencing reactions or subsequent manipulations of the DNA, even though it does precipitate together with the DNA.

To precipitate the labeled DNA, I add 2 vol of 100% ethanol at room temperature, mix, and immediately centrifuge. There is no need to use any other concentration of ethanol. The goal is to get the DNA to precipitate with as little salt coprecipitation as possible. This is accomplished by aspirating *all* the liquid after the ethanol precipitation step, and again after the ethanol wash step. The pellet forms nicely on the side of the tube, and it is easy to put the end of the stretched Pasteur pipet all the way to the bottom of the tube and aspirate all the liquid. Repeated cycles of resuspension and precipitations are *unnecessary* and *inadvisable*. Prolonged incubation at low temperatures is also strongly discouraged. DNA precipitates well at room temperature, but one gets more salt precipitation at low temperature, thus making things worse, not better.

3. The DNA-labeling procedure involving T4-polynucleotide kinase is lengthy and requires the use of a substantial amount of radioactivity and several changes of buffers. Because of these disadvantages, I never use this enzyme for end-labeling for sequencing.

References

1. Maxam, A. M. and Gilbert, W. (1980) Sequencing end-labeled DNA with base-specific chemical cleavages. *Method. Enzymol.* **65,** 499–560.
2. Sambrook, J., Fritsch, E. F., and Maniatis, T. (1989) *Molecular Cloning, A Laboratory Manual.* Cold Spring Harbor Laboratory, Cold Spring Harbor, NY.

CHAPTER 53

DNA Sequencing by the Chemical Method

Eran Pichersky

1. Introduction

The chemical method of sequencing DNA *(1)* has some advantages and some disadvantages compared with the enzymatic method *(2)*. The major disadvantage is that it takes more time to produce the same amount of sequence. This is so for two main reasons. First, the DNA has to be end-labeled and then reisolated prior to the actual chemical sequencing reactions, a process that usually requires an additional day (*see* Chapter 52). Also, because more DNA is used in the reaction and because the lower specific activity of the sequenced DNA requires the use of an intensifying screen in the autoradiography, bands are not as sharp as in the enzymatic method and therefore it is difficult to obtain reliable sequence past about nucleotide 250 (unless very long gels are run).

Nevertheless, the chemical method is often useful for several reasons. It enables one to begin sequencing anywhere in the clone where a restriction site that can be labeled occurs without any further subcloning. The sequence thus obtained can then be used to synthesize oligonucleotide primers for enzymatic sequencing. In addition, in cases of regions that give poor results in the enzymatic reactions (because of secondary structures that inhibit the polymerase enzyme), the chemical method almost always resolves the problem and yields the correct sequence.

The chemical sequencing reaction has acquired a reputation of being difficult. I believe this reputation is undeserved. In my hands, the chemical method is consistently successful in producing results as reliable as those obtained by the enzymatic method. It has been my experience that

From: *Methods in Molecular Biology, Vol. 58: Basic DNA and RNA Protocols*
Edited by: A. Harwood Humana Press Inc., Totowa, NJ

447

many protocols in molecular biology include unnecessary steps. The likely explanation is probably that when researchers encountered difficulties, they added these steps as a solution to the problem, often on the assumption that the additional steps would not hamper the process, even if they did not help. This is clearly not the case here. In developing the method presented here from pre-existing protocols, many steps have been *eliminated*. In general, I have found that the quality of the sequence has improved with the progressive elimination of these steps. It is still possible that some steps included here are not necessary; certainly no additional steps need to be added. And, of course, the end result has been that the protocol as presented here is very short and the entire process of sequencing (starting with end-labeled DNA) and gel electrophoresis can be accomplished in one (long) day.

2. Materials

1. G Buffer: 50 mM sodium cacodylate, pH 8.0.
2. CT/C Stop: 0.3M sodium acetate, 1 mM EDTA, pH 7.0.
3. AG Stop: 0.3M sodium acetate, pH 7.0.
4. G Stop: 1.5M sodium acetate, 1M 2-mercaptoethanol, pH 7.0.
5. 10% Formic acid.
6. Dimethyl sulfate (DMS) (*see* Note 1).
7. Hydrazine (95% anhydrous) (*see* Note 1).
8. 100% ethanol.
9. 5M NaCl.
10. ddH$_2$O (dd = double distilled).
11. Carrier DNA: 1 mg/mL in ddH$_2$O (any DNA will do; we use plasmid DNA).
12. 10 mg/mL tRNA in ddH$_2$O (any tRNA).
13. The DNA fragment to be sequenced, end-labeled at one end only, in ddH$_2$O (*see* Chapter 52).
14. 10% piperidine (dilution prepared on the day of the experiment).
15. Loading buffer: 100% formamide, 0.1% (w/v) bromophenol blue, 0.1% (w/v) xylene cyanol FF.

3. Method

1. For each DNA fragment to be sequenced, mark four 1.5-mL Eppendorf "reaction" tubes and add the following solutions:
    "G" tube: 1 µL of carrier DNA, 200 µL of G buffer, 5 µL of labeled DNA.
    "AG" tube: 1 µL of carrier DNA, 10 µL of ddH$_2$O, 10 µL of labeled DNA.

"CT" tube: 1 µL of carrier DNA, 10 µL of ddH$_2$O, 10 µL of labeled DNA.

"C" tube: 1 µL of carrier DNA, 15 µL of 5*M* NaCl, 5 µL of labeled DNA.

2. Mark four 1.5-mL Eppendorf "stop" tubes and add the following solutions:

    "G Stop" tube: 2 µL of tRNA, 50 µL of "G Stop" solution, 1 mL of ethanol.

    "AG Stop" tube: 2 µL of tRNA, 200 µL of "AG Stop" solution, 1 mL of ethanol.

    "CT Stop" tube: 2 µL of tRNA, 200 µL of "CT/C Stop" solution, 1 mL of ethanol.

    "C Stop" tube: 2 µL of tRNA, 200 µL of "CT/C Stop" solution, 1 mL of ethanol.

3. To start the reactions (*see* Note 2), add the following:

    "G" tube: 1 µL of DMS, mix, and let the reaction proceed for 5 min at room temperature.

    "AG" tube: 3 µL of 10% formic acid and mix (15 min at 37°C).

    "CT" tube: 30 µL of hydrazine and mix (9 min at room temperature).

    "C" tube: 30 µL of hydrazine and mix (11 min at room temperature).

4. Stop each reaction by pipeting the contents of the corresponding stop tube into the reaction tube (use the same Pasteur pipet; slight crosscontamination of stop solutions has no effect, but do not touch the contents of the reaction solutions with the pipet). Cap the reaction tubes, shake briefly but vigorously, and place in a dry ice-ethanol bath (–80°C) for 3–10 min (3 min are enough, but the tubes can be left there for up to 10 min if other reactions are not done yet; do not leave for longer than 10 min) (*see* Note 3).

5. Centrifuge at 4°C for 7 min, discard supernatant, aspirate the rest of the liquid with a drawn Pasteur pipet, and then add 1 mL of 100% ethanol to the tube, invert twice, and centrifuge for 2 min at room temperature. Aspirate as before, and dry in a vacuum for 10 min.

6. To each reaction tube, add 100 µL of the 10% piperidine solution (do not shake the tubes because there is no need to resuspend the DNA) and place the uncapped tubes in a 90°C heat-block. After 15–30 s, cap the tubes and let stand for 30 min.

7. Remove the tubes from the heat-block, let stand at room temperature for 2–5 min, and centrifuge briefly to get the condensation to the bottom. Puncture one hole in the cap with a syringe, then place in dry ice-ethanol bath for 5 min.

8. Place the tubes in a SpeedVac machine and lyophilize for 2 h. Vacuum should be below 100 millitorr.

9. Prior to gel electrophoresis, add 10 µL of loading buffer to each tube, resuspend the sample by shaking and then a brief centrifugation, and place the tubes in the 90°C heat-block for 10 min. Load 1–2 µL per sample (*see* Notes 4 and 5).

4. Notes

1. Quality of chemicals: In general, the standard laboratory grade chemicals should be used. Some chemicals, however, could be the cause of problems when not sufficiently pure or when too old (presumably degradation products are the culprits). I have only had problems with two chemicals *(see below)*: dimethyl sulfate and hydrazine. Note also that these two chemicals, together with piperidine, are hazardous chemicals. In addition to observing the rules pertaining to the handling of radioactive chemicals, all reactions involving these three chemicals should be carried out in the hood.
2. I typically do all chemical reactions together, timing them so that they *end* at the same time. If one is sequencing 5 different fragments, all 20 tubes can be spun together in a single run (we have a microcentrifuge with 20 slots). When stopping all 20 reactions at about the same time, some reactions are invariably going to run longer than the allotted time. This is usually not a problem, because the reaction times indicated above are general, and they can be extended by up to 30% without much noticeable effect.
3. I always use 100% ethanol. There is no need to use any other concentration of ethanol at any step of the process, and 100% ethanol has the advantage because it evaporates fast. The goal is to get the DNA to precipitate with as little salt coprecipitation as possible. This is accomplished by aspirating all the liquid after the ethanol precipitation step, and again after the ethanol wash step. The pellet forms nicely on the side of the tube, and it is easy to put the end of the stretched Pasteur pipet all the way to the bottom of the tube and aspirate all the liquid. Repeated cycles of resuspension and precipitations are *inadvisable*. Prolonged incubation in the dry ice-ethanol bath is also strongly discouraged. DNA precipitates well at room temperature, but one gets more salt precipitation at low temperature, thus making things worse, not better. I almost always precipitate DNA at room temperature; the only reason step 4 calls for incubation at –80°C is to inhibit further reaction with the reactive reagents that at this stage have not yet been removed.
4. Gel electrophoresis: I use a 60 × 40 cm (0.3-mm thick) gel of 6% acrylamide (20:1 acrylamide:bisacrylamide, 50% urea, 50 mM Tris, 50 mM borate, 1 mM EDTA *[3]*). I run the gel at constant power (65 W), with an aluminum plate to disperse the heat. The samples are loaded twice (a "long run" and a "short run"): The second loading is done when the xylene cyanol dye of the first sample is approximately two-thirds of the way down the

Chemical DNA Sequencing

gel, then electrophoresis is halted when the bromophenol blue of the second sample reaches the bottom of the gel. The complete run takes 5–6 h, and it allows us to read, in the "short run," the sequence from about nucleotide 25 to nucleotide 120–150, and in the "long run" the sequence from nucleotide 100 to about 220–250. Additional sequence may be obtained by running longer gels, by loading the sample a third time, or by a variety of other methods if so desired.

5. Troubleshooting:
   G reaction: This reaction is usually very clean, but it is the reaction most sensitive to prolonged incubation and to the quality of the reactive reagent, DMS. If reaction proceeds longer than the allotted time or if old or bad quality dimethyl sulfate is used, excessive and nonspecific degradation of DNA will occur. Also, when several Gs occur in a row, the 3'-most Gs (lower bands if the 3' end was labeled with the Klenow enzyme) may appear weaker.
   AG reaction: This is usually a trouble-free reaction.
   CT and C reactions: The quality of the hydrazine should be good (it does not have to be exceptional), otherwise excessive and nonspecific degradation will occur. Sometimes faint bands will be seen in the C and CT lanes when the base is G (a strong band is then observed in the G lane). The likely explanation is that the pH in the reaction tubes is too low (there is no buffer in the C and CT reaction tubes, but carryover with the DNA sample might cause this to happen). However, these faint bands are not nearly as strong as the signal in the G lane or as bona fide bands of C and T bases. Also, the T bands in the CT lane are often not as strong as the C bands—this is probably caused by inhibition of the reaction by residual salt (in the C reaction, salt is added specifically to obtain complete inhibition).

References

1. Maxam, A. M. and Gilbert, W. (1980) Sequencing end-labeled DNA with base-specific chemical cleavages. *Meth. Enzymol.* **65,** 499–560.
2. Sanger, F., Nicklen, S., and Coulson, A. R. (1977) DNA sequencing with chain-terminating inhibitors. *Proc. Natl. Acad. Sci. USA* **74,** 5463–5467.
3. Sambrook, J., Fritsch, E. F., and Maniatis, T. (1989) *Molecular Cloning, A Laboratory Manual*. Cold Spring Harbor Laboratory, Cold Spring Harbor, NY.

PART VII

SITE DIRECTED MUTAGENESIS AND PROTEIN SYNTHESIS

CHAPTER 54

Site-Directed Mutagenesis of Double-Stranded Plasmids, Domain Substitution, and Marker Rescue by Comutagenesis of Restriction Enzyme Sites

Jac A. Nickoloff, Win-Ping Deng, Elizabeth M. Miller, and F. Andrew Ray

1. Introduction

Oligonucleotide-directed mutagenesis can be used to introduce specific alterations in cloned DNA, including base-pair substitutions, insertions and deletions. It is an important procedure in studies of gene expression and protein structure/function relationships. A variety of protocols have been developed to mutate specific bases in plasmid DNA, all of which involve annealing oligonucleotide primers with specific mutations to target sequences (*1–4;* reviewed in *5*). Primer-directed DNA synthesis creates a DNA strand containing the desired mutation that may segregate from the wild-type template strand during DNA replication following transformation into an *E. coli* host. Although the theoretical yield of mutant products is 50%, the actual yield is usually much lower, necessitating strategies for enriching for products derived from the mutant strand. Two successful strategies for mutant enrichment involved the in vivo degradation of wild-type strands containing uracil bases in place of thymidine (*3,6; see* Chapter 55) and selection of mutant strands carrying a second mutation that creates a functional selectable marker, such as the β-lactamase gene *(7)*. Systems that involve secondary select-

From: *Methods in Molecular Biology, Vol. 58: Basic DNA and RNA Protocols*
Edited by: A. Harwood Humana Press Inc., Totowa, NJ

able mutations depend on coupling of two primers during DNA synthesis, and are improved by the use of T4 DNA polymerase rather than the Klenow fragment of *E. coli* DNA polymerase, since the T4 enzyme does not displace the mutagenic primers *(8,9)*. The efficiency of coupled-primer mutagenesis is also improved if mismatch repair-deficient (mut S) hosts are used *(10)*, since this increases the probability that the selectable and nonselectable (desired) mutations cosegregate during the first round of DNA replication.

Although these earlier procedures produce mutations at high efficiency, they impose limitations on targets and primer design, and require steps to prepare single-stranded templates. Described here are two related procedures involving comutagenesis of restriction sites that are also efficient, but that offer increased speed and greater flexibility with respect to target choice, primer design, and strategies for introducing multiple mutations *(11,12;* Fig. 1). Both procedures are effective with virtually any plasmid, including double- and single-stranded plasmids. Single-stranded templates are produced by heat denaturation of double-stranded plasmids, and mutagenic primers are annealed that eliminate nonessential restriction site(s) (selectable mutations) and that introduce desired, nonselectable mutations. Selection against nonmutant plasmids occurs when a mixture of mutant and nonmutant plasmid DNA is treated with the restriction enzyme that recognizes the target nonessential site before transformation into *E. coli*. Such treatment reduces nonmutant plasmids and reduces their transformation efficiency 10- to 1000-fold below that of mutant plasmids, which are resistant to digestion and remain circular. Desired mutations are obtained when selected mutations are linked to desired mutations during primer-directed DNA synthesis, which normally occurs at frequencies >80%. The two procedures differ in the way the mutagenic primers are linked in vitro. In the original procedure, termed Unique Site Elimination (USE) mutagenesis, mutagenic primers complementary to one strand are linked during second-strand synthesis *(11)*. In the second procedure, primers complementary to opposite strands are linked during PCR; subsequently, the denatured PCR product (a "long primer") directs second-strand synthesis (LP-USE mutagenesis; *12*).

Selection for the elimination of restriction sites is both simple to perform and highly general. Most plasmids carry one or more restriction sites that are suitable targets for introducing selectable mutations (USE sites), or suitable USE sites may be engineered. Because circular single-

Oligonucleotide-Directed Mutagenesis

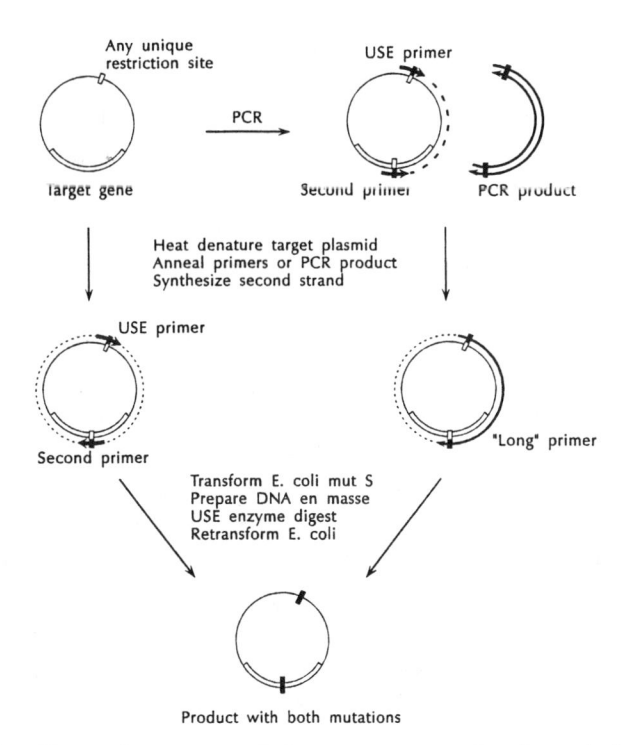

Fig. 1. Site-directed mutagenesis by comutagenesis of restriction sites. The target plasmid with a unique, nonessential restriction site (open bar) is shown at top. USE and LP-USE mutagenesis are diagrammed on the left and right, respectively. Mutagenic primers are shown by black arrows and mutations by black bars. Target DNA is heat-denatured, and primer-directed DNA synthesis (dashed lines) links two mutagenic primers or extends PCR product with linked primers. Mutations introduced by USE primers are selectable. DNA is transformed into an *E. coli* mut S strain to increase cosegregation of the two mutations. DNA prepared from mut S transformants grown *en masse* is digested with the restriction enzyme that recognizes the unique site and transformed into a standard *E. coli* strain. Recovered products usually have both mutations. Selection power may be increased by using additional USE primers targeted to different sites.

stranded templates can be produced simply by heat denaturation of double-stranded plasmid DNA, virtually any double-stranded plasmid can be efficiently mutated with these procedures. Circular single-stranded DNAs produced from M13 phage or phagemids are also suitable templates. Since target genes can be mutagenized without being subcloned, these procedures are well suited to mutagenesis of genes in expression vectors. An additional benefit of starting with double-stranded plasmids is that the mutagenic primers can be complementary to either strand. This feature facilitates the design of multipurpose primers for use in mutagenesis, DNA sequencing, hybridization, and PCR amplification. Furthermore, primer pairs can be either complementary to one strand *(11)* or they may be complementary to opposite strands, if PCR is performed first *(12)*.

General selection primers are designed to eliminate restriction sites in sequences shared by most vectors. With this strategy, a small set of selection primers is required for mutagenizing most plasmids. USE sites can be present outside or within genes, including essential genes, such as those conferring antibiotic resistance. If target sites are within an essential gene, phenotypically silent mutations are introduced by changing bases in the third position of a codon or by creating inframe insertion mutations. Another valuable feature of some selection primers is that they eliminate a restriction site by converting it to another site. If the newly created site is also unique, multiple mutations can be introduced into target genes sequentially by "cycling" between unique sites. Table 1 lists a set of USE primers with these features. Alternatively, multiple mutations can be introduced simultaneously with USE mutagenesis simply by adding more mutagenic primers to synthesis reactions. The frequency of simultaneous multiple mutagenesis is approximately equal to the product of individual mutagenesis frequencies. Although it is possible to introduce multiple mutations simultaneously with the PCR-based procedure (by adapting a PCR-based mutagenesis procedure; *see* ref. *13*), it is simpler to introduce simultaneous mutations with the standard procedure.

These procedures were originally termed USE mutagenesis since the simplest versions involved selecting for the loss of a single, unique restriction site. However, additional flexibility and selective power can be gained if multiple sites are eliminated. It is possible to select for the elimination of two (or more) identical or nonidentical sites. When multiple sites are mutagenized, selection pressure against nonmutants is increased, a factor that may be important if very large plasmids (>10 kbp) are employed.

Restriction site comutagenesis procedures can be adapted to transfer mutations from a gene in any genome to a cloned copy of the gene or from one plasmid context to another. In these adaptations, mutations are

Oligonucleotide-Directed Mutagenesis

Table 1
General Unique Site Elimination Primers

Target site	New site	Sequence	Vectors
1a. AatII	SalI	GTGCCACCTG**TCGA**CTAAGAAACC	1,2,4
1b. SalI	AatII	GTGCCACCTG**ACGT**CTAAGAAACC	1,2,4
2a. AflIII	BglII	CAGGAAAGAA**GATCT**GAGCAAAAG	1–6
2b. BglII	AflIII	CAGGAAAGAA**CATGT**GAGCAAAAG	1–6
3a. ScaI	MluI	CTGTGACTGGTGA**CGCG**TCAACCAAGTC	1–6
3b. MluI	ScaI	CTGTGACTGGTGA**GTACT**CAACCAAGTC	1–6
4a. SspI	EcoRV	CTTCCTTTTTC**GATAT**CATTGAAGCATTT	1,2,4
4b. EcoRV	SspI	CTTCCTTTTTC**AATAT**TATTGAAGCATTT	1,2,4
5. AlwNI	PvuII	CTGGCAGCAG Δ **CTG**GTAACAG	1–6
6. XmnI	None	TTG**G**AAAACGCTCTTCGGGGCG	1,3–6
7. BsaI	None	TGATACCGC**GGG**ACCCACGCTC	1–6
8. ScaI	None	CTGGTGA**GTA**TTCAACCAAGTC	1–6
9. EcoRI	EcoRV	CGGCCAGTGA**TATC**GAGCTCGG	1
10. HindIII	MluI	CAGGCATGCA**CGCG**TGGCGTAATC	1

Target restriction sites and newly created sites are given for intergenic and polylinker target sites. Target codons are given for sites in the β-lactamase gene (lower-case letters). Sequences are complementary to the sense strand of the β-lactamase gene. Newly formed restriction sites (or mutant target restriction sites if no new site is formed) are in bold type, and mutant bases are underlined. Vectors: 1, pUC19; 2, pBR322; 3, pBluescript and pBluescriptII; 4, pSP6/T3 and pSP6/T7-19; 5, pT7/T3α-19; 6, pTZ19R. Primer pairs (i.e., 1a and 1b) cycle new mutant and wild-type sites for sequential procedures. Primer 1b will work with indicated vectors if wild-type SalI sites are destroyed. Primer 2b will work with pSP6/T3 only if the wild-type BglII site is destroyed. Primer 4b will work with pBR322 if the wild-type EcoRV site is destroyed. The AlwNI primer deletes the nucleotides CCA (Δ). EcoRI and HindIII primers eliminate polylinker sites and cannot be used if DNA is inserted into sites present in mutagenic primers.

transferred without fragment subcloning so there are no requirements for natural or engineered restriction sites in the DNA from which mutations are to be transferred. The three adaptations of LP-USE mutagenesis described here represent powerful systems, applicable to any organism, for analyzing the function of specific genes or domains within genes. For example, starting with a cloned copy of a gene, specific mutations may be produced by USE or LP-USE mutagenesis (Fig. 1). Mutant genes may be transferred to a host cell to assay for phenotypic effects of the mutation (i.e., by transfection), and the presence of the mutation is then confirmed by using the first adaptation of LP-USE mutagenesis (Fig. 2). The functional analysis of specific domains in related genes in gene families or from different species (e.g., ref. *14*) is facilitated by the second adaptation of LP-USE to effect domain substitution. Domain substitution is

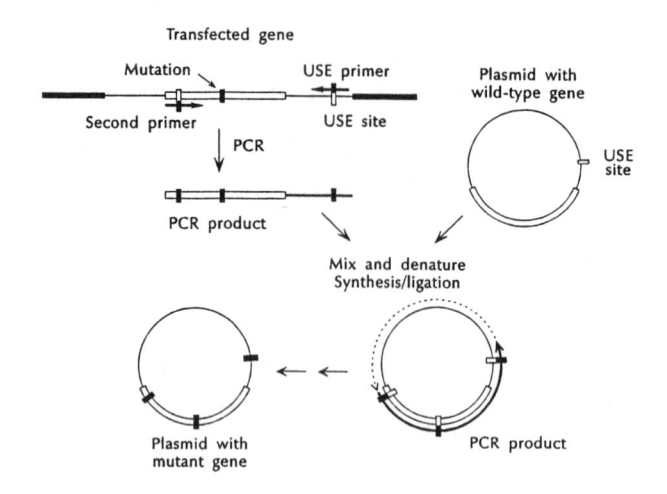

Fig. 2. Confirming mutations in transfected genes. A plasmid carrying a mutant gene (thin lines) is shown integrated into genomic DNA (heavy lines). PCR is performed with a USE primer and a second primer flanking the mutation, producing a fragment with three mutations (bars). The PCR product then directs second-strand synthesis with a plasmid carrying the wild-type gene as a template. Remaining steps are as shown in Fig. 1. The second primer may also carry a mutation that creates a new restriction site. Products that have lost the USE site and gained the new restriction site will carry the mutation originally present in the transfected gene, which is identified by DNA sequence analysis.

also convenient for transferring mutations from one plasmid context to another. Thus, mutations in a gene may be transferred from an expression vector with a constitutive promoter to one with an inducible promoter (Fig. 3). The third adaptation of LP-USE may be used to transfer unknown mutations efficiently in any gene for which a cloned copy of the gene or cDNA is available (marker rescue). In this procedure, mutations are copied from genomic DNA (or mRNA) by PCR amplification using primers flanking the gene of interest. A cloned copy of the gene or cDNA is engineered with USE sites flanking the region of interest. These USE sites are not present in the genomic DNA or mRNA, and will be absent in PCR products. Mutagenesis of the cloned gene with the PCR product directing DNA synthesis removes both USE sites and transfers to the cloned gene any mutations present in the mutant gene at high efficiency (Fig. 4). Mutations are then identified by DNA sequence analysis. Since cloned copies of genes may be present in expression plasmids, the

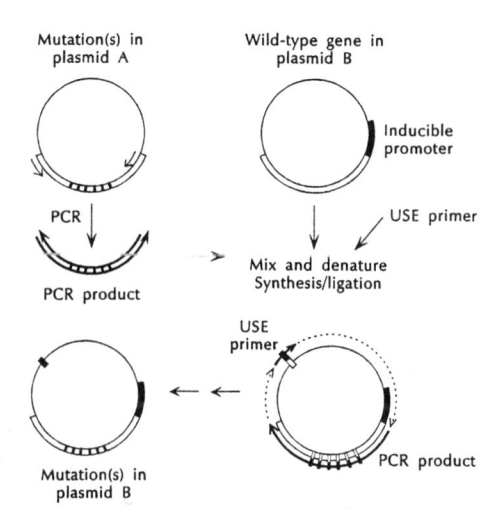

Fig. 3. Domain replacement. Mutations (black bars) are transferred from a copy of a gene in one type of plasmid ("A") to another type ("B"), which in this case is an expression vector with an inducible promoter. The denatured PCR amplification product of the mutant domain is linked to a USE primer during second-strand synthesis with circular single-stranded plasmid B as template. Remaining steps are as shown in Fig. 1.

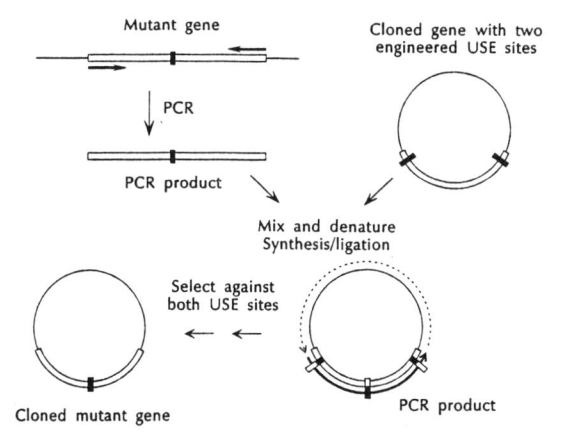

Fig. 4. Marker rescue. A mutation in a gene (black bar) is amplified with two wild-type PCR primers. The PCR product directs second-strand synthesis with a plasmid template carrying a cloned copy of the wild-type gene engineered with USE sites at either end of the gene (black bars). The USE sites may be created by inserting linkers into natural sites, or they may be created *de novo* by site-directed mutagenesis. Selection against both USE sites (as shown in Fig. 1) produces plasmids with the genomic mutation.

resulting mutant genes may be reintroduced into cells to confirm their phenotype. If a gene has many exons and a cDNA is not available, mutations in specific exons for which clones are available are first identified by SSCP analysis *(15)*. Exons with mutations are then amplified, and the mutations are transferred to a cloned copy of that exon and sequenced. Although mutations may be directly identified by sequencing PCR products, marker rescue has the advantage that rescued mutations may then be reintroduced into cells for phenotypic analysis.

2. Materials
2.1. Phosphorylating Mutagenic Primers
1. 10X kinase buffer: 700 m*M* Tris-HCl, pH 7.6, 100 m*M* MgCl$_2$, 50 m*M* dithiothreitol.
2. 10 m*M* ATP: Store at –20°C.
3. Mutagenic oligonucleotide primers: 100 pmol of oligonucleotides in distilled H$_2$O or PCR product with linked mutagenic primers (approx 0.005–1.5 pmol). (*see* Note 1)
4. T4 polynucleotide kinase: 10 U/µL, store at –20°C.

2.2. Site-Directed Mutagenesis
5. Plasmid DNA with one or more nonessential restriction sites.
6. *E. coli* mut S strain, e.g., BMH 71–18 mut S *(10)*.
7. 10X annealing buffer: 200 m*M* Tris-HCl, pH 7.5, 100 m*M* MgCl$_2$, 500 m*M* NaCl.
8. 10X synthesis/ligation buffer: 100 m*M* Tris-HCl, pH 7.5, 5 m*M* each of dATP, dTTP, dGTP, dCTP, 10 m*M* ATP, 20 m*M* dithiothreitol.
9. T4 DNA polymerase: (10 U/µL).
10. T4 DNA ligase: (400 U/µL).
11. 10X stop mix: 50% glycerol, 0.5% SDS, 10 m*M* EDTA, 0.25% bromphenol blue, 0.25% xylene cyanol FF.
12. Sepharose CL-6B: *see* Chapter 27.
13. Reagents and equipment for preparation and transformation of competent cells, preparation and restriction digestion of plasmid DNA, and spin-column chromatography.

3. Methods
3.1. Phosphorylation of Oligonucleotide Primers or PCR Product (see Note 2)
1. To a 1.5-mL tube, add 2 µL of 10X kinase buffer, 2 µL of 10 m*M* ATP, 100 pmol of oligonucleotide (about 0.2 µg of a 21-base oligonucleotide) or 0.005–1.5 pmol of PCR product, and H$_2$O to 20 µL.

Oligonucleotide-Directed Mutagenesis

2. Add 10 U of T4 polynucleotide kinase, and incubate for 2 h at 37°C.

There is no need to purify phosphorylated oligonucleotides or PCR products before use in second-strand synthesis.

3.2. Site-Directed Mutagenesis

The procedure is performed in three phases:

a. In vitro second-strand/ligation reaction, transformation of *E. coli* mut S host, and overnight growth en masse,
b. Plasmid DNA purification, restriction enzyme digestion, and transformation of a normal *E. coli* host, and
c. Purification and identification of mutant plasmids.

Controls important for assessing the efficiency of individual steps are included in the following description.

1. To anneal the mutagenic primers to the target plasmid, add to a 1.5-mL microcentrifuge tube 5 µL of phosphorylated primers (25 pmol) or 5 µL of phosphorylated PCR product, 0.025 pmol of plasmid DNA (about 0.1 µg), 2 µL of 10X annealing buffer, H_2O to 20 µL (*see* Note 3).
2. Mix solution, and heat to 100°C for 3 min. Quickly place tube in an ice bath for 1 min, and then spin for 5 s in a microcentrifuge to collect condensation. Incubate at room temp for 30 min.
3. Add to the annealing mixture: 3 µL of 10X synthesis/ligation buffer, 3 U of T4 DNA polymerase, 400 U of T4 DNA ligase, H_2O to 30 µL (add enzymes last). Mix solution and incubate for 30 min at 37°C if oligonucleotide primers are used, or 5 min at 37°C if PCR products >1000 bp are used (*see* Note 4).
4. Stop the reaction by adding 3 µL of 10X stop mix, vortex briefly, and incubate for 5 min at 65°C. Purify DNA by passing the solution through a Sepharose CL-6B spin column.
5. Transform *E. coli* strain BMH 71–18 mut S (or other appropriate mut S strain) by electroporation or another standard procedure (*see* Note 5). Suspend cells in 1 mL of SOC, and incubate for 1 h at 37°C. Assay transformation efficiency by inoculating 100 µL of the cell suspension on an LB agar dish containing an appropriate antibiotic, and incubate overnight at 37°C. To the remaining cells, add 5 mL of LB containing antibiotic and incubate overnight at 37°C with vigorous shaking. One-tenth of the transformation mixture is spread on an agar dish to determine the number of primary transformants in the liquid culture. This assay is important because the number of primary transformants defines the maximum complexity of the subsequent plasmid DNA preparation (*see* Note 6).

6. Isolate plasmid DNA from the 5-mL culture (*see* Chapters 31–33).
7. Digest 0.25 µg of plasmid DNA with the restriction enzyme(s) that recognizes the targeted USE site(s) in a volume of 20 µL. Monitor the digestion by electrophoresis of 15 µL and comparison to an equivalent amount of undigested DNA on an agarose gel. The digested DNA sample should show little or no circular forms (*see* Notes 7 and 8).
8. Transform an aliquot of the digested DNA into a normal *E. coli* host (i.e., HB101 or DH5α). Isolate plasmid DNA from individual transformants and screen for the loss of USE sites. Among those with mutations in USE selection sites, approx 80% will carry the desired, nonselectable mutations, which are normally identified by DNA sequence analysis (*see* Note 9).

Typically, 50–100% of secondary transformants will be mutant at USE sites (*see* Notes 10 and 11). When desired mutations are introduced by oligonucleotide-directed second-strand synthesis, about 80% of the products that are mutant at USE sites will carry desired, nonselected mutations (*see* Notes 12 and 13). If two or more nonselected mutations are introduced using multiple primers, mutations are produced independently, each with about 80% efficiency. With the PCR-based procedure, desired mutations are often present in 100% of isolates with mutations in USE sites (*see* Notes 13–17).

4. Notes

1. Multiple nonselected mutations are introduced by using additional primers, each of which must anneal to the same strand as the selectable USE primer. For LP-USE mutagenesis, the two primers must be complementary to opposite strands to effect PCR amplification.
2. The ligation reaction in Section 3.2., step 3 requires phosphorylated primers or PCR products. Mutants can be recovered without ligation. However, this is not recommended, since even though strand displacement in vitro is avoided by using T4 DNA polymerase *(8,9)*, strand displacement may occur in vivo.
3. The success of these procedures depends on the cosegregation in vivo of two mutations that are linked in vitro. Several factors influence the formation and maintenance of linked mutations. First, linkage of mutagenic primers during second-strand synthesis requires that both primers anneal to a single template molecule and direct synthesis of a complete second strand, with ligation sealing the nicks. Annealing of both primers is aided by high primer:template ratios. Typically, ratios of 1000:1 are used, but adjustments may be required depending on the number and types of mismatches formed when primers anneal to the target. Ratios of PCR product to template are less critical, since PCR links selected and nonselected

mutagenic primers before second-strand synthesis. Effective molar ratios of PCR product:template range from 0.1–100:1.
4. T4 DNA polymerase synthesizes a complete second-strand, and DNA ligase seals the nick producing a covalently closed double-stranded circular plasmid. T4 DNA polymerase is more processive if T4 gene 32 protein is added to the synthesis/ligation reaction; this optional component may be helpful for some primer–template combinations. Use 1.5 µL of a fresh 10X dilution (in TE; about 15 pmol) for 0.025 pmol of a 5–6 kbp template. This achieves about 50% saturation, since one gene 32 monomer binds to 10 nucleotides. DNA synthesis may be affected by template quality and self-annealing of template or primers.
5. *E. coli* mut S strains are used because they are defective in mismatch repair, increasing the probability that mismatched bases cosegregate at the first round of plasmid replication in the host. However, mut S strains are mutator strains and may spontaneously acquire undesired traits, including reversion to mismatch repair proficiency, which will reduce mutagenesis efficiencies. Therefore, competent mut S cells should be prepared from frozen stocks instead of from serially passaged cultures. Normal *E. coli* strains should not be used for the initial transformation, since mutagenesis efficiencies are reduced dramatically *(7)*.
6. It is important to determine the number of primary transformants expanded in liquid culture (Section 3.2., step 5), because this number defines the maximum complexity of the subsequent plasmid DNA preparation. Since the percentage of mutants at this stage can vary from 5 to <1%, cultures expanded from fewer than 100 primary transformants may have few or no mutants. Note that even a single primary transformant will yield a stationary-phase culture during an overnight incubation. Therefore, culture expansion to stationary phase cannot be used to judge the number of primary transformants. Although any procedure that yields a sufficient quantity of transformants may be used, transformation efficiencies are usually highest with electroporation.
7. The selection against nonmutant plasmids depends on the complete digestion of DNA isolated from primary transformants (or secondary transformants if a second round of selection is used; *see* Note 10). Digesting large amounts of DNA is not recommended, since this increases the chance of incomplete digestion and may produce an unacceptably high background of nonmutant plasmids. Incomplete digestions may be avoided by using a large excess of restriction enzyme. Since the percentage of mutant plasmids among primary transformants is usually low (1–5%), nearly all the DNA should be converted to linear molecules. Therefore, the mutant (circular) species is usually not visible on ethidium bromide-stained agarose gels.

8. Selection and mutagenesis efficiencies have not been found to be influenced by the type of restriction site eliminated (i.e., blunt-end, 5'- or 3'-extensions) or by the distance separating primers.
9. Screening plasmids for loss of restriction sites is a rapid and convenient way to estimate the mutagenesis efficiency before proceeding with DNA sequence analysis.
10. If fewer than 30% of isolates fail to lose the USE site, enrichment is possible with a second round of selection. This can be done by repeating steps 6–8 with plasmid DNA isolated from secondary transformant colonies (Section 3.2., step 8) suspended in 5 mL of LB medium (without further growth). A second round of selection works best if two conditions are met: (a) At least 200 primary transformants were expanded in liquid culture (Section 3.2., step 5), as determined in control plating assays, and (b) plasmid DNA is prepared from a pool of at least 200 secondary transformants (Section 3.2., step 8). As the number of primary or secondary transformants decreases, the risk of isolating siblings increases, and additional rounds of selection may only serve to enrich for siblings that have lost the USE site, but do not contain desired mutations. Secondary selections are effective because plasmids mutant at USE sites are enriched by the first selection, and because plasmids prepared from standard *E. coli* strains are of higher quality than plasmids from mut S strains and are therefore easier to digest to completion.
11. Another way to decrease the wild-type background frequency is to apply USE selection pressure twice: before the first transformation into mut S cells and, as outlined in Fig. 1, before the second transformation.
12. If selected mutations are recovered with high efficiency, but nonselected mutations are recovered rarely or not at all, increase the nonselected primer:selected primer ratio.
13. As with any primer-directed synthesis reaction, inefficient annealing or primer extension may cause problems. When circular species with nicks in one strand are denatured, the desired single-stranded circular templates are produced. Therefore, it is not necessary to start with highly supercoiled preparations. However, single-stranded circular templates are not produced if nicks are present in both strands, so highly nicked preparations do not produce suitable templates.
14. When PCR products are used, it is important to sequence the entire amplified region because PCR may generate unwanted mutations *(16)*. This problem can be diminished by using thermostable enzymes with high fidelity and by minimizing the number of PCR cycles. Occasionally, unwanted mutations are found near or within mutagenic primer sequences, and these may arise from residual error-prone repair in mut S strains *(12)*. However,

such mutations are easily identified during DNA sequence analysis. Unwanted mutations are not common elsewhere.
15. When confirming the presence of a mutation in transfected genes, design one PCR primer to be complementary to vector sequences if there is a possibility that endogenous genes will be amplified (Fig. 2).
16. The efficiency of linking PCR products lacking a selectable mutation to a USE primer (Fig. 3) depends on the PCR product:USE primer:template ratios. In this case PCR product:USE primer ratios ≥1 are recommended.
17. If many mutations are present in a single mutagenic primer, products may contain some, but not all of the desired mutations, possibly because many closely spaced mismatches may provide a strong signal to recruit residual mismatch repair enzymes in mut S strains.

References

1. Zoller, M. J. and Smith, M. (1987) Oligonucleotide-directed mutagenesis: a simple method using two oligonucleotide primers and a single-stranded DNA template. *Methods Enzymol.* **154,** 329–350.
2. Kramer, W. and Fritz, H.-J. (1987) Oligonucleotide-directed construction of mutations via gapped duplex DNA. *Methods Enzymol.* **154,** 350–367.
3. Kunkel, T. A., Roberts, J. D., and Zakour, R. A. (1987) Rapid and efficient site-specific mutagenesis without phenotypic selection. *Methods Enzymol.* **154,** 367–382.
4. Carter, P. (1987) Improved oligonucleotide-directed mutagenesis using M13 vectors. *Methods Enzymol.* **154,** 382–403.
5. Smith, M. (1985) In vitro mutagenesis. *Annu. Rev. Genet.* **19,** 423–462.
6. Kunkel, T. A. (1985) Rapid and efficient site-specific mutagenesis without phenotypic selection. *Proc. Natl. Acad. Sci. USA* **82,** 488–492.
7. Lewis, M. K. and Thompson, D. V. (1990) Efficient site directed *in vitro* mutagenesis using ampicillin selection. *Nucleic Acids Res.* **18,** 3439–3443.
8. Masumune, Y. and Richardson, C. A. (1971) Strand displacement during deoxyribonucleic acid synthesis at single strand breaks. *J. Biol. Chem.* **246,** 2692–2701.
9. Nossal, N. G. (1974) DNA synthesis on a double-stranded DNA template by the T4 bacteriophage DNA polymerase and the T4 gene 32 DNA unwinding protein. *J. Biol. Chem.* **249,** 5668–5676.
10. Zell, R. and Fritz, H.-J. (1987) DNA mismatch-repair in *Escherichia coli* counteracting the hydrolytic deamination of 5-methyl-cytosine residues. *EMBO J.* **6,** 1809–1815.
11. Deng, W. P. and Nickoloff, J. A. (1992) Site-directed mutagenesis of virtually any plasmid by eliminating a unique site. *Anal. Biochem.* **200,** 81–88.
12. Ray, F. A. and Nickoloff, J. A. (1992) Site-specific mutagenesis of almost any plasmid using a PCR-based version of unique-site elimination. *Biotechniques* **13,** 342–346.
13. Higuchi, R. (1989) Using PCR to engineer DNA, in *PCR Technology* (Erlich, H. A., ed.), Stockton, New York, pp. 61–70.

14. Rubin, J. B., Verselis, V. K., Bennett, M. V. L., and Bargiello, T. A. (1992) A domain substitution procedure and its use to analyze voltage dependence of homotypic gap junctions formed by connexins 26 and 32. *Proc. Natl. Acad. Sci. USA* **89,** 3820–3824.
15. Orita, M., Suzuki, Y., Sekiya, T., and Hayashi, K. (1989) Rapid and sensitive detection of point mutations and DNA polymorphisms using the polymerase chain reaction. *Genomics* **5,** 874–879.
16. Saiki, R. K., Gelfand, D. H., Stoffel, S., Scharf, S. J., Higuchi, R., Horn, G. T., Mullis, K. B., and Erlich, H. A. (1988) Primer–directed enzymatic amplification of DNA with a thermostable DNA polymerase. *Science* **239,** 487–491.

CHAPTER 55

A Protocol for Site-Directed Mutagenesis Employing a Uracil-Containing Phagemid Template

Michael K. Trower

1. Introduction

Site-directed mutagenesis is a powerful technique by which specific changes can be generated in a target DNA molecule for either structure-function investigations or for creating and deleting endonuclease restriction sites. There have been several oligonucleotide-directed mutagenesis procedures described in the literature that allow for selection against the wild-type DNA, thereby giving enrichment in those sequences carrying the desired mutation *(1–4)*. Collectively, these methods suffer from a number of drawbacks, including multi-enzymatic steps, the necessity for DNA precipitation, the need for convenient restriction sites, and a requirement for specialized vectors. An elegant method for achieving site-directed mutagenesis has been described by Kunkel et al. *(5,6)* that avoids these problems and at the same time produces a high percentage of DNA molecules with the designed mutation. This process is schematically outlined in Fig. 1. It relies on the observation that an *E. coli* strain lacking two key enzymes, dUTPase *(dut)* and uracil N-glycosylase *(ung)*, will synthesize DNA with a small number of uracil bases substituted for thymine. Uracil containing single-stranded DNA (ssDNA) prepared from this *dut⁻, ung⁻* host strain can be used as a template for a mutagenic oligonucleotide to prime in vitro DNA synthesis of a complementary strand. When this heteroduplex is transformed into a wild-type *(dut⁺, ung⁺) E.*

From: *Methods in Molecular Biology, Vol. 58: Basic DNA and RNA Protocols*
Edited by: A. Harwood Humana Press Inc., Totowa, NJ

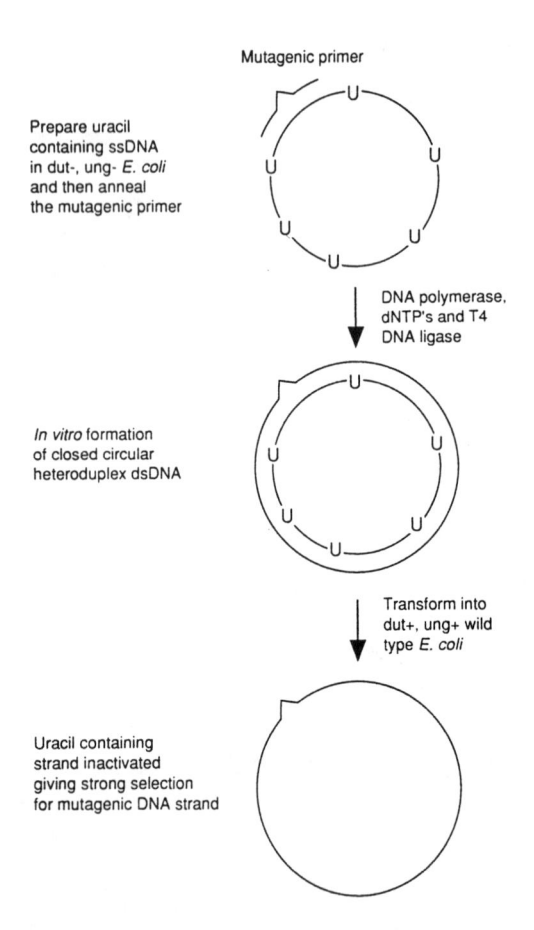

Fig. 1. Site-directed mutagenesis based on the Kunkel method. A uracil-containing ssDNA is prepared by passage through a dut^-, ung^- E. coli host strain. The template is primed in vitro with the mutagenic oligonucleotide and then a dsDNA heteroduplex is generated following the addition of a DNA polymerase and T4 DNA ligase. The heteroduplex is transformed into a dut^+, ung^+ wild-type E. coli that selects for the newly synthesized mutagenic DNA strand through preferential inactivation of the uracil-containing DNA strand.

coli the uracil template strand is selectively inactivated, resulting in a majority of the progeny having the designed mutation. Described below is a protocol that is based on the Kunkel method, but uses a phagemid as the source of single-stranded DNA rather than the more usual M13 bacteriophage. This has the distinct advantage that it reduces the amount of subcloning required during the site-directed mutagenesis process.

2. Materials

Solutions are stored at room temperature unless stated otherwise.

1. *E. coli* strains:
   a. CJ236, a *dut⁻, ung⁻* strain containing the F' episome or other host with the same genotypic features.
   b. Wild-type *dut⁺, ung⁺* strain, such as DH5α, JM109, NM522, TG1, or XL1–Blue. The wild-type host need not have a F' episome.
2. T4 polynucleotide kinase (10 U/µL).
3. 10X T4 polynucleotide kinase buffer: 0.5M Tris-HCl, pH 7.5, 0.1M MgCl$_2$, 50 mM DTT, 0.5 mg/mL nuclease-free bovine serum albumin (BSA). Store at –20°C.
4. Klenow: the large fragment of DNA polymerase I (5 U/µL).
5. 10X Klenow polymerase buffer: 0.5M Tris-HCl, pH 7.4, 0.1M MgCl$_2$, 10 mM DTT, 0.25 mg/mL nuclease-free BSA. Store at –20°C.
6. dNTP stock: 4 mM each of dATP, dCTP, dGTP, and dTTP. Store at –20°C.
7. ATP: 10 mM solution. Store at –20°C.
8. T4 DNA ligase: 400 U/µL.
9. Mutagenic primer (*see* Note 1): a synthetic oligonucleotide at 0.5 µM.

3. Methods

For mutagenesis, a uracil-containing ssDNA template must be prepared. The *E. coli* strain CJ236 is an ideal host because it is both *dut⁻, ung⁻* and contains the F' episome, allowing the helper phage to infect the host cells.

1. Transform 10–50 ng of phagemid plasmid DNA, containing the insert with the target DNA sequence, into the strain CJ236 and select on an appropriate medium (*see* Note 2). Prepare phagemid ssDNA using the protocol described in Chapter 44 (*see* Note 3).
2. Mix 1.0 µL of 10X T4 polynucleotide kinase buffer, 1.0 µL of ATP, 1.0 µL of mutagenic primer, 6.5 µL of distilled water, and 0.5 µL of T4 polynucleotide kinase. Incubate for 60 min at 37°C and then inactivate the kinase at 70°C for 5 min.
3. While the phosphorylated oligonucleotide is still at 70°C, add 2 µL of the uracil containing template (approx 20–50 ng DNA) and place the reaction tube into 200–300 mL of water at 70°C. Allow to cool to room temperature.
4. Add 2.0 µL of 10X Klenow polymerase buffer, 1.0 µL of dNTP stock, 2.0 µL of ATP, 3.0 µL of distilled water, 1.0 µL of Klenow enzyme, and 1.0 µL of T4 DNA ligase. Incubate at 16°C overnight (*see* Note 4).
5. Transform 1–10 µL of the extension/ligation reaction into a competent wild-type *dut⁺, ung⁺ E. coli* host strain and plate out on selective media.

A   NcoI recognition site

5'GAGGATCGTTTC**C**CATG**G**TTGAACAAGATG 3'

                AphA-2 gene start codon

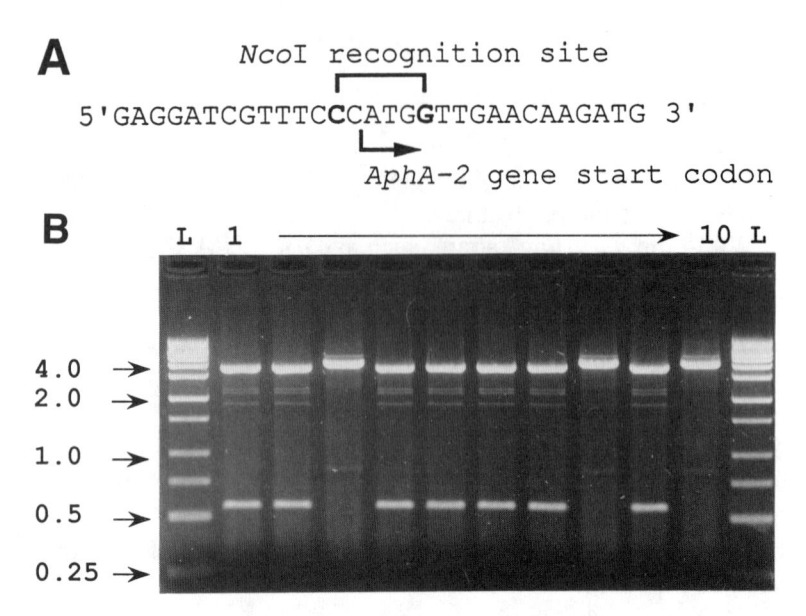

B   L   1 ——————————→ 10 L

4.0 →
2.0 →
1.0 →
0.5 →
0.25 →

Fig. 2. **(A)** Design of the mutant oligonucleotide to introduce a *Nco*I restriction enzyme site (CCATGG) at the start codon of the Tn5 *AphA-2* kanamycin phosphotransferase gene. This generates two changes from the wild-type sequence (highlighted in bold); a G to C at −2 and A to G at +4 with respect to the *AphA-2* start codon. Reading 5' to 3', the 30-mer has 12 bases flanking each side of the two point mutations. The arrow on the figure indicates direction of transcription of the *AphA–2* structural gene. **(B)** Site-directed mutagenesis was carried out and plasmid DNA was prepared from ten of the several hundred colonies on the transformation plates. Seven of the ten digested with *Nco*I to generate two fragments of 3.75 kbp and 560 bp (lanes 1, 2, 4–7, 9). This is consistent with the introduction of a second *Nco*I recognition sequence into the vector; in clones lacking the designed mutation a single linearized 4.3-kb band of the plasmid DNA is formed (lanes 3, 8, 10). L, 1-kb ladder (Stratagene).

Following transformation it is common to find ten to several hundred colonies present on the plates. Typically the efficiency of mutagenic site incorporation is >50% (*see* Notes 5–8), which allows direct screening of a small number of clones by DNA sequencing to identify those carrying the designed mutation. Moreover, if the site-directed changes create or delete a restriction site, then these can initially be investigated by endonuclease restriction analysis, as shown in Fig. 2. All site-directed changes, including those involving restriction sites, should be confirmed by DNA sequencing (*see* Note 9).

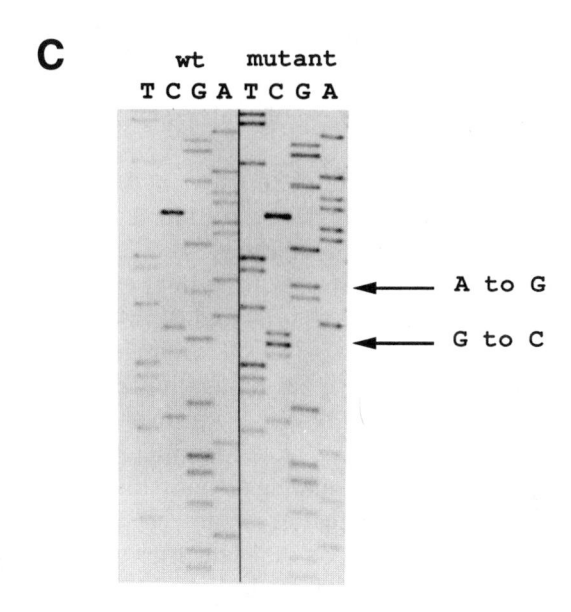

Fig. 2. (C) Sequencing of wild-type and mutated plasmid DNA. A 17-mer oligonucleotide that hybridizes 55 bases upstream of the *AphA-2* start codon was used as a primer to obtain the sequences for both wild-type and *Nco*I mutated vectors.

4. Notes

1. A mutagenic primer should consist of a region of mismatch flanked by sequences complementary to the template to ensure the oligonucleotide anneals efficiently to the specified target. Mismatches can involve insertions, deletions, and substitutions, or a combination of these. If a single or double change is introduced, 10 flanking bases on either side of the mutation site are usually sufficient. This can be extended up to 15 bases and more if the flanking regions are either rich in A + T residues or if substantial alterations to the target sequence are to be incorporated. Whenever possible, include one or more G or C residues at the 5' terminus of the primer to protect the oligonucleotide from displacement following extension by the DNA polymerase, even if this extends the primer length. Avoid primers that include inverted repeats because this can lead to primer duplex formation preventing their hybridization to the target sequence. Do not be afraid to design primers that loop out significant regions of DNA. For instance, using the protocols described, a 237-bp region of the human cytomegalovirus IE$_1$ coding sequence was specifically deleted with a 30-mer containing 15 bp flanking each side of the deletion site (Stephen Walker, personal communication).

2. Competent cells of CJ236 and wild-type host strain can be prepared by the Frozen Storage Buffer (FSB) method of Hanahan (*7; see* Chapter 29). This method is convenient since it enables aliquots of competent cells to be stored at –70°C until required.
3. Kunkel *(5,6)* incorporates a uridine supplement (0.25 µg/mL) in his protocol for the preparation of M13 ssDNA, but I have found this addition to be unnecessary with phagemid preparations.
4. Klenow is the DNA polymerase used for primer extension in this protocol. It lacks ssDNA 5' exonuclease activity and therefore is unable to degrade the mutagenic primer. T4 and T7 DNA polymerases may also be used. Both have the advantage of higher processivities compared to Klenow, and their use can considerably reduce the primer extension reaction time-period. In my hands, however, I have found Klenow to be a much more reliable enzyme in ultimately isolating clones with the desired site-directed changes and it is this enzyme that I recommend.
5. A lack of any colonies present on the plate after transformation is usually attributable to low competency of the host strain. For all transformations a control plasmid should be transformed at the same time to confirm that the *E. coli* strain used is competent.
6. Lack of colonies may also arise from failure to form heteroduplex DNA. In these circumstances, first check that dsDNA is being generated from the ssDNA template by comparing a sample of the oligonucleotide extension/ ligation to a control sample of the ssDNA template on a 1% agarose gel. The phagemid ssDNA band should have disappeared from the extension/ ligation sample and a new band of greater intensity formed that runs significantly slower than the ssDNA template. This band should be either the closed circular relaxed dsDNA (form IV) heteroduplex, or through failure of the ligation reaction the nicked circular dsDNA (form III).

The presence of only a phagemid ssDNA band in the extension/ligation reaction sample indicates that no heteroduplex DNA has been formed. This may be caused by one of a number of problems: inactive DNA polymerase or the presence of inhibitory contaminants, failure of the mutagenic primer to anneal, or the presence of secondary structures, such as hairpin loops, in the template. In the first instance use a new batch of Klenow, check the dilution of the mutagenic primer, and reprecipitate the ssDNA template to remove any contaminants. Confirmation that the mutant oligonucleotide is hybridizing to the correct site on the template can also be demonstrated by using it as a primer in a sequencing reaction with the uracil-containing phagemid ssDNA containing the cloned target sequence. The DNA sequence obtained should be located just downstream of the mutagenic priming site.

7. Mutant colonies should still be isolated, although probably at a lower efficiency, in a situation where the T4 DNA ligase is working inefficiently and not removing the nick between the newly synthesized strand and the 5' end of the mutagenic oligonucleotide. The DNA repair systems of the host strain will complete circularization of the mutant strand in vivo.
8. A low frequency of mutagenesis is likely to result from lack of uracil in the ssDNA template so that it is not inactivated following transformation into the wild-type host strain, mutant oligonucleotide displacement by the advancing DNA polymerase, or DNA impurities in the phagemid ssDNA preparation that prime the template for DNA synthesis.

   The incorporation of uracil into the template can be tested by transformation of 10 ng of the ssDNA template into both CJ236 and a wild-type dut^+, ung^+ *E. coli*. A $>10^5$-fold reduction would be expected between CJ236 and the wild-type strains, because a template containing uracil should be selectively degraded.

   Displacement of the primer may be reduced by increasing the number of template complementary G/C base pairs in the 5' end of the primer. Alternatively, either T4 or native T7 DNA polymerase (not Sequenase) may be used as a replacement for Klenow since these enzymes lack strand displacement activities.

   Priming on the second strand because of DNA contaminants can be identified by carrying out the extension/ligation step in the absence of mutagenic oligonucleotide. If the phagemid ssDNA is converted to a dsDNA product, fresh template should be prepared.
9. The whole site-directed mutagenesis procedure, including sequencing to identify clones with the designed mutation, can be completed in a week. Furthermore, since only 1/25 of the uracil-containing ssDNA is consumed per reaction, the same template can be used to generate other mutations in the cloned insert, thereby reducing the time window by a day.

Acknowledgments

I would like to thank Sydney Brenner of the Department of Medicine, University of Cambridge, for providing me with the opportunity to devise this protocol, and I was supported by a fellowship from E. I. du Pont de Nemours and Co. Inc., Wilmington, DE.

References

1. Vandeyar, M. A., Weiner, M. P., Hutton, C. J., and Batt, C. A. (1988) A simple and rapid method for the selection of oligodeoxynucleotide-directed mutants. *Gene* **65,** 129–133.

2. Stanssens, P., Opsomer, C., McKeown, Y. M., Kramer, W., Zabeau, M., and Fritz, H.-J. (1989) Efficient oligonucleotide-directed construction of mutations in expression vectors by the gapped duplex DNA method using alternating selectable markers. *Nucleic Acids Res.* **17,** 4441–4454.
3. Lewis, M. K. and Thompson, D. V. (1990) Efficient site directed *in vitro* mutagenesis using ampicillin selection. *Nucleic Acids Res.* **18,** 3439–3443.
4. Sayers, J. R., Schmidt, W., Wendler, A., and Eckstein, F. (1988) Strand specific cleavage of phosphorthioate containing DNA by reaction with restriction endonucleases in the presence of ethidium bromide. *Nucleic Acids Res.* **16,** 803–814.
5. Kunkel, T. A. (1985) Rapid and efficient site-specific mutagenesis without phenotypic selection. *Proc. Natl. Acad. Sci. USA* **82,** 488–492.
6. Kunkel, T. A., Roberts, J. D., and Zakour, R. A. (1987) Rapid and efficient site-specific mutagenesis without phenotypic selection. *Meth. Enzymol.* **154,** 367–382.
7. Hanahan, D. (1985) Techniques for transformation of *E. coli,* in *DNA Cloning: A Practical Approach,* vol. 1 (Glover, G. M., ed.), IRL, Oxford, pp. 109–135.

CHAPTER 56

In Vitro Translation of Messenger RNA in a Rabbit Reticulocyte Lysate Cell-Free System

Louise Olliver and Charles D. Boyd

1. Introduction

The identification of specific messenger RNA molecules and the characterization of the proteins encoded by them has been greatly assisted by the development of in vitro translation systems. These cell-free extracts comprise the cellular components necessary for protein synthesis, i.e., ribosomes, tRNA, rRNA, amino acids, initiation, elongation and termination factors, and the energy-generating system *(1)*. Heterologous mRNAs are faithfully and efficiently translated in extracts of HeLa cells *(2)*, Krebs II ascites tumor cells *(2)*, mouse L cells *(2)*, rat and mouse liver cells *(3)*, Chinese hamster ovary (CHO) cells *(2)*, and rabbit reticulocyte lysates *(2,4)*, in addition to those of rye embryo *(5)* and wheat germ *(6)*. Translation in cell-free systems is simpler and more rapid (60 min vs 24 h) than the in vivo translation system using *Xenopus* oocytes.

The synthesis of mRNA translation products is detected by their incorporation of radioactively labeled amino acids, chosen specifically to be those occurring in abundance in the proteins of interest. Analysis of translation products usually involves specific immunoprecipitation *(7)*, followed by polyacrylamide gel electrophoresis *(8)* and fluorography *(9)* (*see* Fig. 1).

In vitro translation systems have played important roles in the identification of mRNA species and the characterization of their products, the investigation of transcriptional and translational control, and the cotrans-

From: *Methods in Molecular Biology, Vol. 58: Basic DNA and RNA Protocols*
Edited by: A. Harwood Humana Press Inc., Totowa, NJ

93,000
68,000
65,000
63,000

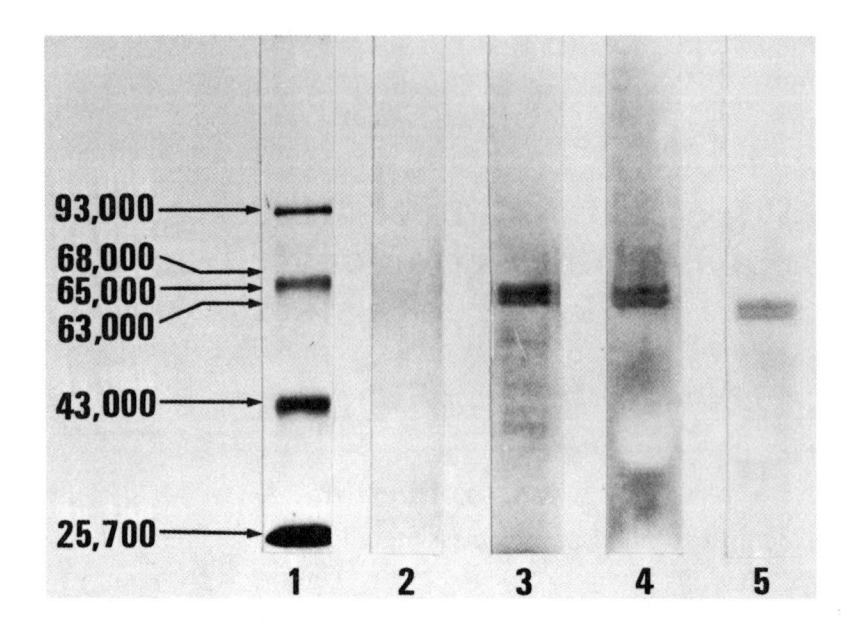

43,000

25,700
    1   2   3   4   5

Fig. 1. SDS polyacrylamide gel electrophoretic analysis of in vitro translation products. In vitro translation products were derived from exogenous mRNA in an mRNA-dependent reticulocyte lysate cell-free system. Following electrophoresis on 8% SDS polyacrylamide gels, radioactive protein products were analyzed by fluorography. Lane 1: [^{14}C]-labeled proteins of known molecular weights, i.e., phosphorylase A (93K), bovine serum albumin (68K), ovalbumin (43K), α-chymotrypsinogen (25.7K). Lanes 2–5 represent [^{3}H]-proline-labeled translation products of the following mRNAs: Lane 2: endogenous reticulocyte lysate mRNA, Lane 3: 0.3 µg of calf nuchal ligament polyadenylated RNA. Lane 4: 0.3 µg of calf nuchal ligament polyadenylated RNA, and immunoprecipitated with 5 µL of sheep antiserum raised to human tropoelastin, Lane 5: 0.3 µg calf nuchal ligament polyadenylated RNA and cotranslationally processed by 0.3 A$_{260}$ nm microsomal membranes.

lational processing of secreted proteins by microsomal membranes added to the translation reaction *(10,11)*. This chapter describes the rabbit reticulocyte lysate system for in vitro translation of mRNA.

Although the endogenous level of mRNA is lost in reticulocyte lysates, it may be further reduced in order to maximize the dependence of translation on the addition of exogenous mRNA. This reduction is achieved by treatment with a calcium-activated nuclease that is thereafter inactivated by the addition of EGTA *(4)*. The system is thus somewhat disrupted with respect to the in vivo situation and is particularly sensitive to the presence of calcium ions. The resulting lysate, however, is the most efficient in vitro translation system with respect to the exogenous mRNA-stimulated incorporation of radioactive amino acids into translation products. It is therefore particularly appropriate for the study of translation products. The system is sensitive, however, to regulation by a number of factors, including hemin, double-stranded RNA, and depletion of certain metabolites. The effects of these factors on regulation of translation of various mRNAs may therefore be investigated. Despite the efficiency of reticulocyte lysates, the competition for initiation of translation by various mRNA species may differ from the in vivo situation. Products therefore may not be synthesized at in vivo proportions; the wheat germ extract cell-free system (see Chapter 57) reflects the in vivo situation more faithfully. Nuclease-treated rabbit reticulocyte lysate cell-free systems are available as kits from a number of commercial suppliers.

2. Materials

All in vitro translation components are stored at −70°C. Lysates, microsomal membranes, and [^{35}S]-labeled amino acids are particularly temperature-labile and therefore should be stored in convenient aliquots at −70°C; freezing and thawing cycles must be minimized. Solutions are quick-frozen on dry ice or in liquid nitrogen prior to storage.

1. Folic acid; 1 mg/mL folic acid, 0.1 mg/mL vitamin B$_{12}$, 0.9% (w/v) NaCl, pH 7.0; filtered through a 0.45-µm filter and stored in aliquots at −20°C.
2. 2.5% (w/v) phenylhydrazine, 0.9% (w/v) sodium bicarbonate, pH 7.0 (with NaOH). Stored no longer than 1 wk at −20°C in single dose aliquots. Thawed unused solution must be discarded. Hydrazine degrades to darken the straw color.
3. Physiological saline: 0.14M NaCl, 1.5 mM MgCl$_2$, 5 mM KCl. Stored at 4°C.
4. 1 mM hemin.
5. 0.1M CaCl.
6. 7500 U/mL micrococcal nuclease in sterile distilled water. Stored at −20°C.
7. Rabbit reticulocyte lysate. This is prepared essentially as described by Pelham and Jackson (4). Rabbits are made anemic by intramuscular injection of 1 mL folic acid solution on d 1, followed by six daily injections of 0.25 mL/kg body weight of 2.5% phenylhydrazine solution (see Note 1). At a reticulocyte count of at least 80%, blood is collected on d 7 or 8 by cardiac puncture into a 200-mL centrifuge tube containing approx 3000 U of heparin, and mixed well. Preparation should continue at 24°C.

a. Blood is centrifuged at 120g, 12 min, 2°C, and plasma removed by aspiration.
b. Cells are resuspended in 150 mL *ice cold* saline and washed at 650g for 5 min. Washing is repeated three times.
c. The final pellets are rotated gently in the bottle, then transferred to Corex tubes (which are only half-filled). An equal volume of saline is added, the cells gently suspended, then pelleted at 1020g for 15 min at 2°C. The leukocytes (buffy coat) are then removed by aspiration with a vacuum pump.
d. In an ice bath, an equal volume of ice-cold sterile deionized distilled water is added and the cells lysed by vigorous vortexing for 30 s (*see* Note 2). The suspension is then immediately centrifuged at 16,000g for 18 min at 2°C.
e. At 4°C the supernatant is carefully removed from the pellet of membranes and cell debris. This lysate is then quick frozen in liquid nitrogen in aliquots of approx 0.5 mL.

The optimum hemin concentration is determined by varying its concentration from 0–1000 µM during the micrococcal nuclease digestion. Lysate (477.5 µL), 5 µL of 0.1M $CaCl_2$, and 5 µL of nuclease (75 U/mL final concentration) is mixed. A 97.5 µL volume of this is incubated with 2.5 µL of the relevant hemin concentration at 20°C for 20 min. A 4 µL 0.05M solution of EGTA is added to stop the digestion (*see* Note 3). The optimum hemin concentration is that allowing the greatest translational activity (incorporation of radioactive amino acids) in a standard cell-free incubation (*see* Section 3.). A quantity of 25 µM is generally used to ensure efficient chain initiation.

Lysates are extremely sensitive to ethanol, detergents, metals, and salts, particularly calcium. Stored at −70°C, reticulocyte lysates remain active for more than 6 mo.

8. L-[^{3}H]- or L-[^{35}S]-amino acids. A radioactive amino acid, labeled to a high specific activity (140 Ci/mmol tritiated, or approx 1 Ci/mmol [^{35}S]-labeled amino acids), is added to the translation incubation to enable detection of the translation products. An amino acid known to be abundant in the protein of interest is chosen. Radioactive solutions should preferably be aqueous; those of low pH should be neutralized with NaOH; ethanol should be removed by lyophilization, and the effect of solvents on lysate activity should be tested. [^{35}S] degrades rapidly to sulfoxide and should be aliquoted and stored at −70°C to prevent interference by sulfoxides.
9. Messenger RNA. Total RNA may be extracted from various tissues by a number of methods (*see* Chapters 1 and 16). RNA stored in sterile dH_2O at −70°C is stable for more than a year. Contamination by ions, metals, and detergents should be avoided.

Reticulocyte Lysate Cell-Free System

Phenol may be removed by chloroform:butanol (4:1) extractions; salts are removed by precipitation of RNA in 0.4M potassium acetate, pH 6.0, in ethanol. Ethanol should be removed by lyophilization. Convenient stock concentrations for translation are 1.5 mg/mL total RNA or 150 µg/mL polyA$^+$ RNA.

10. Translation cocktail: 250 mM HEPES, pH 7.2, 400 mM KCl, 19 amino acids at 500 mM each (excluding the radioactive amino acid), 100 mM creatine phosphate.
11. 20 mM magnesium acetate, pH 7.2.
12. 2.0M potassium acetate, pH 7.2.
13. Sterile distilled H$_2$O.

Sterile techniques are used; RNase contamination is avoided by heat-treatment of glassware (250°C, 12 h) or by treatment of heat-sensitive materials with diethylpyrocarbonate, followed by rinsing in distilled water. Sterile gloves are worn throughout the procedure.

3. Methods

In vitro translation procedures are best carried out in autoclaved plastic microfuge tubes (1.5 mL); a dry incubator is preferable to waterbaths for provision of a constant temperature. All preparations are performed on ice.

1. Prepare (on ice) the following reaction mix (per inculcation): 0.7 µL of dH$_2$O, 1.3 µL of 2.0M potassium acetate (see Note 4), 5 µL (10–50 µCi) of radioactive amino acid (see Note 5), and 3 µL of translation cocktail. Components are added in the above order, vortexed, and 10 µL is aliquoted per incubation tube on ice (see Note 6).
2. Add 10 µL (300 mg) of total mRNA (see Notes 7 and 8). A control incubation without the addition of exogenous mRNA detects translation products of residual endogenous reticulocyte mRNA.
3. A 10 µL volume of lysate is added last to initiate translation. If required, 0.5 A$_{260}$ nm U of microsomal membranes are also added at this point for cotranslational processing of translation products (see Notes 9 and 10).
4. The mixture is vortexed gently prior to incubation at 37°C for 60 min. The reaction is stopped by placing the tubes on ice (see Note 6).
5. Detection of mRNA-directed incorporation of radioactive amino acids into translation products is performed by determination of acid-precipitable counts
   At the initiation and termination of the incubation, 5-µL aliquots are spotted onto glass fiber filters that are then air-dried. Filters are then placed into 10 mL/filter of the following solutions:
   a. 10% (v/v) cold trichloroacetic acid (TCA) for 10 min on ice.
   b. 5% (v/v) boiling TCA for 15 min, to degrade primed tRNAs.
   c. 5% (v/v) cold TCA for 10 min on ice.

The filters are then washed in 95% (v/v) ethanol, then in 50% (v/v) ethanol-50% (v/v) acetone, and finally in 100% (v/v) acetone. The filters are dried at 80°C for 30 min. TCA-precipitated radioactivity is determined by immersing the filters in 5 mL of toluene-based scintillation fluid and counting in a scintillation counter.

Exogenous mRNA-stimulated translation can be expected to result in a five- to 30-fold increase over background of incorporation of [^{3}H]- or [^{35}S]-labeled amino acids, respectively, into translation products.

6. An equal volume of 2% (w/v) SDS, 20% (w/v) glycerol, 0.02% (w/v) bromophenol blue, $1M$ urea is added to the remaining 20 µL of translation mixture. This is made $0.1M$ with respect to dithiothreitol, heated at 95°C for 6 min, and slowly cooled to room temperature prior to loading onto a polyacrylamide gel of appropriate concentration (between 6 and 17%). After electrophoresis, radioactive areas of the gel are visualized by fluorography (Fig. 1).

4. Notes

1. Maximum anemia may be achieved by reducing the dose of phenylhydrazine on d 3, then increasing it on following days. The reticulocyte count is determined as follows:
   a. 100 µL of blood is collected in 20 µL of 0.1% heparin in saline.
   b. 50 µL of blood heparin is incubated at 37°C for 20 min with 50 µL of 1% (w/v) brilliant cresyl blue, 0.6% (w/v) sodium citrate, 0.7% (w/v) sodium chloride.
   c. Reticulocytes appear under the microscope as large, round, and with blue granules. Erythrocytes are small, oval, and agranular.
2. The volume of water (in mL) required to lyse the reticulocyte preparation is equal to the weight of the pellet in the tube.
3. Endogenous mRNAs of lysates are degraded by a calcium-activated nuclease that is inactivated by EGTA. Lysates are therefore sensitive to calcium ions, the addition of which must be avoided to prevent degradation of added mRNAs by this activated nuclease.
4. Optimum potassium concentrations may vary from 30–100 mM depending on mRNAs used and should be determined prior to definitive translations. Similarly, specific mRNAs may require altered magnesium concentrations, although a concentration of 0.6–1.0 mM is generally used.
5. Specific activities greater than those described in Section 2. may result in depletion of the amino acid concerned, with subsequent inhibition of translation.
6. Vigorous vortexing decreases efficiency of translation, therefore do so gently when preparing the reaction mix.
7. The optimum mRNA concentration should be determined prior to definitive experiments by varying the mRNA concentrations while keeping other

variables constant. Care should be taken to avoid excess mRNA; polyadenylated RNA in excess of 1 µg has been noted to inhibit translation.
8. Heating of mRNA at 70–80°C for 1 min followed by quick cooling in an ice bath, prior to addition to the incubation mixture, has been shown to increase the efficiency of translation of GC-rich mRNA; for example, heating elastin mRNA at 70–80°C prior to translation resulted in a 100% increase, compared with unheated mRNA, of incorporation of radioactivity into translation products *(12)*.
9. Cotranslational processing of translation products may be detected by the addition of dog pancreas microsomal membranes to the translation incubation. These may be prepared as described by Jackson and Blobel *(11)* or may be ordered with a commercial translation kit. Microsomal membranes should be stored in aliquots of approx 5 A_{260} nm U in 20 mM HEPES, pH 7.5, at −70°C. Repeated freezing and thawing must be avoided.
10. The addition of spermidine at approx 0.4 mM has been noted to increase translation efficiency in certain cases *(12)*, possibly by stabilizing relevant nucleic acids. However, this effect may also be lysate-dependent and should be optimized if necessary for individual lysate preparations.

References

1. Lodish, H. F. (1976) Translational control of protein synthesis. *Annu. Rev. Biochem.* **45**, 39–72.
2. McDowell, M. J., Joklik, W. K., Villa-Komaroff, L., and Lodish, H. F. (1972) Translation of reovirus messenger RNAs synthesized in vitro into reovirus polypeptides be several mammalian cell-free extracts. *Proc. Natl. Acad. Sci. USA* **69**, 2649–2653.
3. Sampson, J., Mathews, M. B., Osborn, M., and Borghetti, A. F. (1972) Hemoglobin messenger ribonucleic acid translation in cell-free systems from rat and mouse liver and Landschutz ascites cells. *Biochemistry* **11**, 3636–3640.
4. Pelham, H. R. B. and Jackson, R. J. (1976) An efficient mRNA-dependent translation system from reticulocyte lysates. *Eur. J. Biochem.* **67**, 247–256.
5. Carlier, A. R. and Peumans, W. J. (1976) The rye embryo system as an alternative to the wheat-system for protein synthesis in vitro. *Biochem. Biophys. Acta* **447**, 436–444.
6. Roberts, B. E. and Paterson, B. M. (1973) Efficient translation of tobacco mosaic virus RNA and rabbit globin 9S RNA in a cell-free system from commercial wheat germ. *Proc. Natl. Acad. Sci. USA* **70**, 2330–2334.
7. Kessler, S. W. (1981) Use of protein A-bearing staphylococci for the immunoprecipitation and isolation of antigens from cells, in *Methods in Enzymology* (Langone, J. J. and Van Vunakis, H., eds.), Academic, New York, pp. 441–459.
8. Laemmli, U. K. (1970) Cleavage of structural proteins during the assembly of the head of bacteriophage T4. *Nature* **227**, 680–685.
9. Banner, W. M. and Laskey, R. A. (1974) A film detection method for tritium-labelled proteins and nucleic acids in polyacrylamide gels. *Eur. J. Biochem.* **46**, 83–88.

10. Shields, D. and Blobel, G. (1978) Efficient cleavage and segregation of nascent presecretory proteins in a reticulocyte lysate supplemented with microsomal membranes. *J. Biol. Chem.* **253**, 3753–3706.
11. Jackson, R. C. and Blobel, G. (1977) Post-translational cleavage of presecretory proteins with an extract of rough microsomes, from dog pancreas, with signal peptidase activity. *Proc. Natl. Acad. Sci. USA* **74**, 5598–5602.
12. Karr, S. R., Rich, C. B., Foster, J. A., and Przybyla, A. (1981) Optimum conditions for cell-free synthesis of elastin. *Coll. Res.* **1**, 73–81.

CHAPTER 57

In Vitro Translation of Messenger RNA in a Wheat Germ Extract Cell-Free System

Louise Olliver, Anne Grobler-Rabie, and Charles D. Boyd

1. Introduction

The wheat germ extract in vitro translation system has been used widely for faithful and efficient translation of viral and eukaryotic messenger RNAs in a heterologous cell-free system *(1–9)*. With respect to the yield of translation products, the wheat germ extract is less efficient than most reticulocyte lysate cell-free systems. There are advantages, however, of using wheat germ extracts:

1. The in vivo competition of mRNAs for translation is more accurately represented, making the wheat germ system preferable for studying regulation of translation *(1)*.
2. Particularly low levels of endogenous mRNA and the endogenous nuclease activity *(10)* obviate the requirement for treatment with a calcium-activated nuclease. There is, therefore, less disruption of the in vivo situation and contamination with calcium ions is less harmful. The identification of all sizes of exogenous mRNA-directed translation products is facilitated because of the low levels of endogenous mRNA present.
3. There is no posttranslational modification of translation products; primary products are therefore investigated, although processing may be achieved by the addition of microsomal membranes to the translation reaction.
4. The ionic conditions of the reaction may be altered to optimize the translation of large or small RNAs *(2)* (*see* Note 1).

Translational activity is optimized by the incorporation of an energy-generating system of ATP, GTP, creatine phosphate, and creatine kinase *(3)*. Wheat germ is inexpensive and commercially available (*see* Note 2); preparation of the extract is rapid and simple, resulting in high yields. Wheat germ extract cell-free system kits are also commercially available.

2. Materials

Components of the wheat germ in vitro translation system are heat-labile and must be stored in aliquots of convenient volumes at −70°C. Freeze-thaw cycles must be minimized. Sterile techniques are used throughout. RNase contamination is prevented by heat-sterilization (250°C, 8 h) of glassware and tips, and so on, or by diethyl pyrocarbonate treatment of glassware, followed by thorough rinsing of equipment in sterile distilled water.

1. Wheat germ extract: This is prepared essentially as described by Roberts and Paterson *(4)*. The procedure must be carried out at 4°C, preferably in plastic containers since initiation factors stick to glass. Fresh wheat germ (approx 5 g) (*see* Note 2) is ground with an equal weight of sand and 28 mL of 20 mM HEPES, pH 7.6, 100 mM KCl, 1 mM magnesium acetate, 2 mM CaCl$_2$, and 6 mM 2-mercaptoethanol, added gradually. This mixture is then centrifuged at 28,000g for 10 min at 2°C, pH 6.5. This pH prevents the release of endogenous mRNA from polysomes and therefore removes the requirement for a preincubation to allow polysome formation *(4,5)*. The supernatant (S-28) is then separated from endogenous amino acids and plant pigments that are inhibitory to translation, by chromatography through Sephadex G-25 (coarse) in 20 mM HEPES, pH 7.6, 120 mM KCl, 5 mM magnesium acetate, and 6 mM 2-mercaptoethanol. Reverse chromatography will prevent the loss of amino acids. Fractions of more than 20 A$_{260}$ nm/mL are pooled before being stored in aliquots at a concentration of approx 100 A$_{260}$ nm/mL, at −70°C. The extract remains translationally active for a year or more.
2. L-[^{3}H]- or L-[^{35}S]-amino acids: 10–50 µCi of an appropriate amino acid (abundant in the protein[s] of interest) is added to the reaction to allow detection of translation products. Convenient specific activities are 140 Ci/mmol tritiated, or 1 Ci/mmol [^{35}S]-amino acids, respectively (*see* Note 3). Aqueous solutions should be used since ethanol, salts, detergents, and various solvents interfere with translation. Ethanol should be removed by lyophilization and the effects on translation of other solutions should be determined prior to their use. [^{35}S]-labeled amino acids must be stored in

small aliquots at −70°C where they remain stable for up to 6 mo, after which time sulfoxide products of degradation inhibit translation.
3. Messenger RNA: The extraction of both total and polyadenylated RNA has been described by a number of authors *(10–12)*. Total RNA (1.5 mg/mL) or 150 µg/mL polyadenylated RNA (in sterile distilled water) are convenient stock concentrations. RNA is stable for more than a year at −70°C. Contamination with potassium (*see* Note 1), phenol, and ethanol must be prevented by 70% (v/v) ethanol washes, chloroform:butanol (4:1) extractions, and lyophilization respectively.
4. 10X energy mix: 10 mM ATP, 200 µM GTP, 80 mM creatine phosphate. Potassium salts of the nucleotide triphosphates should be used and the final pH adjusted (if necessary) to 7.4–7.6 with sodium hydroxide.
5. 0.5–1.0M potassium acetate (*see* Note 1), 25 mM magnesium acetate.
6. 20 mM dithiothreitol.
7. 0.6–1.2 mM spermine or 4.0–8.0 mM spermidine (*see* Note 4).
8. 0.2M HEPES, pH 7.4–7.6 (*see* Note 5).
9. 200–500 µg/mL creatine kinase (*see* Note 6).

3. Method

All preparations are carried out on ice. After use, components are quick-frozen on dry ice. Reactions are carried out in sterile plastic microfuge tubes.

1. Mix the following solutions (all components are v/50 µL): 5 µL of energy mix, 5 µL of potassium and magnesium acetate, 5 µL of dithiothreitol, 5 µL of HEPES, 5 µL of spermine, 10 µL of 0.3–8.0 µg mRNA in dH$_2$O, (*see* Note 7), 10 µL of wheat germ extract, 10 µL of creatine kinase (0.8–1.0 A$_{260}$ U), and 5 µL of creatine kinase. If a number of incubations are to be made, a master mix of the first five solutions may be prepared and 25 µL aliquoted into each reaction tube. Creatine kinase is added last to ensure that no energy is wasted. The solutions are mixed by tapping the tube or by gentle vortexing. Microsomal membranes (0.5 A$_{260}$ U) may be added before the creatine kinase to detect cotranslational modification of translation products (*see* Note 8).
2. Incubate at 28°C for 1 h (*see* Note 9). The reaction is terminated by placing the tubes at 4°C.
3. Incorporation of radioactive amino acids into mRNA-derived translation products is detected by TCA precipitation of an aliquot of the reaction (*see* Chapter 56 and Note 10). Incorporation of radioactivity into translation products is generally not as well-stimulated by mRNA added to wheat germ extracts as it is in described reticulocyte lysates.
4. The remaining in vitro translation products may be analyzed further by standard techniques, including tryptic mapping and ion-exchange chroma-

tography, but specific products may be analyzed by immunoprecipitation followed by SDS-polyacrylamide gel electrophoresis.

4. Notes

1. Wheat germ extract translational activity is particularly sensitive to variation in the concentration of potassium ions. At concentrations lower than 70 mM, small mRNAs are preferentially translated, whereas larger mRNAs are completely translated at potassium acetate concentrations of 70 mM or greater *(2,5)*. Polypeptides of up to 200 kDa are synthesized under correct ionic conditions *(9)*. Furthermore, chloride ions appear to inhibit translation such that potassium acetate should preferably be used *(5)*. In this context, residual potassium should be removed from RNA preparations, by 70% (v/v) ethanol washes.
2. Inherent translational activity varies with the batch of wheat germ. Israeli mills (for example "Bar-Rav" Mill, Tel Aviv) supply wheat germ, the extracts of which are usually active.
3. Most of the endogenous amino acids are removed by chromatography through Sephadex G-25 (coarse). Depending on the batch of wheat germ extract, addition of amino acids (to 25 µM) and/or tRNA (to 58 µg/mL) may be necessary to optimize translational activity. Wheat germ extract is particularly sensitive to amino acid starvation; use of radioactive amino acids at specific activities greater than those suggested may result in inhibition of translation because of amino acid starvation.
4. The use of either spermine or spermidine generally stimulates translation, and is essential for the synthesis of larger polypeptides *(5)*, probably by stabilizing longer mRNAs. Omission of either compound will increase the optimum magnesium acetate concentration to 4.0–4.3 mM.
5. HEPES has been shown to buffer the wheat germ extract in vitro translation system more effectively than Tris-acetate *(4)*. Use of the latter will alter the optimum potassium and magnesium concentration.
6. Commercial preparations of creatine kinase differ with respect to the levels of nuclease contamination. This must be considered when larger amounts of the enzyme are to be used.
7. Heating of large mRNAs at 70°C for 1 min followed by rapid cooling on ice increases the efficiency of their translation in wheat germ extract in vitro translation systems.
8. Cotranslational processing of translation products may be detected by the addition of dog pancreas microsomal membranes to the translation incubation. They may be prepared as described by Jackson and Blobel *(12)* or may be ordered with a commercial translation kit. Microsomal membranes should be stored in aliquots of approx 5 A_{260} nm U in 20 mM HEPES, pH 7.5, at –70°C. Repeated freezing and thawing must be avoided.

9. mRNA-stimulated incorporation of radioactive amino acids into translation products is linear, after a 5 min lag, for 50 min and is complete after 90 min. The system is labile at temperatures >30°C; optimum activity is achieved at 25–30°C depending on the batch of wheat germ extract. An incubation temperature of 28°C is generally used.
10. In order to obtain maximum translational activity, it is necessary to determine the optima for each preparation of wheat germ extract; mRNA concentration, potassium and magnesium concentrations, and incubation temperature. Take into account the concentration of salts in the wheat germ extract column eluate.

References

1. Steward, A. G., Lloyd, M., and Arnstein, H. R. V. (1977) Maintenance of the ratio of α and β globin synthesis in rabbit reticulocytes. *Eur. J. Biochem.* **80**, 453–459.
2. Benveniste, K., Wilczek, J., Ruggieri, A., and Stern, R. (1976) Translation of collagen messenger RNA in a system derived from wheat germ. *Biochemistry* **15**, 830–835.
3. Huntner, A. R., Farrell, P. J., Jackson, R. J., and Hunt, T. (1977) The role of polyamines in cell-free protein in the wheat germ system. *Eur. J. Biochem.* **75**, 149–157.
4. Roberts, B. E. and Paterson, B. M. (1973) Efficient translation of tobacco mosaic virus RNA and rabbit globin 9S RNA in a cell-free system from commercial wheat germ. *Proc. Natl. Acad. Sci. USA* **70**, 2330–2334.
5. Davies, J. W., Aalbers, A. M. J., Stuik, E. J., and van Kammen, A. (1977) Translation of cowpea mosaic RNA in cell-free extract from wheat germ. *FEBS Lett.* **77**, 265–269.
6. Boedtker, H., Frischauf, A. M., and Lehrach, H. (1976) Isolation and translation of calvaria procollagen messenger ribonucleic acids. *Biochemistry* **15**, 4765–4770.
7. Patrinou-Georgoulas, M. and John, H. A. (1977) The genes and mRNA coding for the theory chains of chick embryonic skeletal myosin. *Cell* **12**, 491–499.
8. Larkins, B. A., Jones, R. A., and Tsai, C. Y. (1976) Isolation and in vitro translation of zein messenger ribonucleic acid. *Biochemistry* **15**, 5506–5511.
9. Schroder, J., Betz, B., and Hahlbrock, K. (1976) Light-induced enzyme synthesis in cell suspension cultures of *petroselinum*. *Eur. J. Biochem.* **67**, 527–541.
10. Pelham, H. R. B. and Jackson, R. J. (1976) An efficient mRNA-dependent translation system from reticloctye lysates. *Eur. J. Biochem.* **67**, 247–256.
11. Darnbrough, C. H., Legon, S., Hunt, T., and Jackson, R. J. (1973) Initiation of protein synthesis: evidence for messenger RNA-independent binding of methionyl-transfer RNA to the 40S ribosomal subunit. *J. Mol. Biol.* **76**, 379–403.
12. Jackson, R. C. and Blobel, G. (1977) Post-translational cleavage of presecretory proteins with an extract of rough microsomes, from dog pancreas, with signal peptidase activity. *Proc. Natl. Acad. Sci. USA* **74**, 5598–5602.

CHAPTER 58

One-Step Purification of Recombinant Proteins with the 6xHis Tag and Ni-NTA Resin

Joanne Crowe, Brigitte Steude Masone, and Joachim Ribbe

1. Introduction

The 6xHis/Ni-NTA system is a fast and versatile tool for the affinity purification of recombinant proteins and antigenic peptides. It is based on the high-affinity binding of six consecutive histidine residues (the 6xHis tag) to immobilized nickel ions, giving a highly selective interaction that allows purification of tagged proteins or protein complexes from <1% to >95% homogeneity in just one step *(1,2)*. The tight association between the tag and the Ni-NTA resin allows contaminants to be easily washed away under stringent conditions, yet the bound proteins can be gently eluted by competition with imidazole, or a slight reduction in pH. Moreover, because the interaction is independent of the tertiary structure of the tag, 6xHis-tagged proteins can be purified even under the strongly denaturing conditions required to solubilize inclusion bodies.

The six histidine residues that comprise the 6xHis tag can be attached at either end of the recombinant protein, are uncharged at physiological pH, and are very poorly immunogenic in all species except some monkeys. Consequently, the 6xHis tag very rarely affects the structure or function of the tagged protein, and need not be removed after purification *(3)*. Its small size makes it ideal for incorporation into any expression system.

From: *Methods in Molecular Biology, Vol. 58: Basic DNA and RNA Protocols*
Edited by: A. Harwood Humana Press Inc., Totowa, NJ

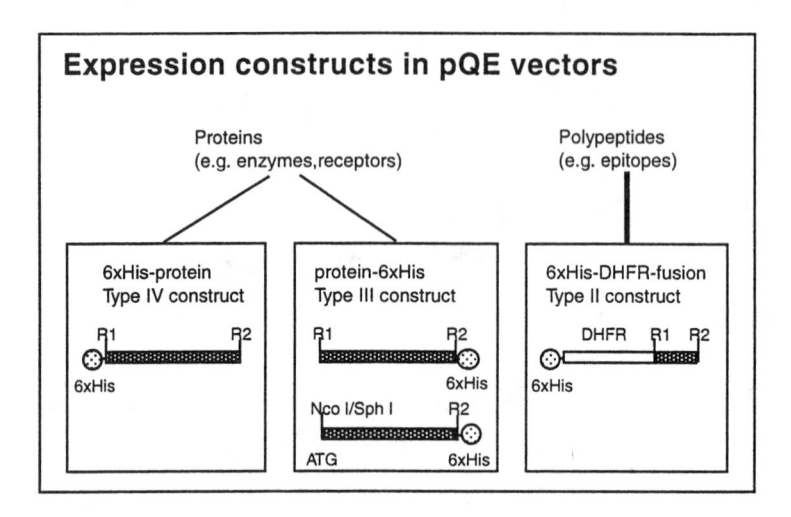

Fig. 1. Expression constructs available with pQE vectors.

The advantages of 6xHis/Ni-NTA purification have been combined with a high level bacterial expression system to create an elegant yet simple strategy, allowing protein purification whether the expressed protein is at low or high levels, denatured, or associated with other proteins, DNA, or RNA. It is currently used in a wide variety of applications, ranging from the large scale purification of proteins for antibody production, to the purification of antibodies, subunits, and substrates through their interactions with the tagged proteins.

1.1. Expression of Proteins Using the pQE Expression Vectors

The pQE expression vectors provide high level expression in *E. coli* of proteins or peptides containing a 6xHis affinity tag. The tag may be placed at the N-terminus of the protein to create a Type IV construct, at the C-terminus of the protein to create a Type III construct, or at the C-terminus of a protein utilizing its original ATG start codon to create a Type ATG construct (pQE-60) (Fig. 1). If small peptides are being synthesized, they can be fused to mouse DHFR to create a Type II construct, where the poorly immunogenic DHFR stabilizes the peptide during expression, and enhances its antigenicity.

The pQE plasmids were derived from plasmids pDS56/RBSII and pDS781/RBSII-DHFRS *(1)*. They contain the following elements as shown for two typical vectors pQE-30 (Type IV) and pQE-40 (Type II) (Fig. 2):

6xHis / Ni-NTA

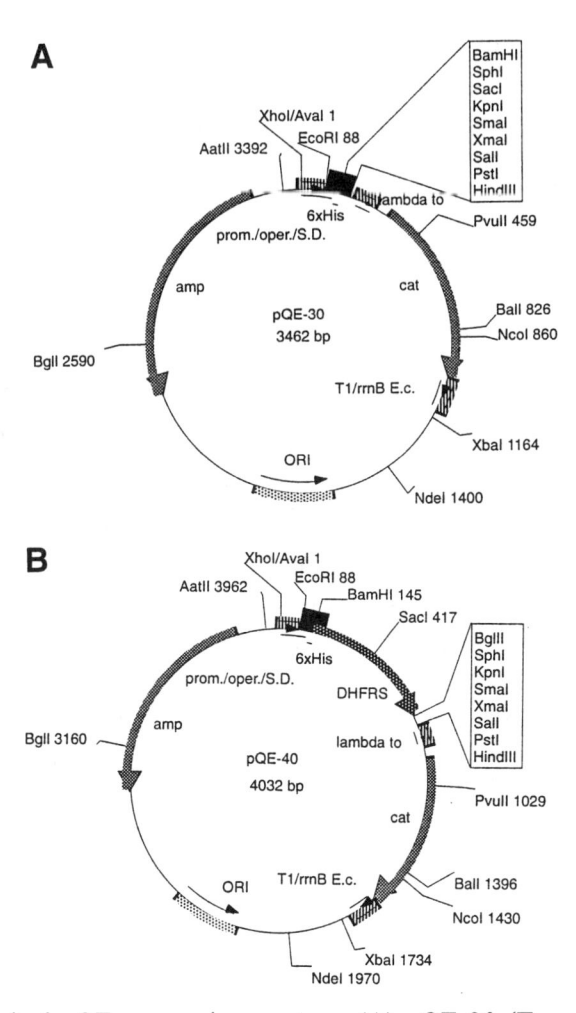

A

[Plasmid map: pQE-30, 3462 bp. Features: AatII 3392, XhoI/AvaI 1, EcoRI 88, BamHI, SphI, SacI, KpnI, SmaI, XmaI, SalI, PstI, HindIII, lambda to, 6xHis prom./oper./S.D., PvuII 459, amp, cat, BalI 826, NcoI 860, BglI 2590, T1/rrnB E.c., XbaI 1164, ORI, NdeI 1400]

B

[Plasmid map: pQE-40, 4032 bp. Features: AatII 3962, XhoI/AvaI 1, EcoRI 88, BamHI 145, SacI 417, 6xHis prom./oper./S.D., DHFRS, BglII, SphI, KpnI, SmaI, XmaI, SalI, PstI, HindIII, lambda to, PvuII 1029, amp, cat, BglI 3160, BalI 1396, NcoI 1430, T1/rrnB E.c., XbaI 1734, NdeI 1970, ORI]

Fig. 2. Typical pQE expression vectors. (**A**) pQE-30 (Type IV construct). The polycloning region is directly 3' to the 6xHis tag sequence. (**B**) pQE-40 (Type II construct) contains a DHFRS sequence between the 6xHis tag and the polycloning region. DHFRS stabilizes short protein sequences.

1. An optimized, regulatable promoter/operator element N250PSN250P29, consisting of the *E. coli* phage T5 promoter (recognized by *E. coli* RNA polymerase) containing two *lac* operator sequences for tight regulation;
2. A synthetic ribosome binding site, RBSII, designed for optimal recognition and binding;
3. Optimized 6xHis affinity tag coding sequence;
4. The mouse DHFR coding sequence (in some vectors only);

5. A multicloning site (available in all reading frames);
6. Translation stop codons in all reading frames;
7. The transcriptional terminator "t_0" from phage lambda;
8. The nontranslated open reading frame for chloramphenicol acetyltransferase;
9. The transcriptional terminator t_1 of the *E. coli* rrnB operon; and
10. The replication region and the gene for β-lactamase of plasmid pBR322.

The *E. coli* host cells M15[pREP4] and SG13009[pREP4] contain multiple copies of the plasmid pREP4. This carries the gene for neomycin phosphotransferase (NEO) conferring kanamycin resistance, and the *lacI* gene encoding the *lac* repressor. The multiple copies of pREP4 present in the host cells ensure high levels of *lac* repressor and tight regulation of protein expression. The plasmid is maintained in *E. coli* cells in the presence of kanamycin at a concentration of 25 µg/mL *(1)*. Expression from pQE vectors is rapidly induced by the addition of IPTG, which inactivates the repressor. The level of IPTG used for induction can be varied to control the level of expression.

The *E. coli* host strains M15[pREP4] or SG13009[pREP4] carrying the repressor (pREP4) plasmids are recommended for the production of recombinant proteins. *E. coli* strains that contain the *lacI*^{q} gene, such as XL1-Blue, JM109, and TG1, are suitable for storing and propagating the pQE plasmids, since they produce enough *lac* repressor to block expression without carrying the pREP4 plasmid. They may also be used as expression hosts; however, expression will not be as tightly regulated as in strains carrying the pREP4 plasmid, which may lead to problems with "toxic" proteins.

The affinity tag for purification on the Ni-NTA resin consists of just six consecutive histidine residues. This small size means that there is minimal addition of extra amino acids to the recombinant protein. It is very poorly immunogenic or nonimmunogenic in all species except some monkeys and, being uncharged at physiological pH, rarely affects the secretion or folding of the protein to which it is attached. In the hundreds of proteins purified using this system, the 6xHis tag has almost never been found to interfere with the structure or function of the purified protein. Several proteins carrying the 6xHis tag have now been crystallized and shown to have the same structure as the nontagged protein *(4)*.

The pQE vectors do not include a protease cleavage site because it is seldom necessary to remove the 6xHis tag after purification, and in many cases the presence of the tag can be an advantage—the purified, functional proteins still carrying the 6xHis tag can be immobilized on Ni-NTA resin and used in a number of downstream applications, such as antibody purification and protein interaction studies *(5–7)*. Most protease cleavage sites consist of highly charged amino acid residues, and once incorporated into recombinant protein, cleavage (which can be tedious and inefficient) will often be necessary to avoid problems in downstream applications. However, if it is necessary to remove the 6xHis tag from the protein after purification, an appropriate cleavage site can be incorporated into the construct during cloning. The rTEV protease (Life Technologies, Gaithersburg, MD) carries a 6xHis tag and can be easily separated from the cleaved protein (along with 6xHis tag fragments and any uncleaved proteins) by adsorption to the Ni-NTA column *(8)*.

The small size of the 6xHis tag makes it ideally suited for inclusion in a variety of other expression systems. It works well in prokaryotic, mammalian, yeast, baculovirus, and other eukaryotic systems. The six histidine residues can be easily inserted into the expression construct, at the C- or N-terminus of the protein, by PCR, mutagenesis, or ligation of a small synthetic fragment.

1.2. The Use of Ni-NTA Resin to Purify the Expressed Proteins

Immobilized metal chelate affinity chromatography was first used to purify proteins in 1975 *(9)* and has become a widely used technique because of its efficiency and ease of use. Until the development of NTA (nitrilo-tri-acetic acid), the chelating ligand iminodiacetic acid (IDA) was charged with metal ions, such as Zn^{2+} and Ni^{2+}, and then used to purify a variety of different proteins and peptides *(10)*. However, IDA, which has only three chelating groups, does not tightly bind these metal ions with six coordination sites. As a consequence, the ions may be washed out of the resin on loading with mildly chelating proteins and peptides, or during the washing of the bound proteins, resulting in low binding capacity, low yields, and impure products.

NTA is a novel chelating adsorbent that was developed in order to overcome these problems (Fig. 3). The NTA occupies four of the ligand binding sites in the coordination sphere of the Ni^{2+} ion (leaving two sites free to interact with the 6xHis tag), and consequently binds the metal ions more stably *(11)*. As a result, Ni-NTA resin binds proteins about

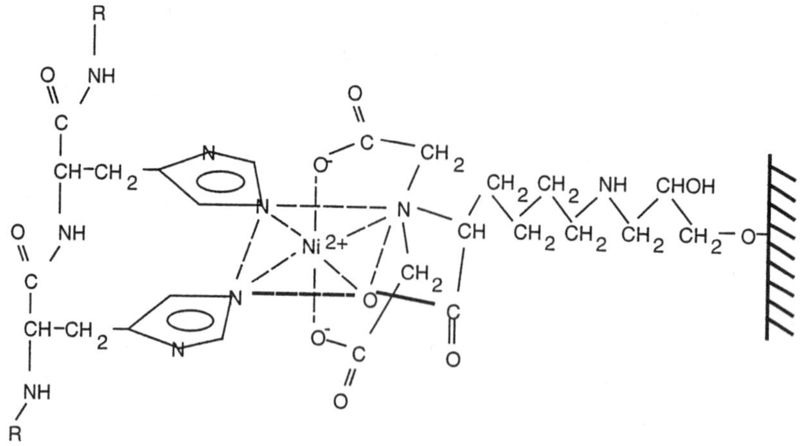

Fig. 3. Model for the binding of neighboring His residues to Ni-NTA resin.

100–1000 times more tightly than Ni-IDA, allowing the purification of proteins constituting <1% of total cellular protein to >95% homogeneity in just one step *(2)*.

Ni-NTA resins are composed of a high surface concentration of NTA ligand attached to supports, such as Sepharose CL-6B (NTA agarose) and silica. The Ni-NTA agarose binding capacity ranges from 5–10 mg 6xHis-tagged protein/mL of resin (i.e., 1 mL of Ni-NTA agarose typically binds 300–400 nmol of protein), depending on the size, properties, and purification conditions for the given protein. Ni-NTA agarose is very stable and easy to handle, and can be stored at room temperature. Ni-NTA spin columns contain Ni-NTA coupled to silica in a convenient microspin format for protein minipreps. Each spin column can purify up to 100 µg of 6xHis-tagged protein from cellular lysates in just 15 min. Protein minipreps are an ideal way to rapidly optimize expression and purification conditions, concentrate poorly expressed proteins for visualization on gels, and screen engineered constructs for function.

Proteins containing 6xHis tags, located at either the amino or carboxyl terminus of the protein, bind to the Ni-NTA resin with an affinity far greater than the affinity between most antibodies and antigens, and enzymes and substrates. As a consequence, the background of proteins that bind to the resin because of the presence of naturally occurring neighboring histidine residues can be easily washed away under relatively stringent conditions without affecting the binding of the tagged proteins. The high

Table 1
Reagents that Normally Do Not Affect Binding
of the 6xHis Tag to Ni-NTA Resin

6M GuHCl	8M urea
2% Triton® X-100	2% Tween-20
1% CHAPS	20 mM β-mercaptoethanol
50% glycerol	30% ethanol
2M NaCl	5 mM CaCl$_2$
25 mM Tris-HCl	20 mM imidazole

binding constant also allows proteins in very dilute solutions, such as those expressed at low levels or secreted into the media, to be efficiently bound to the resin and purified.

The binding of tagged proteins to the resin does not require any functional protein structure and is thus unaffected by strong denaturants, such as 6M guanidine hydrochloride (GuHCl) or 8M urea. This means that, unlike purification systems that rely on antigen/antibody or enzyme/substrate interactions, Ni-NTA can be used to purify almost all proteins—even those that are insoluble under native (nondenaturing) conditions (such as hydrophobic proteins and other proteins that form inclusion bodies) and that must be denatured prior to purification. In addition, *E. coli* proteins that could copurify because of the formation of disulfide bonds can be easily removed by the addition of low levels of β-mercaptoethanol to the loading buffer.

The presence of low levels of detergents, such as Triton X-100 and Tween-20, or high salt concentrations (Table 1), also has no effect on the binding, allowing the complete removal of proteins that would normally copurify because of nonspecific hydrophobic or ionic interactions. Nucleic acids that might associate with certain DNA and RNA binding proteins can also be efficiently removed without affecting the recovery of the tagged protein.

Elution of the tagged proteins from the column can be achieved by several methods. Reducing the pH will cause the histidine residues to become protonated, which allows them to dissociate from the Ni-NTA ligand. Monomers are generally eluted at around pH 5.9, whereas aggregates and proteins that contain more than one tag elute at around pH 4.5. Elution can also be achieved by competition with imidazole buffer, which binds to the Ni-NTA and displaces the tagged protein. Low levels of

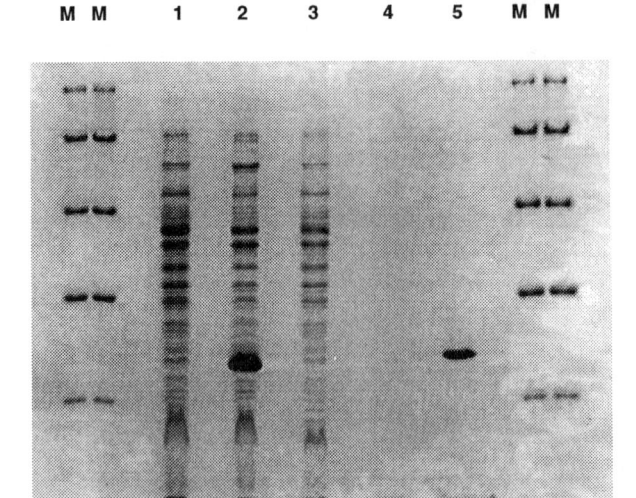

Fig. 4. Purification of DHFR from bacteria (denaturing conditions). Lane 1: uninduced cells; lane 2: induced cells; lane 3: column flow-though, pH 8.0; lane 4: wash fraction, pH 6.3; lane 5: pure eluted protein, pH 5.9; lane M: molecular weight markers.

imidazole can also be used to selectively elute contaminants that bind less strongly to the resin, or to prevent binding of contaminants to the resin during loading *(2)*. A typical result of Ni-NTA purification is shown in Fig. 4.

1.3. General Considerations

This chapter describes a protein miniprep procedure to enable the investigator to confirm correct protein expression, and protocols for bulk purification of proteins from *E. coli* under both nondenaturing and denaturing conditions. Although each procedure works very well for most proteins, some modifications may be necessary if host systems other than *E. coli* are used. The purification power of the 6xHis Ni-NTA system will be enhanced if the conditions are optimized for each individual protein. Possible modifications are considered in Notes 1–5.

Proteins purified under denaturing conditions can be used directly for antibody induction, refolded in solution by dialysis, or refolded on the Ni-NTA column and eluted or used as immobilized ligands for further studies (*see* Note 1).

2. Materials

The pQE-vectors, *E. coli* host strains M15[pREP4] and SG13009-[pREP4], Ni-NTA agarose, and Ni-NTA spin columns are available exclusively from Qiagen GmbH (Hilden, Germany), Qiagen Inc. (Chatsworth, CA), Qiagen Ltd. (Dorking, UK), Qiagen AG (Basel, Switzerland), and their distributors.

2.1. Rapid Screening of Mini-Expression Cultures

1. Culture media: Use LB-medium and its modifications, 2X YT or Super Broth, containing 100 µg/mL ampicillin and 25 µg/mL kanamycin for growth of M15 cells containing pQE expression and pREP4 repressor plasmids (*see* Note 6). LB medium: 10 g of bacto-tryptone, 5 g of bacto-yeast extract, and 5 g of NaCl/L. 2X YT medium: 16 g of bacto-tryptone, 10 g of bacto-yeast extract, and 5 g of NaCl/L. Super medium: 25 g of bacto-tryptone, 15 g of bacto-yeast extract, and 5 g of NaCl/L.
2. IPTG: Stock concentration 1M.
3. Buffer A: 6M guanidine hydrochloride (GuHCl), 0.1M NaH$_2$PO$_4$, 0.01M Tris-HCl, pH adjusted to 8.0 with NaOH.
4. Buffer B: 8M urea, 0.1M NaH$_2$PO$_4$, 0.01M Tris-HCl, pH adjusted to 8.0 with NaOH. Owing to the dissociation of urea, the pH has to be adjusted immediately before use.
5. Buffer C: Same composition as Buffer B, but pH adjusted to 6.3 with HCl. Owing to the dissociation of urea, the pH has to be adjusted immediately before use.
6. Buffer E: Same composition as Buffer B, but pH adjusted to 4.5 with HCl. Owing to the dissociation of urea, the pH has to be adjusted immediately before use.
7. 5X SDS-PAGE sample buffer: 15% β-mercaptoethanol, 15% SDS, 1.5% bromophenol blue, and 50% glycerol.
8. 12.5% polyacrylamide gels containing 0.2% SDS *(12)*.

2.2. Native Purification of Soluble Proteins

9. Sonication buffer: 50 mM NaH$_2$PO$_4$, 300 mM NaCl, 10 mM imidazole, 1 mM PMSF.
10. Lysozyme: Stock concentration 10 mg/mL.
11. RNase: Stock concentration 200 mg/mL.
12. DNase: Stock concentration 60 mg/mL.
13. Wash buffer: 50 mM NaH$_2$PO$_4$, 300 mM NaCl, 20 mM imidazole, 1 mM PMSF.
14. Elution buffer: 50 mM NaH$_2$PO$_4$, 300 mM NaCl, 250 mM imidazole.

2.3. Denaturing Purification of Insoluble Proteins

15. Buffer D: Same composition as Buffer B, but pH adjusted to 5.9 with HCl. Owing to the dissociation of urea, the pH has to be adjusted immediately before use.

3. Methods

3.1. Rapid Screening of Mini-Expression Cultures

The following is a basic protocol for the expression and screening of small cultures by purification of 6xHis-tagged proteins on Ni-NTA spin columns. Purification is performed under denaturing conditions, which will lead to the isolation of any tagged protein, independent of its solubility within the cell. In addition, denaturing the 6xHis-tagged protein fully exposes the 6xHis tag, leading to improved binding kinetics and increased yields in comparison with native purification on Ni-NTA spin columns (see Note 1).

Lysing cells in Buffer B will solubilize most proteins and inclusion bodies, and allows the lysate to be analyzed directly by SDS-PAGE. If the cells or protein do not solubilize in Buffer B, however (e.g., very hydrophobic proteins, and some membrane proteins), then Buffer A must be used. GuHCl is a more efficient solubilization reagent than urea, and may be necessary to solubilize inclusion bodies. It may also be necessary to add nonionic detergents (see Note 7 for ways to treat samples in Buffer A before SDS-PAGE).

For proteins expressed at very high levels (>10 mg/L, ca. 8% of total protein), the lysate should be no more than 25 times concentrated. For lower expression levels (2–5 mg/L), the lysate may be up to 50 times concentrated. Expression levels lower than 1 mg/L require the lysate to be concentrated at least 50 times in order to be able to detect any protein by Coomassie staining (see Note 8 and Table 2).

Some proteins may be subject to degradation during cell harvest, lysis, or even during growth after induction. In these cases, addition of PMSF (0.1–1 mM) or other protease inhibitors is recommended. PMSF treatment during cell growth may result, however, in reduced expression levels. All culture media should contain ampicillin at 100 µg/mL and kanamycin at 25 µg/mL.

1. Inoculate 10 mL of LB broth containing 100 µg/mL ampicillin and 25 µg/mL kanamycin with a fresh colony of M15[pREP4] containing the pQE expression plasmid. Grow at 37°C overnight.

Table 2
Examples of the Recommended Cell Culture Volume
for Use with Ni-NTA Spin Columns[a]

His-tagged protein, mg/L	Expression level, %	Culture volume, mL	His-tagged protein in 600 µL lysate, µg	Concentration factor
Denaturing conditions				
50	40	3	90	3X
10	8	10	60	10X
2	1.6	25	30	25X
0.5	0.4	50	15	50X
0.1	0.08	100	6	100X
Nondenaturing conditions				
>1	>1	50	>30	50X
<1	<1	100	<60	100X

[a]In relation to the expected expression level and purification conditions for a cell lysis volume of 1 mL.

2. Dilute 200 µL of uninduced overnight culture into 10 mL fresh LB broth (1:50 dilution) containing 100 µg/mL ampicillin and 25 µg/mL kanamycin. Grow at 37°C with vigorous shaking until the A_{600} reaches 0.7–0.9 (see Note 8 and Table 2).
3. Add IPTG to a final concentration of 1 mM, and grow the culture at 37°C for an additional 4 h (see Note 9).
4. Harvest the cells by centrifugation for 10 min at 4000g, and discard supernatants.
5. Resuspend cells in 0.1 vol of Buffer B. Lyse cells by gently vortexing, taking care to avoid frothing, or stir cells for 1 h at room temperature (see Note 10).
6. Centrifuge the lysate at 10,000g for 10 min at room temperature to remove the cellular debris, and transfer the supernatant to a fresh tube. Save the pellets and 20 µL of lysate for SDS-PAGE analysis (see Note 11).
7. Pre-equilibrate a Ni-NTA spin column with 600 µL of Buffer B. Centrifuge for 2 min at ≤700g (approx 2000 rpm) (see Note 12).
8. Load up to 600 µL of the cleared lysate supernatant containing the 6xHis-tagged protein onto the pre-equilibrated Ni-NTA spin column. Centrifuge for 2 min at ≤700g (see Note 13). Save flow-through for SDS-PAGE analysis.
9. Wash the Ni-NTA spin column 2× with 600 µL Buffer C. Centrifuge for 2 min at ≤700g. Save the flow-through for SDS-PAGE analysis (see Note 14).
10. Elute the protein with 2 × 200 µL Buffer E. Centrifuge for 2 min at 2000 rpm, and collect the eluates in separate tubes (see Note 15).

11. Take 20 µL samples of all fractions and add 5 µL of 5X PAGE sample buffer (*see* Note 7). Boil for 7 min at 95°C. Analyze samples on a 12.5% polyacrylamide gel and visualize proteins by staining with Coomassie blue.

3.2. Native Bulk Purification Protocol

This protocol is for use with Ni-NTA agarose, to purify up to 5–10 mg of soluble 6xHis-tagged protein. Before purifying proteins under nondenaturing conditions, it is important to determine how much of the protein is soluble in the cytoplasm, and how much is in insoluble precipitates or inclusion bodies. Therefore, parallel purification under denaturing conditions (Section 3.3.) is recommended.

The amount of purified protein will depend on the expression level. This protocol is designed for an expression level from 1–5 mg/L. For optimal results, the protocol should be scaled up or down according to the expression level of the 6xHis-tagged protein. The flow rate during loading, washing, and elution should not exceed 1 mL/min.

1. Grow and induce a 1-L culture as described in Section 3.1.
2. Harvest the cells by centrifugation at 4000g for 20 min. Resuspend the pellet in sonication buffer at 2–5 vol/g of wet weight (*see* Note 16). Freeze sample in dry ice/ethanol (or overnight at –20°C), and thaw in cold water. Alternatively, add lysozyme to 1 mg/mL, and incubate on ice for 30 min.
3. Sonicate on ice (1 min bursts/1 min cooling/200–300 W) and monitor cell breakage by measuring the release of nucleic acids at A_{260} until it reaches a maximum.
4. If the lysate is very viscous, add RNase A to 10 µg/mL and DNase I to 5 µg/mL, and incubate on ice for 10–15 min. Alternatively, draw the lysate through a narrow-gage syringe needle several times. Centrifuge at >10,000g for 20 min at 4°C and collect the supernatant. Save 20 µL for SDS-PAGE.
5. Add 2 mL of a 50% slurry of Ni-NTA agarose, previously equilibrated in sonication buffer, and stir on ice for 60 min (*see* Note 17).
6. Load the lysate and Ni-NTA agarose into a column and collect the column flow-through for SDS-PAGE. Work at 4°C if possible.
7. Wash with 8 mL sonication buffer, or until the A_{280} of the flow-through is below 0.01 (*see* Note 18). Collect wash fractions for SDS-PAGE.
8. Wash with 2 × 8 mL wash buffer, or until the flow-through A_{280} is below 0.01.
9. Elute the protein with 6 mL elution buffer (*see* Note 19). Collect 500 µL fractions, and analyze 5 µL samples on SDS-PAGE (*see* Notes 7, 20, and 21).

3.3. Denaturing Bulk Purification of Insoluble Proteins

This protocol is for use with Ni-NTA agarose, to purify up to 5–10 mg of 6xHis-tagged protein. Purification under denaturing conditions is often more efficient than purification under native conditions, and is essential when proteins cannot be solubilized without denaturation.

The amount of purified protein will depend on the expression level. This protocol is designed for an expression level ca. 10 mg/L. For optimal results, the protocol should be scaled up or down according to the expression level of the 6xHis-tagged protein. The flow rate during loading, washing, and elution should not exceed 1 mL/min.

1. Grow and induce a 500-mL culture as described in Section 3.1. Harvest the cells by centrifugation at 4000*g* for 10 min. Store at −70°C if desired.
2. Thaw cells for 15 min and resuspend in Buffer B at 5 mL/g wet weight (*see* Notes 10 and 22). Stir cells for 1 h at room temperature. Centrifuge lysate at 10,000*g* for 15 min at room temperature. Collect supernatant and save 20 µL for SDS-PAGE (*see* Note 18).
3. Add 2 mL of a 50% slurry of Ni-NTA agarose, previously equilibrated in Buffer B. Stir at room temperature for 45 min, then load resin carefully into a column. Collect flow-through for SDS-PAGE.
4. Wash with 10 mL of Buffer B. If necessary, wash further until the flow-through A_{280} is below 0.01.
5. Wash with 6 mL of Buffer C, or until the flow-through A_{280} is below 0.01.
6. Elute the protein with 10 mL of Buffer E (*see* Note 23). Collect 500 µL fractions and analyze by SDS-PAGE (*see* Notes 7, 20, and 21).

Problems that may be encountered during purification are discussed in Notes 23–28.

4. Notes

1. Many proteins remain soluble during expression and can be purified in their native form under nondenaturing conditions on Ni-NTA resin; others, however, form insoluble precipitates. Since almost all of these proteins are soluble in 6*M* guanidinium hydrochloride, Ni-NTA chromatography and the 6xHis tag provide a universal system for the purification of recombinant proteins. The decision whether to purify the tagged proteins under denaturing or nondenaturing (native) conditions depends both on the solubility and location of the protein, and on the accessibility of the 6xHis tag. Proteins that remain soluble in the cytoplasm, or are secreted into the periplasmic space, can generally be purified under nondenaturing conditions (but note the exception below). If the protein is insoluble, or located in

inclusion bodies, then it must generally be solubilized by denaturation before it can be purified. Some proteins, however, may be solubilized by the addition of detergents, and it is worth experimenting with different solubilization techniques if it is important to retain the native configuration of the protein. Many proteins that form inclusion bodies are also present at some level in the cytoplasm, and may be efficiently purified in their native form, even at very low levels, on Ni-NTA resin.

In rare cases the 6xHis tag is hidden by the tertiary structure of the native protein, so that soluble proteins require denaturation before they can be bound to Ni-NTA resin. If denaturation of the protein is undesirable, the problem is usually solved by moving the tag to the opposite terminus of the protein. Proteins that have been purified under denaturing conditions can either be used directly or refolded in dilute solution by dialyzing away the denaturants in the presence of reduced and oxidized glutathione *(13,14)*. It is also possible to renature proteins on the Ni-NTA column *(2,15,16)*.

2. Proteins may be purified on Ni-NTA resin in either a batch or a column procedure. The batch procedure entails binding the protein to the Ni-NTA resin in solution, decanting the supernatant, and then packing the protein/resin complex into a column for the washing and elution steps. This promotes more efficient binding, particularly under native conditions, and reduces the amount of debris that is loaded onto the column. In the column procedure, the Ni-NTA column is packed and washed, and the cell lysate is applied slowly to the column.

3. Background contamination arises from proteins that contain neighboring histidines, and thus have some affinity for the resin; proteins that copurify because they are linked to the 6xHis-tagged protein by disulfide bonds; proteins that associate nonspecifically with the tagged protein; and nucleic acids that associate with the tagged protein. All of these contaminants can be easily removed by washing the resin under the appropriate conditions. Proteins that contain neighboring histidines in the primary sequence are not common in bacteria, but are quite abundant in mammalian cells. These proteins bind to the resin much more weakly than proteins with a 6xHis tag, and can be easily washed away, even when they are much more abundant than the tagged protein *(2)*. (*See* Note 4 for additional information.) The addition of 10–20 mM β-mercaptoethanol to the loading buffer will reduce background owing to crosslinked proteins. Do not use >1 mM DTT, since higher concentrations may reduce Ni^{2+} ions.

Proteins that are associated with the tagged protein or the resin owing to nonspecific interactions and nucleic acids that copurify can be removed by washing with low levels of detergent (up to 2% Triton X-100 or 0.5%

sarkosyl); increasing the salt concentration up to $2M$ NaCl; or including 30% ethanol or 50% glycerol to reduce hydrophobic interactions. The optimum levels of any of these reagents should be determined empirically for different proteins.

4. Removal of background proteins and elution of tagged proteins from the column may be achieved by either lowering the pH in order to protonate the histidine residues or by the addition of imidazole, which competes with tagged proteins for binding sites on the Ni-NTA resin. Although both methods are equally effective, the imidazole is somewhat milder, and is recommended in cases where the protein would be damaged by a reduction in pH (e.g., tetrameric aldolase) *(3)*.

In bacterial expression systems, it is seldom necessary to wash the bound protein under very stringent conditions, since proteins are expressed to high levels and the background is low. In mammalian systems, however, or under native conditions where many more neighboring histidine residues will be exposed to the resin, it may be necessary to increase the stringency of the washing considerably. This can be done by gradually decreasing the pH of the wash buffer, or by slowly increasing the concentration of imidazole. The pH or imidazole concentration that can be tolerated before elution begins will vary slightly for each protein.

In situations where the tagged protein is very dilute and the background is likely to be high (such as in mammalian expression systems), it is also useful to bind the 6xHis-tagged protein to the resin under conditions in which the background proteins do not compete for the binding sites, i.e., at a slightly lower pH or in the presence of low levels of imidazole. Likewise, the purification process will be optimized if the amount of tagged protein is closely matched to the capacity of the of the resin used, i.e., if the amount of resin is minimized (H. Stunnenberg, personal communication). Since the 6xHis-tagged protein has a higher affinity for the Ni-NTA resin than do the background proteins, it can fill all the available binding sites and very few background proteins will be retained on the resin.

5. Do not use strong reducing agents, such as DTT or DTE, on the column, because they will reduce the Ni^{2+} ions and cause them to elute from the resin. In most situations, β-mercaptoethanol can be used at levels up to 20 mM, but even these low levels might cause problems occasionally when the protein itself has a strongly reducing nature. Use any reducing agent with care, and if in doubt, test it out on a small amount of Ni-NTA resin first. Strong chelating agents will chelate the Ni, and also cause it to elute from the NTA resin. Do not use EDTA, EGTA, or any other chelating agents. (Note however, that there are examples where 1 mM EDTA has been used successfully in buffers.)

6. We suggest that expression should be tried in all three media in parallel, and a time-course taken of expression after induction. There are often striking differences noted between the level of expression in different media at different times.
7. On minigels it is usually sufficient to analyze 5 µL samples of each fraction in an equal volume of SDS-PAGE loading buffer, with or without 3% β-mercaptoethanol. Since the fractions that contain GuHCl will precipitate with SDS, they must either be very dilute (1:6), dialyzed before analysis, or separated from the guanidinium hydrochloride by TCA precipitation. Dilute samples to 100 µL; add an equal volume of 10% TCA; leave on ice 20 min; spin for 15 min in a microfuge; wash pellet with 100 µL of ice-cold ethanol; dry; and resuspend in sample buffer. If there is any guanidinium hydrochloride present, samples must be loaded immediately after boiling for 7 min at 95°C.
8. The required volume of expression culture is mainly determined by the expression level, cellular location of the protein, and purification conditions. For purification of poorly expressed proteins, the minimum cell culture volume should be 30 mL. A 50X concentrated cell lysate should be loaded onto the Ni-NTA spin column to increase the amount of 6xHis-tagged protein, the viscosity and, therefore, the yield. For highly expressed proteins, particularly under denaturing conditions, the situation is different (*see* Table 2): For proteins that are expressed at very high levels (>10 mg/L, i.e., equivalent to an expression level of approx 8% of total cellular protein) the cell lysate should be no more than 10X concentrated. At an expression level of 10 mg/L, 600 µL of the 10X concentrated cell lysate in Buffer B contains approx 60 µg of 6xHis-tagged protein. For lower expression levels (2–5 mg/L), 25X concentrated cell lysates (600 µL cell lysate = 30–75 µg) should be loaded onto the Ni-NTA spin column. For expression levels lower than 1 mg/L, the cell lysate should be concentrated 50–100X.
9. For proteins that are very sensitive to protein degradation, the induction time should be reduced and a time-course of expression should be determined. In some cases, addition of 0.1–1 m*M* PMSF after induction is recommended to inhibit PMSF-sensitive proteases. PMSF treatment can result, however, in reduced expression levels.
10. The solution should become translucent when lysis is complete. It is preferable to lyse the cells in Buffer B so that the cell lysate can be analyzed directly by SDS-PAGE. If the cells or the protein do not solubilize in Buffer B, then Buffer A must be used. *See* Note 7 for ways to treat samples in Buffer A before loading onto SDS-PAGE gels.
11. The supernatant will contain all solubilized proteins. Any insoluble proteins will be pelleted. Retain the pellet for analysis by SDS-PAGE.

12. It is important not to exceed 2000 rpm (approx 700g) when centrifuging Ni-NTA spin columns. At higher speeds, NTA silica particles become compressed, leading to high flow rates (channeling), and inefficient binding.
13. In situations where the binding kinetics may be poor, for example under native conditions, reloading the column flow-through is recommended.
14. Wash the Ni-NTA spin column with Buffer C even if Buffer A was used to initially solubilize the protein. Most proteins will remain soluble in Buffer C. It may not be necessary to repeat the Buffer C wash. The number of wash steps required to obtain highly pure protein is determined primarily by the expression level of the 6xHis-tagged protein. When the expression level is high, two wash steps are usually sufficient for removal of contaminants. For very low expression levels or highly concentrated lysates, three wash steps may be required to achieve high purity.
15. Most of the 6xHis-tagged protein (>80%) should elute in the first 200 µL eluate, particularly when proteins smaller than 30 kDa are purified. The remainder will elute in the second 200 µL. If dilution of the protein is undesirable, do not combine the eluates or, alternatively, elute in 100–150-µL aliquots to increase protein concentration.
16. The composition of the sonication, wash, and elution buffers can be modified to suit the particular application, e.g., by adding low levels of imidazole, 1–2% Tween, 5–10 mM β-mercaptoethanol, 1 mM PMSF, or increased NaCl or glycerol concentrations.
17. If the 6xHis-tagged protein does not bind under these conditions the concentration of imidazole in the sonication buffer should be reduced to 1–5 mM. *See also* Note 1 for information about "hidden" tags.
18. Discolored or impure reagents may affect optical density readings, and imidazole will absorb light at 280 nm.
19. The protein may be eluted by a number of different means. If preferred, elution can be performed using pH, either as a continuous or step gradient decreasing from pH 8.0 to 4.5. Most proteins will be efficiently eluted by wash buffer at pH 4.5, although many (particularly monomers) can be eluted at a higher pH.
20. Where possible, follow the chromatography by A_{280} and collect pools rather than fractions.
21. Do not boil a sample that contains imidazole before SDS-PAGE, since it will hydrolyze acid labile bonds. Heat the sample for a few minutes at 37°C immediately before loading the gel.
22. The purification is performed in 8M urea, since 6M guanidinium hydrochloride precipitates in the presence of SDS, making SDS-PAGE analysis of samples difficult. Otherwise, both urea and GuHCl or combinations thereof can be used throughout the whole purification procedure.

23. Alternative elution procedures may be used. Monomers can usually be eluted in Buffer D, whereas multimers, aggregates, and proteins with two 6xHis tags will generally elute in Buffer E. Elution can also be carried out using a pH 4.0–6.5 gradient in $8M$ urea, $0.1M$ NaH_2PO_4, or $0.01M$ Tris-HCl. If elution at a higher pH is desired, most proteins can be eluted with Buffer B containing 100–250 mM imidazole.
24. If expression levels are too high, one or more of the following may help. Using less IPTG (0.01 mM) will reduce the expression level 10–15-fold. Alternatively, reduce time of induction and/or lower temperature; try the induction at higher cell densities (0.8 A_{600}); start with the nondenaturing (native) protocol, followed by the denaturing protocol for the residual, nondissolved cellular debris and inclusion bodies.
25. In case of precipitation during purification, check for aggregates of purified proteins and try Tween or Triton additives (up to 2%), adjust to 10–20 mM β-mercaptoethanol, and check for stabilizing cofactor requirements (e.g., Mg^{2+}). Make sure that the salt concentration is at least 300 mM NaCl. Check room temperature (>20°C) for the denaturing protocol.
26. In case of an insufficient binding of 6xHis to Ni-NTA resin, check for the presence of chelating agents (EDTA/EGTA) or high concentrations of electron-donating groups (NH_4), ionic detergents, and components, such as glycine, histamine, or zinc, and repeat the binding step with new buffers. In cases where the protein is purified from culture media, it may be necessary to dialyze before binding to remove such components. If the 6xHis tag is hidden in the native protein structure, improve the exposure by adding small concentrations of urea or detergents to the nondenaturing sample preparation buffer. Slower binding kinetics can be compensated by longer contact times with NTA, preferably under batch-binding conditions. Alternatively, try 6xHis at the opposite terminus, or use completely denaturing conditions (Buffer A with 10 mM β-mercaptoethanol). Avoid any Ni^{2+} complexing reagents.
27. Background binding can be suppressed by adjusting the amount of Ni-NTA agarose according to the 6xHis protein expression level. Try to match the binding capacity to no more than 2–5 times the amount of tagged protein.
28. Partially hidden 6xHis tags in native proteins can cause earlier elution characteristics.
29. Do not determine the size of the recombinant product by SDS-PAGE. Adding or replacing amino acids can shift protein bands, suggesting a molecular weight several kDa higher than expected. Therefore, the addition of the 6xHis tag cannot be sized this way.

Further Reading

Reviews

Hochuli, E. (1989) Genetically designed affinity chromatography using a novel metal chelate adsorbent. *Biol. Act. Mol.* 217–239.
Hochuli, E., and Piesecki, S. (1992) Interaction of hexahistidine fusion proteins with nitrilotriacetic acid-chelated Ni2+ ions. *Methods: A Companion to Methods in Enzymology* **4**, 68–72.
Kaslow, D. and Shiloach, J. (1994) Production, purification and immunogenicity of a malaria transmission-blocking vaccine candidate: TBV25H expressed in yeast and purified using Ni-NTA agarose. *Bio/Technology* **12**, 494–499.

Selected Examples

Abate, C., Luk, D., Gentz, R., Rauscher, F. J., III, and Curran, T. (1990) Expression and purification of the leucine zipper and DNA-binding domains of Fos and Jun: both Fos and Jun contact DNA directly. *Proc. Natl. Acad. Sci. USA* **87**, 1032–1036.
Bush, G. L., Tassin, A., Friden, H., and Meyer, D. I. (1991) Purification of a translocation-competent secretory protein precursor using nickel ion affinity chromatography. *J. Biol. Chem.* **266**, 13,811–13,814.
Gentz, R., Certa, U., Takacs, B. J., Matile, H., Dobeli, H., Pink, R., Mackay, M., Bone, N., and Scaife, J. G. (1988) Major surface antigen pl90 of *Plasmodium falciparum*: detection of common epitopes present in a variety of plasmodia isolates. *EMBO J.* **7**, 225–230.
Gentz, R., Chen, C., and Rosen, C. A. (1989) Bioassay for trans-activation using purified immunodeficiency virus tat-encoded protein: trans-activation requires mRNA synthesis. *Proc. Natl. Acad. Sci. USA* **86**, 821–824.
Le Grice, S. F. J. and Grueninger-Leitch, F. (1990) Rapid purification of homodimer HIV-I reverse transcriptase by metal chelate affinity chromatography. *Eur. J. Biochem.* **187**, 307–314.
Stüber, D., Bannwarth, W., Pink, J. R. L., Meloen, R. H., and Matile, H. (1990) New B cell epitopes in the plasmodium falciparum malaria circumsporozoite protein. *Eur. J. Immunol.* **20**, 819–824.
Takacs, B. J. and Girard, M.-F. (1991) Preparation of clinical grade proteins produced by recombinant DNA technologies. *J. Immunol. Meth.* **143**, 231–240.

References

1. Stüber, D., Matile, H., and Garotta, G. (1990) System for high-level production in *Escherichia coli* and rapid purification of recombinant proteins: application to epitope mapping, preparation of antibodies, and structure-function analysis, in *Immunological Methods*, vol. IV (Lefkovits, I. and Pernis, B., eds.), Academic, New York, pp. 121–152.
2. Janknecht, R., de Martynoff, G., Lou, J., Hipskind, R. A., Nordheim, A., and Stunnenberg, H. G. (1991) Rapid and efficient purification of native histidine-tagged protein expressed by recombinant vaccinia virus. *Proc. Natl. Acad. Sci. USA* **88**, 8972–8976.

3. Dobeli, H.,Trecziak, A., Gillessen, D., Matile, H., Srivastava, I. K., Perrin, L. H., Jakob, P. E., and Certa, U. (1990) Expression, purification, biochemical characterization and inhibition of recombinant Plasmodium falciparum aldolase. *Mol. Biochem. Parasitol.* **41,** 259–268.
4. Lindner, P., Guth, B., Wülfing, C., Krebber, C., Steipe, B., Müller, F., and Plückthun, A. (1992) Purification of native proteins form the cytoplasm and periplasm of *Escherichia coli* using IMAC and histidine tails: a comparison of proteins and protocols. *Methods: A Companion to Methods in Enzymology* **4(2),** 41–55.
5. Gu, J., et. al. (1994) Recombinant proteins attached to a Ni-NTA column: use in affinity purification of antibodies. *BioTechniques* **17(2),** 257–262.
6. Gamer, J., Bujard, H., and Bukau, B. (1992) Physical interaction between heatshock proteins DnaK, DnaJ, and GpE and the bacterial heat-shock transcription factor σ^{32}. *Cell* **69,** 833–842.
7. Sporeno, E., Paonessa, G., Salvati, A. L., Graziani, R., Delmastro, P., Ciliberto, G., and Toniatti, C. (1994) Oncostatin *M* binds directly to gp130 and behaves as interleukin-6 antagonist on a cell line expressing gp130 but lacking functional oncostatin *M* receptors. *J. Biol Chem.* **269(15),** 10,991–10,995.
8. Polayes, D., Goldstein, A., Ward, G., and Hughes, A. (1994) TEV protease, recombinant: a site-specific protease for cleavage of affinity tags from expressed proteins. *Focus (Life Technologies, Inc.)* **16(1),** 2–5.
9. Porath, J., Carlsson, J., Olsson, I., and Belfrage, G. (1975) Metal chelate affinity chromatography, a new approach to protein fractionation. *Nature* **258,** 598,599.
10. Sulkowski, E. (1985) Purification of proteins by IMAC. *Trends Biotechnol.* **3,** 1–7.
11. Hochuli, E., Dobeli, H., and Schacher, A. (1987) New metal chelate adsorbent selective for proteins and peptides containing neighboring histidine residues. *J. Chromatog.* **411,** 177–184.
12. Takacs, B. J. (1979) in *Immunological Methods,* vol. 1 (Lefkovits, I. and Pernis, B., eds.), Academic, New York, p. 81.
13. Thanos, D. and Maniatis, T. (1992) The high mobility group protein HMG I(Y) is required for NF-κB-dependent virus induction of the human IFN-β gene. *Cell* **71,** 623–635.
14. Fischer, B., Perry, B., Summer, I., and Goodenough, P. (1992) A novel sequential procedure to enhance the renaturation of recombinant protein from *Escherichia coli* inclusion bodies. *Prot. Eng.* **5(6),** 593–596.
15. Urban, S. and Hildt, E. (1994) Use of Ni-NTA resin for isolation of cellular proteins binding to HBV transactivator proteins HBx and MHB[St]. *QIAGEN News* **1,** 14–16.
16. Sinha, D., et. al. (1994) Ligand binding assays with recombinant proteins refolded on an affinity matrix. *BioTechniques* **17(3),** 509–514.

Index

A
Agarose gel loading buffer, 12, 18
Alkaline phosphatase detection, 41, 44, 46–48, 55–57, 60, 163
Antidigoxigenin antibodies, 41–51
Antifluoroscein antibodies, 59, 60
Automated sequencing,
  dye primer method, 428, 429
  dye terminator method, 430
  gel electrophoresis, 432, 434
  preparation of gel, 431, 432
  using Taq polymerase, 429, 430

B
Bacterial growth medium,
  CY, 180
  L broth, 180
  SOB, 242
  SOC, 250
  Superbroth, 499
  TB, 258
  TBG, 364
  2xTY269, 360
Bacterial hosts,
  λ bacteriophage, 173, 179, 201
  M13 bacteriophage, 342, 359
  plasmid, 471
  phagemid, 364
  for site-directed mutagenesis,
    mutS strains, 456, 462, 465
    dut⁻/ung⁻ strains, 469, 471
Blue/white selection, 232
Blunt ending DNA, 231, 232
  in gel, 355

C
Calf intestinal alkaline phosphatase (CIP), 106–108, 110, 231, 346
cDNA synthesis,
  PCR template, 284
  RNaseH method, 191–197
Compatible cohesive ends, 227
Competition hybridization, 33–35
cos sites, 171,172

D
100X Denhardt's solution, 33
DEPC treatment of water, 4, 15
Dephosphorylation,
  for endlabeling, 108
  of λ arms, 206
  during subcloning, 231
Dextran sulfate, 34
DNA,
  preparation of genomic DNA, 1–9, 280
  partial digestion, 184, 185

E
ECL detection, 69, 72
Electroelution, 134
Endlabeling DNA,
  using T4 kinase, 105–108
  using Klenow, 105–108, 444
  using T4 DNA polymerase, 110
Ethidium bromide, 18
Exonuclease III, 352, 354

F
Formamide, deionization, 130, 134
Formamide dye mix, 394, 406

511

G

Gel electrophoresis,
  agarose, 23–25
  denaturing agarose, 117, 121, 122
  denaturing polyacrylamide, see Sequence Gels)
  native polyacrylamide, 94, 95
Gel purification,
  by agarose gel electrophoresis, 239
  by PAGE, 95, 444
Genomic DNA, see DNA
Glass beads, preparation, 238, 239
Glycerol in sequence reactions, 375

H

6xHis tagging of proteins, 491–498
Hybridization,
  colony, 217
  M13 bacteriophage, 218
  Northern, 123–125
  plaque, see λ Libraries
  Southern, 31–39, 46, 47, 58, 59, 71, 72

I

Immunoscreening, see λ Libraries
In situ hybridization (ISH), 161–163
In vitro packaging of λ bacteriophage, 171, 172, 174
In vitro translation,
  using rabbit reticulocyte lysate, 481–483
  using wheat germ extract, 487–489

K

Klenow fragment of DNA polymerase I, 105–108

L

λ Libraries amplification, 186
  immunoscreening, 211–215
  plate lysate, 215
  plating, 186, 207
  screening by hybridization, 214
  screening by PCR, 335–337
  secondary and tertiary screens, 215
  size calculation, 189
λ miniprep, 216
λ packaging mixes,
  preparation, 173, 174
  use of 174, 207
λ phage, preparation, EMBL4, 181, 182
  λgt10, 203, 204
  λgt11, 204, 205
  purification,
    using CsCl gradient, 181, 182, 204
    using PEG, 208
λ vectors, 177–179, 199–201
Ligation, 232, 346
  in gel, 355
Linkers,
  addition to cDNA, 195–197
  addition during subcloning, 232

M

M13 bacteriophage,
  preparation of RF form, 344, 345
  preparation of single-stranded DNA, 359–362
M13 helper phage, 364, 365
Maxam and Gilbert sequencing, see Sequencing, chemical method

N

Ni-NTA spin column, 496
Northern blotting, 117, 118, 121–123

O

Oligonucleotides,
  design of PCR primers, 298, 307, 318–320
  fluoroscein labeling, 78, 79
  hybridization, 79, 80
  radiolabeling, 150, 406, 421, 462, 463, 471

P

Phenol preparation, 5, 276, 360
Phosphate-buffered saline (PBS), 4

Index

Phagemid, preparation of single-
stranded DNA, 363–365
Plasmid preparation,
  alkaline lysis method, 259–261
  rapid boiling method, 266
Plasmid, purification,
  by CsCl gradient, 261, 262, 271
  by diatomaceous earth, 269–272
  by glass beads, 369
Plasmid vectors, pGEM, 88, 365
  pBluescriptII, 88, 317, 365
Polymerase chain reaction (PCR),
  basic reaction, 284, 285
  cloning, see T-vectors
  colony screening by PCR, 330, 331
  degenerate PCR, 303–310
  digoxigenin labeling, 45, 46
  inverse PCR (IPCR), 293–297
  nested PCR, 287, 297
  preparation of sequence templates, 434, 435
  preparation of riboprobe templates, 86, 87
  primers, see Oligonucleotides
  radiolabeling, 290, 325–328
Primer extension analysis, 137–145
Probe labeling,
  fluoroscein-labeled riboprobe, 157–161
  HRP conjugation, 44–46, 68, 71
  radiolabeled riboprobe, 87
  random prime method, 27–29, 43, 45, 58
Protein purification, see 6xHis tagging of proteins
Pyrophosphatase, 375, 378

R

Rabbit reticulocyte lysate,
  preparation, 479, 480
Restriction enzyme buffers, 13
Restriction enzyme digestion, 11–15
Reverse transcriptase,
  AMV, 191, 193, 194

MMLV, 278, 284
RNA,
  preparation of total cellular RNA, 1–9, 118–120
  preparation of polyA$^+$ RNA, 120–121
  purification using CsCl, 281–283
  purification by RNAzol™, 283, 284
RNA polymerase, 83, 90
RNAse protection assay, 131–136
RNasin, 144, 278

S

S1 mapping, 147–153
S1 nuclease, 147, 151
S1 nuclease treatment after Exo III digestion, 352, 355
Sephadex G25 spin column, preparation, 143
Sephadex G50 spin column, preparation, 408
Sepharose CL-6B spin column, preparation, 229–231
Sequenase v. 2.0 T7 DNA polymerase, 373–375, 378
Sequence gel compressions, 383
Sequence gels, 390, 391
  buffer gradient, 390–392
  glycerol-tolerant gels, 376
  reading, 397–399
  running, 395, 396
Sequencing,
  chemical method, 448–450
  double-stranded templates, 369, 370, 376, 377, 381
  PCR products, 403–411
  single-stranded templates, 380
  thermal cycle sequencing, 413–422
  using labeled primers, 407, 417
Silver staining, 97–103
Site-directed mutagenesis,
  2 site method, 463, 464

USE, 456, 457
LP-USE, 457
using uracil containing templates, 469–474
Slot blotting, 129, 130
Southern blotting, 23–25
20X SSC, 33

T

T-vectors, 314–318
T4 DNA polymerase, 353, 456
T4 gene 32 protein, 465
T4 polynucleotide kinase, 106–108, 406
Terminal transferase, 78, 79
Top agar, 180
Transformation of *E. coli*,
  by calcium method, 243–245
  by electroporation, 251, 252
  by the Hanahan method, 244, 245
  with M13 bacteriophage, 245

U

UV,
  crosslinking, 123
  damage, 85, 127
  decontamination, 279
  shadowing, 96

V

Vent$_R$™(exo⁻) DNA polymerase, 413, 415

W

Wheat germ extract, preparation, 486